# Studies in Classification, Data Analysis, and Knowledge Organization

T0180767

For further volumes:
http://www.springer.com/series/1564

Wolfgang Gaul • Andreas Geyer-Schulz
Lars Schmidt-Thieme • Jonas Kunze
Editors

# Challenges at the Interface of Data Analysis, Computer Science, and Optimization

Proceedings of the 34th Annual Conference
of the Gesellschaft für Klassifikation e. V.,
Karlsruhe, July 21 - 23, 2010

 Springer

*Editors*

Prof. Dr. Wolfgang Gaul
Karlsruhe Institute of Technology (KIT)
Institute of Decision Theory
and Operations Research
Kaiserstr. 12
76128 Karlsruhe
Germany
wolfgang.gaul@kit.de

Prof. Andreas Geyer-Schulz
Karlsruhe Institute of Technology (KIT)
Insitute for Information Systems
and Management (IISM)
Kaiserstr. 12
76131 Karlsruhe Baden-Württemberg
Germany
andreas.geyer-schulz@kit.edu

Prof. Dr. Dr. Lars Schmidt-Thieme
University of Hildesheim
Institute of Computer Science
Marienburger Platz 22
31141 Hildesheim
Hildesheim
Germany
schmidt-thieme@ismll.de

Jonas Kunze
Karlsruhe Institute of Technology (KIT)
Institute for Information Systems
and Management (IISM)
Kaiserstraße 12
76128 Karlsruhe
Germany
jonas.kunze@kit.edu

ISSN 1431-8814
ISBN 978-3-642-24465-0     e-ISBN 978-3-642-24466-7
DOI 10.1007/978-3-642-24466-7
Springer Heidelberg Dordrecht London New York

Library of Congress Control Number: 2012930643

Printed on acid-free paper

Springer is part of Springer Science+Business Media (www.springer.com)

# Preface

Revised versions of selected papers presented at the 34th Annual Conference of the German Classification Society (GfKl (Gesellschaft für Klassifikation), a member of IFCS (International Federation of Classification Societies)), held at KIT (Karlsruhe Institute of Technology) in July 2010, are contained in this volume of "Studies in Classification, Data Analysis, and Knowledge Organization".

One aim of the conference was to provide a platform for discussions on results concerning the interface that data analysis has in common with other areas such as, e.g., computer science, operations research, and statistics from a scientific perspective, as well as with various application areas when "best" interpretations of data that describe underlying problem situations need knowledge from different research directions.

Practitioners and researchers – interested in data analysis in the broad sense – had the opportunity to discuss recent developments and to establish cross-disciplinary cooperation in their fields of interest. More than 200 persons attended the conference, 158 talks (including plenary and semiplenary lectures) were presented. The audience of the conference was quite international with about 90 contributions from participants from abroad with the largest groups of foreign presenters from Italy, Japan, and Poland. Additionally, a program for librarians (14 presentations) was organized. Parallel to the conference a German-Japanese workshop sponsored by GfKl and JCS (Japanese Classification Society) took place with additional 24 talks.

Sixty of the papers presented at the conference are contained in this volume (The contributions given at the German-Japanese workshop will be published elsewhere.). As an unambiguous assignment of topics addressed in single papers is sometimes difficult the contributions are grouped in a way that the editors found appropriate. Within (sub)chapters the presentations are listed in alphabetical order with respect to the authors' names. At the end of this volume an index is included that, additionally, should help the interested reader.

Last but not least, we would like to thank all participants of the conference for their interest and various activities which, again, made the 34th annual GfKl conference and this volume an interdisciplinary possibility for scientific discussion, in particular all authors and all colleagues who reviewed papers, chaired sessions or were otherwise involved. Here, the teams of Prof. Gaul, Prof. Geyer-Schulz, and Prof. Schmidt-Thieme have provided support in a way that outside persons cannot fully appreciate. Additionally, we gratefully take the opportunity to acknowledge support by Deutsche Forschungsgemeinschaft (DFG) as well as the Fakultät für Wirtschaftswissenschaften of KIT.

This volume was put together at Lehrstuhl "Information Services & Electronic Markets", and, at least, Jonas Kunze should be explicitly mentioned for the final editing tasks. As always we thank Springer Verlag, Heidelberg, especially Dr. Martina Bihn, for excellent cooperation in publishing this volume.

Karlsruhe and Hildesheim (Germany)                                    *Wolfgang Gaul*
                                                                    *Andreas Geyer-Schulz*
                                                                     *Lars Schmidt-Thieme*
                                                                          *Jonas Kunze*

# Conference Organization

*Local Organizers*

Prof. Dr. Wolfgang Gaul (Chair)
Prof. Dr. Andreas Geyer-Schulz
Prof. Dr. Martin E. Ruckes
Prof. Dr. Detlef Seese
Prof. Dr. Karl-Heinz Waldmann
Fakultät für Wirtschaftswissenschaften
Kollegium am Schloss, Universität Karlsruhe (TH), KIT


*Scientific Program Committee*

Y. Baba (Tokyo, Japan)
D. Baier (Cottbus)
H.-H. Bock (Aachen)
A.-L. Boulesteix (München)
M.P. Brito (Porto, Portugal)
J.M. Buhmann (Zurich, Switzerland)
A. Cerioli (Parma, Italy)
R. Decker (Bielefeld)
L. de Raedt (Leuven, Belgium)
W. Esswein (Dresden)
W. Gaul (Karlsruhe)
M. Greenacre (Barcelona, Spain)
P.J.F. Groenen (Rotterdam, The Netherlands)
M. Grötschel (Berlin)
Ch. Hennig (London, Great Britain)
K. Hornik (Vienna, Austria)
O. Hudry (Paris, France)
R. Klein (Augsburg)

P. Kuntz (Nantes, France)
G. McLachlan (Brisbane, Australia)
M. Mizuta (Sapporo, Japan)
A. Okada (Tokyo, Japan)
S.T. Rachev (Karlsruhe)
M. Schader (Mannheim)
L. Schmidt-Thieme (Hildesheim, Chair)
W. Seidel (Hamburg)
M. Spiliopoulou (Magdeburg)
M. Vichi (Rome, Italy)
C. Weihs (Dortmund)

# Contents

## Part II   Quantification Theory

## Part III   Analysis of m-Mode n-Way and Asymmetric Data

## Part IV   Analysis of Visual, Spatial, and Temporal Data

## Part VI   Text Mining

## Part VII   Dimension Reduction

## Part VIII   Statistical Musicology

## Part IX   Data Analysis in Banking and Finance

**Part XIII   Analysis of Tourism Data**

# Part I
# Classification, Cluster Analysis, and Multidimensional Scaling

# Fuzzification of Agglomerative Hierarchical Crisp Clustering Algorithms

**Mathias Bank and Friedhelm Schwenker**

**Abstract** User generated content from fora, weblogs and other social networks is a very fast growing data source in which different information extraction algorithms can provide a convenient data access. Hierarchical clustering algorithms are used to provide topics covered in this data on different levels of abstraction. During the last years, there has been some research using hierarchical fuzzy algorithms to handle comments not dealing with one topic but many different topics at once. The used variants of the well-known fuzzy $c$-means algorithm are nondeterministic and thus the cluster results are irreproducible. In this work, we present a deterministic algorithm that fuzzifies currently available agglomerative hierarchical crisp clustering algorithms and therefore allows arbitrary multi-assignments. It is shown how to reuse well-studied linkage metrics while the monotonic behavior is analyzed for each of them. The proposed algorithm is evaluated using collections of the RCV1 and RCV2 corpus.

## 1 Introduction

In recent years, the interest in user generated data out of fora, weblogs, social networks and recommendation systems has increased significantly. Many different methods are applied to extract relevant information out of this a priori unstructured

M. Bank (✉)
University of Ulm, Faculty for Mathematics and Economics, Germany
e-mail: mathias.bank@gmail.com

F. Schwenker
University of Ulm, Institute of Neural Information Processing, Germany
e-mail: friedhelm.schwenker@uni-ulm.de

W. Gaul et al. (eds.), *Challenges at the Interface of Data Analysis, Computer Science, and Optimization*, Studies in Classification, Data Analysis, and Knowledge Organization, DOI 10.1007/978-3-642-24466-7_1, © Springer-Verlag Berlin Heidelberg 2012

data. One of the main tasks consists in topic detection. To group a large set of data –
in this case user comments – according to underlying structures, cluster analysis
techniques are applied. Especially hierarchical clustering algorithms have shown
many advantages because they generate different levels of abstraction in which the
user can decide on his own which one is the best for his individual information
request.

Generally, user generated comments are not only dealing with one topic.
Therefore, it must be possible to assign each document to more than one cluster.
In literature, different methods have been proposed to generate hierarchical clusters
with the possibility of multiple assignment. Next to pyramidal clusters (Diday
1987), which are not flexible enough due to order limitations of the data elements,
different variations of the well known fuzzy-c-means algorithm have been suggested
(Mendes-Rodrigues and Sacks 2005; Torra 2005; Bordogna and Pasi 2009). These,
however, suffer from the drawbacks of partitioning clustering algorithms. They are
neither deterministic nor do they guarantee to create local optimal clusters due to
random initialization. Additionally, it is necessary to predict the number of possible
clusters.

In this work, we propose a new clustering method that fuzzifies well-known
agglomerative hierarchical crisp clustering algorithms. The deterministic algorithm
generates locally optimized clusters while well-known linkage methods can be
reused with small modifications. The degree of branching can be specified with
a fuzzifier $f$ that is directly applied to the similarity matrix. It is shown that the
generated clusters can still be monotonic depending on the used linkage measure
even though the induced dissimilarity measures are no longer ultrametrics. Using
the pairwise merged clusters, an additional shrinking process is proposed to generate
topic related groups with more than two cluster elements.

The overall quality of the proposed clustering algorithm is analyzed using
a cosine quality measure that indicates how well each element fits into the
corresponding clusters. It is applied to text collections created out of the RCV1
and RCV2 (German) corpus.

## 2  Generalization of Agglomerative Crisp Clustering Algorithms

### 2.1  Overall Algorithm

The following listing presents an abstract overview of the complete algorithm. Each
step is discussed in the following in detail. As the algorithm is based on a symmetric
similarity matrix, the discussion and the algorithm are limited to the upper triangular
matrix.

Input: similarity matrix $S$, linkage measure $s$, min. similarity $\varepsilon$
While ($\exists i, j : s_{ij} \geq \varepsilon$)
  Select $C_i$ and $C_j$ with $s_{ij} = \max(s_{kl}) \forall k, l, k \neq l$
  Update $s_{ij} := 0$
  Apply fuzzifier $f$ to $C_i$ and $C_j$
  Insert $C_{ij}$ into $S$ according to linkage measure $s$
Calculate Fuzzy Membership
Apply Shrinking Process for Topic Detection

### 2.1.1 Basic Concept

Similar to the crisp clustering process, the proposed agglomerative clustering algorithm starts with a symmetrical similarity matrix which is initially filled with all similarities between data points (singletons). At each iteration, the algorithm looks for the highest similarity value $s_{ij}, i \neq j$ to select two clusters $C_i$ and $C_j$, which are merged in the next step. An agglomerative hierarchical crisp clustering algorithm would delete all similarity entries $s_{il}$ and $s_{jl}$ of these clusters. In contrast the proposed algorithm does not delete any entry but updates only the similarity value $s_{ij}$ to 0, while the newly created cluster $C_{ij} = C_i \cup C_j$ is added to the similarity matrix. Thus, the matrix dimensions grow by 1 in each iteration.

With the proposed method it is possible to reuse the remaining similarity values of $C_i$ and $C_j$ for multiple cluster assignments. However, it is no longer possible to use well-known linkage measures to calculate the similarity between the new cluster $C_{ij}$ and an existing cluster $C_k$. It is necessary to extend these measures to deal with common subgroups. To ensure that $C_i$ and $C_j$ will not be remerged with cluster $C_{ij}$, the following conditions must be satisfied:

$$s_{(ij)i} = 0 \tag{1a}$$

$$s_{(ij)j} = 0 \tag{1b}$$

Generally, if a cluster $C_k \subset C_{ij}$, the similarity will be set to 0. In the following, this condition is called the *Special Subgraph Property*. Extending well-known linkage measures in this way, a finite hierarchical clustering algorithm is ensured.

In addition to complete subgroups, it is also possible to analyze clusters with common elements: $C_k \not\subset C_{ij}, C_k \cap C_{ij} \neq \emptyset$. This special situation is only possible in fuzzy mode. It depends on the used linkage measure whether it is useful to distinguish between common and uncommon data points in both clusters:

- If Group Average, Minimum Variance or Centroid based linkage metrics are used, it will make sense to use all data points for similarity calculation because outliers are tolerated.
- If Single Linkage is used, it will be helpful not to use common data points. Otherwise, the algorithm would prefer to merge clusters having common nodes.

If the algorithm is wanted to disregard common data points for similarity calculation, virtual clusters $C'_k$ and $C'_{ij}$ will be needed that only consist of uncommon data points. The linkage measures have to be modified to use these virtual clusters for similarity calculation which means that the algorithm follows a non-combinatorial strategy. It already fulfills the Special Subgraph Property because the similarity will be 0 if there are no uncommon data points. In the following, this method is called the *General Subgraph Property*.

### 2.1.2 Fuzzifier

In most application domains, there are many non-zero similarity entries. It is not appropriate to use all of them to generate a hierarchical cluster graph. Thus, it is very important to limit the degree of multiple assignment. Whether a similarity value of one cluster $C_i$ to another cluster $C_k$ is relevant depends on all available similarity values of cluster $C_i$. To take account of high similarity values only, a *fuzzifier* $f \in [0, 1]$ is multiplied by all similarity values (except for self similarity). Applying this similarity modification for each iteration, a similarity value over time is given:

$$s_{il}(t) = s_{il}(0) \cdot f^\gamma \quad \forall l \neq i , \tag{2a}$$

$$s_{jl}(t) = s_{jl}(0) \cdot f^\gamma \quad \forall l \neq j . \tag{2b}$$

In these equations $\gamma$ denotes how many times the cluster $C_i$ and $C_j$ have been selected for combination. If we additionally define a minimum similarity $\epsilon$, we will even improve this method by defining:

$$\forall l : s_{il}(0) \cdot f^\gamma < \epsilon \Rightarrow s_{il} = 0 , \tag{3a}$$

$$\forall l : s_{jl}(0) \cdot f^\gamma < \epsilon \Rightarrow s_{jl} = 0 . \tag{3b}$$

With the help of the *fuzzifier* it is possible to specify the degree of multiple assignment. A fuzzifier $f = 0$ creates crisp clusters. Allowing different fuzzifiers for each cluster level, it is even possible to adjust the degree of multiple assignment more flexibly.

### 2.1.3 Fuzzy Membership

The proposed clustering algorithm creates a directed acyclic graph providing a binary membership information whether a node is the child of the other node. Storing the similarity value $s_{ij}$ for each merging step to the corresponding edge in the cluster graph, a weighted graph is generated. It can be used to calculate a fuzzy membership degree directly after the end of the merging process. In the following, $C_i$ is a cluster node, $C_{ij}$ is the parent cluster generated of $C_i$ and $C_j$. $C_p, p = 1, \ldots, m$ are all clusters that have been merged with $C_i$ to other parents

$C_{ip}$. The membership of $C_i$ to its parent $C_{ij}$ can be calculated in this way:

$$\mu_{C_i.C_{ij}} = \frac{s_{ij}}{\sum_p s_{ip}} \tag{4}$$

## 2.2 Monotonic Behavior

Johnson and Milligan showed that hierarchical crisp clustering algorithms are monotonic except for centroid and median based linkage measures (Johnson 1967; Milligan 1979). In case of hierarchical crisp clustering this means that all these linkage measures induce an ultrametric (Johnson 1967). Pyramidal clustering – introduced by Diday – no longer induces an ultrametric but a Robinson metric while creating monotonic clusters (Diday 1987). The proposed clustering algorithm does not induce any metric due to the fuzziness. It is easy to show that in the fuzzy case the triangular inequality is no longer valid. Nevertheless a dissimilarity is induced.

The following analyzes whether the created hierarchical fuzzy clusters are monotonic for different linkage measures. Centroid and median based linkage methods are not analyzed because these measures are not monotonic in non-overlapping clusters (Milligan 1979). Like in crisp clustering algorithms, the similarity $s_{(ij)i}$ for $C_i \subset C_{ij}$ has to be 0. Thus the distinction has only to deal with overlapping clusters because otherwise the recurrence relation introduced by Lance and Williams (Lance and Williams 1966, 1967) could be used. To prove monotonic clusters, we use in the following a dissimilarity measure $d$ to make comparison to the literature easier. The General Subgraph Property and the Special Subgraph Property are handled separately.

### 2.2.1 Special Subgraph Property

Applying only the Special Subgraph Property, all elements are taken into account calculating the dissimilarity to other clusters. Therefore, the recurrence relations (Lance and Williams 1966, 1967) can be reused for Single Linkage, Complete Linkage and Weighted Average which therefore create monotonic clusters (Milligan 1979). The recurrence relations for Group Average and Minimum Variance (also known as Ward method) have to be adapted because the formulas do not take care of common elements which only appear once in the new cluster. The number of common elements has to be removed in $\alpha_1$, $\alpha_2$ and $\beta$ ($\gamma = 0$). Assuming non-empty clusters, it is sure that $\alpha_1 > 0$, $\alpha_2 > 0$ and $\alpha_1 + \alpha_2 + \beta \geq 1$ for both measures. This also means that $\gamma \geq -min(\alpha_1, \alpha_2)$. Hence, the created measures are monotonic (based on (Batagelj 1981)).

### 2.2.2  General Subgraph Property

Disregarding common elements, it is not possible to use the recurrence relation (Lance and Williams 1966, 1967) for overlapping clusters because $d_{ik}$, $d_{jk}$ and $d_{ij}$ may change. In this situation, the algorithm follows a non-combinatorial strategy so that the original data must be retained.

To analyze monotonic behavior, it has to be checked if $d_{ij} \leq d_{(ij)k}$ for each created cluster. If this is true, the algorithm will create monotonic clusters. The analysis has only to care about dissimilarities $d_{(ij)k}$ with $C_{ij} \cap C_k \neq \emptyset$.

- *Single Linkage*: The clustering algorithm ensures that $d_{ij} \leq d_{ik}$ and $d_{ij} \leq d_{jk}$. Without loss of generality assume that $e \in C_k$ is the nearest element to cluster $C_i$ and that $e$ is also element of cluster $C_j$. This means that $d_{(ij)k} > d_{ik}$ if common elements are not used. Otherwise if $e \notin C_j$, the dissimilarity $d_{(ij)k} = \min(d_{ik}, d_{jk}) \geq d_{ij}$. Therefore it is sure that $d_{ij} \leq d_{(ij)k}$ and the created cluster is monotonic.
- *Complete Linkage*: The clustering algorithm ensures that $d_{ij} \leq d_{ik}$ and $d_{ij} \leq d_{jk}$. Assuming that $e \in C_k$ is the furthest element to cluster $C_i$ and that $e$ is also element of cluster $C_j$. Than $d_{ik} \leq d_{ij}$ which is only possible if both dissimilarities are equal. Otherwise if $e \notin C_j$, the dissimilarity $d_{(ij)k} = \max(d_{ik}, d_{jk}) \geq d_{ij}$. Thus $d_{(ij)k} \geq d_{ij}$ and the created cluster is monotonic.
- *Group Average*, *Weighted Average* and *Minimum Variance*: Every measure uses all elements available in the cluster. Using only a part of each cluster for calculating dissimilarity values, it cannot be ensured to get monotonic clusters anymore.

Except for Single Linkage and Complete Linkage, the created fuzzy clusters are not necessarily monotonic any more.

## 2.3  Topic Groups

In most cases, the proposed binary merging process does not represent real data structures in terms of topic groups. This is why many different researchers have used partitioning clustering algorithms for hierarchical clustering tasks (Torra 2005; Mendes-Rodrigues and Sacks 2005; Bordogna and Pasi 2009). To take advantage of the deterministic algorithm proposed in this work, we introduce a different approach that generates topic groups with more than two elements.

Let $h$ be the maximum distance in the created directed acyclic graph. Comparing the similarity values of all nodes, it is possible to calculate the similarity differences for all edges in the created cluster graph. While $h$ is greater than a predefined level, merge these clusters with the smallest similarity differences. As maximum level, we used $h \in [5; 10]$ which is nearer to real world hierarchy structures and thus easier to interpret. This process is called the *Shrinking Process* in the following.

# 3 Evaluation

## 3.1 Data

Based on the selected similarity measure, the proposed algorithm was designed to create locally optimized clusters. Therefore, it does not make sense to evaluate the generated clusters with any internal criterion. Instead we use externally labeled data to analyze the fuzzy behavior of the proposed algorithm. As data source, seven different document collections have been randomly generated out of the RCV1 and RCV2 (German) corpora[1]. Each one consists of between $9,350$ and $26,501$ documents. The fuzzy behavior is analyzed for $f \in \{0.0, 0.1, \ldots, 0.8\}$ for level $1, 0$ else (in this scenario, it is only necessary to multiply assign documents, not cluster structures). To generate monotonic clusters, the algorithm uses Group Average with the Special Subgraph Property. All clusters are limited to level size $h = 5$ to take topic generation into account.

## 3.2 Methodology

Next to defuzzification methods to reuse crisp evaluation measures, many different specialized fuzzy evaluation methods have been proposed (e.g. Bäck and Hussain 1996; Nuovo and Catania 2007; Mendes and Sacks 2003) that have been applied to evaluate hierarchical clusters (e.g. Mendes-Rodrigues and Sacks 2005; Bordogna and Pasi 2009) although they only analyze partitions. Each of them assumes that each cluster can be assigned to one label which is not valid for the discussed application scenario with multi-labeled topic groups. Therefore a different quality measure is necessary.

To take care of multi-labeled clusters, each document label is propagated according to its membership value to the corresponding cluster parents. Representing all labels in vector form, it is possible to calculate the cosine for each document to the corresponding clusters. In this way, the *cosine quality measure* calculates how well the element fits into the cluster: The more the cosine value is to 1, the better the element fits into the cluster.

In addition to the quality itself, it is necessary to check if all relevant elements are assigned to a cluster. Unfortunately, it is quite difficult in a hierarchical cluster to define whether an element belongs to a cluster (e.g. Fig. 1). Therefore we analyze the fuzziness behavior by counting the number of multiple assignments per document.

---

[1] http://trec.nist.gov/data/reuters/reuters.html

**Fig. 1** Label propagation for
quality measure in a crisp
cluster example with multi
labeled clusters

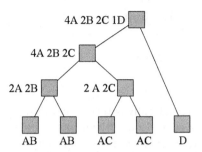

Table 1 Evaluation results: The cosine quality for all documents, the cosine quality related only to multi-assigned documents and the fuzziness is analyzed via minimum, quartiles, median and maximum

| f | Overall Quality | | | | | Multi assigned Quality | | | | | Fuzziness | | | | |
|---|---|---|---|---|---|---|---|---|---|---|---|---|---|---|---|
| | min | q1 | med | q2 | max | min | q1 | med | q2 | max | min | q1 | med | q2 | max |
| 0.0 | 0.01 | 0.91 | 0.98 | 1.00 | 1.00 | – | – | – | – | – | 1 | 1 | 1 | 1 | 1 |
| 0.1 | 0.01 | 0.91 | 0.98 | 1.00 | 1.00 | 0.00 | 0.90 | 0.98 | 1.00 | 1.00 | 1 | 1 | 1 | 2 | 2 |
| 0.2 | 0.00 | 0.90 | 0.98 | 1.00 | 1.00 | 0.00 | 0.87 | 0.98 | 1.00 | 1.00 | 1 | 1 | 1 | 2 | 3 |
| 0.3 | 0.00 | 0.90 | 0.98 | 1.00 | 1.00 | 0.00 | 0.87 | 0.98 | 1.00 | 1.00 | 1 | 1 | 1 | 2 | 3 |
| 0.4 | 0.00 | 0.90 | 0.98 | 1.00 | 1.00 | 0.00 | 0.88 | 0.98 | 1.00 | 1.00 | 1 | 1 | 1 | 3 | 4 |
| 0.5 | 0.00 | 0.89 | 0.98 | 1.00 | 1.00 | 0.00 | 0.89 | 0.98 | 1.00 | 1.00 | 1 | 1 | 1 | 3 | 5 |
| 0.6 | 0.00 | 0.89 | 0.98 | 1.00 | 1.00 | 0.00 | 0.89 | 0.98 | 1.00 | 1.00 | 1 | 1 | 1 | 3 | 7 |
| 0.7 | 0.00 | 0.88 | 0.98 | 1.00 | 1.00 | 0.00 | 0.90 | 0.98 | 1.00 | 1.00 | 1 | 1 | 1 | 3 | 10 |
| 0.8 | 0.00 | 0.88 | 0.98 | 1.00 | 1.00 | 0.00 | 0.90 | 0.98 | 1.00 | 1.00 | 1 | 1 | 1 | 3 | 15 |

## 3.3   Results

Table 1 shows the cosine quality and the fuzziness behavior for different fuzzifiers $f$. As it can be seen, the overall quality for all documents does not change much, nearly 75% of the data is not affected by quality changes. For 25% of the data the cosine quality decreases slightly. Analyzing only these elements, which have been multiply assigned, this quality reduction cannot be explained. Therefore the Shrinking Process must cause these quality changes: Higher fuzzifier values create larger clusters before the Shrinking Process is applied. More clusters have to be merged to limit the cluster size, which causes only very small quality differences. The number of multiple assignments increases with higher fuzzifier values: 25% of the data benefits from the possibility to be multiply assigned.

## 4   Conclusion

In this work, we have shown a fuzzification method for agglomerative hierarchical crisp clustering algorithms. The deterministic algorithm creates reproducible acyclic cluster graphs and therefore the possibility for arbitrary multiple assignments.

Well-known linkage methods can be reused, if the measure takes the Special Subgraph Property into account. This ensures a finite clustering algorithm and monotonic clusters except for centroid and median based linkage measures. Disregarding overlapping regions to calculate similarity values, monotonic clusters cannot be ensured except for Single Linkage and Complete Linkage. The data benefits from the multiple assignment while the cluster quality is hardly affected. The Shrinking Process was shown not to influence the cluster quality significantly.

It is obvious that with increasing fuzzifier $f$ the cluster process needs more computing time. Therefore, future work has to deal with speed improvements especially with the help of parallelization.

# References

Bäck C, Hussain M (1996) Validity measures for fuzzy partitions. In: Bock HH, Polasek W (eds) Data analysis and information systems. Springer, Berlin, pp 114–125

Batagelj V (1981) Note on ultrametric hierarchical clustering algorithms. Psychometrika 46:351–352

Bordogna G, Pasi G (2009) Hierarchical-hyperspherical divisive fuzzy c-means (h2d-fcm) clustering for information retrieval. In: WI-IAT '09: Proceedings of the 2009 IEEE/WIC/ACM International Joint Conference on Web Intelligence and Intelligent Agent Technology, IEEE Computer Society, pp 614–621

Diday E (1987) Orders and overlapping clusters by pyramids. Tech. Rep. RR-0730, INRIA, URL http://hal.inria.fr/inria-00075822/en/

Johnson SC (1967) Hierarchical clustering schemes. Psychometrika 32:241–254

Lance GN, Williams WT (1966) A generalized sorting strategy for computer classifications. Nature 212:218–219, DOI 10.1038/212218a0

Lance GN, Williams WT (1967) A general theory of classificatory sorting strategies 1. hierarchical systems. Comput J 9(4):373–380

Mendes MES, Sacks L (2003) Evaluating fuzzy clustering for relevance-based information access. In: IEEE International Conference on Fuzzy Systems, pp 648–653

Mendes-Rodrigues MES, Sacks L (2005) A scalable hierarchical fuzzy clustering algorithm for text mining. In: The 5th International Conference on Recent Advances in Soft Computing, URL http://lesacks.googlepages.com/rasc2004.pdf

Milligan GW (1979) Ultrametric hierarchical clustering algorithms. Psychometrika 44:343–346

Nuovo AGD, Catania V (2007) On external measures for validation of fuzzy partitions. In: Foundations of Fuzzy Logic and Soft Computing. Springer, Berlin, pp 491–501

Torra V (2005) Fuzzy c-means for fuzzy hierachical clustering. In: FUZZ '05: The 14th IEEE International Conference on Fuzzy Systems, pp 646–651

# An EM Algorithm for the Student-*t* Cluster-Weighted Modeling

Salvatore Ingrassia, Simona C. Minotti, and Giuseppe Incarbone

**Abstract** Cluster-Weighted Modeling is a flexible statistical framework for modeling local relationships in heterogeneous populations on the basis of weighted combinations of local models. Besides the traditional approach based on Gaussian assumptions, here we consider Cluster Weighted Modeling based on Student-*t* distributions. In this paper we present an EM algorithm for parameter estimation in Cluster-Weighted models according to the maximum likelihood approach.

## 1 Introduction

The functional dependence between some input vector **X** and output variable $Y$ based on data coming from a heterogeneous population $\Omega$, supposed to be constituted by $G$ homogeneous subpopulations $\Omega_1, \ldots, \Omega_g$, is often estimated using methods able to model local behavior like Finite Mixtures of Regressions (FMR) and Finite Mixtures of Regressions with Concomitant variables (FMRC), see e.g. Frühwirth-Schnatter (2005). As a matter of fact, the purpose of these models is to identify groups by taking into account the local relationships between some response variable $Y$ and some $d$-dimensional explanatory variables $\mathbf{X} = (X_1, \ldots, X_d)$. Here we focus on a different approach called *Cluster-Weighted Modeling* (CWM) proposed first in the context of media technology in

S. Ingrassia · G. Incarbone
Dipartimento di Impresa, Culture e Società, Università di Catania Corso Italia 55, - 95129 Catania, Italy
e-mail: s.ingrassia@unict.it; gincarbo@diit.unict.it

S.C. Minotti (✉)
Dipartimento di Statistica, Università di Milano-Bicocca Via Bicocca degli Arcimboldi 8 - 20126 Milano, Italy
e-mail: simona.minotti@unimib.it

W. Gaul et al. (eds.), *Challenges at the Interface of Data Analysis, Computer Science, and Optimization*, Studies in Classification, Data Analysis, and Knowledge Organization, DOI 10.1007/978-3-642-24466-7_2, © Springer-Verlag Berlin Heidelberg 2012

order to recreate a digital violin with traditional inputs and realistic sound, see Gershenfeld et al. (1999).

From a statistical point of view, CWM can be regarded as a flexible statistical framework for capturing local behavior in heterogeneous populations. In particular, while FMR considers the conditional probability density $p(y|\mathbf{x})$, the CWM approach models the joint probability density $p(\mathbf{x}, y)$ which is factorized as a weighted sum over $G$ clusters. More specifically, in CWM each cluster contains an input distribution $p(\mathbf{x}|\Omega_g)$ (i.e. a local model for the input variable $\mathbf{X}$) and an output distribution $p(y|\mathbf{x}, \Omega_g)$ (i.e. a local model for the dependence between $\mathbf{X}$ and $Y$). Some statistical properties of the Cluster-Weighted (CW) models have been established by Ingrassia et al. (2010); in particular it is shown that, under suitable hypotheses, CW models generalize FMR and FMRC. In the literature, CW models have been developed under Gaussian assumptions; moreover, a wider setting based on Student-$t$ distributions has been introduced and will be referred to as the Student-$t$ CWM. In this paper, we focus on the problem of parameter estimation in CW models according to the likelihood approach; in particular to this end we present an EM algorithm.

The rest of the paper is organized as follows: the Cluster-Weighted Modeling is introduced in Sect. 2; the EM algorithm for the estimation of the CWM is described in Sect. 3; in Sect. 4 we provide conclusions and ideas for further research.

## 2   The Cluster-Weighted Modeling

Let $(\mathbf{X}, Y)$ be a pair of a random vector $\mathbf{X}$ and a random variable $Y$ defined on $\Omega$ with joint probability distribution $p(\mathbf{x}, y)$, where $\mathbf{X}$ is the $d$-dimensional input vector with values in some space $\mathcal{X} \subseteq \mathbb{R}^d$ and $Y$ is a response variable having values in $\mathcal{Y} \subseteq \mathbb{R}$. Thus $(\mathbf{x}, y) \in \mathbb{R}^{d+1}$. Assume that $\Omega$ can be partitioned into $G$ disjoint groups, say $\Omega_1, \ldots, \Omega_G$, that is $\Omega = \Omega_1 \cup \cdots \cup \Omega_G$. *Cluster-Weighted Modeling* (CWM) decomposes the joint probability of $(\mathbf{X}, Y)$ as:

$$p(\mathbf{x}, y; \boldsymbol{\theta}) = \sum_{g=1}^{G} p(y|\mathbf{x}, \Omega_g)\, p(\mathbf{x}|\Omega_g)\, \pi_g, \tag{1}$$

where $\pi_g = p(\Omega_g)$ is the mixing weight of group $\Omega_g$, $p(\mathbf{x}|\Omega_g)$ is the probability density of $\mathbf{x}$ given $\Omega_g$ and $p(y|\mathbf{x}, \Omega_g)$ is the conditional density of the response variable $Y$ given the predictor vector $\mathbf{x}$ and the group $\Omega_g$, $g = 1, \ldots, G$, see Gershenfeld et al. (1999). Vector $\boldsymbol{\theta}$ denotes the set of all parameters of the model. Hence, the joint density of $(\mathbf{X}, Y)$ can be viewed as a mixture of local models $p(y|\mathbf{x}, \Omega_g)$ weighted (in a broader sense) on both the local densities $p(\mathbf{x}|\Omega_g)$ and the mixing weights $\pi_g$. Throughout this paper we assume that the input-output relation can be written as $Y = \mu(\mathbf{x}; \boldsymbol{\beta}) + \varepsilon$, where $\mu(\mathbf{x}; \boldsymbol{\beta})$ is a given function of $\mathbf{x}$ (depending also on some parameters $\boldsymbol{\beta}$) and $\varepsilon$ is a random variable with zero mean and finite variance.

Usually, in the literature about CWM, the local densities $p(\mathbf{x}|\Omega_g)$ are assumed to be multivariate Gaussians with parameters $(\boldsymbol{\mu}_g, \boldsymbol{\Sigma}_g)$, that is $\mathbf{X}|\Omega_g \sim N_d(\boldsymbol{\mu}_g, \boldsymbol{\Sigma}_g)$, $g = 1, \ldots, G$; moreover, also the conditional densities $p(y|\mathbf{x}, \Omega_g)$ are often modeled by Gaussian distributions with variance $\sigma_{\varepsilon,g}^2$ around some deterministic function of $\mathbf{x}$, say $\mu_g(\mathbf{x}; \boldsymbol{\beta})$, $g = 1, \ldots, G$. Thus $p(\mathbf{x}|\Omega_g) = \phi_d(\mathbf{x}; \boldsymbol{\mu}_g, \boldsymbol{\Sigma}_g)$, where $\phi_d$ denotes the probability density of a $d$-dimensional multivariate Gaussian and $p(y|\mathbf{x}, \Omega_g) = \phi(y; \mu(\mathbf{x}; \boldsymbol{\beta}_g), \sigma_{\varepsilon,g}^2)$, where $\phi$ denotes the probability density of a uni-dimensional Gaussian. Such model will be referred to as the Gaussian CWM. In the simplest case, the conditional densities are based on linear mappings $\mu(\mathbf{x}; \boldsymbol{\beta}_g) = \mathbf{b}_g'\mathbf{x} + b_{g0}$, with $\boldsymbol{\beta} = (\mathbf{b}_g', b_{g0})'$, $\mathbf{b}_g \in \mathbb{R}^d$ and $b_{g0} \in \mathbb{R}$, yielding the linear Gaussian CWM:

$$p(\mathbf{x}, y; \boldsymbol{\theta}) = \sum_{g=1}^{G} \phi(y; \mathbf{b}_g'\mathbf{x} + b_{g0}, \sigma_{\varepsilon,g}^2)\, \phi_d(\mathbf{x}; \boldsymbol{\mu}_g, \boldsymbol{\Sigma}_g)\, \pi_g. \qquad (2)$$

Under suitable assumptions, one can show that model (2) leads to the same posterior probability of Finite Mixtures of Gaussians (FMG), Finite Mixtures of Regressions (FMR) and Finite Mixtures of Regressions with Concomitant variables (FMRC). In this sense, we shall say that CWM contains FMG, FMR and FMRC, as summarized in the following table, see Ingrassia et al. (2010) for details:

|       | $p(\mathbf{x}|\Omega_g)$ | $p(y|\mathbf{x}, \Omega_g)$ | assumption |
|-------|--------------------------|------------------------------|------------|
| FMG   | Gaussian | Gaussian | linear relationship between $\mathbf{X}$ and $Y$ |
| FMR   | none     | Gaussian | $(\boldsymbol{\mu}_g, \boldsymbol{\Sigma}_g) = (\boldsymbol{\mu}, \boldsymbol{\Sigma})$, g=1,…,G |
| FMRC  | none     | Gaussian | $\boldsymbol{\Sigma}_g = \boldsymbol{\Sigma}$ and $\pi_g = \pi$, g=1,…,G |

In order to provide more realistic tails for real-world data with respect to the Gaussian models and rely on a more robust estimation of parameters, the CWM with Student-t components has been introduced by Ingrassia et al. (2010). Let us assume that in model (1) both $p(\mathbf{x}|\Omega_g)$ and $p(y|\mathbf{x}, \Omega_g)$ are Student-$t$ densities. In particular, we assume that $\mathbf{X}|\Omega_g$ has a multivariate $t$ distribution with location parameter $\boldsymbol{\mu}_g$, inner product matrix $\boldsymbol{\Sigma}_g$ and $\nu_g$ degrees of freedom, that is $\mathbf{X}|\Omega_g \sim t_d(\boldsymbol{\mu}_g, \boldsymbol{\Sigma}_g, \nu_g)$, and $Y|\mathbf{x}, \Omega_g$ has a $t$ distribution with location parameter $\mu(\mathbf{x}; \boldsymbol{\beta}_g)$, scale parameter $\sigma_g^2$ and $\zeta_g$ degrees of freedom, that is $Y|\mathbf{x}, \Omega_g \sim t(\mu(\mathbf{x}; \boldsymbol{\beta}_g), \sigma_g^2, \zeta_g)$, so that

$$p(\mathbf{x}, y; \boldsymbol{\theta}) = \sum_{g=1}^{G} t(y; \mu(\mathbf{x}; \boldsymbol{\beta}_g), \sigma_g^2, \zeta_g)\, t_d(\mathbf{x}; \boldsymbol{\mu}_g, \boldsymbol{\Sigma}_g, \nu_g)\, \pi_g. \qquad (3)$$

This implies that, for $g = 1, \ldots, G$,

$$\mathbf{X}|\boldsymbol{\mu}_g, \boldsymbol{\Sigma}_g, \nu_g, U_g \sim N_d\left(\boldsymbol{\mu}_g, \frac{\boldsymbol{\Sigma}_g}{u_g}\right),$$

$$Y|\mu(\mathbf{x},\boldsymbol{\beta}_g),\sigma_g,\zeta_g,W_g \sim N\left(\mu(\mathbf{x},\boldsymbol{\beta}_g),\frac{\sigma_g}{w_g}\right),$$

where $U_g$ and $W_g$ are independent random variables such that

$$U_g|\boldsymbol{\mu}_g,\boldsymbol{\Sigma}_g,\nu_g \sim \Gamma\left(\frac{\nu_g}{2},\frac{\nu_g}{2}\right) \quad \text{and} \quad W_g|\mu(\mathbf{x},\boldsymbol{\beta}_g),\sigma_g,\zeta_g \sim \Gamma\left(\frac{\zeta_g}{2},\frac{\zeta_g}{2}\right). \quad (4)$$

Model (3) will be referred to as the Student-$t$ CWM; the special case in which $\mu(\mathbf{x};\boldsymbol{\beta}_g)$ is some linear mapping will be called the *linear t-CWM*. In particular, we remark that the linear $t$-CWM defines a wide family of densities which includes mixtures of multivariate $t$-distributions and mixtures of regressions with Student-$t$ errors.

Obviously, the two models have a different behavior. An example is illustrated using the NO dataset, which relates the concentration of nitric oxide in engine exhaust to the equivalence ratio, see Hurn et al. (2003). Data have been fitted using both Gaussian and Student-$t$ CWM. The two classifications differ by four units, which are indicated by circles around them (two units classified in either group 1 or group 2; two units classified in either group 2 or group 4), see Fig. 1. In particular, there are two units that the Gaussian CWM classifies in group 4 but which are a little bit far from the other points of the same group; instead, such units are classified in group 2 by means of the Student-$t$ CWM. If we consider the following *index of weighted model fitting* (IWF) $\mathscr{E}$ defined as:

$$\mathscr{E} = \left(\frac{1}{N}\sum_{n=1}^{N}\left[y_n - \left(\sum_{g=1}^{G}\mu(\mathbf{x}_n;\boldsymbol{\beta}_g)p(\Omega_g|\mathbf{x}_n,y_n)\right)\right]^2\right)^{1/2}, \quad (5)$$

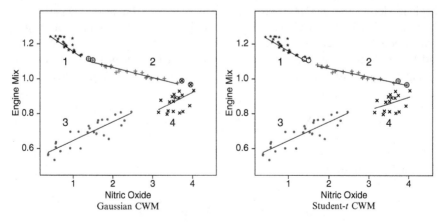

**Fig. 1** Gaussian and Student-$t$ CW models for the NO dataset

we get $\mathscr{E} = 0.108$ in the Gaussian case and $\mathscr{E} = 0.086$ in the Student CWM. Thus the latter model showed a slightly better fit than the Gaussian CWM.

## 3   CWM Parameter Estimation Via the EM Algorithm

In this section we present the main steps of the EM algorithm in order to estimate the parameters of CWM according to the likelihood approach. We illustrate here the Student-*t* distribution case; similar ideas can be easily developed in the Gaussian case. Let $(\mathbf{x}_1, y_1), \ldots, (\mathbf{x}_N, y_N)$ be a sample of $N$ independent observation pairs drawn from (3) and set $\mathbf{X} = (\mathbf{x}_1, \ldots, \mathbf{x}_N)$, $Y = (y_1, \ldots, y_N)$. Then, the likelihood function of the Student-*t* CWM is given by

$$L_0(\boldsymbol{\psi}; \underset{\sim}{\mathbf{X}}, \mathbf{Y}) = \prod_{n=1}^{N} p(\mathbf{x}_n, y_n; \boldsymbol{\psi}) =$$

$$\prod_{n=1}^{n} \left[ \sum_{g=1}^{G} p(y_n|\mathbf{x}_n; \boldsymbol{\beta}_g, \zeta_g) \, p_d(\mathbf{x}_n; \boldsymbol{\theta}_g, v_g) \, f(w; \zeta_g) \, f(u; v_g) \, \pi_g \right], \tag{6}$$

where $f(w; \zeta_g)$ and $f(u; v_g)$ denote the density functions of $W_g$ and $U_g$ respectively given in (4). Maximization of $L_0(\boldsymbol{\psi}; \underset{\sim}{\mathbf{X}}, \mathbf{Y})$ with respect to $\boldsymbol{\psi}$, for given data $(\mathbf{X}, \mathbf{Y})$, yields the estimate of $\boldsymbol{\psi}$. If we consider fully categorized data $\{(\mathbf{x}_n, y_n, \mathbf{z}_n, u_n, w_n) : n = 1, \ldots, N\}$, then the complete-data likelihood function can be written in the form

$$L_c(\boldsymbol{\psi}; \underset{\sim}{\mathbf{X}}, \mathbf{Y}) = \prod_{n,g} p(y_n|\mathbf{x}_n; \boldsymbol{\beta}_g, \zeta_g)^{z_{ng}} \, p_d(\mathbf{x}_n; \boldsymbol{\theta}_g, v_g)^{z_{ng}} \, f(w; \zeta_g)^{z_{ng}} \, f(u; v_g)^{z_{ng}} \, \pi_g^{z_{ng}},$$

where $z_{ng} = 1$ if $(\mathbf{X}_n, Y_n)$ comes from the $g$-th population and $z_{ng} = 0$ elsewhere.

Let us take the logarithm, then after some algebra we get:

$$\mathscr{L}_c(\boldsymbol{\psi}; \underset{\sim}{\mathbf{X}}, \mathbf{Y}) = \ln L_c(\boldsymbol{\psi}; \underset{\sim}{\mathbf{X}}, \mathbf{Y}) = \mathscr{L}_{1c}(\boldsymbol{\beta}) + \mathscr{L}_{2c}(\boldsymbol{\theta}) + \mathscr{L}_{3c}(\zeta) + \mathscr{L}_{4c}(v) + \mathscr{L}_{5c}(\pi)$$

where

$$\mathscr{L}_{1c}(\boldsymbol{\beta}) = \sum_{n=1}^{N} \sum_{g=1}^{G} z_{ng} \frac{1}{2} \left[ -\ln 2\pi + \ln w_n - \ln \sigma_{\epsilon,g}^2 - w_n \frac{(y_n - \mathbf{b}_g' \mathbf{x}_n - b_{0g})^2}{\sigma_{\epsilon,g}^2} \right]$$

$$\mathcal{L}_{2c}(\boldsymbol{\theta}) = \sum_{n=1}^{N}\sum_{g=1}^{G} z_{ng}\frac{1}{2}\left[-p\ln 2\pi + p\ln u_n - \ln|\boldsymbol{\Sigma}_g| - u_n(\mathbf{x}_n-\boldsymbol{\mu}_g)'\boldsymbol{\Sigma}_g^{-1}(\mathbf{x}_n-\boldsymbol{\mu}_g)\right]$$

$$\mathcal{L}_{3c}(\boldsymbol{\zeta}) = \sum_{n=1}^{N}\sum_{g=1}^{G} z_{ng}\left[-\ln\Gamma\left(\frac{\zeta_g}{2}\right) + \left(\frac{\zeta_g}{2}\right)\ln\left(\frac{\zeta_g}{2}\right) + \frac{\zeta_g}{2}(\ln w_n - w_n) - \ln w_n\right]$$

$$\mathcal{L}_{4c}(\boldsymbol{v}) = \sum_{n=1}^{N}\sum_{g=1}^{G} z_{ng}\left[-\ln\Gamma\left(\frac{v_g}{2}\right) + \left(\frac{v_g}{2}\right)\ln\left(\frac{v_g}{2}\right) + \frac{v_g}{2}(\ln u_n - u_n) - \ln u_n\right]$$

$$\mathcal{L}_{5c}(\boldsymbol{\pi}) = \sum_{n=1}^{N}\sum_{g=1}^{G} z_{ng}[\ln\pi_g].$$

The **E-step** on the $(k+1)$-th iteration of the EM algorithm requires the calculation of the conditional expectation of the complete-data loglikelihood function $\ln L_c(\boldsymbol{\psi};\mathbf{X},\mathbf{Y})$, say $Q(\boldsymbol{\psi};\boldsymbol{\psi}^{(k)})$, evaluated using the current fit $\boldsymbol{\psi}^{(k)}$ for $\boldsymbol{\psi}$. To this end, the quantities $\mathbb{E}_{\boldsymbol{\psi}^{(k)}}\{Z_{ng}|\mathbf{x}_n,y_n\}$, $\mathbb{E}_{\psi^{(k)}}\{U_n|\mathbf{x}_n,\mathbf{z}_n\}$, $\mathbb{E}_{\psi^{(k)}}\{\ln U_n|\mathbf{x}_n,\mathbf{z}_n\}$, $\mathbb{E}_{\psi^{(k)}}\{W_n|y_n,\mathbf{z}_n\}$ and $\mathbb{E}_{\psi^{(k)}}\{\ln W_n|y_n,\mathbf{z}_n\}$ have to be evaluated, for $n = 1,\ldots,N$ and $g = 1,\ldots,G$. It follows that

$$\mathbb{E}_{\boldsymbol{\psi}^{(k)}}\{Z_{ng}|\mathbf{x}_n,y_n\} = \tau_{ng}^{(k)} = \frac{\pi_g^{(k)} p(y_n|\mathbf{x}_n;\boldsymbol{\beta}_g^{(k)},\zeta_g^{(k)})p_d(\mathbf{x}_n;\boldsymbol{\theta}_g^{(k)},v_g^{(k)})}{\sum_{j=1}^{G}\pi_j^{(k)} p(y_n|\mathbf{x}_n;\boldsymbol{\beta}_j^{(k)},\zeta_j^{(k)})p_d(\mathbf{x}_n;\boldsymbol{\theta}_j^{(k)},v_j^{(k)})}$$

is the posterior probability that the $n$-th observation $(\mathbf{x}_n,y_n)$ belongs to the $g$-th component of the mixture. Further, we have that

$$\mathbb{E}_{\psi^{(k)}}\{U_n|\mathbf{x}_n,z_n\} = u_{ng}^{(k)} = \frac{v_g^{(k)}+d}{v_g^{(k)} + \left(\mathbf{x}_n-\boldsymbol{\mu}_g^{(k)}\right)'\boldsymbol{\Sigma}_g^{-1(k)}\left(\mathbf{x}_n-\boldsymbol{\mu}_g^{(k)}\right)}$$

$$\mathbb{E}_{\psi^{(k)}}\{\ln U_n|\mathbf{x}_n,z_n\} = \ln u_{ng}^{(k)} + \left\{\psi\left(\frac{v_g^{(k)}+d}{2}\right) - \ln\left(\frac{v_g^{(k)}+d}{2}\right)\right\}$$

$$\mathbb{E}_{\psi^{(k)}}\{W_n|y_n,z_n\} = w_{ng}^{(k)} = \frac{\zeta_g^{(k)}+1}{\zeta_g^{(k)} + \frac{\left(y_n - \mathbf{b}_g'^{(k)}\mathbf{x}_n - b_{g0}^{(k)}\right)^2}{\sigma_{\epsilon,g}^{2(k)}}}$$

$$\mathbb{E}_{\psi^{(k)}}\{\ln W_n|y_n,z_n\} = \ln w_{ng}^{(k)} + \left\{\psi\left(\frac{\zeta_g^{(k)}+1}{2}\right) - \ln\left(\frac{\zeta_g^{(k)}+1}{2}\right)\right\},$$

where $\psi(s) = \{\partial\Gamma(s)/\partial s\}/\Gamma(s)$.

In the **M-step**, on the $(k+1)$-th iteration of the EM algorithm, we maximize the conditional expectation of the complete-data loglikelihood $Q$, with respect to

$\boldsymbol{\psi}$. The solutions for the posterior probabilities $\pi_g^{(k+1)}$ and the parameters $\boldsymbol{\mu}_g^{(k+1)}$, $\boldsymbol{\Sigma}_g^{(k+1)}$ of the local input densities $p_d(\mathbf{x}_n|\boldsymbol{\theta}_g, v_g)$, for $g = 1,\ldots,G$, exist in a closed form and coincide with the case of mixtures of multivariate $t$ distributions (Peel and McLachlan 2000), that is:

$$\pi_g^{(k+1)} = \frac{1}{N}\sum_{n=1}^{N}\tau_{ng}^{(k)}$$

$$\boldsymbol{\mu}_g^{(k+1)} = \frac{\sum_{n=1}^{N}\tau_{ng}^{(k)}u_{ng}^{(k)}\mathbf{x}_n}{\sum_{n=1}^{N}\tau_{ng}^{(k)}u_{ng}^{(k)}}$$

$$\boldsymbol{\Sigma}_g^{(k+1)} = \frac{\sum_{n=1}^{N}\tau_{ng}^{(k)}u_{ng}^{(k)}(\mathbf{x}_n - \boldsymbol{\mu}_g^{(k+1)})(\mathbf{x}_n - \boldsymbol{\mu}_g^{(k+1)})'}{\sum_{n=1}^{N}\tau_{ng}^{(k)}u_{ng}^{(k)}}.$$

The updates $\mathbf{b}_g^{(k+1)}, b_{g0}^{(k+1)}$ and $\sigma_{\epsilon,g}^{2(k+1)}$ of the parameters for the local output densities $p(y_n|\mathbf{x}_n; \boldsymbol{\beta}_g, \zeta_g)$, for $g = 1,\ldots,G$, are obtained by the solution of the equations

$$\frac{\partial\mathbb{E}_{\psi^{(k)}}\{\mathcal{L}_c(\boldsymbol{\psi}|\mathbf{x}_n, y_n)\}}{\partial\mathbf{b}_g'} = \mathbf{0}'$$

$$\frac{\partial\mathbb{E}_{\psi^{(k)}}\{\mathcal{L}_c(\boldsymbol{\psi}|\mathbf{x}_n, y_n)\}}{\partial b_{g0}} = 0$$

$$\frac{\partial\mathbb{E}_{\psi^{(k)}}\{\mathcal{L}_c(\boldsymbol{\psi}|\mathbf{x}_n, y_n)\}}{\partial\sigma_{\epsilon,g}^2} = 0$$

and it can be proved that they are given by

$$\mathbf{b}_g'^{(k+1)} = \left(\frac{\sum_{n=1}^{N}\tau_{ng}^{(k)}w_{ng}^{(k)}y_n\mathbf{x}_n'}{\sum_{n=1}^{N}\tau_{ng}^{(k)}w_{ng}^{(k)}} - \frac{\sum_{n=1}^{N}\tau_{ng}^{(k)}w_{ng}^{(k)}y_n}{\sum_{n=1}^{N}\tau_{ng}^{(k)}w_{ng}^{(k)}}\frac{\sum_{n=1}^{N}\tau_{ng}^{(k)}w_{ng}^{(k)}\mathbf{x}_n'}{\sum_{n=1}^{N}\tau_{ng}^{(k)}w_{ng}^{(k)}}\right)\cdot$$

$$\cdot\left(\frac{\sum_{n=1}^{N}\tau_{ng}^{(k)}w_{ng}^{(k)}\mathbf{x}_n\mathbf{x}_n'}{\sum_{n=1}^{N}\tau_{ng}^{(k)}w_{ng}^{(k)}} - \left(\frac{\sum_{n=1}^{N}\tau_{ng}^{(k)}w_{ng}^{(k)}\mathbf{x}_n'}{\sum_{n=1}^{N}\tau_{ng}^{(k)}w_{ng}^{(k)}}\right)^2\right)^{-1}$$

$$b_{g0}^{(k+1)} = \frac{\sum_{n=1}^{N}\tau_{ng}^{(k)}w_{ng}^{(k)}y_n}{\sum_{n=1}^{N}\tau_{ng}^{(k)}w_{ng}^{(k)}} - \mathbf{b}_g^{(k+1)}\frac{\sum_{n=1}^{N}\tau_{ng}^{(k)}w_{ng}^{(k)}\mathbf{x}_n'}{\sum_{n=1}^{N}\tau_{ng}^{(k)}w_{ng}^{(k)}}$$

$$\sigma_{\epsilon,g}^{2(k+1)} = \frac{\sum_{n=1}^{N}\tau_{ng}^{(k)}w_{ng}^{(k)}[y_n - (\mathbf{b}_g^{(k+1)})'\mathbf{x}_n' + b_{g0}^{(k+1)})]^2}{\sum_{n=1}^{N}\tau_{ng}^{(k)}w_{ng}^{(k)}}.$$

The updates $v_g^{(k+1)}$ and $\zeta_g^{(k+1)}$ for the degrees of freedom $v_g$ and $\zeta_g$ need to be computed iteratively and are given by the solutions of the following equations, respectively:

$$-\psi\left(\frac{v_g}{2}\right) + \ln\left(\frac{v_g}{2}\right) + 1 + \frac{1}{\sum_{n=1}^{N}\tau_{ng}^{(k)}} \cdot \sum_{n=1}^{N}\tau_{ng}^{(k)}\left(\ln u_{ng}^{(k)} - u_{ng}^{(k)}\right) +$$

$$+ \psi\left(\frac{v_g^{(k)}+d}{2}\right) - \ln\left(\frac{v_g^{(k)}+d}{2}\right) = 0$$

and

$$-\psi\left(\frac{\zeta_g}{2}\right) + \ln\left(\frac{\zeta_g}{2}\right) + 1 + \frac{1}{\sum_{n=1}^{N}\tau_{ng}^{(k)}} \cdot \sum_{n=1}^{N}\tau_{ng}^{(k)}\left(\ln w_{ng}^{(k)} - w_{ng}^{(k)}\right) +$$

$$+ \psi\left(\frac{\zeta_g^{(k)}+1}{2}\right) - \ln\left(\frac{\zeta_g^{(k)}+1}{2}\right) = 0$$

where $\psi(s) = \{\partial\Gamma(s)/\partial s\}/\Gamma(s)$.

## 4 Concluding Remarks

CWM is a flexible approach for clustering and modeling functional relationships based on data coming from a heterogeneous population. Besides traditional modeling based on Gaussian assumptions, in order to provide more realistic tails for real-world data and rely on a more robust estimation of parameters, we considered also CWM based on Student-$t$ distributions. Here we focused on a specific issue concerning parameter estimates according to the likelihood approach. For this aim, in this paper we presented an EM algorithm which has been implemented in Matlab and R language for both the Gaussian and the Student-$t$ case.

Our simulations showed that the initialization of the algorithms is quite critical, in particular the initial guess has been chosen according to both a preliminary clustering of data using a $k$-means algorithm and a random grouping of data, but our numerical studies pointed out that there is no overwhelming strategy. Similar results have been obtained by Faria and Soromenho (2010) in the area of mixtures of regressions, where the performance of the algorithm depends essentially on the configuration of the true regression lines and the initialization of the algorithms.

Finally we remark that, in order to reduce such critical aspects, suitable constraints on the eigenvalues of the covariance matrices could be implemented. This provides ideas for future work.

# References

Faria S, Soromenho G (2010) Fitting mixtures of linear regressions. J Stat Comput Simulat 80:201–225

Frühwirth-Schnatter S (2005) Finite mixture and markov switching models. Springer, Heidelberg

Gershenfeld N, Schöner B, Metois E (1999) Cluster-weighted modeling for time-series analysis. Nature 397:329–332

Hurn M, Justel A, Robert CP (2003) Estimating mixtures of regressions. J Comput Graph Stat 12:55–79

Ingrassia S, Minotti SC, Vittadini G (2010) Local statistical modeling via the cluster-weighted approach with elliptical distributions. ArXiv: 0911.2634v2

Peel D, McLachlan GJ (2000) Robust mixture modelling using the $t$ distribution. Stat Comput 10:339–348

## References

# Analysis of Distribution Valued Dissimilarity Data

Masahiro Mizuta and Hiroyuki Minami

**Abstract** We deal with methods for analyzing complex structured data, especially, distribution valued data. Nowadays, there are many requests to analyze various types of data including spatial data, time series data, functional data and symbolic data. The idea of symbolic data analysis proposed by Diday covers a large range of data structures. We focus on distribution valued dissimilarity data and multidimensional scaling (MDS) for these kinds of data. MDS is a powerful tool for analyzing dissimilarity data. The purpose of MDS is to construct a configuration of the objects from dissimilarities between objects. In conventional MDS, the input dissimilarity data are assumed (non-negative) real values. Dissimilarities between objects are sometime given probabilistically; dissimilarity data may be represented as distributions. We assume that the distributions between objects $i$ and $j$ are non-central chi-square distributions $\chi^2(p, \delta_{ij}/\gamma_{ij})$ multiplied by a scalar (say $\gamma_{ij}$), i.e. $s_{ij} \sim \gamma_{ij} \chi^2(p, \delta_{ij}/\gamma_{ij})$. We propose a method of MDS under this assumption; the purpose of the method is to construct a configuration; $x_i \sim N(\mu_i, \alpha_i^2 I_p), i = 1, 2, \cdots, n$.

## 1 Introduction

Most methods for data analysis assume that the data are sets of numbers with structures. For example, typical multivariate data are identified as a set of $n$ vectors of $p$ real numbers and dissimilarity data on pairs of $n$ objects are represented as a $n$ by $n$ matrix. However, there is an increasing need for the analysis of various types of data. We must analyze huge and complex data sets. Conventional methods for data analysis cannot be applied to these data sets directly. In such a case, an approach

M. Mizuta (✉) · H. Minami
Hokkaido University, Japan
e-mail: mizuta@iic.hokudai.ac.jp; min@iic.hokudai.ac.jp

W. Gaul et al. (eds.), *Challenges at the Interface of Data Analysis, Computer Science, and Optimization*, Studies in Classification, Data Analysis, and Knowledge Organization, DOI 10.1007/978-3-642-24466-7_3, © Springer-Verlag Berlin Heidelberg 2012

is to classify the observations into distinct groups and to analyze the groups. But, the groups are not ordinary observations but sets, distributions, intervals etc.

In the 1980s, Diday proposed *SDA (Symbolic Data Analysis)* to deal with these kinds of data (Diday and Bock 1987). We can learn SDA from many good textbooks (Bock and Diday 2000; Diday and Noirhomme-Fraiture 2008; Billard and Diday 2007 *etc.*) From another viewpoint, SDA is a method for second-level objects, i.e. classes, categories and concepts. In conventional data analysis, we assume that we can get observations (e.g., a dog i.e. individual or first-level object). If we would like to analyze a species of dogs, it is appropriate for us to analyze second-level objects as they are.

There are many ways for descriptions of symbolic objects (or concepts) including interval values, modal interval values, and distribution values. Among them, the use of interval values is a simple extension of the real values. But, it is impossible to investigate the status for individuals in the interval. The use of distribution values is one solution to consider them.

We sometimes get dissimilarities between objects and have to analyze them. It is nothing unusual to have dissimilarities probabilistically. We can assume that the dissimilarities are distributions. There are two major approaches to analyze dissimilarity data: cluster analysis and multidimensional scaling (MDS).

We focus on distribution valued data and MDS for these kinds of data. The most fundamental and important issue is a family of distributions. We adopt non-central chi-square distributions for dissimilarity data and propose a method for MDS.

## 2 Distribution Valued Data and Multidimensional Scaling

In SDA, targets of the analysis are called *concepts*. A set of individuals is a typical concept in SDA. From the view point of data analysis, the description of concept is very important. In order to simplify, we assume that targets are sets of individuals and individuals are measured by $p$ dimensional vectors. A basic method for description of a concept in this case is a $p$ dimensional interval value with the maximum value and the minimum value among the individuals belonging to the concept for each dimension. But, the description with an interval value is not so robust for outliers. Another description is a distribution: e.g. a normal, chi-square, multidimensional normal distribution, etc.

We quickly review a conventional MDS for preparation. It is easy to calculate a distance between two points which have coordinates. But the inverse problem is not easy. When we get dissimilarities between objects, the purpose of MDS is to construct a configuration of the object in $p$ dimensional space. One of the simple MDS methods is Torgerson's method (Mardia et al. 1979 *etc.*) The method is described here. For dissimilarity data $S = \{s_{ij}\}$, we calculate

$$b_{ij} := -\frac{1}{2}(s_{ij}^2 - s_{i.}^2 - s_{.j}^2 + s_{..}^2),$$

where $s_{i.}^2 = \frac{1}{n} \sum_{j=1}^n s_{ij}^2, s_{.j}^2 = \frac{1}{n} \sum_{i=1}^n s_{ij}^2$, and $s_{..}^2 = \frac{1}{n^2} \sum_{i,j}^n s_{ij}^2$.

The matrix $B = (b_{ij})$ is decomposed by a singular value decomposition; $B = T\Lambda T^T$, where $\Lambda = diag(l_1, l_2, \cdots, l_n)$ and $T = (t_1, \cdots, t_n)$. We can get a configuration $X = (t_1, \cdots, t_r)diag(\sqrt{l_1}, \cdots, \sqrt{l_r})$. The criterion of Torgerson's method is to minimize the differences between the input dissimilarities and the distances between objects in the constructed configuration. Other famous method of MDS is Kruskal's method. The criterion of it is to minimize the differences of *orders* of the input and configuration.

## 3   MDS for Distribution Valued Data

In this section, we propose a method of MDS for distribution valued dissimilarity data. It is natural that the configurations as the result of MDS are distributions when the input dissimilarities are distributions. For preparation, we discuss *distance* between two normal distributions. We assume that the $i$-th distribution is a $p$ dimensional normal distribution $\xi_i \sim N(\mu_i, \alpha_i^2 I_p)$, where $n$ is the size of objects $(i = 1, 2, \cdots, n)$.

Property 1.
Because $x_i \sim N(\mu_i - \mu_j, (\alpha_i^2 + \alpha_j^2) I_p)$, $(x_i - x_j)^T ((\alpha_i^2 + \alpha_j^2) I_p)^{-1}(x_i - x_j)$ follows a non-central chi-square distribution with $p$ degrees of freedom and noncentrality parameter $\frac{1}{\alpha_i^2 + \alpha_j^2}(\mu_i - \mu_j)^T (\mu_i - \mu_j)$. $\|x_i - x_j\|^2$ follows a non-central chi-square distribution $\chi^2(p, \frac{1}{\alpha_i^2 + \alpha_j^2}(\mu_i - \mu_j)^T (\mu_i - \mu_j))$ multiplied by a constant $\alpha_i^2 + \alpha_j^2$.

Property 2.
Mean and variance of the non-central chi-square distribution $\chi^2(m, \delta)$ are $m + \delta$ and $2(m + 2\delta)$, respectively.

The outline of the proposed method is here; after the input dissimilarities are approximated with noncentral chi-square distributions, we derive set of normal distributions $x_i \sim N(\mu_i, \alpha_i^2 I_p)$ corresponding to objects.

At a first step, we approximate distributions of dissimilarities with noncentral chi-square distributions. We adopt moment methods; we estimate the parameters of the noncentral chi-square distribution $\delta$ and $\gamma_{ij}$ using means and variances of input dissimilarities by Property 2.

We put the estimated dissimilarity distributions as $s_{ij} = \gamma_{ij} \chi^2(p, \delta_{ij})$ for object $i$ and object $j$ and put configuration of the $i$-th object as $x_i \sim N(\mu_i, \alpha_i^2 I_p)$. We get two kinds of distributions:

1. Distributions of the input dissimilarities between the $i$-th and $j$-th objects: $s_{ij} = \gamma_{ij} \chi^2(p, \delta_{ij})$
2. Distributions of the squared distances between the $i$-th and $j$-th normal distributions : $(\alpha_i^2 + \alpha_j^2)\chi^2(p, \frac{1}{\alpha_i^2 + \alpha_j^2}(\mu_i - \mu_j)^T (\mu_i - \mu_j))$.

The goal of the proposed method is to find out parameters $\alpha_i, \mu_i; i = 1, 2, \cdots, n$ for which the two distributions (1) and (2) are closely similar.

Because both distributions are non-central chi-square distributions, we can compare them with parameters:

$$\begin{cases} \gamma_{ij} \approx \alpha_i^2 + \alpha_j^2 \\ \delta_{ij} \approx \frac{1}{\alpha_i^2 + \alpha_j^2}(\mu_i - \mu_j)^T(\mu_i - \mu_j). \end{cases}$$

We can calculate $\mu_i$ with ordinal MDS method, for example Torgerson's method, because $\delta_{ij}, \gamma_{ij}$ are known. $\frac{1}{2}(\gamma_{ij} - \gamma_{kj} + \gamma_{ik}) \sim \frac{1}{2}(\alpha_i^2 + \alpha_j^2 - \alpha_k^2 - \alpha_j^2 + \alpha_i^2 + \alpha_k^2) = \alpha_i^2$ for any $j, k$. From the averages of them

$$\frac{1}{2n^2}\sum_{j=1}^{n}\sum_{k=1}^{n}(\gamma_{ij} - \gamma_{kj} + \gamma_{ik}) \sim \alpha_i^2,$$

we get $\alpha_i$. $N(\mu_i, \alpha_i^2 I_p)$ is the configuration of the objects.

## 4   Actual Example

We will apply an actual data set to the proposed method. The Internet is indispensable for many people. We can usually access any site quickly. But sometimes we must wait for a long time. This is a problem of turn around time or round trip time (RTT). Round trip time is determined by many factors. It is natural that the round trip time follows a distribution.

The Resilient Project in Japan placed around 30 servers in many prefectures in Japan and collected ICMP RTT data every 5 min over years. We use a part of the data: 9 sites, 1728 trials for each pair of sites. Figure 1 is a histogram of RTT between site 4 and site 8. We can get configurations of the 9 sites as normal distributions. Figure 2 shows the configurations for the results, where the centers of circles are average vectors, and the radiuses represent 3 sigma.

The configuration shows the relations among sites; actually the sites 3, 5 and 6 are located at Tokyo and relatively stable compared with the other sites. The geographic distance between site 4 and site 8 is small. But, they belong to different Internet providers.

## 5   Concluding Remarks

In this paper, we deal with the analysis of distribution valued dissimilarity data. Cluster analysis and MDS are effective methods for dissimilarity data in general.

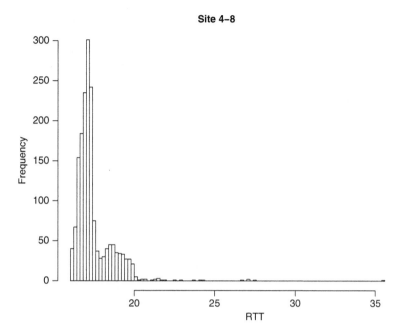

**Fig. 1** A Histogram of Round Trip Time

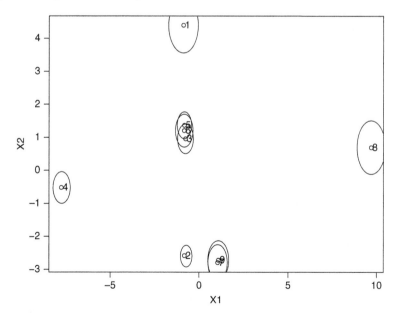

**Fig. 2** Configuration of Sites with the Proposed MDS Method

We focused on MDS and proposed a method for dealing with distribution valued dissimilarity data.

For the concluding remarks, interval valued data and distribution valued data will be compared. In the case that individuals are described as $p$ dimensional vectors, $2p$ parameters are needed for interval valued data and $p(p+3)/2$ parameters for distribution valued data. When $p = 1$ or 2, the difference between $2p$ and $p(p+3)/2$ is small. But, as $p$ increases, $p(p+2)/2$ increases rapidly. It is reasonable that dissimilarity is represented by one dimensional value or distribution. At least for dissimilarity data, distribution valued data are useful for data analysis.

We have reported MDS for functional data (Mizuta 2000, 2003). MDS for functional dissimilarity valued data is a future work.

# References

Billard L, Diday E (2007) Symbolic data analysis. Wiley, Chichester, UK

Bock HH, Diday E (2000) Analysis of symbolic data. Springer, Berlin

Diday E, Bock HH (1987) The symbolic approach in clustering and related methods of data analysis. In: Classification and related methods of data analysis. North-Holland, Amsterdam, pp 673–684

Diday E, Noirhomme-Fraiture M (2008) Symbolic data analysis and the SODAS software. Wiley, Chichester, UK

Mardia KV, Kent JT, Bibby JM (1979) Multivariate analysis. Academic, London, San Diego

Mizuta M (2000) Functional multidimensional scaling. In: Proceedings of the Tenth Japan and Korea Joint Conference of Statistics, pp 77–82

Mizuta M (2003) Multidimensional scaling for dissimilarity functions with continuous argument. J Jpn Soc Comput Stat 15(2):327–333

# An Overall Index for Comparing Hierarchical Clusterings

## I. Morlini and S. Zani

**Abstract** In this paper we suggest a new index for measuring the distance between two hierarchical clusterings. This index can be decomposed into the contributions pertaining to each stage of the hierarchies. We show the relations of such components with the currently used criteria for comparing two partitions. We obtain a similarity index as the complement to one of the suggested distances and we propose its adjustment for agreement due to chance. We consider the extension of the proposed distance and similarity measures to more than two dendrograms and their use for the consensus of classification and variable selection in cluster analysis.

## 1 Introduction

In cluster analysis, one may be interested in comparing two or more hierarchical clusterings obtained for the same set of $n$ objects. Indeed, different clusterings may be obtained by using different linkages, different distances or different sets of variables. In the literature the most popular measures have been proposed for comparing two partitions obtained by cutting the trees at a certain stage of the two hirarchical procedures (Rand (1971); Fowlkes and Mallows (1983); Hubert and Arabie (1985); Meila (2007); Youness and Saporta (2010)). Less attention has been devoted to the comparison of the global results of two hierarchical classifications, i.e. two dendrograms obtained for the same set of objects. Sokal and Rohlf (1962) have

I. Morlini (✉)
Department of Economics, University of Modena and Reggio Emilia, Via Berengario 51, 41100 Modena, Italy
e-mail: isabella.morlini@unimore.it

S. Zani
Department of Economics, University of Parma, Via Kennedy 6, 43100 Parma, Italy
e-mail: sergio.zani@unipr.it

W. Gaul et al. (eds.), *Challenges at the Interface of Data Analysis, Computer Science, and Optimization*, Studies in Classification, Data Analysis, and Knowledge Organization, DOI 10.1007/978-3-642-24466-7_4, © Springer-Verlag Berlin Heidelberg 2012

introduced the so-called cophenetic correlation coefficient (see also Rohlf 1982 and Lapointe and Legendre 1995). Baker (1974) has proposed the rank correlation between stages where pairs of objects combine in the tree for measuring the similarity between two hierarchical clusterings. Reilly et al. (2005) have discussed the use of Cohen's kappa in studying the agreement between two classifications.

In this work we suggest a new index for measuring the dissimilarity between two hierarchical clusterings. This index is a distance and can be decomposed into the contributions pertaining to each stage of the hierarchies. In Sect. 2 we define the new index for two dendrograms. We then present its properties and its decomposition with reference to each stage. Section 3 shows the relations of each component of the index with the currently used criteria for comparing two partitions. Section 4 considers the similarity index obtained as the complement to one of the suggested distances and shows that its single components obtained at each stage of the hierarchies can be related to the measure $B_k$ suggested by Fowlkes and Mallows (1983). This section also deals with the adjustment of the similarity index for agreement due to chance. Section 5 considers the extension of the overall index to more than two clusterings. Section 6 gives some concluding remarks.

## 2  The Index and Its Properties

Suppose we have two hierarchical clusterings of the same number of objects, $n$. Let us consider the $N = n(n - 1)/2$ pairs of objects and let us define, for each non trivial partition in $k$ groups ($k = 2, \ldots, n - 1$), a binary variable $X_k$ with values $x_{ik} = 1$ if objects in pair $i$ ($i = 1, \ldots, N$) are classified in the same cluster in partition in $k$ groups and $x_{ik} = 0$ otherwise. A binary ($N \times (n - 2)$) matrix $\mathbf{X}_g$ for each clustering $g$ ($g = 1, 2$) may be derived, in which the columns are the binary variables $X_k$. A global measure of dissimilarity between the two clusterings may be defined as follows:

$$Z = \frac{\| \mathbf{X}_1 - \mathbf{X}_2 \|}{\| \mathbf{X}_1 \| + \| \mathbf{X}_2 \|} \tag{1}$$

where $\| \mathbf{A} \| = \sum_i \sum_k \| a_{ik} \|$ is the $L_1$ norm of the matrix $\mathbf{A}$. In expression (1), since the matrices involved take only binary values, the $L_1$ norm is equal to the square of the $L_2$ norm.

Index $Z$ has the following properties:

- It is bounded in [0,1].
- $Z = 0$ if and only if the two hierarchical clusterings are identical and $Z = 1$ when the two clusterings have the maximum degree of dissimilarity, that is when for each partition in $k$ groups and for each $i$, objects in pair $i$ are in the same group in clustering 1 and in two different groups in clustering 2 (or vice versa).
- It is a distance, since it satisfies the conditions of non negativity, identity, symmetry and triangular inequality (Zani (1986)).

- The complement to 1 of $Z$ is a similarity measure, since it satisfies the conditions of non negativity, normalization and symmetry.
- It does not depend on the group labels since it refers to pairs of objects.
- It may be decomposed in $(n - 2)$ parts related to each pair of partitions in $k$ groups since:

$$Z = \sum_k Z_k = \sum_k \sum_i \frac{|x_{1ik} - x_{2ik}|}{\| \mathbf{X}_1 \| + \| \mathbf{X}_2 \|} \qquad (2)$$

The plot of $Z_k$ versus $k$ shows the distance between the two clusterings at each stage of the procedure.

## 3  The Comparison of Two Partitions in $k$ Groups

Let us consider the comparison between two partitions in $k$ groups obtained at a certain stage of the hierarchical procedures. The measurement of agreement between two partitions of the same set of objects is a well-known problem in the classification literature and different approaches have been suggested (see, i.e., Brusco and Steinley 2008; Denoeud 2008). In order to highlight the relation of the suggested index with the ones proposed in the literature, we present the so-called matching matrix $M_k = [m_{fj}]$ where $m_{fj}$ indicates the number of objects placed in cluster $f$ $(f = 1, \ldots, k)$ according to the first partition and in cluster $j$ $(j = 1, \ldots, k)$, according to the second partition (Table 1). Information in Table 1 can be collapsed in a $(2 \times 2)$ contingency table, showing the cluster membership of the object pairs in each of the two partitions (Table 2).

The number of pairs which are placed in the same cluster according to both partitions is

$$T_k = \sum_{f=1}^{k} \sum_{j=1}^{k} \binom{m_{fj}}{2} = \frac{1}{2} \left[ \sum_{f=1}^{k} \sum_{j=1}^{k} m_{fj}^2 - n \right] \qquad (3)$$

**Table 1** Matching matrix $M_k$

|       | 1        | $\ldots$ | $j$      | $\ldots$ | $k$      | Total     |
|-------|----------|----------|----------|----------|----------|-----------|
| 1     | $m_{11}$ | $\ldots$ | $\ldots$ | $\ldots$ | $m_{1k}$ | $m_{1.}$  |
| 2     | $m_{21}$ | $\ldots$ | $\ldots$ | $\ldots$ | $m_{2k}$ | $m_{2.}$  |
| $\vdots$ | $\vdots$ | $\vdots$ | $\vdots$ | $\vdots$ | $\vdots$ | $\vdots$ |
| $f$   | $\ldots$ | $\ldots$ | $m_{fj}$ | $\ldots$ | $\ldots$ | $m_{f.}$  |
| $\vdots$ | $\vdots$ | $\vdots$ | $\vdots$ | $\vdots$ | $\vdots$ | $\vdots$ |
| $k$   | $m_{k1}$ | $\ldots$ | $m_{kj}$ | $\ldots$ | $m_{kk}$ | $m_{k.}$  |
| Total | $m_{.1}$ | $\ldots$ | $m_{.j}$ | $\ldots$ | $m_{.k}$ | $n$       |

**Table 2** Contingency table of the cluster membership of the $N$ object pairs

| First clustering ($g = 1$) | Second clustering ($g = 2$) | | Sum |
|---|---|---|---|
| | Pairs in the same cluster | Pairs in different clusters | |
| Pairs in the same cluster | $T_k$ | $P_k - T_k$ | $P_k$ |
| Pairs in different clusters | $Q_k - T_k$ | $U_k = N + T_k - P_k - Q_k$ | $N - P_k$ |
| Sum | $Q_k$ | $N - Q_k$ | $N = n(n-1)/2$ |

The counts of pairs joined in each partition are:

$$P_k = \sum_{f=1}^{k} \binom{m_{f.}}{2} = \frac{1}{2}\left[\sum_{f=1}^{k} m_{f.}^2 - n\right] \tag{4}$$

$$Q_k = \sum_{j=1}^{k} \binom{m_{.j}}{2} = \frac{1}{2}\left[\sum_{j=1}^{k} m_{.j}^2 - n\right] \tag{5}$$

The numerator of formula (2) with reference to the two partitions in $k$ groups can be expressed as a function of the previous quantities:

$$\sum_{i=1}^{N} |x_{1ik} - x_{2ik}| = P_k + Q_k - 2T_k \tag{6}$$

The well-known Rand index (Rand 1971) computed for two partitions in $k$ groups is given by (see Warrens 2008, for the derivation of the Rand index in terms of the quantities in Table 2):

$$R_k = \frac{N - P_k - Q_k + 2T_k}{N} \tag{7}$$

Therefore, the numerator of $Z_k$ in (2) can be expressed as a function of the Rand index:

$$\sum_{i=1}^{N} |x_{1ik} - x_{2ik}| = N(R_k - 1) \tag{8}$$

The information in Table 2 can also be summarized by a similarity index, e.g. the simple matching coefficient (Sokal and Michener 1958):

$$_{SM}I_k = \frac{T_k + (N + T_k - P_k - Q_k)}{N} = \frac{N + 2T_k - P_k - Q_k}{N} \tag{9}$$

If the Rand index is formulated in terms of the quantities in Table 2 it is equivalent to the simple matching coefficient and can be written as:

$$\sum_{i=1}^{N} |x_{1ik} - x_{2ik}| = N(_{SM}I_k - 1) \tag{10}$$

## 4 The Complement of the Index

Since $\| \mathbf{X}_1 \| = \sum_k Q_k$ and $\| \mathbf{X}_2 \| = \sum_k P_k$, the complement to 1 of $Z$ is:

$$S = 1 - Z = \frac{2 \sum_k T_k}{\sum_k Q_k + \sum_k P_k} \qquad (11)$$

Also the similarity index $S$ may be decomposed in $(n - 2)$ parts $V_k$ related to each pair of partitions in $k$ groups:

$$S = \sum_k V_k = \sum_k \frac{2 T_k}{\sum_k Q_k + \sum_k P_k} \qquad (12)$$

The components $V_k$, however, are not similarity indices for each $k$ since they assume values $< 1$ even if the two partitions in $k$ groups are identical. For this reason, we consider the complement to 1 of each $Z_k$ in order to obtain a single similarity index for each pair of partitions:

$$S_k = 1 - Z_k = \frac{\sum_{j=2}^{n-1} P_j + \sum_{j=2}^{n-1} Q_j - P_k - Q_k + 2 T_k}{\sum_{j=2}^{n-1} P_j + \sum_{j=2}^{n-1} Q_j} \qquad (13)$$

Expression (13) can be written as:

$$S_k = \frac{\sum_{j \neq k} P_j + \sum_{j \neq k} Q_j + 2 T_k}{\sum_j P_j + \sum_j Q_j} \qquad (14)$$

The index suggested by Fowlkes and Mallows (1983) for two partitions in $k$ groups in our notation is given by:

$$B_k = \frac{2 T_k}{\sqrt{2 P_k 2 Q_k}} = \frac{T_k}{\sqrt{P_k Q_k}} \qquad (15)$$

The statistics $B_k$ and $S_k$ may be thought of as resulting from two different methods of scaling $T_k$ to lie in the unit interval. Furthermore, in $S_k$ and $B_k$ the pairs $U_k$ (see Table 2), which are not joined in either of the clusterings, are not considered as indicative of similarity. On the contrary, in the Rand index, the pairs $U_k$ are considered as indicative of similarity. With many clusters, $U_k$ must necessarily be large and the inclusion of this count makes $R_k$ tending to 1, for large $k$. How the treatment of the pairs $U_k$ may influence so much the values of $R_k$ for different $k$ or the values of $R_k$ and $B_k$, for the same $k$, is illustrated in Wallace (1983).

A similarity index between two partitions may be adjusted for agreement due to chance (Hubert and Arabie 1985; Albatineh et al. 2006; Warrens 2008). With reference to formula (13) the adjusted similarity index $AS_k$ has the form:

$$AS_k = \frac{S_k - E(S_k)}{max(S_k) - E(S_k)} \qquad (16)$$

Under the hypothesis of independence of the two partitions, the expectation of $T_k$ in Table 2 is:

$$E(T_k) = P_k Q_k / N \qquad (17)$$

Therefore, the expectation of $S_k$ is given by:

$$E(S_k) = \frac{\sum_{j \neq k} P_j + \sum_{j \neq k} Q_j + 2 P_k Q_k / N}{\sum_j P_j + \sum_j Q_j} \qquad (18)$$

Given that $max(S_k) = 1$, we obtain:

$$AS_k = \frac{\frac{\sum_{j \neq k} P_j + \sum_{j \neq k} Q_j + 2 T_k - \sum_{j \neq k} P_j - \sum_{j \neq k} Q_j - 2 P_k Q_k / N}{\sum_k P_k + \sum_k Q_k}}{\frac{\sum_k P_k + \sum_k Q_k - \sum_{j \neq k} P_j - \sum_{j \neq k} Q_j - 2 P_k Q_k / N}{\sum_k P_k + \sum_k Q_k}} \qquad (19)$$

Simplifying terms, this reduces to:

$$AS_k = \frac{2 T_k - 2 P_k Q_k / N}{P_k + Q_k - 2 P_k Q_k / N} \qquad (20)$$

The adjusted Rand index for two partitions in $k$ groups is given by (Warrens 2008):

$$AR_k = \frac{2(N T_k - P_k Q_k)}{N(P_k + Q_k) - 2 P_k Q_k} \qquad (21)$$

and so $AS_k$ is equal to the Adjusted Rand Index.

## 5   Extension to More than Two Clusterings

When a set of $G$ $(G > 2)$ hierarchical clusterings for the same set of objects is available, we may be interested to gain insights into the relations of the different classifications. The index $Z$ defined in (1) may be applied to each pair of clusterings in order to produce a $G \times G$ distance matrix:

$$\mathbf{Z} = [Z_{gh}], \qquad g, h = 1, \dots, G. \qquad (22)$$

Furthermore, considering the index $S$ defined in (11) for each pair of dendrograms, we obtain a $G \times G$ similarity matrix:

$$\mathbf{S} = [S_{gh}], \qquad g, h = 1, \dots, G \qquad (23)$$

that displays the proximities between each pair of classifications. Usually, the $G$ clusterings are obtained applying different algorithms to the same data set. In this case, matrices $Z$ and $S$ may be useful in the context of the "consensus of classifications", i.e. the problem or reconciling clustering information coming from different methods (Gordon and Vichi 1998; Krieger and Green 1999). Clusterings with high distances (or low similarities) from all the others can be deleted before computing the single (consensus) clustering.

Indexes $Z$ and $S$ can also be used for variable selection in cluster analysis (Fowlkes et al. 1988; Fraiman et al. 2008; Steinley and Brusco 2008). The inclusion of "noisy" variables can actually degrade the ability of clustering procedures to recover the true underlying structure. For a set of $p$ variables and a certain clustering method, we suggest different approaches.

First we may obtain the $p$ one dimensional clustering with reference to each single variable and then compute the $p \times p$ similarity matrix $S$. The pairs of variables reflecting the same underlying structure show high similarity and can be used to obtain a multidimensional classification. On the contrary, the noisy variables should present a similarity with the other variables near to the expected value for chance agreement. We may select a subset of variables that best explains the classification into homogeneous groups. These variables help us to better understand the multivariate structure and suggest a dimension reduction that can be used in a new data set for the same problem (Fraiman et al. 2008).

A second approach consists in finding the similarities between clusterings obtained with subsets of variables (regarding, for example, different features). This approach is helpful in finding aspects that lead to similar partitions and subsets of variables that, on the contrary, lead to different clusterings.

A third way to proceed consists in finding the similarities between the "master" clustering obtained by considering all variables and the clusterings obtained by eliminating each single variable in turn, in order to highlight the "marginal" contribution of each variable to the master structure.

# 6 Concluding Remarks

In this paper we have introduced a new index to compare two hierarchical clusterings. This measure is a distance and it is appealing since it does summarize the dissimilarity by one number and can be decomposed in contributions relative to each pair of partitions. This "additive" feature is necessary for comparisons with other indices and for interpretability purposes. The complement to 1 of the suggested measure is a similarity index and it also can be expressed a sum of the components with reference to each stage of the clustering procedure.

The new distance is a measure of dissimilarity of two sequences of partitions of $n$ objects into $2, 3, \ldots, n-2, n-1$ groups. The fact that these partitions came from successive cutting of two hierarchical trees is irrelevant. The partitions could also come from a sequence of non hierarchical clusterings (obtained, i.e., by $k$-means methods with a different number of groups).

Further studies are needed in order to illustrate the performance of the suggested indices on both real and simulated data sets.

# References

Albatineh AN, Niewiadomska-Bugaj M, Mihalko D (2006) On similarity indices and correction for chance agreement. J Classification 23:301–313

Baker FB (1974) Stability of two hierarchical grouping techniques. Case I: Sensitivity to data errors. JASA 69:440–445

Brusco MJ, Steinley D (2008) A binary integer program to maximize the agreement between partitions. J Classification 25:185–193

Denoeud L (2008) Transfer distance between partitions. Adv Data Anal Classification 2:279–294

Fowlkes EB, Mallows CL (1983) A method for comparing two hierarchical clusterings. JASA 78:553–569

Fowlkes EB, Gnanadesikan R, Kettenring JR (1988) Variable selection in clustering. J Classification 5:205–228

Fraiman R, Justel A, Svarc M (2008) Selection of variables for cluster analysis and classification rules. JASA 103:1294–1303

Gordon AD, Vichi M (1998) Partitions of partitions. J Classification 15:265–285

Hubert LJ, Arabie P (1985) Comparing partitions. J Classification 2:193–218

Krieger AM, Green PE (1999) A generalized Rand-index method for consensus clusterings of separate partitions of the same data base. J Classification 16:63–89

Lapointe FJ, Legendre P (1995) Comparison tests for dendrograms: A comparative evaluation. J Classification 12:265–282

Meila M (2007) Comparing clusterings – an information based distance. J Multivariate Anal 98(5):873–895

Rand WM (1971) Objective criteria for the evaluation of clustering methods. JASA 66:846–850

Reilly C, Wang C, Rutherford M (2005) A rapid method for the comparison of cluster analyses. Statistica Sinica 15:19–33

Rohlf FJ (1982) Consensus indices for comparing classifications. Math Biosci 59:131–144

Sokal RR, Michener CD (1958) A statistical method for evaluating systematic relationships. Univ Kans Sci Bull 38:1409–1438

Sokal RR, Rohlf FJ (1962) The comparison for dendrograms by objective methods. Taxon 11:33–40

Steinley D, Brusco MJ (2008) Selection of variables in cluster analysis: An empirical comparison of eight procedures. Psychometrika 73:125–144

Wallace DL (1983) Comment on the paper "A method for comparing two hierarchical clusterings". JASA 78:569–578

Warrens MJ (2008) On the equivalence of Cohen's Kappa and the Hubert-Arabie adjusted Rand index. J Classification 25:177–183

Youness G, Saporta G (2010) Comparing partitions of two sets of units based on the same variables. Adv Data Anal Classification 4,1:53–64

Zani S (1986) Some measures for the comparison of data matrices. In: Proceedings of the XXXIII Meeting of the Italian Statistical Society, Bari, Italy, pp 157–169

# An Accelerated $K$-Means Algorithm Based on Adaptive Distances

**Hans-Joachim Mucha and Hans-Georg Bartel**

**Abstract** Widely-used cluster analysis methods such as $K$-means and spectral clustering require some measures of (pairwise) distance on the multivariate space. Unfortunately, distances are often dependent on the scales of the variables. In applications, this can become a crucial point. Here we propose an accelerated $K$-means technique that consists of two steps. First, an appropriate weighted Euclidean distance is established on the multivariate space. This step is based on univariate assessments of the importance of the variables for the cluster analysis task. Here, additionally, one gets a crude idea about what the number of clusters $K$ is at least. Subsequently, a fast $K$-means step follows based on random sampling. It is especially suited for the purpose of data reduction of massive data sets. From a theoretical point of view, it looks like MacQueen's idea of clustering data over a continuous space. However, the main difference is that our algorithm examines only a random sample in a single pass. The proposed algorithm is used to solve a segmentation problem in an application to ecology.

## 1 Introduction

The $K$-means method is a well-known cluster analysis technique. Already more than 40 years ago, MacQueen (1967) pointed out: "Also, the $k$-means procedure is easily programmed and is computationally economical, so that it is feasible to process very large samples on a digital computer.". However, to our knowledge, the

H.-J. Mucha (✉)
Weierstrass Institute, Mohrenstraße 39, 10117 Berlin, Germany
e-mail: mucha@wias-berlin.de

H.-G. Bartel
Department of Chemistry, Humboldt University Berlin, Brook-Taylor-Straße 2, 12489 Berlin, Germany
e-mail: hg.bartel@yahoo.de

W. Gaul et al. (eds.), *Challenges at the Interface of Data Analysis, Computer Science, and Optimization*, Studies in Classification, Data Analysis, and Knowledge Organization, DOI 10.1007/978-3-642-24466-7_5, © Springer-Verlag Berlin Heidelberg 2012

original idea dates back to Steinhaus in the fifties (Steinhaus 1956). Nowadays real huge piles of data are collected everywhere, especially in finance, remote sensing, genomics. The increasing size of data sets poses challenges to their statistical treatment. Cluster analysis methods such as the proposed accelerated $K$-means can help to summarize large data sets, and thus to reduce their size and complexity.

A general principle of dealing with large data sets is clustering based on sub-samples. It dates back to Kaufman and Rousseeuw (1986) (see for a modified version in Kaufman and Rousseeuw 1990). The remaining observations (not belonging to the sub-sample) are classified subsequently into the clusters. Concerning $K$-means clustering as a data mining technique, Faber et al. (1994) and Faber (1994) analysed real massive data sets obtained by the Landsat system (satellite images). It was a pioneering work, to our knowledge. The main aim was a rigorous reduction of the size of the dataset without loosing the essential information. For the Landsat data in (Faber et al. 1994), a "practical" $K = 256$ has been chosen because 256 is the greatest number of distinct colours that can be stored as an "inexpensive" byte. Figure 1 (at the left hand side) illustrates such a kind of practical motivated data reduction without looking for a native cluster structure.

Here we propose a fast $K$-means algorithm based on random sampling. It clusters the data set in a single pass. From a theoretical point of view, it looks like MacQueen's idea for clustering data over a continuous space into $K$ regions $R_k$ (MacQueen 1967)

$$E_K = \sum_{k=1}^{K} E_k = \sum_{k=1}^{K} \int_{x \in R_k} p(x)\|x - z_k\|^2 dx \tag{1}$$

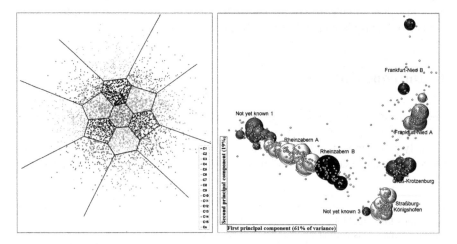

**Fig. 1** Voronoi tesselation of 2-dimensional no-structure data into 15 clusters by $K$-means clustering (left). Archaeometry (Mucha et al. 2003): Principal components analysis (PCA) plot showing the points, $K$-means cores (bubbles), and the final clusters (bubbles of same grey scale value)

Formally, we have to minimize the total error measure $E_K$ (which is given by the sum of the $E_k$'s, the within-cluster error measures) with respect to $K$ and to each region $R_k$ (clusters). Here $z_k$ is the centroid of the cluster $k$, and $p(x)$ is a density function defined in the continuous space. Before starting $K$-means, a weighted squared Euclidean distance (4) has to be established on the multivariate space. It will be based on univariate assessments of the variables. Usually, there are many variables describing an object, but often their contribution to cluster analysis is different. Above all, the weighted distance (4) allows different scales of the variables.

## 2  $K$-Means Clustering and (Weighted) Squared Euclidean Distance

Let us consider $J$-dimensional observations $\mathbf{x}_1, \mathbf{x}_2, \ldots, \mathbf{x}_I$ that are Gaussian i.i.d. The corresponding data matrix, consisting of $I$ rows and $J$ columns (variables), is denoted by $\mathbf{X} = (x_{ij})$. Further, let $\mathscr{C} = \{1, \ldots, i, \ldots, I\}$ denote the finite set of the $I$ observations. The main task of the $K$-means method is finding a partition of $\mathscr{C}$ in $K$ non-empty clusters (subsets) $\mathscr{C}_k, k = 1, 2, \ldots, K$, such that the sum of squares (SS) criterion

$$V_K = \sum_{k=1}^{K} \operatorname{tr}(\mathbf{W}_k) \tag{2}$$

has to be minimized with respect to a fixed $K$. Herein $\mathbf{W}_k = \sum_{i \in \mathscr{C}_k} (\mathbf{x}_i - \bar{\mathbf{x}}_k)(\mathbf{x}_i - \bar{\mathbf{x}}_k)^T$ is the sample cross-product matrix for the cluster $\mathscr{C}_k$. The reference point (or centroid) $\bar{\mathbf{x}}_k$ is the usual maximum likelihood estimate of the expectation value in cluster $\mathscr{C}_k$. The $K$-means method is leading to the well-known Voronoi tessellation, where the objects have minimum distance to their centroid and thus, the borderlines between clusters are hyperplanes. At the left hand side, Fig. 1 shows the result of $K$-means clustering of 4,000 random generated points (Gaussian).

Meanwhile there are several modifications of the original $K$-means method in use (Jain and Dubes 1988). In applications, often a combination of $K$-means clustering with other cluster analysis methods is used (Mucha 1992). For example, the plot at the right hand side of Fig. 1 illustrates the usage of the $K$-means method for constructing a large number of prototype clusters (or cores) which can be subsequently combined into a few final clusters using the hierarchical Ward's method. Instead of Ward's method, a more sophisticated model-based approach can be used in order to find the final clusters.

An equivalent form of the SS criterion (2) without centroids (and additionally with positive masses of observations $m_i$ and weights of variables $q_j$) is

$$V_K = \sum_{k=1}^{K} \frac{1}{M_k} \sum_{i \in \mathscr{C}_k} m_i \sum_{l \in \mathscr{C}_k, l > i} m_l d_Q(\mathbf{x}_i, \mathbf{x}_l). \tag{3}$$

Here, $M_k$ is the mass of the cluster $\mathscr{C}_k$. Now, an estimation of the expectation values of clusters $\mathscr{C}_k$ is no longer necessary. The integration of masses allows more general observations such as reference points of micro-clusters or of cores, see Fig. 1 (right). A native mass of a reference point is the cardinality of the micro-cluster. In (3), $d_Q$ is the weighted squared Euclidean distance between the observations $i$ and $l$

$$d_{il} = d_Q(\mathbf{x}_i, \mathbf{x}_l) = (\mathbf{x}_i - \mathbf{x}_l)^T \mathbf{Q}(\mathbf{x}_i - \mathbf{x}_l), \tag{4}$$

and $\mathbf{Q}$ is diagonal with $q_{jj} = q_j > 0$. In doing so, at least scaling problems can be handled fashionably without an additional data preprocessing step such as the standardization of variables.

**Example.** Iris flower dataset ($I = 150$, $J = 4$), see (Fisher 1936). The standard weights of the variables $q_j = 1/s_j$ (inverse variances) and of the observations $w_i = 1$ are used. The $s_j$ are the diagonal elements of $\mathbf{Q} = (I - 1)^{-1}\mathbf{Y}\mathbf{Y}^T$ with $\mathbf{Y} = \mathbf{X} - I^{-1}\mathbf{1}\mathbf{1}^T\mathbf{X}$. Figure 2 shows the result of the $K$-means method (3).

Much better results can be obtained by using adaptive weights $q_j$ that can be estimated during the iteration cycles of $K$-means clustering (Mucha 1992, 1995; Mucha and Sofyan 2000). The use of a more general positive definite (adaptive) matrix $\mathbf{Q}$ in cluster analysis dates back to Späth (1985). He recommended the use of the $K$ inverse within-cluster covariance matrices $\mathbf{Q}_k$ instead of $\mathbf{Q}$.

Why on the earth adaptive weights? We assume that each variable has its own specific contribution in finding clusters. A variable with more than one mode in its density should be an important one for clustering. The more important a variable $j$ is the higher should be the weight $q_j$ in (4) in the case of same scale of the variables. This approach is different to clustering based on subsets of variables as proposed by Friedman and Meulman (2004).

**Example.** Iris data. Basic adaptive weights are $\mathbf{Q} = (\mathrm{diag}(\overline{\mathbf{S}}))^{-1}$ with $q_j = 1/\overline{s}_j$ (inverse pooled within-cluster variances), where $\overline{\mathbf{S}} = (I - K)^{-1}\sum_{k=1}^{K}\mathbf{W}_k$ is the pooled covariance matrix with pooled within-cluster variances in the diagonal. Usually, the estimates are obtained in an iterative manner, for further details see

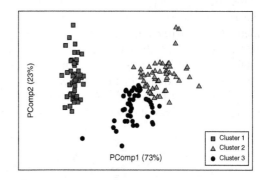

**Fig. 2** PCA plot showing the result of $K$-means clustering based on standard weights (which is the same as $K$-means clustering using z-scores). Here 22 out of 150 observations are classified into the wrong species (that gives an error rate of 15%)

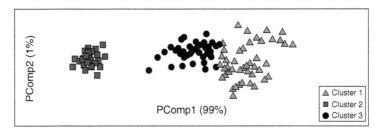

**Fig. 3** Adaptive PCA plot showing the result of an adaptive $K$-means clustering

(Mucha and Klinke 1993). In doing so, only 6 errors are counted using the adaptive $K$-means method, see Fig. 3.

# 3   An Accelerated $K$-Means Method Based on Adaptive Distances

We propose the following accelerated $K$-means algorithm (abbreviated as A$K$MA) based on adaptive distances that works on random sub-samples. First, finding appropriate adaptive weights by univariate investigation of the variables. (Alternatively, one can extend the investigation also to bivariate and to multivariate inspections.) Thereupon, establishing an appropriate squared weighted Euclidean distance (4) on the multivariate space. Then, in the $K$-means step, initialise the clusters by a random sub-sample of size $K$, i.e. set up the so-called seed points of clusters. The clustering algorithm itself starts by examining also only a random sample once. If the dataset is very large and the sample is representative of the dataset, the proposed procedure should converge similar to an algorithm that examines every observation in sequence. Such a sample yields a reasonable estimate of the error measure without using all the observations in the original (huge) dataset. Both, working on sub-samples and no iterations greatly accelerate the $K$-means clustering.

**First step of A$K$MA:** univariate investigations. Now let's look at the proposed algorithm in more detail. The univariate assessment of the variables is based on how likely it is that the underlying distribution is heterogeneous with several different populations. For example, there are different statistical tests for finding the presence of more than one mode (Hartigan and Hartigan 1985; Hennig 2009).

**Example.** Sub-sample of the Iris data (15 observations). The univariate nonparametric density estimation of the variable "Petal Length" suggests two modes (bandwidth $= 1$) in Fig. 4 (left). Estimating the density of the squared Euclidean distances (105 pairwise distance values, bandwidth $= 2$) is another possibility, see the plot at the right hand side of Fig. 4. Obviously, if there are two clusters (two well-separated modes) of nearly the same cardinality in the variable, then one could expect three modes in the density estimation of the corresponding pairwise distances. Apparently, one could extend the investigation of the values of variables

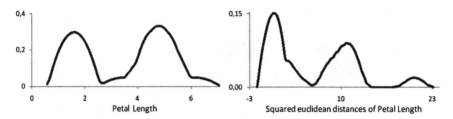

**Fig. 4** Nonparametric density estimations based on a sub-sample of the Iris data

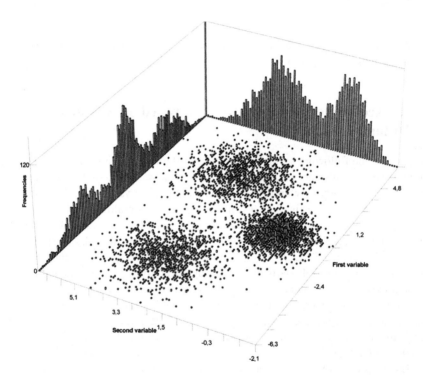

**Fig. 5** Histograms of three-class data (mixture of Gaussians)

also to distances (for further reading concerning pairwise distances, see Murtagh (2009)). However, it seems to become more difficult the higher the number of clusters is. Therefore, our focus will be on univariate assessments of the original values of the variables.

In addition to the univariate assessment, one gets a crude idea about what the number of clusters is at least in the multivariate setting. This is a kind of "minimax" rule. For example, the univariate histograms in Fig. 5 suggest that there are three modes in the first variable and two modes in the second, respectively. So, the maximum number of clusters based on univariate assessments is three. Obviously,

at the same time, these three clusters are the minimum number of clusters that can
be expected when going to the multivariate setting. Therefore, the main focus is on
finding adaptive (data-driven) weights for 3 clusters at least. The data behind Fig. 5
is a 4,000×2 data matrix **X** consisting of the following three spherical, randomly
generated Gaussian classes of sizes 1,100, 1,600, and 1,300, respectively, with
the following different mean values $(-3, 3)$, $(0, 0)$, $(3, 3)$ and different standard
deviations $(1, 1)$, $(0.7, 0.7)$, and $(1.2, 1.2)$. For further details see Mucha (2009).

How to obtain the weights $q_j$ based on the univariate density? For example,
the estimated $p$-values of the dip test of unimodality (Hartigan and Hartigan 1985)
can be used. Beside statistical tests, univariate cluster analysis can be an useful
practical tool for an univariate investigation of the variables via decomposition of
the heterogeneity of their variances. We recommend separate (univariate) Ward's
clustering that has to be examined for each single variable. As a result, a vector **h** of
so-called criterion values is obtained for every variable that contains the incremental
sum of squares when aggregating the data points. Figure 6 shows the logarithmic
scaled criterion values versus the number of clusters (left hand side) for a sub-
sample of the three class data of Fig. 5. A quite simple univariate assessment of
a variable $j$ is the use of the ratios between subsequent criterion values

$$r_k^{(j)} = \frac{\log(h_k + 1)}{\log(h_{k+1} + 1)}. \tag{5}$$

Here the superscript indicates the variable $j$. Considering a special $K$ for a variable
$j$, the weights $q_j = r_K^{(j)}/s_j$ are used in the subsequent $K$-means clustering. Such
a proposal looks similar to the well-known scree test in hierarchical cluster analysis
(Späth 1980). In Fig. 6, at the right hand side, the plot of the ratios versus the number
of clusters is presented. Considering $K = 3$ the following ratios will be used later
on in the clustering step: $r_3^{(1)} = 1.692$, $r_3^{(2)} = 1.098$. Obviously, the ratios and
weights, respectively, change with respect to the number of clusters $K$. Let's look
at our sub-sample of Iris data (15 observations). In Fig. 7, each line corresponds to

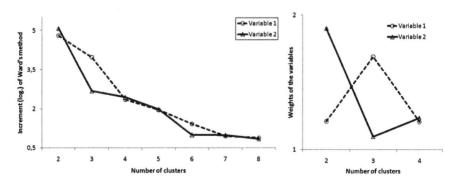

**Fig. 6** Comparative view of both the logarithmic criterion values (left) and the ratios (5) (right) of
the two univariate Ward's clustering of the data of Fig. 5 based on a sub-sample of size 200

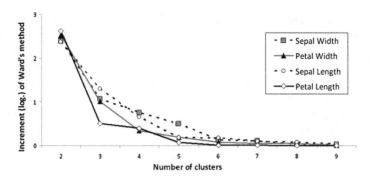

**Fig. 7** Comparative view of the criterion values of the four univariate Ward's clusterings based on a sub-sample of the Iris data. One gets similar results by drawing other samples

one variable. Again, we used the ratio of two subsequent criterion values for finding the weights $q_j$ (which are usually different for different $k = 2, 3, \ldots, 9$).

**Second step of AKMA:** Establishing a distance on the multivariate space. Generally, the multivariate distance is composed in an additive way by

$$\mathbf{D} = \sum_{j=1}^{J} q_j \mathbf{D}_j. \tag{6}$$

It dates back to Gower (1971). Here $\mathbf{D}_j$ is the matrix of squared Euclidean distances based on the variable $j$. The cluster analysis itself starts with an initialisation of clusters by random seed points, and then followed by a "running" $K$-means technique. The latter means that whenever a randomly drawn observation is classified, the corresponding reference point is updated immediately. The clustering algorithm itself examines usually only a random sub-sample (see Bradley et al. 1998). If the dataset is very large and the sample is representative of the dataset, the proposed procedure should converge similar to an algorithm that examines every observation in sequence and that has many iteration steps. This can be confirmed in the case of the three class data (Fig. 5) using the above given weights $q_j$. We examine several AKMA runs on different sub-samples. In any case, the size of the sub-samples is 200. The median of the error rates is about 1.46%. This is only slightly higher than 1.38% (for the usual $K$-means clustering of all 4,000 points) and than 1.25% (Ward's method), see Mucha (2009). Moreover, the estimates of the (true) reference points are of similar quality.

**Example.** The same sub-sample of the Iris data as above (15 observations). At one go, Ward's hierarchical cluster analysis method based on variable-wise distances investigates a variable at different number of clusters $k = 2, 3, \ldots, I$. Figure 7 shows the criterion values of all four separate Ward's clustering runs, one for each single variable. Figure 8 shows the final result of AKMA, where all 150 objects are used in the "running mean" $K$-means step. Here pooled weights are used that

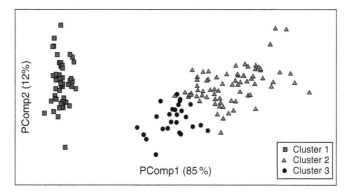

**Fig. 8** Adaptive PCA plot showing the result of the accelerated *K*-means clustering

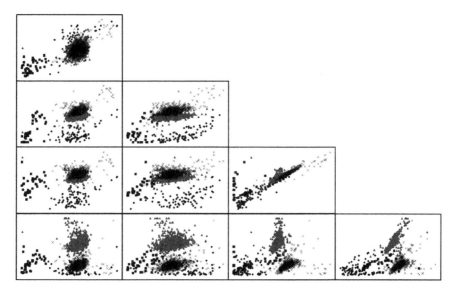

**Fig. 9** Scatter plot matrix showing the result of the accelerated *K*-means clustering

were averaged over the weights for different $k$, $k = 2, 3, 4$. A total of 26 errors are counted.

## 4 Application to Ecology

The flow cytometry measurements of lake Müggelsee in Berlin were taken on the 8th of July 2002. The five variables measure the luminosity of fluorescence of pigments at a certain wavelength (for details, see Mucha et al. 2002). The number of observations is about 23,000. The aim of clustering is to discover different kinds of algae and separate other things from them like dead or anorganic particles. The proposed accelerated $K$-means method A$K$MA is based on random sub-sampling of 10% of the original data. In between, at sample size equals 500 observations one $K$-means optimisation step was included in order to repair "bad" seed points. Figure 9 shows the final result. From the experts point of view the clusters represent reasonable results that can be confirmed to a high degree.

## 5 Summary

The proposed accelerated $K$-means method is based on random sampling. A squared weighted Euclidean distance on the multivariate space is established that is based on univariate assessment of the variables. A$K$MA is a quite fast algorithm that gives good results (error rates and estimates of the reference points). The stability of results has to be investigated yet. Beside univariate assessment of the importance of variables, bivariate investigations seems to be promising, see Fig. 5: The bivariate density gives much better separated regions. Random seed points may cause problems in applications in the case of a small number of clusters. Also, it seems to be difficult to discover small clusters.

## References

Bradley P, Fayyad U, Reina C (1998) Scaling clustering algorithms to large databases. Tech. rep., Microsoft Research

Faber V (1994) Clustering and the continuous $k$-means algorithm. Los Alamos Sci 22:138–144

Faber V, Hochberg JG, Kelly PM, Thomas TR, White JM (1994) Concept extraction. A data-mining technique. Los Alamos Sci 22:123–137, 145–149

Fisher RA (1936) The use of multiple measurements in taxonomic problems. Ann Eugenics 7: 179–188

Friedman JH, Meulman JJ (2004) Clustering objects on subsets of attributes. J Royal Stat Soc B 66(4):815–849, URL http://www-stat.stanford.edu/~jhf/ftp/cosa.pdf

Gower JC (1971) A general coefficient of similarity and some of its properties. Biometrics 27: 857–871

Hartigan JA, Hartigan PM (1985) The dip test of unimodality. Ann Statist 13:70–84

Hennig C (2009) Merging Gaussian mixture components - an overview. In: Mucha HJ, Ritter G (eds) Classification and clustering: Models, software and applications. Report 26, WIAS, Berlin, pp 80–89

Jain AK, Dubes RC (1988) Algorithms for clustering data. Prentice Hall, New Jersey

Kaufman L, Rousseeuw PJ (1986) Clustering large data sets. In: Gelsema ES, Kanal LN (eds) Pattern recognition in Practice II (with discussion). Elsevier/North-Holland, pp 425–437

Kaufman L, Rousseeuw PJ (1990) Finding groups in data. Wiley, New York

MacQueen J (1967) Some methods for classification and analysis of multivariate observations. In: Le Cam LM, Neyman J (eds) Proc. 5th Berkeley Symp. Math. Statist. Prob., Univ. California Press, Berkley, vol 1, pp 281–297

Mucha HJ (1992) Clusteranalyse mit mikrocomputern. Akademie Verlag, Berlin

Mucha HJ (1995) Clustering in an interactive way. discussion paper no. 13. Tech. Rep. Sfb 373, Humboldt-Universität, Berlin

Mucha HJ (2009) ClusCorr98 for Excel 2007: clustering, multivariate visualization, and validation. In: Mucha HJ, Ritter G (eds) Classification and clustering: Models, software and applications. WIAS, Berlin, 26, pp 40–40

Mucha HJ, Klinke S (1993) Clustering techniques in the interactive statistical computing environment XploRe. Tech. Rep. 9318, Institut de Statistique, Université Catholique de Louvain, Louvain-la-Neuve

Mucha HJ, Sofyan H (2000) Cluster analysis. discussion paper no. 49. Tech. Rep. Sfb 373, Humboldt-Universität, Berlin

Mucha HJ, Simon U, Brüggemann R (2002) Model-based cluster analysis applied to flow cytometry data of phytoplankton. Tech. Rep. 5, WIAS, Berlin, URL http://www.wias-berlin.de/

Mucha HJ, Bartel HG, Dolata J (2003) Core-based clustering techniques. In: Schader M, Gaul W, Vichi M (eds) Between data science and applied data analysis. Springer, Berlin, pp 74–82

Murtagh F (2009) The remarkable simplicity of very high dimensional data: application of model-based clustering. J Classification 26:249–277

Späth H (1980) Cluster analysis algorithms for data reduction and classification of objects. Ellis Horwood, Chichester

Späth H (1985) Cluster dissection and analysis. Ellis Horwood, Chichester

Steinhaus H (1956) Sur la division des corps matériels en parties. Bull de l'Académie Polonaise des Sci IV(12):801–804

# Bias-Variance Analysis of Local Classification Methods

Julia Schiffner, Bernd Bischl, and Claus Weihs

**Abstract** In recent years an increasing amount of so called local classification methods has been developed. Local approaches to classification are not new. Well-known examples are the $k$ nearest neighbors method and classification trees (e.g. CART). However, the term 'local' is usually used without further explanation of its particular meaning, we neither know which properties local methods have nor for which types of classification problems they may be beneficial. In order to address these problems we conduct a benchmark study. Based on 26 artificial and real-world data sets selected local and global classification methods are analyzed in terms of the bias-variance decomposition of the misclassification rate. The results support our intuition that local methods exhibit lower bias compared to global counterparts. This reduction comes at the price of an only slightly increased variance such that the error rate in total may be improved.

## 1 Introduction

Lately the amount of literature on local approaches to classification is increasing. The probably best-known example is the $k$ nearest neighbors method. Many more local approaches have been developed, most of them can be considered localized versions of standard classification techniques such as linear discriminant analysis (LDA), logistic regression, naïve Bayes, support vector machines (SVM), neural networks, boosting etc. The main idea of local classification methods is as follows: Since it is often difficult to find a single classification rule that is suitable for the whole population, rather concentrate on subsets of the population and calculate several individual rules that are only valid for single subsets. Two questions immediately arise: What is the effect of this localization and in which situations

J. Schiffner (✉) · B. Bischl · C. Weihs
Department of Statistics, TU Dortmund University, 44221 Dortmund, Germany
e-mail: schiffner@statistik.tu-dortmund.de

W. Gaul et al. (eds.), *Challenges at the Interface of Data Analysis, Computer Science, and Optimization*, Studies in Classification, Data Analysis, and Knowledge Organization, DOI 10.1007/978-3-642-24466-7_6, © Springer-Verlag Berlin Heidelberg 2012

are local methods especially appropriate? Many authors when proposing a new local classification method just demonstrate superior performance over standard methods on selected data sets. To our knowledge there are only few theoretical results regarding the performance of local methods and no extensive studies that compare several types of local methods across many data sets. A useful concept to gain deeper insight into the behavior of learning algorithms is the bias-variance decomposition of prediction error. It was originally introduced for quadratic loss functions, but generalizations to the misclassification rate have been developed. In a benchmark study on real-world and synthetic data we assess the bias-variance decomposition of the misclassification rate for different types of local methods as well as global methods. For a start in Sect. 2 a short introduction to local approaches to classification is given and different types of local methods are described. The bias-variance decomposition of the misclassification rate and its specific properties are explained in Sect. 3. The benchmark study and the results are presented in Sect. 4. Finally, in Sect. 5 a summary and an outlook to future work are given.

## 2 Local Approaches to Classification

Due to space limitations we can only give a short introduction. We do not go into details here, but rather point to references on certain topics. In classification it is assumed that each object $\omega$ in the population $\Omega$ belongs to one and only one class $y = Y(\omega) \in \mathcal{Y}$, with $\mathcal{Y}$ denoting the set of class labels. Additionally, on each object measurements $x = X(\omega) \in \mathcal{X}$ are taken. Both, $X$ and $Y$, are assumed to be random variables. The aim is to find a classification rule or classifier $D : \mathcal{X} \to \mathcal{Y}$ that based on the measurements predicts the class labels. The set $\mathcal{X}$ is usually called predictor space and often $\mathcal{X} \subset \mathcal{R}^d$. A local classifier is specialized on subsets of the population $\Omega$. A local classification method induces one or more local classifiers and aggregates them if necessary. According to which subsets of the population are addressed several types of local approaches can be distinguished:

*Observation-Specific Methods* For each object in the population an individual classification rule is built based on training observations near the trial point $x$. The best-known example is $k$NN. But in principle every classification method can be localized this way, for example a localized form of LDA is described in Czogiel et al. (2007). While for $k$NN locally constant functions are fitted to the data, in case of LLDA locally linear functions are used. A review paper concerned with observation-specific methods is Atkeson et al. (1997).

*Partitioning Methods* These methods partition the predictor space $\mathcal{X}$. Examples are CART, mixture-based approaches like mixture discriminant analysis (MDA, Hastie and Tibshirani 1996) and other multiple prototype methods like learning vector quantization (LVQ, Hastie et al. 2009). Strictly speaking $k$NN belongs to both groups because it generates a Voronoi tessellation of $\mathcal{X}$.

There are some more local approaches like *multiclass to binary* strategies (Allwein et al. 2000) and *discriminant-adaptive approaches* (Hand and Vinciotti 2003) that are beyond the scope of this paper. Since in local classification model assumptions need only be valid for subsets of the population instead of the whole population they are relaxed. For this reason localized methods exhibit more flexibility than their global counterparts and are expected to give good results in the case of irregular class boundaries. Localization is only one way to obtain flexible classifiers. Other global ways are e.g. using polynomials of higher degrees and/or kernel methods.

# 3 Bias-Variance Decomposition of the Misclassification Rate

The bias-variance decomposition of prediction error was originally introduced for quadratic loss. The two main concerns when generalizing it beyond quadratic loss are finding reasonable definitions of bias and variance on the one hand and deriving the decomposition of prediction error on the other hand (e.g. James 2003).

*Definitions of Noise, Bias and Variance* Let $\mathscr{P}_{X,Y}$ denote the joint distribution of $X$ and $Y$ and let $P(Y = y \mid x)$ be the class posterior probabilities. In classification noise at a fixed trial point $x$ is the irreducible or Bayes error

$$Var(Y \mid x) = E_Y\big[L_{01}\big(Y, S(Y \mid x)\big)\mid x\big] = 1 - \max_y P(Y = y \mid x), \text{ where} \quad (1)$$

$$S(Y \mid x) = \arg\min_y E_Y\big[L_{01}(Y, y)\mid x\big] = \arg\max_y P(Y = y \mid x) \quad (2)$$

is the Bayes prediction at $x$ and $L_{01} : \mathscr{Y} \times \mathscr{Y} \to \{0, 1\}$ denotes the zero-one loss function. Let $\hat{Y} = D(x)$ denote the prediction at $x$. Since $D$ is calculated based on training data that result from a random draw from $(\mathscr{X}, \mathscr{Y})$ according to $P_{\mathscr{X},\mathscr{Y}}$, $\hat{Y}$ is not fixed. Bias and variance are defined as

$$bias(\hat{Y} \mid x) = L_{01}\big(S(Y \mid x), S(\hat{Y} \mid x)\big) = I\big(S(Y \mid x) \neq S(\hat{Y} \mid x)\big), \quad (3)$$

$$Var(\hat{Y} \mid x) = E_{\hat{Y}}\big[L_{01}\big(\hat{Y}, S(\hat{Y} \mid x)\big)\mid x\big] = 1 - \max_y P(\hat{Y} = y \mid x), \text{ where} \quad (4)$$

$$S(\hat{Y} \mid x) = \arg\min_y E_{\hat{Y}}\big[L_{01}(\hat{Y}, y)\mid x\big] = \arg\max_y P(\hat{Y} = y \mid x) \quad (5)$$

is usually called the main prediction and $I$ is the indicator function. Bias measures the systematic deviation of the main prediction from $S(Y \mid x)$, while $Var(\hat{Y} \mid x)$ indicates the random variation of $\hat{Y}$ around the main prediction. $S$ is an operator that reveals the systematic parts of $Y$ and $\hat{Y}$. The definitions given here are natural generalizations of those in the quadratic case, i.e. if the quadratic loss is used instead of $L_{01}$ they reduce to the standard definitions of noise, squared bias and variance.

*Decomposition of the Prediction Error*  In case of quadratic loss the prediction error can be decomposed into the sum of noise, bias and variance and thus both, high bias and high variance, are detrimental to prediction accuracy. In case of zero-one loss the role of variance is completely different. In James (2003) it is shown that in the two-class case the misclassification rate can be decomposed as follows

$$E_{Y,\hat{Y}}\left[L_{01}(Y,\hat{Y})|\,x\right] = Var(Y|\,x) + bias(\hat{Y}|\,x) + Var(\hat{Y}|\,x)$$
$$-2Var(Y|\,x)bias(\hat{Y}|\,x) - 2Var(Y|\,x)Var(\hat{Y}|\,x)$$
$$-2bias(\hat{Y}|\,x)Var(\hat{Y}|\,x) + 4Var(Y|\,x)bias(\hat{Y}|\,x)Var(\hat{Y}|\,x). \tag{6}$$

In contrast to the quadratic case the decomposition additionally contains interactions. The negative interaction effect of bias and variance indicates that variance corrects the prediction in case of bias. If the number of classes is larger than two, this correction does not occur for sure, but with a certain probability, which makes the decomposition even more complicated.

*Systematic and Variance Effects*  In order to obtain a simpler decomposition, James (2003) distinguishes between bias and variance as measures of systematic deviance and random variation on the one hand and the effects of bias and variance on the prediction error on the other hand. He defines systematic and variance effects as

$$SE(Y,\hat{Y}|\,x) = E_{Y}\left[L_{01}\left(Y,S(\hat{Y}|\,x)\right) - L_{01}\left(Y,S(Y|\,x)\right)|\,x\right], \tag{7}$$

$$VE(Y,\hat{Y}|\,x) = E_{Y,\hat{Y}}\left[L_{01}(Y,\hat{Y}) - L_{01}\left(Y,S(\hat{Y}|\,x)\right)|\,x\right]. \tag{8}$$

The systematic effect is the change in prediction error if instead of the Bayes prediction $S(Y|\,x)$ the main prediction $S(\hat{Y}|\,x)$ is used. The variance effect measures the change in prediction error due to random variation of $\hat{Y}$ around the main prediction. While $SE(Y,\hat{Y}|\,x) \geq 0$ the variance effect for the reasons explained above can also take negative values. Under squared loss bias and variance coincide with their respective effects. Generally, an estimator with zero bias also has zero systematic effect and an estimator with zero variance has zero variance effect. With systematic and variance effects an additive decomposition of the misclassification rate is obtained as follows

$$E_{Y,\hat{Y}}\left[L_{01}(Y,\hat{Y})|\,x\right] = Var(Y|\,x) + SE(Y,\hat{Y}|\,x) + VE(Y,\hat{Y}|\,x). \tag{9}$$

## 4  Benchmark Study

The aim of our study is to get more insight into the properties of local methods and the effect of localization. As explained in Sect. 2 we assume that local methods in general exhibit rather low bias and decreased bias compared to their global counterparts. This reduction probably comes at the price of an increased variance. As explained in Sect. 3 variance needs not be detrimental, but in conjunction with low bias it is likely to be. Questions of interest are: Is the error rate in total increased or decreased and is bias or variance the main contributor to prediction error when using local methods? Moreover, it would be useful to know if there are differences between distinct types of local methods. We expect this since, for example, CART is known for high variance, whereas $k$NN is reported as stable by Breiman (1996). Finally, since we justify our assumption of bias reduction with the increased flexibility of local methods, we would like to know if there are differences between local methods and global approaches of similar flexibility.

*Data Sets* We consider 26 data sets, both artificial and real-world data. Most data sets are taken from the UCI repository (Frank and Asuncion 2010) and the mlbench R-package (Leisch and Dimitriadou 2010). The threenorm, twonorm and waveform data were used in Breiman (1996). The crystal and encoded data sets are described in more detail in Szepannek et al. (2008), the mixture data are taken from Bishop (2006). The hvdata set is an artificial data set described in Hand and Vinciotti (2003). The orange and South-African-heart-disease (SAheart) data are taken from Hastie et al. (2009) and the crabs data are available in the MASS R-package (Venables and Ripley 2002). In Table 1 a survey of the data sets used in the study is given. On the left hand the real-world and on the right hand the artificial data sets are shown. The crabs and wine data sets as well as the hvdata and twonorm data sets pose relatively easy problems since the decision boundary is linear. The circle, orange, ringnorm, subclasses and threenorm data sets are quadratic and the xor and subclasses2 data sets are more complex classification problems.

*Classification Methods* We consider ten classification methods, three global and seven local methods.

- Global methods: We apply two linear methods, LDA and multinomial regression (MNR), and as a more flexible method a SVM with polynomial kernel (poly SVM). The degree of the polynomial is tuned in the range of 1 to 3.
- Partitioning methods: We consider CART, learning vector quantization (LVQ1) and mixture discriminant analysis (MDA, Hastie and Tibshirani 1996).
- Observation-specific methods: We use $k$NN, localized LDA (LLDA, Czogiel et al. 2007) and SVMs with RBF kernel (RBF SVM).
- Moreover, we consider random forests (RF), which can be regarded as a local classification method in conjunction with an ensemble method. Since several classification trees are combined RF normally exhibits lower variance than CART.

*Measuring Noise, Bias and Variance* In case of artificial data the noise level is usually known. When dealing with real-world data noise can be estimated by means of a consistent classifier. We follow the proposition of James (2003) and employ the 3NN method where the neighbors are weighted by a Gaussian kernel. In order to calculate bias and variance we need to estimate the distribution

**Table 1** Characteristics of the 26 data sets used in the benchmark study: Number of observations, number of classes, dimensionality, number of numeric and number of categorical predictors

| Data Set | #Obs | #Class | Dim | #Num | #Categ | Data Set | #Obs | #Class | Dim | #Num | #Categ |
|---|---|---|---|---|---|---|---|---|---|---|---|
| Breastcancer | 683 | 2 | 9 | 0 | 9 | Circle | 1,000 | 2 | 4 | 4 | 0 |
| Car | 1,728 | 4 | 6 | 0 | 6 | Cuboids | 1,002 | 4 | 3 | 3 | 0 |
| Crabs | 200 | 2 | 5 | 5 | 0 | Hvdata | 1,000 | 2 | 6 | 6 | 0 |
| Credit-g | 1,000 | 2 | 20 | 7 | 13 | Mixture | 200 | 2 | 2 | 2 | 0 |
| Crystal | 2,746 | 3 | 37 | 37 | 0 | Orange | 1,000 | 2 | 10 | 10 | 0 |
| Encoded | 794 | 2 | 6 | 6 | 0 | Ringnorm | 1,000 | 2 | 4 | 4 | 0 |
| Glass | 214 | 6 | 9 | 9 | 0 | Spirals | 1,000 | 2 | 2 | 2 | 0 |
| Ionosphere | 351 | 2 | 34 | 34 | 0 | Subclasses | 300 | 3 | 5 | 5 | 0 |
| Pima | 768 | 2 | 8 | 8 | 0 | Subclasses2 | 990 | 3 | 5 | 5 | 0 |
| SAheart | 462 | 2 | 9 | 8 | 1 | Threenorm | 1,000 | 2 | 4 | 4 | 0 |
| Sonar | 208 | 2 | 60 | 60 | 0 | Twonorm | 1,000 | 2 | 4 | 4 | 0 |
| Soybean | 562 | 15 | 35 | 0 | 35 | Waveform | 1,000 | 3 | 21 | 21 | 0 |
| Vowel | 990 | 11 | 9 | 9 | 0 | Xor | 1,000 | 8 | 4 | 4 | 0 |

of $\hat{Y}$. For this purpose we use a nested resampling strategy. In the outer loop we employ subsampling (4/5 splits with 100 repetitions). The subsample is used for training while based on the remaining data error, bias, variance and their effects are estimated. In the inner loop we use 5-fold cross-validation for parameter tuning which is required for all employed classification methods except LDA and MNR. A more detailed description of how bias, variance etc. can be estimated can be found in James (2003). All calculations were carried out in R 2.11.1 (R Development Core Team 2009) using the mlr R-package (Bischl 2010).

**Results.** While space limitations preclude a full description of the results, some of the main observations are reported here. First, we assess if there are significant performance differences between the classification methods. In Fig. 1 for each classification method a boxplot of the misclassification rates on the 26 data sets is shown. Moreover, a ranking of the methods based on their performance is obtained. For this purpose for each data set a linear mixed effects model is fitted to the misclassification rates obtained on the individual subsamples. Then pairwise differences between the classification methods are tested based on Tukey contrasts and thus an order among the algorithms is established. Finally, a partial consensus ranking of the methods over all data sets is obtained (Eugster et al. 2008). This ranking is not unique. One of eight possible rankings is shown on the right side of Fig. 1. From bottom to top the performance of the classification methods increases. An edge indicates that the error rate of the method displayed on the lower level is significantly smaller than that of the method above. In all eight consensus rankings RBF SVM and polynomial SVM show best performance. Moreover, there are no methods that beat LLDA and MDA. Both linear methods and LVQ1 exhibit the highest error rates. Next, in order to explain the differences in error rates we consider the bias-variance decomposition. Figure 2 shows the average misclassification rates as well as systematic and variance effects over all 26 data sets. For all classification methods the systematic effects are considerably larger than the variance effects. Differences in error rates are mainly caused by changes in systematic effect. Both methods with the lowest error rates, polynomial and RBF SVMs, exhibit the smallest systematic effects. Their variance effects are hardly larger than that of

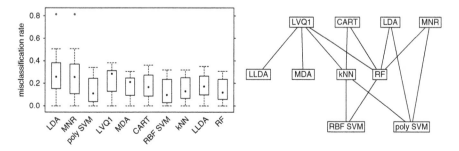

**Fig. 1** Left: Boxplot of the error rates on the 26 data sets. Right: Consensus ranking of the classification methods

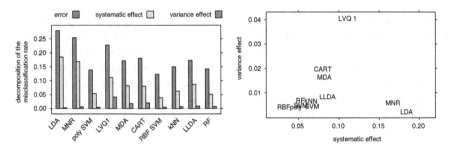

**Fig. 2** Left: Error rates, systematic and variance effects averaged over the 26 data sets. Right: Variance effect versus systematic effect

LDA which is minimal under all classification methods. The localized versions of LDA, LLDA and MDA, both exhibit considerably lower systematic effects, but only slightly increased variance effects. LVQ1 and as expected CART exhibit the largest variance effects. Compared to CART random forests (RF) shows a smaller average variance effect, but as well a smaller systematic effect. In order to visualize which classification methods show a similar behavior with respect to the bias-variance decomposition we plot the variance effect against the systematic effect. We can recognize two clusters, one formed by the linear methods LDA and MNR, the other one formed by highly flexible methods $k$NN, RF as well as RBF and polynomial SVM. Neither global and local methods nor the different types of local methods we mentioned in Sect. 2 can be distinguished based on this plot. The reason may be that the bias-variance decomposition only reflects the degree of flexibility of classification methods and that the way how it is obtained, by localization or otherwise, is not relevant.

## 5 Summary and Outlook

In order to gain insight into the performance of local classification methods we assessed the bias-variance decomposition of the error rate for local and global classification methods based on real-world and synthetic data. The results support our intuition that localized approaches exhibit considerably lower bias or rather systematic effect than global counterparts. Contrary to our assumptions most local methods under consideration, except LVQ1 and CART, only have a slightly increased variance effect, the main contributors to prediction error are noise or systematic effect. In terms of the bias-variance decomposition neither global and local methods nor the different types of local methods could be distinguished. In the future we would like to apply more classification methods. Some types mentioned in Sect. 2 like multiclass to binary have not been included yet. Moreover, we plan to relate the results to characteristics of different types of local classification methods. All conclusions were drawn based on averages over the 26 data sets in the study, differences between distinct data sets are not taken into account. In order to gain deeper insight and to get hints in what situation which method will probably perform well, it would be useful to relate the results to characteristics of the data sets.

# References

Allwein EL, Shapire RE, Singer Y (2000) Reducing multiclass to binary: A unifying approach for margin classifiers. J Mach Learn Res 1:113–141

Atkeson CG, Moore AW, Schaal S (1997) Locally weighted learning. Artif Intell Rev 11(1-5): 11–73

Bischl B (2010) mlr: Machine learning in R. URL http://mlr.r-forge.r-project.org

Bishop CM (2006) Pattern recognition and machine learning. Springer, New York

Breiman L (1996) Bias, variance, and arcing classifiers. Tech. Rep. 460, Statistics Department, University of California at Berkeley, Berkeley, CA, URL www.stat.berkeley.edu

Czogiel I, Luebke K, Zentgraf M, Weihs C (2007) Localized linear discriminant analysis. In: Decker R, Lenz HJ (eds) Advances in data analysis, Springer, Berlin Heidelberg, Studies in classification, data analysis, and knowledge organization, vol 33, pp 133–140

Eugster MJA, Hothorn T, Leisch F (2008) Exploratory and inferential analysis of benchmark experiments. Tech. Rep. 30, Institut für Statistik, Ludwig-Maximilians-Universität München, Germany, URL http://epub.ub.uni-muenchen.de/4134/

Frank A, Asuncion A (2010) UCI machine learning repository. University of California, Irvine, School of Information and Computer Sciences, URL http://archive.ics.uci.edu/ml

Hand DJ, Vinciotti V (2003) Local versus global models for classification problems: Fitting models where it matters. American Statistician 57(2):124–131

Hastie T, Tibshirani R (1996) Discriminant analysis by Gaussian mixtures. J Royal Stat Soc B 58(1):155–176

Hastie T, Tibshirani R, Friedman J (2009) The elements of statistical learning: data mining, inference, and prediction, 2nd edn. Springer, New York

James GM (2003) Variance and bias for general loss functions. Mach Learn 51(2):115–135

Leisch F, Dimitriadou E (2010) mlbench: Machine learning benchmark problems. R package version 2.0-0

R Development Core Team (2009) R: A Language and Environment for Statistical Computing. R Foundation for Statistical Computing, Vienna, Austria, URL http://www.R-project.org

Szepannek G, Schiffner J, Wilson J, Weihs C (2008) Local modelling in classification. In: Perner P (ed) Advances in data mining. Medical applications, e-commerce, marketing, and theoretical aspects, Springer, Berlin Heidelberg, LNCS, vol 5077, pp 153–164

Venables WN, Ripley BD (2002) Modern applied statistics with S, 4th edn. Springer, New York, URL http://www.stats.ox.ac.uk/pub/MASS4

# Effect of Data Standardization on the Result of $k$-Means Clustering

Kensuke Tanioka and Hiroshi Yadohisa

**Abstract** In applying clustering to multivariate data, in which there are some large-scale variables, clustering results depend on the variables more than the user's needs. In such cases, we should standardize the data to control the dependency.

For high-dimensional data, Doherty et al. (Appl Soft Comput 7:203–210, 2007) argued numerically that data standardization by variable range leads to almost the same results regardless of the kinds of norms, although Aggarwal et al. (Lect Notes Comput Sci 1973:420–434, 2001) showed theoretically that a fraction norm reduces the effect of the curse of high dimensionality for $k$-means result more than the Euclidean norm does. However, they have not considered the effects of standardization and factors properly.

In this paper, we verify the effects of six data standardization methods with various norms and examine factors that affect the clustering results for high-dimensional data. As a result, we show that data standardization with the fraction norm reduces the effect of the curse of high dimensionality and gives a more effective result than data standardization with the Euclidean norm and not applying data standardization with the fraction norm.

K. Tanioka (✉)
Graduate School of Culture and Information Science, Doshisha University Kyoto, 610-0313, Japan
e-mail: dik0012@mail4.doshisha.ac.jp

H. Yadohisa
Department of Culture and Information Science, Doshisha University Kyoto 610-0313, Japan
e-mail: hyadohis@mail.doshisha.ac.jp

W. Gaul et al. (eds.), *Challenges at the Interface of Data Analysis, Computer Science, and Optimization*, Studies in Classification, Data Analysis, and Knowledge Organization, DOI 10.1007/978-3-642-24466-7_7, © Springer-Verlag Berlin Heidelberg 2012

# 1 Introduction

$K$-means clustering is a non-hierarchical clustering method that classifies objects into groups with multivariate data. If there exist some variables, that have a large scale or great variability, these variables strongly affect the clustering result. In these cases, data standardization would be used to control the scale or variability.

Nonetheless, no single approach to data standardization has been shown to be the best. Milligan and Cooper (1988) showed the effect of standardization methods on hierarchical clustering results and concluded that standardization by variable range is the most effective. For $k$-means clustering, Steinley (2004) indicated that standardization by the maximum of variables is also effective.

For high-dimensional data, Aggarwal et al. (2001) showed theoretically that clustering by using the fraction norm provides more distinct results than by using the Euclidean norm. In contrast, Doherty et al. (2007) argued numerically that data standardization resulted in the $k$-means clustering with the Euclidean norm outperforming $k$-means with the fraction norm. Moreover, despite their differences, these papers have not considered the effects of:

(a) Factors such as noise dimensions, outlier conditions and cluster size.
(b) Other standardization methods on the results of $k$-means clustering by using a non-Euclidean norm.

As a result, in this paper, we examine the effect of six data standardization methods on the result of $k$-means clustering through Monte Carlo simulation using a non-Euclidean norm because for real data, we can't verify whether the factor affects the clustering result or not. The purpose of this simulation is to compare how data standardizations with the fraction norm affect various factors such as error conditions, variance of variables, cluster configurations and so on.

# 2 Fraction Norm

Here we introduce the fraction norm for $k$-means clustering. The Minkowski norm, $L_p$, which is a family of distance measures, is described as:

$$L_p(\boldsymbol{x}, \boldsymbol{y}) = \left\{ \sum_{i=1}^{d} |x_i - y_i|^p \right\}^{1/p}, \quad (p \geq 1) \tag{1}$$

where $d$ is the number of dimension of vectors. Aggarwal et al. (2001) extended this concept to $p \in (0, 1)$ for reducing the effect of the curse of high dimensionality. Figure 1 represents a unit length from the origin in the Euclidean plane with $L_p$ norms ($p = 0.3, 0.5, 1, 2$). $L_{0.3}$ shows an inwards-curved loci, although $L_2$ traces an outwards-curved loci. The properties of the fraction norm are as follows:

**Fig. 1** Unit length loci from the origin with $L_p (p = 2, 1, 0.5, 0.3)$norms

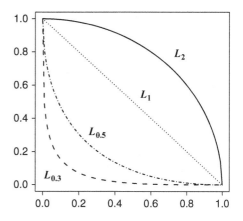

**Table 1** Norms of vectors $a$ and $b$ from origin measured with $L_p$

| Norm | $L_p(a)$ | $L_p(b)$ | $L_p(b)/L_p(a)$ |
|---|---|---|---|
| $p = 2$ | 8.83 | 50.29 | 5.69 |
| $p = 1/3$ | 238.62 | 506.95 | 2.12 |

(1) the fraction measures defined by $L_p$ with $p \in (0, 1)$ do not satisfy the triangle inequality; (2) it reduces the effect of outlier values compared with Euclidean measures.

For example, consider the two vectors $a = (2, 3, 4, 7)$ and $b = (2, 3, 4, 50)$, and let $L_p(x)$ be the norm between vector $x$ and the origin. Table 1 shows the length of vectors $a$, $b$ and the ratio $a$ to $b$ measured with the $L_2$ and $L_{1/3}$ norm. The ratio of the $L_{1/3}(a)$ to $L_{1/3}(b)$ is less than the ratio of the $L_2(a)$ to $L_2(b)$.

## 3 Methods of Standardization

In this section, we present the six standardization methods and describe their properties using Monte Carlo simulation. First, we define

$$X = (x_1, x_2, \cdots, x_n), \text{ the } \mathbf{N} \times \mathbf{n} \text{ data matrix, where } x_j \text{ is the column}$$
$$\text{vector;}$$

$$\mathbf{1} = \text{an } \mathbf{N} \times 1 \text{ vector of ones;}$$

$$\bar{x}_j = \text{mean of variable } x_j \text{ } (\mathbf{j} = 1, 2, \cdots \mathbf{n});$$

$$\sigma_j = \text{standard deviation of variable } x_j \text{ } (\mathbf{j} = 1, 2, \cdots, \mathbf{n});$$

$$\max(x_j) = \text{maximum of variable } x_j \text{ } (\mathbf{j} = 1, 2, \cdots, \mathbf{n});$$

$$\min(x_j) = \text{minimum of variable } x_j \text{ } (\mathbf{j} = 1, 2, \cdots, \mathbf{n});$$

Rank$(x_j)$ = within − variable ranking of variable $x_j$ ($j = 1, 2, \cdots, n$);

$$x_j^q = x_{(a)j} + b(x_{(a+1)j} - x_{(a)j}) \quad (0 \le q \le 1; \; j = 1, 2, \cdots, n);$$

where $x_{(\cdot)}$ is an order statistic and $\mathbf{a}$ and $\mathbf{b}$ are defined by:

$$a = [q(n + 1)], \quad b = q(n + 1) - a.$$

In short, $x_j^q$ is the $100q$th percentile.

The original data and the six standardization functions are as follows:

$$Z_0(x_j) = x_j, \tag{2}$$

$$Z_1(x_j) = (x_j - \bar{x}_j \mathbf{1}')/\sigma_j, \tag{3}$$

$$Z_2(x_j) = x_j / \max(x_j), \tag{4}$$

$$Z_3(x_j) = x_j / (\max(x_j) - \min(x_j)), \tag{5}$$

$$Z_4(x_j) = x_j / N \bar{x}_j, \tag{6}$$

$$Z_5(x_j) = \text{Rank}(x_j), \tag{7}$$

and

$$Z_6(x_j) = x_j / (x_j^{0.975} - x_j^{0.025}). \tag{8}$$

$Z_0$ represents the original data. $Z_1$, $Z_2$, $Z_3$, $Z_4$ and $Z_5$ were examined by Milligan and Cooper (1988) and $Z_6$ was recommended by Mirkin (2005).

Although Milligan and Cooper (1988) concluded that $Z_3$ is the most effective compared to the other standardizations, $Z_3$ is affected by outliers. In particular, under the outlier condition, applying $Z_3$ to each variable in the data matrix causes the scales of those variables, that contain outliers to become smaller than the scale of those variables that do not . However, $Z_6$ provides a more equal scale for the variables than $Z_3$.

## 4 Simulation Design

We use a Monte Carlo simulation to verify the effects of data standardization on the result of $k$-means clustering. The method for evaluating clustering results is to generate data with a known cluster structure and compare the recovered cluster structure with the known cluster structure on several data structures. The design of this simulation is based on Milligan (1985), Milligan and Cooper (1988), Steinley (2004), and Steinley and Henson (2005), with some modifications. For the data generation procedure, overlap between clusters is adjusted on the first dimension of the variable space (Steinley and Henson 2005). For all other dimensions, clusters

**Table 2** Factors in the simulation

| Factor No. | Factor name | Factor No. | Factor name |
|---|---|---|---|
| Factor 1 | Number of clusters | Factor 6 | Distribution of variables |
| Factor 2 | Number of variables | Factor 7 | Variance of variables |
| Factor 3 | Number of observations | Factor 8 | Error conditions |
| Factor 4 | Cluster densities | Factor 9 | Dissimilarities |
| Factor 5 | Initial seeds | Factor 10 | Probability that clusters overlap |

are allowed to either overlap or not and the maximum range of the data is limited to be two-thirds of the range of the first dimension. In this simulation, we use 10 factors to retain the validity of the simulation and use the adjusted Rand index (ARI) to evaluate the clustering results (Hubert and Arabie 1985).

Next, we describe the levels of the 10 factors shown in Table 2.

**Number of clusters:** The number of clusters has three levels, 5,10 and 20 (Steinley and Brusco 2007).

**Number of variables:** The number of variables has two levels, 25 and 50 (Steinley and Brusco 2007).

**Number of observations:** The number of observations is 200 because the number of observations hardly influences the recovery of the known cluster structure (Steinley and Brusco 2007).

**Cluster densities:** This factor has three levels:

(a) All clusters have the same number of observations.
(b) Three clusters have 50% of the observations while the remaining observations are evenly divided among the remaining clusters.
(c) Three clusters have 20% of the observations while the remaining observations are evenly divided among the remaining clusters.

**Initial seeds:** $K$-means clustering depends on the initial seeds. Thus, we perform $k$-means clustering 10 times, which is the number of cluster updates, and on each implementation randomly generate the initial seeds. The number of times the initial seeds are generated is 200. The classification result is selected according to the criterion that minimizes the overall within cluster variance.

**Distribution of variables:** The distribution of each variable is selected randomly from a normal distribution, uniform distribution and triangular distribution (Steinley and Henson 2005).

**Variance of variables:** This factor has two levels:

(a) The variance of variables is not manipulated.
(b) Half of the number of variables are multiplied by random numbers between 5 and 10 (Steinley 2004).

**Error conditions:** This factor has three levels:

(a) Error-free: Neither outliers nor noise dimensions are added to the generated data.
(b) Outliers: Outliers (20% of the observations) are added to the error-free data.
(c) Noise dimensions: Half of the number of dimensions of the error-free data is replaced with noise dimensions where the noise dimensions are distributed from uniform distributions. The range of the noise dimensions is equal to the first dimension.

**Dissimilarities:** This factor has two levels: Euclidean norm, $p = 2$ and fraction norm, $p = 0.3$.

**Probability that clusters overlap:** On the first dimension, the probability that two clusters overlap is manipulated by five levels, 0.01, 0.1, 0.2, 0.3 and 0.4 (Steinley and Henson 2005).

## 5  Results

In this section, we show the results of the Monte Carlo simulation from two perspectives, error conditions and probability that clusters overlap.

Table 3 shows the recovery results for the data standardization methods under various error conditions and dissimilarities. Each cell in the table represents the average value of ARI. Under any error condition, the data standardizations with the fraction norm provide more effective results on the $k$-means methods than data standardizations with the Euclidean norm. Using the fraction norm reduces the difference among the recoveries from the various data standardization methods more than the Euclidean norm.

In the error-free conditions, $Z_5$ with a Euclidean norm produces a higher recovery than any other standardization method with a Euclidean norm. When using the fraction norm, $Z_2$ and $Z_3$ are also effective for the result of $k$-means clustering.

As is well known, clustering results of $k$-means methods are affected by outliers. In the outlier conditions, $Z_5$ shows the highest recovery of any other standardization

**Table 3** Effect of the error conditions and kinds of norm on the recovery for the six standardization methods

| Error conditions | Error-free | | Outliers | | Noise dimensions | |
|---|---|---|---|---|---|---|
| Norm | $L_2$ | $L_{0.3}$ | $L_2$ | $L_{0.3}$ | $L_2$ | $L_{0.3}$ |
| $Z_0$ | 0.25 | 0.45 | 0.12 | 0.32 | 0.11 | 0.51 |
| $Z_1$ | 0.29 | 0.53 | 0.11 | 0.39 | 0.43 | 0.64 |
| $Z_2$ | 0.37 | 0.59 | 0.25 | 0.48 | 0.46 | 0.68 |
| $Z_3$ | 0.37 | 0.58 | 0.24 | 0.49 | 0.47 | 0.67 |
| $Z_4$ | 0.32 | 0.52 | 0.19 | 0.38 | 0.40 | 0.66 |
| $Z_5$ | 0.55 | 0.66 | 0.51 | 0.64 | 0.56 | 0.64 |
| $Z_6$ | 0.34 | 0.56 | 0.12 | 0.38 | 0.47 | 0.66 |

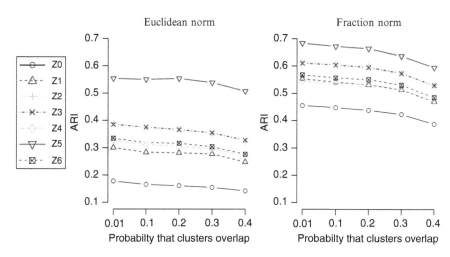

**Fig. 2** Effect of probability that clusters overlap on the recovery for the standardization methods

methods regardless of norm, although the recoveries of the other standardization methods are lower than those under any error conditions as a whole. $Z_6$ gives a lower recovery than $Z_2$ and $Z_3$, even though $Z_6$ provides more equal scale of variables than $Z_2$ and $Z_3$ in the error condition.

High-dimensional data usually contains many noise variables. In the noise dimension conditions, $Z_0$ with the Euclidean norm is affected by noise dimensions, although the other methods gave stable recoveries. The reason is that the data values in the noise dimension is uniformly-distributed in the range and does not affect the $k$-means results, although the scale affects the $k$-means result.

Figure 2 represents the recoveries with respect to the probability that clusters will overlap on the first dimension. Data standardizations with the fraction norm provide higher recoveries than those with a Euclidean norm regardless of the probability. As a result, $Z_5$ provides the most effective recovery on the $k$-means method regardless of norm. The reason is quite understandable: $Z_5$ gives the highest recovery of any method in the outliers condition.

## 6   Conclusion

We can obtain two conclusions from this simulation. Firstly, as a whole, the recoveries of data standardization methods with the fraction norm reduce the effect of the curse of high dimensionality more than with the Euclidean norm, although Doherty et al. (2007) argued that data standardization leads to the same clustering results regardless of norm. However, this simulation partly supports the results of Doherty et al. (2007) that for the fraction norm, the differences of the recoveries between standardization methods are smaller than those for the Euclidean norm.

Secondarily, the result in our simulation indicates that $Z_5$ is the most effective for $k$-means clustering of all standardization methods we tested, whereas Steinley (2004) concluded that $Z_2$ and $Z_3$ are the most effective. The difference occurs because Steinley (2004) only considered variables distributed from normal distributions. In particular, $k$-means clustering postulates that each cluster detected is distributed from a normal distribution. However, real data does not always follow a normal distribution. In this simulation, we also employed two kinds of distributions that are not normal distributions. Thus, the recoveries in this simulation are lower than those in Steinley (2004) and $Z_2$ and $Z_3$ are affected by the effect of the variables that follow non-normal distributions. For example, when variables are distributed as a triangular distribution, which represents a left-or right-skewed distribution, and the probability that clusters overlap is high, $k$-means clustering tends to consider the data around the highest probability of the distribution with the data generated from the neighboring distribution as a cluster. However, $Z_5$ does not consider the data around the highest probability of the distribution, but considers the ranking of the variable.

The results of this study indicate that we should apply $Z_5$ to multivariate data with the fraction norm when we classify the observations by the $k$-means method for high-dimensional data. The reason is quite understandable. The fraction norm is more effective for reducing the curse of dimensionality than the Euclidean norm and $Z_5$ is less affected by the distributions of variables than the other standardization methods.

Although we provide several results in this simulation, the simulation design still requires some modifications. For example, we did not consider the properties of high-dimensional data such as sparsities and masking variables. Thus, the effects of these factors on the clustering results need to be verified through Monte Carlo simulations. Finally, wherever feasible in the future, we would like to prove that data standardization with the fraction norm reduces the effect of the curse of high dimensionality on $k$-means clustering more effectively than data standardization without the fraction norm.

# References

Aggarwal CC, Hinneburg A, Keim DA (2001) On the surprising behavior of distance metrics in high dimensional space. Lecture Notes in Computer Science 1973:420–434

Doherty KAJ, Adams RG, Davey N (2007) Unsupervised learning with normalised data and non-euclidean norms. Applied Soft Computing 7:203–210

Hubert L, Arabie P (1985) Comparing partititions. Journal of Classifications 2:193–218

Milligan GW (1985) An algorithm for generating artificial test clusters. Psychometrika 50:123–127

Milligan GW, Cooper MC (1988) A study of standardization of variables in cluster analysis. J Classification 5:181–204

Mirkin B (2005) Clustering for data mining: A data recovery approach. Chapman and Hall, Boca Raton, FL

Steinley D (2004) Standardizing variables in k-means clustering. In: Banks D, House L, McMorris FR, Arabie P, Gaul W (eds) Classification, clustering, and data mining application. Springer, Berlin, pp 53–60

Steinley D, Brusco MJ (2007) Initializing k-means batch clustering: A critical evaluation of several techniques. J Classification 24:99–121

Steinley D, Henson R (2005) OCLUS: An analytic method for generating clusters with known overlap. J Classification 22:221–250

# A Case Study on the Use of Statistical Classification Methods in Particle Physics

**Claus Weihs, Olaf Mersmann, Bernd Bischl, Arno Fritsch, Heike Trautmann, Till Moritz Karbach, and Bernhard Spaan**

**Abstract** Current research in experimental particle physics is dominated by high profile and large scale experiments. One of the major tasks in these experiments is the selection of interesting or relevant events. In this paper we propose to use statistical classification algorithms for this task. To illustrate our method we apply it to an Monte-Carlo (MC) dataset from the BABAR experiment. One of the major obstacles in constructing a classifier for this task is the imbalanced nature of the dataset. Only about 0.5% of the data are interesting events. The rest are background or noise events. We show how ROC curves can be used to find a suitable cutoff value to select a reasonable subset of a stream for further analysis. Finally, we estimate the $CP$ asymmetry of the $B^{\pm} \rightarrow DK^{\pm}$ decay using the samples extracted by the classifiers.

## 1 Introduction

The field of experimental particle physics is driven by the analysis of large datasets collected by detectors at high energy labs around the world. These experiments are usually designed to achieve two goals. First, they should provide proof for theoretical predictions and secondly the measurements are used to refine estimates of unknown constants. A byproduct of the collected data may be the observation

C. Weihs (✉) · O. Mersmann · B. Bischl · A. Fritsch · H. Trautmann
Statistics Department, TU Dortmund, Germany
e-mail: weihs@statistik.tu-dortmund.de; olafm@statistik.tu-dortmund.de;
bischl@statistik.tu-dortmund.de; fritsch@statistik.tu-dortmund.de;
trautmann@statistik.tu-dortmund.de

T. M. Karbach · B. Spaan
Physics Department, TU Dortmund, Germany
e-mail: moritz.karbach@tu-dortmund.de; spaan@physik.tu-dortmund.de

W. Gaul et al. (eds.), *Challenges at the Interface of Data Analysis, Computer Science, and Optimization*, Studies in Classification, Data Analysis, and Knowledge Organization, DOI 10.1007/978-3-642-24466-7_8, © Springer-Verlag Berlin Heidelberg 2012

of so called new physics. This term is used to describe phenomena currently not covered by one of the established theories.

One large particle physics experiment, which was started in 1994, is the BABAR (Boutigny et al. 1994) experiment at the Stanford Linear Accelerator (SLAC). In this case study, we will focus on this experiment and the data collected by its detector. One of the goals of the researchers working on the BABAR project is to further study the charge parity (CP) violations predicted by the Standard Model of particle physics (c.f. Griffiths (1987)). Of particular interest are the coefficients of the Cabibbo–Kobayashi–Masukawa ($CKM$) mixing matrix which, in layman's terms, describes how quarks transition from one generation to another. It can be described by three angles, $\alpha$, $\beta$ and $\gamma$ as well as a complex phase parameter. The parameter of interest for the BABAR collaboration is $\gamma$. Its value can be constrained by measuring the $CP$ asymmetry of the $B^{\pm}$ decay. To generate these decays, electrons and positrons are collided resulting in a $B^+ B^-$ pair in about 50% of all collisions. The $B$ mesons are very short lived and decay into a myriad of other particles, e.g. into a $D$ meson and a charged kaon $K$. By counting the charge asymmetry of these kaons it is possible to infer constraints for the angle $\gamma$.

In this case study we will show how statistical classification methods can be used in the process of identifying kaon events which can then be used to estimate the $A_{CP}$ coefficient, which in turn is used in determining $\gamma$. This work is based on the methodology used in Karbach (2009) but has a much narrower scope. $A_{CP}$ is defined as

$$A_{CP} = \frac{N_{DK+} - N_{DK-}}{N_{DK+} + N_{DK-}}, \tag{1}$$

where $N_{DK+}$ denotes the number of $B \to DK^+$ decays observed and $N_{DK-}$ denotes the number of $B \to DK^-$ events observed. These cannot be detected directly. Instead, certain characteristics of the decay events are captured by components of the detector. Using the measurements obtained from these components the type of event has to be inferred using constraints motivated by physics as well as statistical techniques. This filtering of events to isolate interesting decays is divided into several steps. First, a triggering system in the detector will decide if an event might be interesting. This system is based on a fixed set of rules which can be evaluated very quickly. In a second stage the raw data is used to reconstruct the event. After this step, the data is divided into several streams. A stream contains similarly characterized events as well as background reactions which are of no interest. Finally, a stream is reduced to a data sample by sophisticated rules to extract as many interesting events as possible while simultaneously suppressing as much of the background as possible. A stream will usually have less than one interesting event for every 100 background events. Lastly the parameter of interest, here $A_{CP}$, is estimated from the data using a suitable method. This step can be further subdivided into two subparts, a classification step, as well as the final estimation of $A_{CP}$. In order to be able to judge the quality of our proposed methods, we used a Monte-Carlo dataset that was obtained by simulating the events in the detector and then processing the resulting raw data as described above. The rest of this article is structured as follows. In Sect. 2 we describe the experimental setup, next we explore the dataset in Sect. 3, and then we analyze the dataset in Sect. 4.

## 2 Experimental Setup

The dataset considered in this case study was created using the Toy Monte-Carlo simulator of the BABAR experiment. Reconstruction and initial preprocessing was done the same way it would have been done for a real dataset of the BABAR experiment. Four subsets from the resulting event database were selected for this analysis. One consisted of interesting $B\pm \rightarrow DK\pm$ decays as well as background reactions and the other three contained only background reactions. The number of event and background observations in each dataset is summarized in Table 1. In order to obtain a realistic dataset as it might have resulted from an actual experiment, the four subsets have to be mixed in certain proportions. These are given by

$$\text{data} = \tfrac{1}{3}\text{dat}_1 + \tfrac{1}{3}\text{dat}_2 + \tfrac{1}{2}\text{dat}_3 + \text{dat}_4. \tag{2}$$

A dataset created using these mixing constants contains, on average, about 152 interesting decays and 36,272 background reactions. The a-priori probability of an interesting event is therefore 0.4%.

Since for the real experiment the type of decay is unknown, we are interested in a classification rule that allows us to separate the interesting decays from the rest. Given the extremely unbalanced nature of the dataset this is a nontrivial task. With such a rule, we can directly estimate $A_{CP}$ by counting the number of positively and negatively charged particles.

## 3 Preprocessing

Data preprocessing is always a crucial step in data analysis and is especially important in this setting, where the classes are extremely imbalanced. Exploratory data analysis was performed to identify anomalies in the variables, interesting

**Table 1** The four datasets provided. Signal denotes the interesting $B\pm \rightarrow DK\pm$ events

| Dataset | Signal (#) | Background (#) | Total |
|---------|-----------|----------------|-------|
| dat$_1$ | 456 | 15,364 | 15,820 |
| dat$_2$ | 0 | 2,844 | 2,844 |
| dat$_3$ | 0 | 11,601 | 11,601 |
| dat$_4$ | 0 | 24,402 | 24,402 |
| Total | 456 | 54,211 | 54,667 |

relationships between variables and extreme observations. The findings were then discussed with the experts from particle physics to judge their relevance for the task at hand.

Initially, we identified technical variables which stem from the Monte Carlo simulation. Summary statistics revealed that roughly 30 variables show hardly any variation, i.e. have a smaller variance than $10^{-9}$. These were deemed to be not useful for the classification and therefore not included in the classification step. Eleven variables contained a large percentage of missing values. Discussions with the physicist revealed that all of these variables were measured by one detector component, the DIRC detector, and that it is entirely plausible to have many missing values there. It turns out that these variables were used in the reconstruction phase to generate a new summary variable containing a meaningful value even if the 11 original variables are missing. It was therefore decided to drop these variables and instead use only the derived variable as well as one new variable which indicates if any of the 11 DIRC variables had been missing. After all of these preprocessing steps, 33 variables are left in the dataset.

# 4 Analysis

After preprocessing, four classification algorithms were used to predict the decay type. These were Classification Trees (Breiman et al. 1984), Logistic Regression (Nelder and Wedderburn 1972), Random Forests (Breiman 2001) and Adaboost (Freund and Schapire 1996). Notably absent from this list are all classifiers with hyperparameters which cannot be chosen sensibly independent of the data (e.g. SVMs and other kernel methods). Their use would have required a nested resampling strategy. Due to the highly unbalanced nature of the data, initial tests showed that training and tuning such a classifier was not possible. On average less than 75 interesting events would have been included in a training sample. We do not believe that learning with such a reduced dataset will accurately reflect the performance of a classifer that is trained on the full sample.

In order to assess their relative performance 500 datasets were generated using the mixing ratios from Formula 2. Then, for each dataset a 10-fold cross-validation was used to obtain decision values for all observations.

## 4.1 Results

Of the 500 independent runs, we initially used 100 to choose cutoff values for each classifier. The remaining 400 datasets were then used to assess the power of the classification rules in predicting $A_{CP}$. Since the classes are highly unbalanced and we are only interested in one class, the usual approach of minimizing the misclassification rate fails. Instead we need to tune each classification rule so that it has a high precision, i.e. it excells at identifying interesting decays. At the same time we want a high recall rate since otherwise the total number of events used in the final estimation procedure will be too small. So our judgement of the performance of a classifier is not symmetric w.r.t. the classes as is usually the case, since we are primarily interested in retrieving one class.

Before we present the initial results of our study, we need to define how we want to use the classifiers to estimate $A_{CP}$. We will use two approaches. The first approach will choose a cutoff value for the classifier and apply the obtained classification rule to the test data. The obtained events can then be used to directly estimate $A_{CP}$ by counting their charge and plugging those values into (1). In the second approach, instead of choosing a cutoff, we interpret the decision values of the classifier as probabilities. We then estimate $A_{CP}$ using the following equation

$$\hat{A}_{CP} = \left(\sum_i \pi_i\right)^{-1} \sum_i \pi_i \text{charge}. \tag{3}$$

Here the variable charge is $+1$ if a positively charged particle was observed (presumably a $B \rightarrow DK^+$ decay) and $-1$ otherwise (presumably a $B \rightarrow DK^-$ decay) and $\pi_i$ is the classifier's estimated probability for the $i$-th observation in the test set to be an interesting event.

To choose the required cutoff value for each classifier, precision vs. recall curves are used. These are shown in Fig. 1. We clearly see that the Classification Tree and Logistic Regression cannot compete with Adaboost or Random Forests. Both of the former are outperformed by the latter, independent of the chosen cutoff value. At the same time, no curve is even close to what we would consider a desired curve,

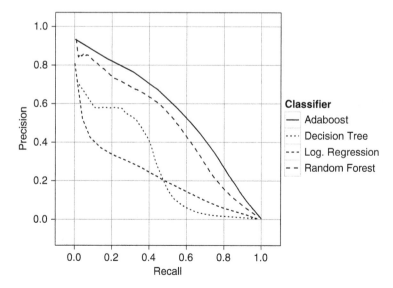

**Fig. 1** Average precision vs. recall curve for the four classifiers used

one that reaches far into the upper right corner. This desired classifier would lead to a high precision while at the same time providing a high recall. Since such a classifier could not be found, the Classification Tree was not included in further analysis. We will keep the Logistic Regression even though it too is dominated by the remaining two classifiers because it is the only method which directly estimates the probabilities $\pi_i$ needed in the second estimation scheme. To study the properties of these classification rules we estimated $A_{CP}$ for each of the 400 remaining mixing-samples. This was done using two different approaches.

Since the curves show that there is no cutoff value which leads to both a high recall as well as a high precision, we will consider three different cutoff values for each classifier. These are chosen so that one cutoff leads to a precision of approximately 80%, one which leads to equal precision and recall and one which leads to an expected recall of about 80%. The reasoning behind this is, that the cutoff value which has a recall rate of 80% could be used as an enhanced filter rule before further processing is applied. Conversely, the classifier with a precision of 80% should lead to a fairly good estimate of $A_{CP}$. The cutoff with equal precision and recall is a good compromise between the other two scenarios. Additionally, true $A_{CP}$ value was also calculated for each mixture. The results of these calculations are summarized in Table 2.

We see that the probability based approach outperforms all other approaches and does especially well using Logistic Regression. On the other hand we saw in the precision vs. recall graph, that logistic regression was the worst classifier. The only explanation we can offer for this is that the probability estimates using the logistic model are probably better than the estimates from the other models.

To further analyze this, kernel density estimates of the $A_{CP}$ estimates for the classifiers were studied. These are shown in Fig. 2. Here we can see that there is considerable bias in the estimates for all three classifiers. This bias decreases as the

**Table 2** Mean, standard deviation (SD), root mean squared error (RMSE) and correlation for $A_{CP}$ estimates. Best estimate for each classifier in bold

| Learner | $A_{CP}$ **Estimate** | | | | |
| | Method | Mean | SD | RMSE | Corr. |
| --- | --- | --- | --- | --- | --- |
| Adaboost | Prob. Est. | −0.078 | 0.085 | 0.079 | 0.659 |
| | 80% Recall | **−0.045** | **0.047** | **0.052** | **0.649** |
| | Precision = Recall | −0.062 | 0.067 | 0.064 | 0.639 |
| | 80% Precision | −0.101 | 0.147 | 0.142 | 0.526 |
| Log. Regression | Prob. Est. | **−0.030** | **0.070** | **0.034** | **0.881** |
| | 80% Recall | −0.006 | 0.051 | 0.043 | 0.866 |
| | Precision = Recall | −0.093 | 0.113 | 0.096 | 0.774 |
| | 80% Precision | 0.028 | 0.396 | 0.401 | 0.074 |
| Random Forest | Prob. Est. | **−0.033** | **0.027** | **0.051** | **0.716** |
| | 80% Recall | −0.019 | 0.030 | 0.057 | 0.550 |
| | Precision = Recall | −0.100 | 0.072 | 0.093 | 0.571 |
| | 80% Precision | −0.103 | 0.198 | 0.199 | 0.345 |

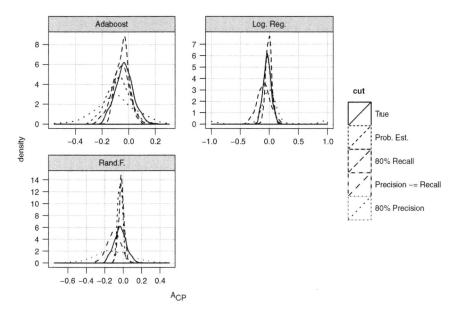

**Fig. 2** Kernel density estimates of the true $A_{CP}$ as well as the estimated $A_{CP}$ using three cutoff values

precision decreases. While initially this may seem counterintuitive, it turns out, that background events have, on average, an $A_{CP}$ of zero, i.e. have no charge asymmetry. Therefore, the more background we include, the more we bias our estimate towards zero, which in this case counters the actual bias of our estimate. We have not found out why the estimates with the high precision data is biased, but we suspect, that the classifier only selects events of a certain structure in this setting which lead to a shift in the charge relation observed.

In a second step, scatterplots of the true $A_{CP}$ versus the $A_{CP}$ estimate were created. These are shown in Fig. 3. We expected the estimates to vary around the depicted diagonal line. Instead we see that there is a systematic trend to deviate from the diagonal while still showing a linear relationship. We cannot fully explain this phenomenon either, but again assume that it is a result of the bias induced by the noise and the classifier. Since this bias appears to be linear, a correction factor could be estimated from the data and used in further experiments. This has, however, not been tried out. Note that this is not observed for the Logistic Regression. Here the points lie around the diagonal as expected except for the case of 80% precision where the classifier breaks down. The erratic behaviour is explained by the fact that this classification rule will only identify one or two interesting events in the test set. Therefore the estimated $A_{CP}$ is either $+1$, $0$ or $-1$ in most of the cases.

Physicists use complex parametric models to estimate $A_{CP}$ using maximum likelihood estimation. This approach was not tried in this case study, but could easily be combined with the classifiers proposed here as a preprocessing step to control the quality of the data that goes into the estimation procedure.

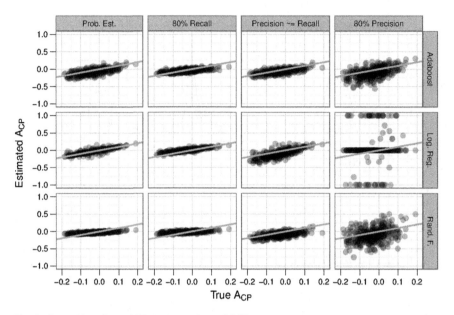

**Fig. 3** Scatterplot of true ACP against estimated ACP

# 5 Outlook and Conclusion

In this case study we have shown how modern classification techniques can be used as filters to select interesting events in particle physics experiments. We employed precision vs. recall graphs to choose a suitable cutoff value for the classifier and then studied the quality of the $A_{CP}$ estimation. This allows us precise control over the amount of background noise in our filtered dataset. This alone provides an enhanced preprocessing step, which allows the physicist to control the quality of the data used for further analysis.

Dealing with such highly unbalanced data in a classification setting is still an open research topic and there is likely much room for improvement here as well. Instead of changing the cutoff value after the training of the classifier, it is likely beneficial to either upweigh the smaller class or choose the cutoff value a-priori to the training phase. This gives the classification algorithm a chance to adapt to the unbalanced cost structure of the problem.

**Acknowledgements** This work was partly supported by the Collaborative Research Center SFB 823 and the Research Training Group "Statistical Modelling" of the German Research Foundation. Due to the computationally intensive nature of this case study, all calculations were performed on the LiDO HPC cluster at the TU Dortmund. We would like to thank the LiDO team for their support.

# References

Boutigny D, et al. (1994) Letter of intent for the study of CP violation and heavy flavor physics at PEP-II. SLAC-0443

Breiman L (2001) Random forests. Mach Learn 45:5–32

Breiman L, Friedman J, Olshen R, Stone C (1984) Classification and regression trees. Wadsworth and Brooks, Monterey, CA

Freund Y, Schapire R (1996) Experiments with a new boosting algorithm. In: Machine Learning: Proceedings of the Thirteenth International Conference, Morgan Kauffman, San Francisco, pp 148–156

Griffiths D (1987) Introduction to elementary particles. Wiley, New York, USA

Karbach TM (2009) Measurement of $CP$ parameters in $B-> D^0_{CP\pm}K$ decays with the BABAR detector. PhD thesis, TU Dortmund, http://hdl.handle.net/2003/26523

Nelder JA, Wedderburn RWM (1972) Generalized linear models. J Royal Stat Soc A 135:370–384

# Problems of Fuzzy c-Means Clustering and Similar Algorithms with High Dimensional Data Sets

Roland Winkler, Frank Klawonn, and Rudolf Kruse

**Abstract** Fuzzy c-means clustering and its derivatives are very successful on many clustering problems. However, fuzzy c-means clustering and similar algorithms have problems with high dimensional data sets and a large number of prototypes. In particular, we discuss hard c-means, noise clustering, fuzzy c-means with a polynomial fuzzifier function and its noise variant. A special test data set that is optimal for clustering is used to show weaknesses of said clustering algorithms in high dimensions. We also show that a high number of prototypes influences the clustering procedure in a similar way as a high number of dimensions. Finally, we show that the negative effects of high dimensional data sets can be reduced by adjusting the parameter of the algorithms, i.e. the fuzzifier, depending on the number of dimensions.

## 1 Introduction

Clustering high dimensional data has many interesting applications: For example, clustering similar music files, semantic web applications, image recognition or biochemical problems. Many tools today are not designed to handle hundreds of dimensions, or in this case, it might be better to call it degrees of freedom. Many clustering approaches work quite well in low dimensions, but especially the

R. Winkler (✉)
German Aerospace Center Braunschweig, German
e-mail: roland.winkler@dlr.de

F. Klawonn
Ostfalia, University of Applied Sciences, German
e-mail: f.klawonn@ostfalia.de

R. Kruse
Otto-von-Guericke University Magdeburg, German
e-mail: kruse@iws.cs.uni-magdeburg.de

W. Gaul et al. (eds.), *Challenges at the Interface of Data Analysis, Computer Science, and Optimization*, Studies in Classification, Data Analysis, and Knowledge Organization, DOI 10.1007/978-3-642-24466-7_9, © Springer-Verlag Berlin Heidelberg 2012

fuzzy c-means algorithm (FCM), (Dunn 1973; Bezdek 1981; Höppner et al. 1999; Kruse et al. 2007) seems to fail in high dimensions. This paper is dedicated to give some insight into this problem and the behaviour of FCM as well as its derivatives in high dimensions.

The algorithms that are analysed and compared in this paper are hard c-means (HCM), fuzzy c-means (FCM), noise FCM (NFCM), FCM with polynomial fuzzifier function (PFCM) and PFCM with a noise cluster (PNFCM) that is an extension of PFCM in the same way like NFCM is an extension of FCM. All these algorithms are prototype based and gradient descent algorithms. Previous to this paper, an analysis of FCM in high dimensions is presented in Winkler et al. (2011) which provides a more extensive view on the high dimension problematic but solely analyses the behaviour of FCM. Not included in this paper is the extension by Gustafson and Kessel (1978) because this algorithm is already unstable in low dimensions. Also not included is the competitive agglomeration FCM (CAFCM) (Frigui and Krishnapuram 1996). The algorithm is not a gradient descent algorithm in the strict sense.

A very good analysis of the influence of high dimensions to the nearest neighbour search is done in Beyer et al. (1999). The nearest neighbour approach can not be applied directly on clustering problems. But the basic problem is similar and thus can be used as a starting point for the analysis of the effects of high dimensional data on FCM as it is presented in this paper.

We approach the curse of dimensionality for the above mentioned clustering algorithms because they seem very similar but perform very differently. The main motivation lies more in observing the effects of high dimensionality rather than producing a solution to the problem. First, we give a short introduction to the algorithms and present a way how to test the algorithms in a high dimensional environment in the next section. In Sect. 3, the effects of a high dimensional data set are presented. A way to use the parameters of the algorithms to work on high dimensions is discussed in Sect. 4. We close this paper with some last remarks in Sect. 5, followed by a list of references.

## 2   The Algorithms and the Test Environment

A cluster is defined as a subset of data objects of the data set $X$ that belong together. The result of a (fuzzy) clustering algorithm is a (fuzzy) partitioning of $X$. All discussed algorithms in this paper proceed by a gradient descent strategy. The method of gradient descent is applied to an objective function with the following form: Let $X = \{x_1, \ldots, x_m\} \subset \mathbb{R}^n$ be a finite set of data objects of the vector space $\mathbb{R}^n$ with $|X| = m$. The clusters are represented by a set of prototypes $Y = \{y_1, \ldots, y_c\} \subset \mathbb{R}^n$ with $c = |Y|$ be the number of clusters. Let $f : [0, 1] \to [0, 1]$ be a strictly increasing function called the fuzzifier function and $U \in \mathbb{R}^{c \times m}$ be the partition matrix with $u_{ij} \in [0, 1]$ and $\forall j : \sum_{i=1}^{c} u_{ij} = 1$. And finally, let $d : \mathbb{R}^n \times \mathbb{R}^n \to \mathbb{R}$ be the Euclidean distance function and $d_{ij} = d(y_i, x_j)$.

The objective function $\boldsymbol{J}$ is defined as

$$J(\boldsymbol{X}, \boldsymbol{U}, \boldsymbol{Y}) = \sum_{i=1}^{c} \sum_{j=1}^{m} f(u_{ij}) d_{ij}^2. \tag{1}$$

The minimisation of $J$ is achieved by iteratively updating the members of $\boldsymbol{U}$ and $\boldsymbol{Y}$ and is computed using a Lagrange extension to ensure the constraints $\sum_{i=1}^{c} u_{ij} = 1$. The iteration steps are denoted by a time variable $t \in \mathbb{N}$ with $t = 0$ as the initialisation step for the prototypes. The algorithms HCM, FCM and PFCM have each a different fuzzifier function. The variances NFCM and PNFCM use a virtual noise cluster with a user specified, constant noise distance to all data objects: $\forall j = 1..m$, $d_{0j} = d_{\text{noise}}$, $0 < d_{\text{noise}} \in \mathbb{R}$. The noise cluster is represented in the objective function as additional cluster with index 0 so that the sum of clusters is extended to $i = 0$ to $c$.

To have a first impression on the behaviour of the algorithms in question, we apply them on a test data set $\boldsymbol{T}_d$. $\boldsymbol{T}_d$ is sampled from a set of normal distributions representing the clusters and 10% of normal distributed noise. In $\boldsymbol{T}_d$, the number of clusters is set to $c = 2n$, in our examples we use values of $n = 2$ and $n = 50$. The performance of a clustering algorithm applied on $\boldsymbol{T}_d$ is measured by the number of correctly found clusters and correctly represented number of noise data objects. A cluster counts as found if at least one prototype is located in the convex hull of the data objects of that cluster.

HCM (Steinhaus 1957) is not able to detect noise and it is not a fuzzy algorithm. The fuzzifier function is the identity: $f_{\text{HCM}}(u) = u$ and the membership values are restricted to $u_{ij} \in \{0, 1\}$. If applied on $\boldsymbol{T}_{50}$, HCM finds around 40 out of 100 clusters.

The fuzzifier function for FCM (Dunn 1973; Bezdek 1981) is an exponential function with $f_{\text{FCM}}(u) = u^{\omega}$ and $1 < \omega \in \mathbb{R}$. In Fig. 1, the prototypes are represented as filled circles, their "tails" represent the way the prototypes took from their initial- to their final location. The devastating effect of a high dimensional data set to FCM is obvious: the prototypes run straight into the centre of gravity of the data set, independently of their initial location and therefore, finding no clusters at all. NFCM

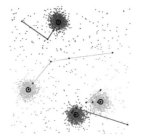 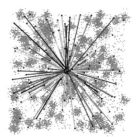

**Fig. 1** FCM, applied on $\boldsymbol{T}_2$ (left) and $\boldsymbol{T}_{50}$ (right)

**Fig. 2** PFCM, applied on $T_2$
(left) and $T_{50}$ (right)

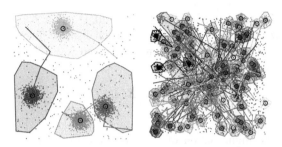

(Dave 1991) is one of the two algorithms considered in this paper that is able to detect noise. The fuzzifier function for NFCM is identical to FCM: $f_{\text{NFCM}} = f_{\text{FCM}}$. Apart from the fact that all data objects have the highest membership value for the noise cluster, the behaviour of the algorithm does not change compared to FCM. PFCM (Klawonn and Höppner 2003) is a mixture of HCM and FCM, as the definition of the fuzzifier function shows: $f_{\text{PFCM}}(u) = \frac{1-\beta}{1+\beta}u^2 + \frac{2\beta}{1+\beta}u$. This fuzzifier function creates an area of crisp membership values around a prototype while outside of these areas of crisp membership values, fuzzy values are assigned. The parameter $\beta$ controls the size of the crisp areas: the low value of $\beta$ means a small crisp area of membership values.

In the 2-dimensional example in Fig. 2, the surrounded and slightly shaded areas represents the convex hull of all data objects with membership value 1. On the right hand side of the figure, it can be seen, that PFCM does not show the same ill behaviour as FCM and NFCM, PFCM finds approximately 90 out of 100 clusters.

PNFCM (presented along with PFCM in Klawonn and Höppner (2003)) is the second algorithm considered in this paper that is able to detect noise. The fuzzifier function of PNFCM is again, identical to PFCM but the objective function is again modified by the noise cluster extension. If PNFCM is applied on $T_{50}$, the result is quite different to PFCM. Due to the specified noise distance, it is very likely that prototypes are initialized so far away from the nearest cluster, that all of its data objects are assigned crisp to the noise cluster. That is true for all prototypes which implies, that no cluster is found and all data objects are assigned as noise. The presented algorithms are in particular interesting because FCM produces useless results, HCM works with a lucky initializian, but their combination PFCM can be applied quite successfully. Since using a noise cluster has at least for PNFCM a negative effect, it is interesting to analyse its influence further.

We want to identify structural problems with FCM (and alike) on high dimensional data sets. Considering real data sets exemplary is not enough to draw general conclusions, therefore, we consider only one but for clustering optimal data set. Let $D = \{x_1, \ldots, x_c\} \subset \mathbb{R}^n$ be a data set that contains of $c > n$ clusters, with one data object per cluster. The clusters (data objects) in $D$ are located on an $n$-dimensional hypersphere surface and are arranged so that the minimal pairwise distance is maximised. The general accepted definition of clustering is: the data set partitioning should be done in such a way, that data objects of the same cluster should be as similar as possible while data objects of different clusters should be as different as possible. $D$ is a perfect data set for clustering because its clusters can be considered infinitely dense and maximally separated. There is one small limitation to that statement: $c$ should not be extremely larger than $n$ ($c < n!$) because the hypersphere surface might be too small for so many prototypes, but that limitation is usually not significant. Algorithms with problems on $D$ will have even more problems on other data sets because there is no "easier" data set than $D$. Especially if more, high-dimensional problems occur like overlapping clusters or very unbalanced cluster sizes.

As the example in Fig. 1-right has shown, the prototypes end up in the centre of gravity for FCM and NFCM. To gain knowledge why this behaviour occurs (and why not in the case of PFCM), the clustering algorithms are tested in a rather artificial way. The prototypes are all initialised in the centre of gravity (COG) and then moved iteratively towards the data objects by ignoring the update procedure indicated by the clustering algorithms. Let $\alpha$ control the location of the prototypes: $\alpha \in [0, 1]$, $x_i \in \mathbb{R}^n$ the $i$'th data object with $cog(D) \in \mathbb{R}^n$ the centre of gravity of data set $D$ implies $y_i : [0, 1] \to \mathbb{R}^n$ with $y_i(\alpha) = \alpha \cdot x_i + (1 - \alpha) \cdot cog(D)$ and finally: $d_{ij}(\alpha) = d(y_i(\alpha), x_j)$. Since the membership values are functions of the distance values and the objective function is a function of membership values and distance values, it can be plotted as a function of $\alpha$.

## 3 The Effect of High Dimensions

In this section, the effects of the high dimensional data sets like $D$ are presented. There are two effects on high dimensional data sets that have strong influence on the membership function: the number of prototypes and the number of dimensions. HCM is not farther analysed because its objective function values do not change due to the number of dimensions or the number of prototypes.

However, for FCM and NFCM, these factors have a strong influence on the objective function. In Fig. 3, the objective functions for these two algorithms are plotted for a variety of dimensions, depending on $\alpha$. For convenience, the objective

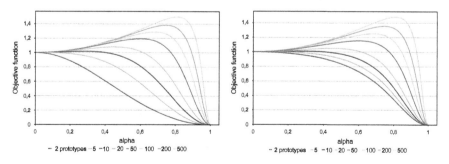

**Fig. 3** Objective function plots for FCM (left) and NFCM (right)

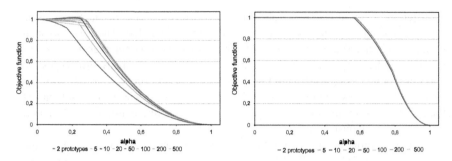

**Fig. 4** Objective function plots for PFCM (left) and PNFCM (right)

function values are normalized to 1 at $\alpha = 0$. The plots show a strong local maximum between $\alpha = 0.5$ and $\alpha = 0.9$. Winkler et al. showed in Winkler et al. (2011) that the number of dimensions effects the objective function by the height of this local maximum. The number of prototypes however influences the location of the maximum: the higher the number of prototypes, the further right the local maximum can be observed. Since these are gradient descent algorithms, the prototypes will run into the centre of gravity if they are initialized left of the local maximum which is exactly what is presented in Fig. 1-right. Since the volume of an $n$-dimensional hypersphere increases exponentially with its radius, it is almost hopeless to initialize a prototype near enough to a cluster so that the prototype converges to that cluster. For this example, in 50 dimensions and with 100 prototypes, the converging hypershere radius is 0.3 times the feature space radius which means, the hypervolume is $7.2 \cdot 10^{-27}$ times the volume of the feature space.

As presented in Fig. 4-left, PFCM does not create such a strong local maximum as FCM, also the local maximum that can be observed is very far left. That is the reason why PFCM can be successfully applied on a high dimensional data set. The situation is quite different for PNFCM, see Fig. 4-right. The fixed noise distance is chosen appropriate for the size of the clusters but the distance of the prototypes to the clusters is much larger. Therefore, all data objects have membership value 0 for the prototypes which explains the constant objective function value.

## 4 How to Exploit the Algorithm Parameters to Increase Their Effectiveness

As the section title indicates, it is possible to exploit the parameters $\omega$ and $d_{noise}$ of FCM, NFCM and PNFCM to tune the algorithms so that they work on high dimensions. The term "exploit" is used because the fuzzifier $\omega$ and the noise distance $d_{noise}$ are chosen dimension dependent and not in order to represent the properties of the data set. Let $\omega = 1 + \frac{2}{n}$ and $d_{noise} = 0.5 \log_2(D)$, the parameter $\beta$ remained at 0.5. Setting the fuzzifier near 1, creates an almost crisp clustering and

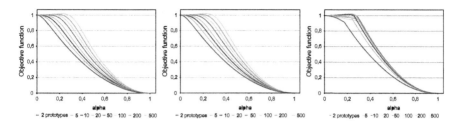

**Fig. 5** Objective function plots for FCM (left), NFCM (middle) and PNFCM (right) with dimension dependent parameters

**Table 1** Performance overview of $T_{50}$ with 100 data objects for each cluster, 1000 noise data objects and each algorithm is applied performed 100 times. The mean value and (sample standard deviation) are displayed

| Algorithm | Found clusters | | Correctly clustered noise | | Incorrect clustered as noise | |
|---|---|---|---|---|---|---|
| HCM | 42.35 | (4.65) | 0 | (0) | 0 | (0) |
| FCM | 0 | (0) | 0 | (0) | 0 | (0) |
| NFCM | 0 | (0) | 1,000 | (0) | 10,000 | (0) |
| PFCM | 90.38 | (2.38) | 0 | (0) | 0 | (0) |
| PNFCM | 0 | (0) | 1,000 | (0) | 10,000 | (0) |
| Adjusted Parameter | | | | | | |
| FCM AP | 88.09 | (3.58) | 0 | (0) | 0 | (0) |
| NFCM AP | 88.5 | (3.37) | 999.77 | (0.58) | 1136.0 | (344.82) |
| PNFCM AP | 92.7 | (2.67) | 995.14 | (3.12) | 96.0 | (115.69) |

by setting the noise distance larger its effect is reduced. That way, FCM and NFCM become similar to HCM and PNFCM becomes similar to PFCM. The results of the adapted parameter are shown in Fig. 5 for FCM, NFCM and PNFCM for $D$ with 100 dimensions. The objective function plots are all very similar to PFCM which would imply that they work just as well.

To test that, we apply each algorithm 100 times on $T_{50}$, the results are presented in Table 1 as mean and sample standard deviation in braces. The found clusters column is the most important one, the other two are just for measuring the performance of recognizing noise data objects. The test clearly shows the improvement by adjusting the parameters according to the number of dimensions.

## 5   Conclusions

The two algorithms HCM and FCM do not work on high dimensions properly. It is very odd therefore that a combination of them in form of PFCM works quite well. We have shown that the reason for this effect is a very small local minimum of PFCM compared to FCM in the COG. We presented, that FCM, NFCM and PNFCM can be tuned in such a way that their objective function shows a similar

behaviour in our test as PFCM, in which case the clustering result is similar on the
test data set $T_{50}$. The question remains why this local minimum occurs. A possible
explanation is presented in Beyer et al. (1999), Durrant and Kabán (2008) as they
identify the effect of distance concentration as being the most problematic in having
a meaningful nearest neighbour search. Further work is needed here for a deeper
understanding of the effect of distance concentration in relation to clustering. It
sounds logical that a clustering algorithm, that is based on the spatial structure
of the data, can not work well if all data objects from all points of view and for
all (variance limited) data distributions seem to be equally distant. But that poses
the question if clustering algorithms in general can produce meaningful results
on arbitrary high dimensional data sets without having some special features that
reduce the complexity of the problem.

The knowledge, gained from the presented experiments is, that prototype based
clustering algorithms seem to need a crisp component. But from the failure of
HCM, it might as well be learned that only crisp assignments of data objects are
not good enough. "Almost crisp" clustering algorithms like the PFCM, however,
perform quite well, at least on $T_{50}$. That is no guaranty that they will work on
high dimensional real data sets, but they are not as hopeless as FCM. However,
this also means that the fuzzifier in the case of FCM and NFCM as well as the
noise distance in case of the NFCM and PNFCM has to be used to counter the high
dimensional effects. This is very unsatisfying as it prevents a suitable modelling of
the data set and a better way would be to adapt the algorithm rather than exploiting
its parameters.

# References

Beyer K, Goldstein J, Ramakrishnan R, Shaft U (1999) When is nearest neighbor meaningful?
    In: Database theory - ICDT'99, Lecture Notes in Computer Science, vol 1540, Springer,
    Berlin/Heidelberg, pp 217–235
Bezdek JC (1981) Pattern recognition with fuzzy objective function algorithms. Plenum, New York
Dave RN (1991) Characterization and detection of noise in clustering. Pattern Recogn Lett
    12(11):657–664
Dunn JC (1973) A fuzzy relative of the isodata process and its use in detecting compact well-
    separated clusters. Cybern Syst Int J 3(3):32–57
Durrant RJ, Kabán A (2008) When is 'nearest neighbour' meaningful: A converse theorem and
    implications. J Complex 25(4):385–397
Frigui H, Krishnapuram R (1996) A robust clustering algorithm based on competitive agglomera-
    tion and soft rejection of outliers. In: IEEE Computer Society Conference on Computer Vision
    and Pattern Recognition, IEEE, pp 550–555
Gustafson DE, Kessel WC (1978) Fuzzy clustering with a fuzzy covariance matrix. IEEE
    17:761–766
Höppner F, Klawonn F, Kruse R, Runkler T (1999) Fuzzy cluster analysis. Wiley, Chichester,
    England
Klawonn F, Höppner F (2003) What is fuzzy about fuzzy clustering? Understanding and improv-
    ing the concept of the fuzzifier. In: Cryptographic Hardware and Embedded Systems - CHES
    2003, Lecture Notes in Computer Science, vol 2779, Springer, Berlin/Heidelberg, pp 254–264

Kruse R, Döring C, Lesot MJ (2007) Advances in fuzzy clustering and its applications. In: Fundamentals of fuzzy clustering. Wiley, pp 3–30

Steinhaus H (1957) Sur la division des corps materiels en parties. Bull Acad Pol Sci, Cl III 4:801–804

Winkler R, Klawonn F, Kruse R (2011) Fuzzy C-Means in High Dimensional Spaces. International Journal of Fuzzy System Applications (IJFSA), 1(1), 1–16. doi:10.4018/IJFSA.2011010101

# Part II
# Quantification Theory

# Reduced Versus Complete Space Configurations in Total Information Analysis

José G. Clavel and Shizuhiko Nishisato

**Abstract** In most multidimensional analyses, the dimension reduction is a key concept and reduced space analysis is routinely used. Contrary to this traditional approach, total information analysis (TIA) (Nishisato and Clavel, Behaviormetrika 37:15–32, 2010) places its focal point on tapping every piece of information in data. The present paper is to demonstrate that the time-honored practice of reduced space analysis may have to be reconsidered as its grasp of data structure may be compromised by ignoring intricate details of data. The paper will present numerical examples to make our point.

## 1 Introduction

The traditional quantification procedure (e.g., Benzécri et al. 1973; Nishisato 2007; 1980; 1984; Greenacre 1984; Gifi 1990) decomposes within-set relations (i.e., the variance-covariance matrix of rows and that of columns) into multiple components, and interprets the first few major components out of many. The joint graphical display of row and column variables is typically handled by symmetric scaling, which is an overlay of the multidimensional configuration of row variables onto that of columns. It is well-known, however, that this symmetric scaling cannot be justified because the space for the row variables is not the same as that for the column variables, rendering the direct overlaying of two different configurations meaningless and uninterpretable. This space discrepancy has been a perennial

J.G. Clavel (✉)
Universidad de Murcia, Spain
e-mail: jjgarvel@um.es

S. Nishisato
University of Toronto, Canada
e-mail: shizuhiko.nishisato@utoronto.ca

W. Gaul et al. (eds.), *Challenges at the Interface of Data Analysis, Computer Science, and Optimization*, Studies in Classification, Data Analysis, and Knowledge Organization, DOI 10.1007/978-3-642-24466-7_10, © Springer-Verlag Berlin Heidelberg 2012

problem in quantification theory, but the practical consideration has forced most researchers to ignore the space discrepancy and to adopt the theoretically incorrect symmetric scaling for joint graphical display. Based on preliminary studies (Nishisato and Clavel 2003; 2008; Clavel and Nishisato 2008), total information analysis or comprehensive dual scaling (Nishisato and Clavel 2010) was proposed to rectify the theoretical problem of symmetric scaling by adopting common space for both row and column variables, which is of greater dimensionality than that of the traditional symmetric scaling, for instance, two-dimensional symmetrically scaled graph requires four-dimensional space. With the expanded dimension of the common space, Nishisato and Clavel (2010) proposed an alternative to the joint graphical display, namely cluster analysis of row and column variables in this total space. This adoption of cluster analysis in total space can be justified because cluster structure in reduced space would almost always be distorted and furthermore graphical display in four dimensions, for example, is hardly practical. While the adoption of the common space for row and column variables may be welcomed by many researchers, there would be a strong objection to the use of total space for clustering. Therefore, the main aim of the current paper is to demonstrate that reduced space analysis can often lead to a distorted image of data structure, that is, relations of variables in total space. In the current paper, reduced space analysis does not mean the adoption of a smaller number of components in the traditional quantification analysis of the contingency table, but the use of the super-distance matrix (the expanded within-set and between-set distance matrix) calculated from a smaller number of solutions (components) associated with the original contingency table than the total number of solutions, the latter being referred to as total information analysis. As for the comparison between symmetric scaling and total information analysis, please see Nishisato and Clavel (2010).

## 2  Total Information Analysis (TIA) or Comprehensive Dual Scaling

Nishisato and Clavel (2003) have shown that the angle between the axis for row variables and that for column variables of component $k$ can be expressed as

$$\theta_k = cos^{-1}\rho_k \tag{1}$$

where $\rho_k$ is the singular value of solution (component) $k$. This means that only when the singular value is one, the angle between the row-variable and the column-variable axes of solution $k$ becomes zero and both row and column variables span the same space. However, when the singular value is one, not only the two axes of the same solution coincide, but also the row variables and the column variables occupy the same points (if the contingency table is $3 \times 5$ there will be at most only three distinct points in the joint graph). Thus, there is no interest for graphical display when the singular value is one. In other words, we are interested in graphical display only when the singular value is not perfect, that is, only when there exists the discrepancy between the two axes. In this case, plotting of the row variables and the column variables for solution $k$ requires two-dimensional space, due to the space discrepancy. Similarly, the familiar two-dimensional symmetric plot actually needs four dimensions to be exact. It is remarkable that when we examine two solutions we are so used to looking at a two-dimensional graph of variables, in spite of the fact that this is actually of a four-dimensional configuration erroneously compressed into two dimensions. What symmetric scaling does is to analyze the within-row variance-covariance matrix and the within-column variance-covariance matrix separately, and plot the two sets of weights from the two matrices in the same space, that is, solution 1 of the row variables and that of column variables using the same axis. This is absolutely wrong because a two-dimensional configuration is represented as a one-dimensional configuration. This erroneous representation, however, has become the standard procedure in most quantification research under the name symmetric scaling. To rectify this current procedure, Nishisato and Clavel (2010) proposed total information analysis (TIA) or comprehensive dual scaling, in which the common space for both row and column variables is used to find the configurations of all the variables in the same space.

In quantification theory, we decompose the data matrix $\mathbf{F}$ as

$$\mathbf{F} = \frac{1}{f_t}\mathbf{D}_n\mathbf{Y}\Phi\mathbf{X}'\mathbf{D}_m \tag{2}$$

where $\mathbf{D}_n$ and $\mathbf{D}_m$ are diagonal matrices of row marginals and column marginals of $\mathbf{F}$, respectively, $\mathbf{Y}$ is the $n \times K$ matrix of eigenvectors associated with the rows, $\mathbf{X}$ is the $m \times K$ matrix of eigenvectors associated with the columns, and $\Phi$ is the $K \times K$ diagonal matrix of singular values. The solution corresponding to the singular value of 1 is called the trivial solution and is discarded from analysis.

The Nishisato–Clavel between-set distance (i.e., the distance between row $i$ and column $j$) is given by:

$$d_{ij} = \sqrt{\sum_{k=1}^{K} \rho_k^2 \left( y_{ik}^2 + x_{jk}^2 - 2\rho_k y_{ik} x_{jk} \right)} \tag{3}$$

Notice that the above distance is calculated by taking into consideration the space discrepancy in the form of the cosine law. The Nishisato–Clavel within-set distance (e.g., the distance between row $i$ and row $i'$) is given by

$$d_{ii'} = \sqrt{\sum_{k=1}^{K} \rho_k^2 \left( y_{ik} - y_{i'k} \right)^2} \tag{4}$$

The within-set distances are identical to the Euclidean distances, calculated for the rows of the $n$-by-$K$ matrix $\mathbf{Y}\Phi$ and computed for the rows of the $m$-by-$K$ matrix $\mathbf{X}\Phi$.

Nishisato and Clavel (2010) define the super-distance matrix $\mathbf{D}^*$, consisting of within-row ($\mathbf{D}_r^*$), within-column ($\mathbf{D}_c^*$), between-row-column ($\mathbf{D}_{rc}^*$) and between-column-row ($\mathbf{D}_{cr}^*$) distance matrices arranged as

$$\mathbf{D}^* = \begin{bmatrix} \mathbf{D}_r^* & \mathbf{D}_{rc}^* \\ \mathbf{D}_{cr}^* & \mathbf{D}_c^* \end{bmatrix} \tag{5}$$

In terms of the formulas, typical elements of this matrix $\mathbf{D}^*$ are

$$\begin{bmatrix} \sqrt{\sum_{k=1}^{K} \rho_k^2 (y_{ik} - y_{i'k})^2} & \sqrt{\sum_{k=1}^{K} \rho_k^2 (y_{ik}^2 + x_{jk}^2 - 2\rho_k y_{ik} x_{jk})} \\ \sqrt{\sum_{k=1}^{K} \rho_k^2 (x_{jk}^2 + y_{ik}^2 - 2\rho_k x_{jk} y_{ki})} & \sqrt{\sum_{k=1}^{K} \rho_k^2 (x_{jk} - x_{j'k})^2} \end{bmatrix} \tag{6}$$

When the original contingency table requires $K$ solutions for exhaustive analysis, the super-distance matrix needs $2K$ solutions. This is the price we must pay to overcome the illogical problem of symmetric scaling and to deal with the common space.

## 3 Problem

With the expanded dimensionality, Nishisato and Clavel (2010) proposed the use of $k$-means clustering of all the variables in total space. This is quite sensible since clusters of variables in reduced space may provide a distorted image of the distribution of variables. The current problem is not on the effects of symmetric scaling, but the question of what happens if the super-distance matrix based on $K$ solutions is substituted with the super-distance matrix calculated from, say two solutions. The latter is referred to as reduced space analysis in this paper. The differences between the reduced space analysis and the analysis based on $K$ solutions, referred here as total information analysis, would depend on data. Therefore, we will look at a few numerical examples and will try to draw some empirical conclusions.

## 4 Reduced Versus Total Space Comparisons

To illustrate detailed steps of analysis, two small examples are used here. As Nishisato (2011) shows, its application to a larger data set is not expected to yield any difficult problems.

## 4.1   Barley Data

The data are from Stebbins (1950). Seeds of six varieties of barley were planted in equal numbers at each of six agricultural stations. At the harvest, 500 randomly chosen seeds were classified into the six varieties at each station. This process was repeated over a number of years. The final results are tabulated in the barley-by-location contingency table.

|     | Arl | Ith | StP | Moc | Mor | Dav |
|-----|-----|-----|-----|-----|-----|-----|
| Co  | 446 | 57  | 83  | 87  | 6   | 362 |
| Ha  | 4   | 34  | 305 | 19  | 4   | 34  |
| Wh  | 4   | 0   | 4   | 241 | 489 | 65  |
| Ma  | 1   | 343 | 2   | 21  | 0   | 0   |
| Ga  | 13  | 9   | 15  | 58  | 0   | 1   |
| Me  | 4   | 0   | 0   | 4   | 0   | 27  |

- Locations: *Arl (Arlington, VA); Ith (Ithaca, NY); StP (St. Paul, MN); Moc (Mocasin, MT); Mor (Moro, OR); Dav (Davis, CA)*
- Varieties: *Co (Coast & Trebi); Ha (Hanchen); Wh (White Smyrna); Man (Manchuria); Ga (Gatami); Me (Meloy)*

Quantification of this table yields in total five solutions, with singular values being 0.85, 0.79, 0.69, 0.24 and 0.13. Table 1 shows the elements of the super-distance matrix based on five solutions (total space) in the lower triangular section and those based on two solutions (reduced space) in the upper triangular section.

As expected, the cross-product matrix obtained from the super-distance matrix (total space) yields ten singular values (8.49, 6.37, 5.28, 1.75, 1.19, 0.92, 0.61, 0.44, 0.33 and 0.11), rather than five, meaning we need ten-dimensional space for the twelve variables.

**Table 1** The super-distance matrices in total space (lower triangle) and reduced space (upper triangle) of barley data

|     | Co   | Ha   | Wh   | Ma   | Ga   | Me   | Arl  | Ith  | StP  | Moc  | Mor  | Dav  |
|-----|------|------|------|------|------|------|------|------|------|------|------|------|
| Co  | 0.00 | 0.39 | 1.80 | 2.52 | 0.80 | 0.21 | 0.50 | 2.23 | 0.62 | 1.19 | 1.93 | 0.45 |
| Ha  | 2.13 | 0.00 | 2.11 | 2.37 | 1.04 | 0.60 | 0.64 | 2.10 | 0.58 | 1.47 | 2.22 | 0.70 |
| Wh  | 1.95 | 2.50 | 0.00 | 2.79 | 1.10 | 1.67 | 1.83 | 2.61 | 2.06 | 0.79 | 0.80 | 1.62 |
| Ma  | 2.56 | 2.89 | 2.80 | 0.00 | 2.18 | 2.66 | 2.55 | 1.17 | 2.43 | 2.41 | 2.88 | 2.51 |
| Ga  | 1.71 | 2.14 | 1.72 | 2.56 | 0.00 | 0.76 | 0.91 | 1.97 | 1.08 | 0.55 | 1.33 | 0.71 |
| Me  | 1.28 | 2.58 | 2.24 | 2.97 | 2.24 | 0.00 | 0.54 | 2.36 | 0.74 | 1.10 | 1.81 | 0.43 |
| Arl | 0.80 | 2.24 | 2.06 | 2.66 | 1.83 | 1.51 | 0.00 | 2.40 | 0.33 | 1.35 | 2.10 | 0.23 |
| Ith | 2.29 | 2.63 | 2.62 | 1.18 | 2.36 | 2.73 | 2.50 | 0.00 | 2.25 | 2.25 | 2.77 | 2.35 |
| StP | 2.00 | 1.28 | 2.46 | 2.89 | 2.12 | 2.46 | 2.28 | 2.74 | 0.00 | 1.58 | 2.35 | 0.53 |
| Moc | 1.45 | 2.14 | 0.96 | 2.47 | 1.37 | 1.89 | 1.69 | 2.33 | 2.17 | 0.00 | 0.79 | 1.12 |
| Mor | 2.09 | 2.61 | 0.85 | 2.91 | 1.90 | 2.37 | 2.35 | 2.80 | 2.66 | 1.12 | 0.00 | 1.86 |
| Dav | 0.70 | 2.04 | 1.76 | 2.56 | 1.65 | 1.39 | 0.62 | 2.40 | 2.05 | 1.44 | 2.03 | 0.00 |

**Table 2** Four clusters of
Barley data in reduced and
total space

| Reduced space | | | |
|---|---|---|---|
| Cluster 1 | Cluster 2 | Cluster 3 | Cluster 4 |
| Ma | Wh | Ga | Co+Ha+Me |
| Ith | Mor | Moc | StP+Arl+Dav |
| Total space | | | |
| Cluster 1 | Cluster 2 | Cluster 3 | Cluster 4 |
| Ma | Wh+Ga | Ha | Co+Me |
| Ith | Moc+Mor | StP | Arl+Dav |

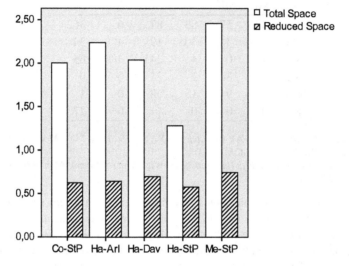

**Fig. 1** Distance comparisons

Four clusters by the $k$-means clustering of the two super-distance matrices are
as shown in Table 2. Clusters formed in reduced space resemble those we see in
symmetric scaling of the first two solutions (not shown here): (Manchuria; Ithaca),
(White Smyrna; Moro), (Gatemi; Mocasin) and (Coast Trebi, Hanchen, Meloy; St.
Paul, Arlington, Davis). In total space, the cluster formation changes. If we look at
Cluster 4 of reduced space, we can look at distances between Hanchen and other
locations, between St. Paul and other varieties, and between Hanchen and St. Paul,
as shown in Fig. 1, which explains why Hanchen and St. Paul form a cluster in total
space. Other new clusters in total space can equally be explained through distance
comparisons between reduced space and total space.

## 4.2 Rorschach Data

Data are from Garmize and Rycklak (1964). Selected Rorschach responses were
collected under six types of experimentally induced moods, fear, anger, depression,
ambition, security and love. Table 3 shows joint frequencies of responses:

**Table 3** The Garmize and Rycklak data

| Rorschach | Induced moods | | | | | |
|-----------|------|-------|------------|----------|----------|------|
| Responses | Fear | Anger | Depression | Ambition | Security | Love |
| Bat       | 33   | 10    | 18         | 1        | 2        | 6    |
| Blood     | 10   | 5     | 2          | 1        | 0        | 0    |
| Butterfly | 0    | 2     | 1          | 26       | 5        | 18   |
| Cave      | 7    | 0     | 13         | 1        | 4        | 2    |
| Clouds    | 2    | 9     | 30         | 4        | 1        | 6    |
| Fire      | 5    | 9     | 1          | 2        | 1        | 1    |
| Fur       | 0    | 3     | 4          | 5        | 5        | 21   |
| Mask      | 3    | 2     | 6          | 2        | 2        | 3    |
| Mountains | 2    | 1     | 4          | 1        | 18       | 2    |
| Rocks     | 0    | 4     | 2          | 1        | 2        | 2    |
| Smoke     | 1    | 6     | 1          | 0        | 1        | 0    |

**Table 4** The super-distance matrices in total space analysis (lower triangular section) and reduced space analysis (upper triangular section)

|       | Bat  | Bloo | Butt | Cave | Clou | Fire | Fur  | Mas  | Mou  | Rock | Smok | Fear | Ange | Depr | Amb  | Secu | Love |
|-------|------|------|------|------|------|------|------|------|------|------|------|------|------|------|------|------|------|
| Bat   | 0.00 | 0.25 | 1.89 | 0.57 | 0.48 | 0.31 | 1.49 | 0.68 | 2.00 | 0.87 | 0.18 | 0.67 | 0.56 | 0.57 | 1.66 | 1.71 | 1.36 |
| Blood | 0.56 | 0.00 | 2.03 | 0.82 | 0.69 | 0.45 | 1.67 | 0.90 | 2.25 | 1.10 | 0.43 | 0.76 | 0.72 | 0.78 | 1.80 | 1.90 | 1.52 |
| Butt  | 1.94 | 2.09 | 0.00 | 1.71 | 1.45 | 1.59 | 0.53 | 1.31 | 2.15 | 1.20 | 1.78 | 1.90 | 1.55 | 1.54 | 1.01 | 1.79 | 0.94 |
| Cave  | 0.85 | 1.34 | 1.84 | 0.00 | 0.40 | 0.60 | 1.22 | 0.41 | 1.43 | 0.51 | 0.40 | 0.79 | 0.57 | 0.41 | 1.49 | 1.36 | 1.15 |
| Cloud | 1.22 | 1.58 | 1.78 | 0.81 | 0.00 | 0.29 | 1.02 | 0.22 | 1.72 | 0.41 | 0.33 | 0.74 | 0.39 | 0.32 | 1.29 | 1.44 | 0.95 |
| Fire  | 1.10 | 0.89 | 1.83 | 1.54 | 1.44 | 0.00 | 1.22 | 0.50 | 1.99 | 0.69 | 0.29 | 0.71 | 0.44 | 0.48 | 1.42 | 1.64 | 1.12 |
| Fur   | 1.66 | 1.96 | 1.16 | 1.53 | 1.48 | 1.70 | 0.00 | 0.84 | 1.68 | 0.71 | 1.35 | 1.52 | 1.15 | 1.10 | 0.89 | 1.43 | 0.65 |
| Mask  | 0.84 | 1.22 | 1.40 | 0.57 | 0.68 | 1.17 | 1.10 | 0.00 | 1.55 | 0.20 | 0.52 | 0.85 | 0.48 | 0.36 | 1.19 | 1.33 | 0.83 |
| Mount | 2.04 | 2.27 | 2.16 | 1.65 | 2.05 | 2.11 | 1.90 | 1.64 | 0.00 | 1.42 | 1.83 | 1.98 | 1.82 | 1.64 | 2.03 | 1.46 | 1.77 |
| Rock  | 1.36 | 1.49 | 1.53 | 1.31 | 1.07 | 0.87 | 1.17 | 0.85 | 1.62 | 0.00 | 0.71 | 0.99 | 0.62 | 0.49 | 1.12 | 1.25 | 0.76 |
| Smoke | 1.65 | 1.49 | 2.32 | 1.93 | 1.66 | 0.72 | 2.06 | 1.60 | 2.25 | 1.00 | 0.00 | 0.65 | 0.47 | 0.43 | 1.55 | 1.58 | 1.23 |
| Fear  | 0.90 | 1.05 | 2.02 | 1.16 | 1.39 | 1.28 | 1.78 | 1.10 | 2.08 | 1.44 | 1.71 | 0.00 | 0.43 | 0.54 | 1.93 | 1.93 | 1.62 |
| Anger | 1.13 | 1.23 | 1.81 | 1.28 | 1.25 | 1.06 | 1.57 | 1.02 | 2.01 | 1.08 | 1.40 | 1.30 | 0.00 | 0.32 | 1.50 | 1.73 | 1.20 |
| Depre | 1.02 | 1.34 | 1.75 | 0.84 | 0.90 | 1.34 | 1.45 | 0.73 | 1.86 | 1.14 | 1.68 | 1.30 | 1.30 | 0.00 | 1.54 | 1.43 | 1.19 |
| Ambi  | 1.77 | 1.95 | 1.15 | 1.68 | 1.64 | 1.73 | 1.31 | 1.32 | 2.10 | 1.45 | 2.14 | 2.04 | 1.81 | 1.79 | 0.00 | 1.87 | 0.39 |
| Secu  | 1.77 | 1.99 | 1.85 | 1.54 | 1.73 | 1.83 | 1.62 | 1.40 | 1.50 | 1.46 | 2.08 | 1.99 | 1.89 | 1.73 | 1.93 | 0.00 | 1.54 |
| Love  | 1.48 | 1.73 | 1.16 | 1.36 | 1.34 | 1.53 | 0.97 | 0.98 | 1.88 | 1.16 | 1.94 | 1.80 | 1.58 | 1.45 | 1.06 | 1.67 | 0.00 |

The table yields five solutions with singular values 0.68, 0.50, 0.41, 0.36 and 0.27, respectively. The super-distance matrices based on five solutions (total space analysis, the lower triangular part) and two solutions (reduced space analysis, the upper triangular part) are given in Table 4. Let us examine k-means clustering results with four clusters (Table 5).

Clusters 1 and 2 are identical in both reduced space and total space. So, let us concentrate on the differences between two sets of clusters. First, we notice that Depression is closer to Anger than to Fear in reduced space, but total space analysis indicates that Depression is closer to Fear than to Anger. Second, Rock is closer to Cave, Cloud and Mask than to Fire and Smoke in reduced space, but in fact total space analysis shows that Rock is closer to Fire and Smoke than to Cave, Cloud and

**Table 5** Four clusters of Barley data in reduced and total space

| Reduced space | | | |
|---|---|---|---|
| Cluster 1 | Cluster 2 | Cluster 3 | Cluster 4 |
| But+Fur | Mount | Bat+Bloo+Fire+Smoke | Cave+Cloud+Mask+Roc |
| Ambi+Love | Secu | Fear | Anger+Depre |
| Total space | | | |
| Cluster 1 | Cluster 2 | Cluster 3 | Cluster 4 |
| But+Fur | Mount | Fire+Roc+Smoke | Bat+Bloo+Cave+Cloud+Mask |
| Ambit+Love | Secu | Anger | Fear+Depre |

**Table 6** Five cluster for barley data in reduced and total space

| Reduced space | | | | |
|---|---|---|---|---|
| Cluster 1 | Cluster 2 | Cluster 3 | Cluster 4 | Cluster 5 |
| Mount | But+Fur | | Cave+Cloud+Mask+Roc | Bat+Bloo+Fire+Smoke |
| Secu | | Ambi+Love | Anger+Depre | Fear |
| Total space | | | | |
| Cluster 1 | Cluster 2 | Cluster 3 | Cluster 4 | Cluster 5 |
| Mount | But+Fur | Fire+Roc+Smoke | Cave+Cloud+Mask | Bat+Bloo |
| Secu | Ambi+Love | Anger | Depre | Fear |

Mask. These are two examples of misleading results when reduced space analysis is used. We can further compare the differences in clusters 3 and 4 between reduced space analysis and total space analysis, noting that the data structure is reflected in total space, not in reduced space. Let us look at five clusters.

Other than Cluster 1, the differences in clusters between reduced space and total space are outstanding. This case, too, strongly suggests that we should look at total space, for reduced space is likely to mislead our conclusions on clusters of variables in data.

# 5   Concluding Remarks

This paper has shown the importance of interpreting data structure in total space, rather than reduced space. In the traditional symmetric scaling, reduced space analysis is typically used and this problematic practice is compounded by a theoretical problem of plotting variables in different coordinates on the common coordinates, calling it joint graphical display of row and column variables. Now that we have refuted the wisdom of symmetric scaling, it is no longer necessary to adopt reduced space analysis, for the super-distance matrix calculated from total space is of the same size as that based on a smaller number of solutions, and clustering of the former matrix is not much more laborious than that of the latter. The only additional step for total space analysis is to extend the computations of within-set and between-set distances to all the solutions (components). The current paper has presented convincing enough examples in favor of total information analysis as a routine procedure. With modern technology, the application of TIA to a larger data set than those used in the current study should not pose too difficult a problem.

# References

Benzécri JP, et al. (1973) L'Analyse des Données: II. L'Analyse des correspondances. Dunod, Paris

Clavel JG, Nishisato S (2008) Joint analysis of within-set and between-set distances. In: Shigemasu K, Okada A, Imaizumi T, Hoshino T (eds) New trends in psychometrics. Universal Academy Press, Tokyo, pp 41–50

Garmize L, Rycklak J (1964) Role-play validation of socio-cultural theory of symbolism. J Consult Psychol 28:107–115

Gifi A (1990) Nonlinear multivariate analysis. Wiley, New York

Greenacre MJ (1984) Theory and applications of correspondence analysis. Academic Press, London

Nishisato A (2007) Multidimensional nonlinear descriptive analysis. Chapman and Hall/CRC, Boca Raton, Florida

Nishisato S (1980) Analysis of categorical data: dual scaling and its applications. The University of Toronto Press, Toronto

Nishisato S (1984) Elements of dual scaling: An introduction to practical data analysis. Lawrence Erlbaum Associates, Hilsdale, NJ

Nishisato S (2011) Quantification theory: Reminiscence and a step forward. In: Gaul W, Geyer-Schulz A, Schmidt-Thieme L, Kunze J (eds) Challenges at the interface of data analysis, computer science, and optimization, studies in classification, data analysis, and knowledge organization. Springer, Heidelberg, Berlin

Nishisato S, Clavel J (2003) A note on between-set distances in dual scaling and correspondence analysis. Behaviormetrika 30:87–98

Nishisato S, Clavel J (2008) Interpreting data in reduced space: A case of what is not what in multidimensional data analysis. In: Shigemasu K, Okada A, Imaizumi T, Hoshino T (eds) New trends in psychometrics. Universal Academy Press, Tokyo, pp 357–366

Nishisato S, Clavel J (2010) Total information analysis: Comprehensive dual scaling. Behaviormetrika 37:15–32

Stebbins CL (1950) Variations and evolution. Columbus University Press, New York

# A Geometrical Interpretation of the Horseshoe Effect in Multiple Correspondence Analysis of Binary Data

Takashi Murakami

**Abstract** When a set of binary variables is analyzed by multiple correspondence analysis, a quadratic relationship between individual scores corresponding to the two largest characteristic roots is often observed. This phenomenon is called the *horseshoe effect*, which is well known as an artifact in the analysis of the perfect scale in Guttman's sense, and also observed in the *quasi scale* as a result of random errors. In addition, although errors are unsystematic and symmetric, scores corresponding to erroneous response patterns lie only inside the horseshoe. This phenomenon, which we will call *filled horseshoe*, is explained by the concept of an affine projection of a hypercube that represents binary data. The image of the hypercube on the plane has the form of a *zonotope*, which is a convex and centrally symmetric polygon, and it is shown that images forming the horseshoe must lie along the vertices of the zonotope, if it exists, and hence, other images must reside inside it.

## 1 Introduction

The perfect scale (Table 1) proposed by Guttman (1944) is an idealized data matrix that guarantees the existence of a unidimensional latent trait behind observed binary variables. However, the multiple correspondence analysis (MCA) of the perfect scale yields apparently spurious dimensions (Guttman 1950). The most conspicuous phenomenon is the quadratic relationship between the scores of individuals corresponding to the first and the second largest characteristic roots (Fig. 1). This phenomenon is called the *horseshoe effect* (e.g., Greenacre 1984).

T. Murakami (✉)
Department of Sociology, Chukyo University, Toyota, 470-0393, Japan
e-mail: tandem06@sass.chukyo-u.ac.jp

W. Gaul et al. (eds.), *Challenges at the Interface of Data Analysis, Computer Science, and Optimization*, Studies in Classification, Data Analysis, and Knowledge Organization, DOI 10.1007/978-3-642-24466-7_11, © Springer-Verlag Berlin Heidelberg 2012

**Table 1** Perfect scale for six variables

|            | Variable |   |   |   |   |   |
|------------|----------|---|---|---|---|---|
| Individual | 1        | 2 | 3 | 4 | 5 | 6 |
| 1          | 0        | 0 | 0 | 0 | 0 | 0 |
| 2          | 1        | 0 | 0 | 0 | 0 | 0 |
| 3          | 1        | 1 | 0 | 0 | 0 | 0 |
| 4          | 1        | 1 | 1 | 0 | 0 | 0 |
| 5          | 1        | 1 | 1 | 1 | 0 | 0 |
| 6          | 1        | 1 | 1 | 1 | 1 | 0 |
| 7          | 1        | 1 | 1 | 1 | 1 | 1 |

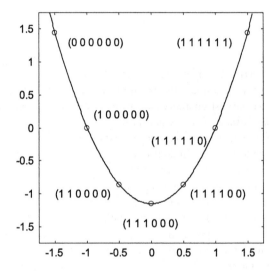

**Fig. 1** Horseshoe effect obtained from the data in Table 1

The reason for the occurrence of the horseshoe in the perfect scale has been proven more or less rigorously. For example, Iwatsubo (1984) solved the simultaneous difference equations, and showed that individual scores corresponding to the $k$-th ($k > 1$) characteristic root are in precise agreement with the $k$-th order polynomial function of the scores corresponding to the first characteristic root. In his theorem, the horseshoe effect is for the case of $k = 2$.

Although the perfect scale would rarely occur for real data, the quasi scale (Suchman 1950) which includes unsystematic errors would be a better model for real data. For example, we constructed a quasi scale by copying each response pattern in Table 1 one thousand times, and changed 25% of ones to zeros, zeros to ones in the obtained 7,000 by 7 data matrix. The scatter diagram of the first and the second scores are shown in Fig. 2. To represent the frequency of each response pattern in the diagram, a random number distributed uniformly with a zero mean and a constant range was added to each score.

A simple question is raised here: Although errors occur unsystematically and symmetrically, why do all the scores corresponding to the response pattern *not* congruent to the perfect scale lie only inside the horseshoe? We will call this

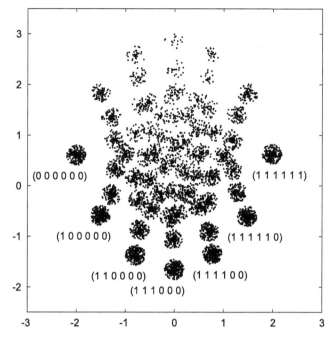

**Fig. 2** A filled horseshoe obtained from an artificial quasi scale

phenomenon a *filled* horseshoe in contrast to an *open* horseshoe given by the perfect scale. There have been limited discussions on this phenomenon in the literature. We will propose its geometrical interpretation.

It is well known (e.g., Yamada and Nishisato 1993) that principal components analysis (PCA) generates exactly the same set of scores for individuals as MCA. Hence, in the following section, we will only consider the results of PCA, and discuss *means* (proportions of ones), characteristic roots of a matrix of *correlations*, and *weights for component scores*.

## 2 Data Cube and Its Affine Projection

### 2.1 Binary Data Represented as a Hypercube

In this section, we will consider the general characteristics of scores obtained from the PCA of binary variables.

Let $\mathbf{X}$ be an $n$-individuals by $p$-variables binary data matrix. A row vector $\mathbf{x}'_i$ is the $i$-th row of $\mathbf{X}$, and is called *response pattern*. Elements of $\mathbf{x}'_i$ are seen as the coordinates of the $p$-dimensional space, and any distinctive response pattern is

**Fig. 3** A datacube for 4
binary variables and the
perfect scale (filled circles)

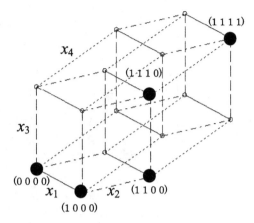

represented as the vertex of a $p$-dimensional hypercube with $2^p$ vertices and $p\,2^{p-1}$
edges of unit length. We will call it a *data cube*. A data cube for four variables is
shown in Fig. 3.

Each edge of a data cube connects two vertices, one of which corresponds to the
value 0 and the other to the value 1 of a variable. There are $2^{p-1}$ parallel edges which
correspond to the two values of a variable for all the different response patterns
of the remaining $p-1$ variables. In other words, $p\,2^{p-1}$ edges are classified into
$p$ classes corresponding to particular variables. Edges belonging to different classes
are drawn by different line types in Fig. 3.

## 2.2 Individual Scores as Affine Projections

The first two scores for individuals in PCA are computed by the following formula

$$\mathbf{F} = (\mathbf{X} - \mathbf{1m'})\mathbf{D}^{-1}\mathbf{K}_2\Lambda_2^{-1/2}, \tag{1}$$

where $\mathbf{F}$ is an $n$ by 2 matrix of scores, $\mathbf{1}$ is an $n$-dimensional vector with all elements
unity, $\mathbf{m}$ is a $p$-dimensional vector of means, $\mathbf{D}$ is a $p$ by $p$ diagonal matrix of
standard deviations, $\Lambda_2$ is a diagonal matrix of the two largest characteristic roots,
and $\mathbf{K}_2$ is a $p$ by 2 matrix of the corresponding characteristic vectors. Hence $\mathbf{f}'_i$,
the $i$-th row of $\mathbf{F}$, is the coordinate of the image of a vertex $\mathbf{x}'_i$ given by the affine
projection

$$\mathbf{f}'_i = \mathbf{x}'_i\mathbf{V} + \mathbf{c}', \tag{2}$$

where

$$\mathbf{V} = \mathbf{D}^{-1}\mathbf{K}_2\Lambda_2^{-1/2},$$

and

$$\mathbf{c}' = -\mathbf{m}'\mathbf{D}^{-1}\mathbf{K}_2\Lambda_2^{-1/2}.$$

The following two properties will be important in the sequel.

**Property 1** All the images of the $2^{p-1}$ edges of a data cube corresponding to a particular variable by an affine projection are mutually parallel line segments of the same length.

This property is trivial because the equity of length and the mutual parallel relationship between line segments, which are also properties of a hypercube, are preserved by an affine projection. It means that the images of the edges of a data cube are also classified into the $p$ classes of line segments with the same length and the same slant in the plane.

**Property 2** The projection of a data cube is a convex polygon.

This property also might be almost trivial because any affine projection of a convex polyhedron is a convex polyhedron as well (e.g., Ziegler 1995), and a hypercube is a convex polyhedron.

We should note that Property 1 and 2 refer to somewhat different aspects of the term *projection*. While Property 1 mentions the projections of all vertices and edges, i.e., the skeleton of a hypercube, Property 2 refers to a shadow of a hypercube as an opaque solid body.

## 2.3 Properties of an Affine Projection of a Data Cube

Property 2 does not show the exact form and properties of the polygon as an affine projection of a hypercube. Now, we will explain how to draw it. It may also be a constructive proof of the existence of the polygonal lines representing a horseshoe. Let us begin with introducing a few concepts.

The *symmetric vertex* of a vertex $\mathbf{x}'_i$ is defined as

$$\tilde{\mathbf{x}}'_i = \mathbf{1}' - \mathbf{x}'_i. \tag{3}$$

For example, the symmetric vertex of $(1 \quad 1 \quad 0 \quad 0)$ is $(0 \quad 0 \quad 1 \quad 1)$.

The *leftmost vertex*, $\mathbf{x}'_L$ is the vertex whose projection gives the attainable minimum of $f_{i1}$. Its $j$-th element is defined as

$$x_{jL} = \begin{cases} 1 & \text{if } v_{j1} < 0, \\ 0 & \text{otherwise,} \end{cases} \tag{4}$$

where $v_{j1}$ is the $j$-th element of the first column of $\mathbf{V}$. The symmetric vertex of $\mathbf{x}'_L$ is the rightmost vertex, $\mathbf{x}'_R$.

The *slope* of the $j$-th variable is defined as

$$s_j = \frac{v_{j2}}{v_{j1}}, \tag{5}$$

which represents the slope of images of edges belonging to the class corresponding to the variable $j$.

Weights and slopes with the means of four variables obtained from an artificial quasi scale, generated by the same way as in Fig. 2, are shown in Table 2. We will explain the procedure for drawing a polygon using Table 2 as an example.

1. Determine $\mathbf{x}'_L$. Because all the elements of the first column of the weight matrix in Table 2 are positive, $\mathbf{x}'_L = (0 \quad 0 \quad 0 \quad 0)$.
2. Sort the slopes in ascending order. It is attained automatically in Table 2.
3. Construct a series of response patterns: Begin with $\mathbf{x}'_L$, append patterns where the values of the variables are changed according to the order determined in the previous step, and reach $\mathbf{x}'_R$. In the case of Table 2, they are $(0 \quad 0 \quad 0 \quad 0)$, $(1 \quad 0 \quad 0 \quad 0)$, $(1 \quad 1 \quad 0 \quad 0)$, $(1 \quad 1 \quad 1 \quad 0)$, and $(1 \quad 1 \quad 1 \quad 1)$, which is the perfect scale.
4. Compute the images of the series of response patterns. Connect the images by line segments, which are actually the images of the edges connecting the two vertices, and the slants of the line segments are equal to the slopes of the corresponding variables. Because the slopes were sorted in an ascending order, the form of the obtained sequential line segments is upward convex.
5. Replicate the same procedure starting from $\mathbf{x}'_R$. The series of response patterns are $(1 \quad 1 \quad 1 \quad 1)$, $(0 \quad 1 \quad 1 \quad 1)$, $(0 \quad 0 \quad 1 \quad 1)$, $(0 \quad 0 \quad 0 \quad 1)$, and $(0 \quad 0 \quad 0 \quad 0)$ in the case of Table 2. This time downward convex line segments are obtained (Figs. 4 and 5).

Intuitively speaking, because we only have $p$ kinds of line segments as materials for drawing a *convex* polygon, the polygon obtained by the procedure described above is unique up to the reflections of the axes. The procedure also demonstrates

**Table 2** Weights and slopes of a four-variable quasi scale

| Variable | Mean | Weight 1 | Weight 2 | Slope |
|---|---|---|---|---|
| 1 | 0.65 | 0.73 | −1.55 | −2.12 |
| 2 | 0.55 | 0.99 | −0.43 | −0.43 |
| 3 | 0.44 | 0.99 | 0.52 | 0.53 |
| 4 | 0.35 | 0.81 | 1.27 | 1.56 |

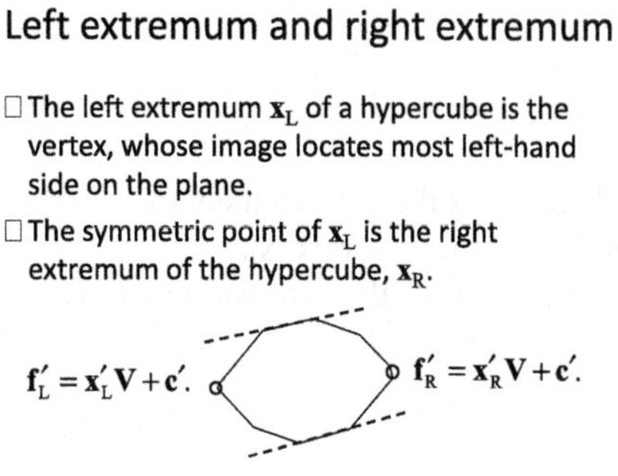

## Left extremum and right extremum

☐ The left extremum $\mathbf{x}_L$ of a hypercube is the vertex, whose image locates most left-hand side on the plane.

☐ The symmetric point of $\mathbf{x}_L$ is the right extremum of the hypercube, $\mathbf{x}_R$.

**Fig. 4** Zonotope as an image of a four-variable quasi scale, and images of the left- and rightmost vertex

$\mathbf{f}'_L = \mathbf{x}'_L \mathbf{V} + \mathbf{c}'.$ $\mathbf{f}'_R = \mathbf{x}'_R \mathbf{V} + \mathbf{c}'.$

**Fig. 5** Images of vertices
and edges of a four-variable
quasi scale

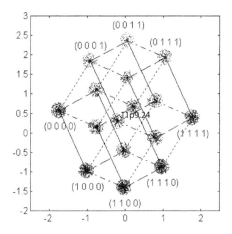

that the polygon has $2p$ vertices, its opposite sides connecting the symmetric vertices are mutually parallel, and is centrally symmetric (Fig. 4). A polygon as the projection of a hypercube is called a zonotope (e.g., Ziegler 1995).

Scores obtained from the data summarized in Table 2 are shown in Fig. 5 in the same manner as Fig. 2. One may see a zonotope including all vertices and edges, and the response patterns belonging to the perfect scale are situated on the vertices of the zonotope. In addition, it is seen as the projection of the hypercube in Fig. 3 from a different angle.

## 3 The Images of Response Patterns in Quasi Scales

### 3.1 Proof of the Filled Horseshoe

We will show that a filled horseshoe is inevitable when PCA is applied to a quasi scale. Let us assume that the images of response patterns belonging to the perfect scale in a quasi scale are arranged on a series of vertices of the zonotope as seen in Fig. 5. It would hold as long as the errors occur unsystematically, and unless the amount of random errors is very large or the number of individual is very small. Then, they form a contour of a horseshoe, and all the remaining images must lie inside it. There cannot be any images outside the horseshoe, because all the images of vertices of a hypercube must be situated inside the zonotope.

No other paths connecting the images of $\mathbf{x}'_L$ and $\mathbf{x}'_R$, usually $\mathbf{0}'$ and $\mathbf{1}'$ have the form of the horseshoe because they must have at least one concave part. This is the answer to the simple question raised in Sect. 1.

## 3.2 Some Additional Properties of the Perfect Scale

One may suspect that the horseshoe of the perfect scale is not necessarily situated on the contour of the zonotope because some series of images that form horseshoe-like shapes reside inside it as in Fig. 2.

However, that is not the case. Let us consider the perfect scale again. It consists of $p+1$ response patterns connecting $\mathbf{0}'$ and $\mathbf{1}'$ changing the value of 0 elements to 1 one by one. Therefore, adjacent images forming the horseshoe must be connected by an edge as seen in Fig. 3, and no horseshoe-like sequence has this property because there are only two convex sequential line segments between the images of an affine projection of a hypercube.

## 3.3 Concluding Remarks

Some anomalous scatter diagrams, which deviate from the linear relationships or the multivariate normal distribution, are often observed in the analysis of real categorical data by PCA and MCA. Some of them probably include the spurious dimensions related to the horseshoe effects. Development of procedures for discriminating the actual multidimensional structures from these artifacts would be a challenging task.

**Acknowledgements** The author wishes to express his deep gratitude to the anonymous reviewers for their valuable comments. This study was partially supported by Special Research Grant of Chukyo University in 2008.

# References

Greenacre MJ (1984) Theory and applications of correspondence analysis. Academic Press, London

Guttman LL (1944) A basis for scaling quantitative data. Am Sociol Rev 9:139–150

Guttman LL (1950) The principal components of scale analysis. In: Stoufer SA, et al. (eds) Measurement and prediction. Princeton University Press, Princeton, pp 312–361

Iwatsubo S (1984) The analytical solutions of Eigenvalue problem in the case of applying optimal scoring to some types of data. In: Diday I, et al. (eds) Data analysis and informatics III. North Holland, Amsterdam, pp 31–40

Suchman EA (1950) The utility of scalogram analysis. In: Stoufer SA, et al. (eds) Measurement and prediction. Princeton University Press, Princeton, pp 122–171

Yamada F, Nishisato S (1993) Several mathematical properties of dual scaling as applied to dichotomous item category data. Jpn J Behaviormetrics 29(1):56–63, (In Japanese)

Ziegler GM (1995) Lectures on polytope. Springer, New York

# Quantification Theory: Reminiscence and a Step Forward

Shizuhiko Nishisato

**Abstract** After a sketch of topics in my life-long work on quantification theory, two robust clustering procedures are proposed to compliment a newly developed scaling procedure, called total information analysis (TIA), with numerical examples.

## 1 Beginning

Factor analysis (Nishisato 1961) and minimum entropy clustering (Nishisato 1966) of binary variables served as a wakening call for the need of a more practical framework for data analysis. A long-range project of data quantification was launched in 1969 (Nishisato 1980). Modern quantification theory goes back to ecologists' work in the early twentieth century (e.g., Gleason 1926; Lenoble 1927; Ramensky 1930), followed almost independently by well-known studies as described in Benzécri et al. (1973), de Leeuw (1973), Lebart, Morineu, and Tabard (1977), Nishisato (1980, 1994, 2007, 2010), Greenacre (1984), Gifi (1990), and Le Roux and Rouanet (2004).

Two desiderata for quantification were targeted: (1) capturing a maximal amount of information in data by transforms which are linear regressions on data (Hirschfeld 1935), and (2) handling as many types of nominal and ordinal categorical data, the main theme of dual scaling, resulting in optimal quantification of incidence and dominance data (Nishisato 1978, 1996, 2007, 2010).

## 2 Forty Years of Work in a Nutshell

This section contains a description of some key topics investigated in Toronto.

S. Nishisato (✉)
9 St. George's Road, Toronto, Ontario M9A 3S9, Canada
e-mail: shizuhiko.nishisato@utoronto.ca

W. Gaul et al. (eds.), *Challenges at the Interface of Data Analysis, Computer Science, and Optimization*, Studies in Classification, Data Analysis, and Knowledge Organization, DOI 10.1007/978-3-642-24466-7_12, © Springer-Verlag Berlin Heidelberg 2012

## 2.1  Quantification of Structured Data

The one-way analysis of variance procedure for non-numerical data (Fisher 1948)
was extended to the general analysis of variance design for rows, or columns, or
both rows and columns simultaneously (Nishisato 1971, 1972; Lawrence 1985;
Nishisato and Lawrence 1989) by considering the quantification of data matrix $\mathbf{F}$
(i.e., frequency and response-pattern tables), expressed in the following general
form:

$$\mathbf{F} = \left(\sum_{j=1}^{n} \mathbf{P}_j\right) \mathbf{F} \left(\sum_{k=1}^{m} \mathbf{Q_k}\right),$$

where $\mathbf{P}_j$ and $\mathbf{Q}_k$ are projection operators such that

$$\sum_{j=1}^{n} \mathbf{P}_j = \mathbf{I} \text{ and } \sum_{k=1}^{m} \mathbf{Q}_k = \mathbf{I}.$$

Optimizing a particular design structure also led to discriminant analy-
sis of categorical data, called forced classification (Nishisato 1984a, 1986;
Nishisato and Gaul  1990;  Nishisato and Baba  1999).  See  also  studies  by
Takane and Shibayama (1991), Takane, Yanai, and Mayekawa (1991), and
Yanai and Maeda (2002).

## 2.2  Robust Quantification

Optimal quantification is highly sensitive to outlier responses, necessitat-
ing robust quantification. The well-known method of reciprocal averages
(Richardson and Kuder 1933; Horst 1935) was replaced by the method of reciprocal
medians (Nishisato 1984b). See also a few other robust procedures (Nishisato 1987;
Sachs 1994).

## 2.3  Missing Responses

Effects of missing responses on quantification results cannot be ignored (e.g.,
Nishisato and Levine 1975; Meulman 1982; Van Buuren and Van Rijckevorsel 1992).
It is popular to optimize a certain statistic as a function of imputed responses, but this
strategy may result in a totally unexpected outcome: the more missing responses the
better the results. To avoid this dilemma, Nishisato and Ahn (1995) considered the
best imputation and the worst imputation, and recommended not to analyze the data
when the two imputations led to significantly different outcomes. This approach is
unique and worth considering for other imputation studies.

## 2.4   Standardizing Multidimensional Space

Information retrieval depends on data designs (Nishisato 1993, 2003a, 2003b). For multiple-choice data, the contribution of response option $p$ of item $j$, $SS(jp)$, and that of item $j$, $SS(j)$ to the total information are:

$$SS(jp) = n(N - f_{jp}), SS(j) = \sum_p SS(jp) = nN(m_j - 1).$$

Nishisato (1991) standardized the contribution of options (i.e., $SS(jp)$=constant for all combinations of $j$ and $p$), and that of items (i.e., $SS(j)$=constant for all $j$). The standardizations reduced the effects of outliers on the results, a noteworthy finding.

## 2.5   Order Constraints on Categories

The imposition of order constraints on categories in quantification was once popular (e.g., Bradley, Katti, and Coons 1962; Nishisato and Arri 1975; Tanaka and Kodake 1980), but can it be justified if we are to extract as much information as possible from data? See Sect. 3.

## 2.6   Down-Grading of Measurement or Desensitization

The importance of down-grading input measurement was noted (Nishisato 2005). If ordinal measurement is left as is, its quantification is restricted to finding only up to monotone relations; if it is down-graded to nominal measurement, however, the quantification is left free from the monotone restriction, allowing category intervals and the order of categories free to vary, thus accommodating nonlinear relations in the results. For quantitative measurement, one can still down-grade it by categorizing each variable with the minimum loss of original information which can then be nonlinearly quantified as nominal measurement. See Eouanzoui (2004), Nishisato's last student in Toronto.

## 2.7   Computational Simplification

Nishisato and Sheu (1980) proposed a piece-wise method of reciprocal averages. Forced classification (Nishisato 1984a, 1986; Nishisato and Gaul 1990; Nishisato and Baba 1999) is another example.

## 2.8 Multidimensional Correlation of Categorical Variables

Pair-wise maximization of correlation typically yields a non-positive-definite matrix. The correlation defined by the first component depends on what variables are in the data set. Nishisato (2006) proposed coefficient $\nu$ as a function of correlations over all possible components. The matrix of $\nu$ is always positive definite or semi-definite and independent of what other variables are included in the data set. Nishisato (2006) proved its equivalence to Cramér's coefficient of association V.

## 2.9 Total Information Analysis (Comprehensive Dual Scaling)

A perennial problem for quantification of a two-way table is that of the discrepancy between row space and column space. The familiar two-dimensional configuration of variables by symmetric scaling, for example, is in fact a 4-dimensional configuration. To overcome this problem, Nishisato and Clavel (2010) proposed total information analysis (TIA), or, comprehensive dual scaling (CDS), based on Nishisato and Clavel (2002, 2008) and Clavel and Nishisato (2008), in which both within-set and between-set distances, computed from all components, are analyzed. For the resultant expanded quantification space, $k$-means clustering is used to summarize relations of row and column variables, rather than by resorting to the problematic joint graphical display of rows and columns in the same space.

## 2.10 A Note on Dual Scaling

In 1976, F. W. Young, G. Saporta, J. de Leeuw and S. Nishisato held a symposium on optimal scaling with J. B. Kruskal as discussant. When J. Ziness questioned appropriateness of the name optimal scaling, Nishisato proposed the name dual scaling, which made its debut in the title of his book (Nishisato 1980). The book was translated into Russian by B. Milkin and S. Adamov, with the publication contract signed by the University of Toronto Press and Finansi Press of Moscow. It was ready to be published when H.H. Bock, W. Gaul, W. Day, H. Bozdogan and S. Nishisato were invited to Moscow in December of 1990. Sadly, the book was never published due to the historic fall of the Soviet Union in January of 1991.

## 3  New Phase: Cluster Analysis for TIA

The traditional quantification theory has a well-known problem that row and column variables cannot be graphed jointly in the same space. TIA is quantification theory based on the common space for both sets of variables, but it must deal with a much larger number of data dimensions, and its results cannot generally be easily mapped graphically. With a larger number of variables, cluster analysis is an ideal tool for summarizing TIA results. Cluster analysis (e.g., Tryon 1939; Sokal and Sneath 1963) flourished after the 1970s (e.g., Bock 1974; Gordon 1981). Rigorous as they may be, they are complex. What if we replace means with medians? Then most observations become irrelevant, and only a small subset of measurements will be used for clustering. This idea is closely related to the robust clustering methods of the PAM Algorithm (Kaufman and Rousseeuw 1987), the forward search method (Cerioli and Rousseeuw 2006) and the trimmed K-means algorithm (Garcia-Escudero et al. 2010). Two simpler procedures than these are proposed here.

## 3.1  Max-Mini Clustering

The data are from Stebbins (1950), a contingency table of six varieties of barley by six agricultural stations, as reported by Nishisato (2010). The table yields five components, and using all of them we calculate the 12-by-12 super-distance matrix of within-set and between-set distances (Nishisato and Clavel 2010) (Table 1). This table now requires ten-dimensional space to accommodate all the variables. Identify the minimum distance in each of the 12 rows (indicated by *) and columns (**), find the maximal value out of these minimal values as the threshold, beyond which all the distances are discarded. In this example, the threshold value is 1.26.

Once those distances larger than the threshold are discarded, the rows and columns of the table are rearranged in such a way that the survivor distances in bold-face are clustered along the diagonal positions of the matrix (Table 2). The inter-cluster distances are average distances of the corresponding sub-matrices. From Table 2, we obtain Cluster I (Hanchen-St. Paul), Cluster II (Manchuria-Ithaca), Cluster III (Coast, Meloy; Davis, Arlington), Cluster IV (White, Gatemi; Mocasin, Moro).

**Table 1** Within-set and between-set distance table

| | | | | | | | | | | | | |
|---|---|---|---|---|---|---|---|---|---|---|---|---|
| Coast(Co) | 0 | 2.13 | 1.95 | 2.56 | 1.71 | 1.28 | 0.62** | 2.32 | 2.04 | 1.47 | 2.13 | **0.57** |
| Hanchen(Ha) | 2.13 | 0 | 2.50 | 2.89 | 2.14 | 2.58 | 2.29 | 2.65 | **0.95** | 2.16 | 2.63 | 2.07 |
| White(Wh) | 1.95 | 2.50 | 0 | 2.80 | 1.72 | 2.24 | 2.10 | 2.64 | 2.48 | 0.90 | **0.65** | 1.79 |
| Manchuria(Ma) | 2.56 | 2.89 | 2.80 | 0 | 2.56 | 2.97 | 2.68 | **0.87** | 2.92 | 2.48 | 2.92 | 2.59 |
| Gatemi(Ga) | 1.71 | 2.14 | 1.72 | 2.56 | 0 | 2.24 | 1.84 | 2.37 | 2.13 | 1.23* | 1.93 | 1.69 |
| Meloy(Me) | 1.28 | 2.58 | 2.24 | 2.97 | 2.24 | 0 | 1.46 | 2.75 | 2.50 | 1.91 | 2.39 | 1.26* |
| Arlington(Ar) | 0.62* | 2.29 | 2.10 | 2.68 | 1.84 | 1.46 | 0 | 2.50 | 2.28 | 1.69 | 2.35 | 0.62 |
| Ithaca(It) | 2.32 | 2.65 | 2.64 | **0.87** | 2.37 | 2.75 | 2.50 | 0 | 2.74 | 2.33 | 2.80 | 2.40 |
| St. Paul(SP) | 2.04 | **0.95** | 2.48 | 2.92 | 2.13 | 2.50 | 2.28 | 2.74 | 0 | 2.17 | 2.66 | 2.05 |
| Mocasin(Mc) | 1.47 | 2.16 | 0.90* | 2.48 | 1.23** | 1.91 | 1.69 | 2.33 | 2.17 | 0 | 1.12 | 1.44 |
| Moro(Mr) | 2.13 | 2.63 | **0.65** | 2.92 | 1.93 | 2.39 | 2.35 | 2.80 | 2.66 | 1.12** | 0 | 2.03 |
| Davis(Da) | **0.57** | 2.07 | 1.79 | 2.59 | 1.69 | 1.26** | 0.62** | 2.40 | 2.05 | 1.44 | 2.03 | 0 |
| | Co | Ha | Wh | Ma | Ga | Me | Ar | It | SP | Mc | Mr | Da |

\* = the minimum row distance; ** = the minimum column distance; boldface = the minimum in both row and column

**Table 2** Re-arranged distance table with clusters (partitions) and inter-cluster distances (means within partitions)

| | | | | | | | | | | | | |
|---|---|---|---|---|---|---|---|---|---|---|---|---|
| Manchuria | 0 | **0.87** | 2.89 | 2.92 | 2.56 | 2.59 | 2.68 | 2.97 | 2.80 | 2.48 | 2.92 | 2.56 |
| Ithaca | *0.87* | 0 | 2.65 | 2.74 | 2.32 | 2.40 | 2.50 | 2.75 | 2.64 | 2.80 | 2.33 | 2.37 |
| Hanchen | | | 0 | **0.95** | 2.13 | 2.07 | 2.29 | 2.58 | 2.50 | 2.13 | 2.16 | 2.14 |
| St. Paul | *2.80* | | *0.95* | 0.00 | 2.04 | 2.05 | 2.28 | 2.50 | 2.48 | 2.66 | 2.17 | 2.13 |
| Coast | | | | 0 | **0.57** | **0.62** | (128) | 1.95 | 2.13 | 1.47 | 1.78 |
| Davis | | | | | 0 | **0.62** | (128) | 179 | 2.03 | 1.44 | 1.69 |
| Arlington | | | | | | 0 | (146) | 2.10 | 2.35 | 1.69 | 1.84 |
| Meloy | *2.60* | | *2.24* | | *0.96* | | 0 | 2.34 | 2.39 | 1.91 | 2.24 |
| White(Wh) | | | | | | | | 0 | **0.65** | **0.99** | (172) |
| Moro | | | | | | | | | 0 | **1.12** | (193) |
| Mocasin | | | | | | | | | | 0 | **1.23** |
| Gatemi | *2.61* | | *2.36* | | *1.94* | | | *1.26* | | | 0 |
| | Ma | It | Ha | SP | Co | Da | Ar | Me | Wh | Mr | Mc | Ga |

**Bold-face** = minimum values, ( ) = originally discarded but reinstated to complete the clusters, *Italics* = mean distances of partitions in the upper triangular part

**Table 3** Between-set distance table

| | Coast | Hanchen | White | Manchuria | Gatemi | Meloy |
|---|---|---|---|---|---|---|
| Arlington | 0.62* | 2.29 | 2.10 | 2.68 | 1.84 | 1.46 |
| Ithaca | 2.32 | 2.65 | 2.64 | 0.87* | 2.37 | 2.75 |
| St. Paul | 2.04 | 0.95* | 2.48 | 2.92 | 2.13 | 2.50 |
| Mocasin | 1.47 | 2.16 | 0.90* | 2.48 | 1.23* | 1.91 |
| Moro | 2.13 | 2.63 | 0.65* | 2.92 | 1.93 | 2.39 |
| Davis | 0.57* | 2.07 | 1.79 | 2.59 | 1.69 | 1.26* |

* = minimum distances in rows and columns

**Table 4** Re-arranged between-set distances

| | Manchuria | Hanchen | Coast | Meloy | White | Gatemi |
|---|---|---|---|---|---|---|
| Ithaca | **0.87** | 2.65 | 2.32 | 2.75 | 2.64 | 2.37 |
| St. Paul | 2.92 | **0.95** | 2.04 | 2.50 | 2.48 | 2.13 |
| Davis | 2.59 | 2.07 | **0.57** | **1.26** | 1.79 | 1.69 |
| Arlington | 2.68 | 2.29 | **0.62** | (1.46) | 2.10 | 1.84 |
| Moro | 2.92 | 2.63 | 2.13 | 2.39 | **0.65** | (1.93) |
| Mocasin | 2.48 | 2.16 | 1.47 | 1.91 | **0.90** | **1.23** |

**Bold-face** = minimum distances in rows and columns, ( ) = members of clusters not from the minimum distance set

## 3.2 Minimum Distance Clustering of Between-Set Distance Matrix

When the table is large, some clusters may consist of only variables from its rows (or columns). But our core interest in contingency table analysis is to find relations between rows and columns. Thus, clustering should be directed only to the between-set distance matrix (Table 3).

**Table 5** Age and suicide between-set distance table

|  | sm | gsH | gsO | hng | drwn | gun | knf | jmp | oth |
|---|---|---|---|---|---|---|---|---|---|
| Male10– | 1.21 | 1.29 | 1.39 | **0.90** | 1.35 | 1.17 | 0.99 | 1.21 | 1.20 |
| 16– | 0.60 | 0.68 | 0.81 | 0.51 | 1.00 | 0.66 | 0.51 | 0.69 | **0.50** |
| 21– | 0.60 | 0.63 | 0.74 | 0.64 | 1.04 | 0.68 | 0.60 | 0.72 | **0.48** |
| 26– | 0.57 | 0.62 | **0.73** | 0.62 | 1.01 | 0.68 | 0.57 | 0.70 | **0.47** |
| 31– | 0.51 | 0.60 | **0.73** | 0.56 | 0.97 | 0.66 | 0.51 | 0.65 | **0.42** |
| 36– | 0.47 | **0.59** | 0.74 | 0.46 | 0.92 | 0.65 | 0.42 | 0.60 | **0.41** |
| 41– | 0.50 | 0.62 | 0.76 | 0.42 | 0.92 | 0.64 | **0.39** | 0.62 | 0.45 |
| 46– | 0.53 | 0.65 | 0.78 | 0.39 | 0.93 | **0.63** | **0.38** | 0.63 | 0.48 |
| 51– | 0.53 | 0.67 | 0.83 | **0.33** | 0.89 | 0.67 | **0.34** | 0.61 | 0.51 |
| 56– | 0.54 | 0.70 | 0.88 | **0.32** | 0.88 | 0.70 | **0.34** | 0.60 | 0.55 |
| 61– | 0.57 | 0.74 | 0.92 | **0.32** | 0.87 | 0.73 | 0.36 | 0.62 | 0.59 |
| 66– | 0.63 | 0.80 | 0.99 | **0.37** | 0.89 | 0.79 | 0.41 | 0.65 | 0.66 |
| 71– | 0.65 | 0.83 | 1.03 | **0.39** | 0.88 | 0.84 | 0.44 | 0.66 | 0.70 |
| 76– | 0.70 | 0.88 | 1.07 | **0.44** | 0.90 | 0.88 | 0.49 | 0.70 | 0.75 |
| 81– | 0.70 | 0.91 | 1.12 | **0.47** | 0.87 | 0.94 | 0.51 | 0.69 | 0.79 |
| 86– | 0.78 | 0.96 | 1.15 | **0.52** | 0.94 | 0.96 | 0.58 | 0.77 | 0.85 |
| 90+ | 0.83 | 1.00 | 1.18 | **0.57** | 0.98 | 0.99 | 0.62 | 0.82 | 0.89 |
| Female10– | **0.57** | 0.81 | 1.04 | 0.73 | 0.94 | 0.98 | 0.67 | 0.61 | 0.66 |
| 16– | **0.62** | 0.90 | 1.12 | 0.91 | 1.00 | 1.12 | 0.83 | 0.71 | 0.75 |
| 21– | **0.62** | 0.91 | 1.15 | 0.92 | 0.96 | 1.16 | 0.83 | 0.70 | 0.77 |
| 26– | **0.52** | 0.80 | 1.05 | 0.81 | 0.90 | 1.04 | 0.72 | 0.60 | 0.65 |
| 31– | **0.48** | 0.80 | 1.06 | 0.77 | 0.85 | 1.04 | 0.67 | 0.56 | 0.64 |
| 36– | **0.42** | 0.75 | 1.01 | 0.71 | 0.82 | 0.99 | 0.62 | 0.52 | 0.58 |
| 41– | **0.35** | 0.72 | 1.01 | 0.59 | 0.73 | 0.95 | 0.50 | **0.42** | 0.56 |
| 46– | **0.36** | 0.74 | 1.02 | 0.59 | 0.72 | 0.96 | 0.50 | 0.43 | 0.58 |
| 51– | **0.37** | 0.75 | 1.04 | 0.55 | 0.68 | 0.96 | 0.47 | **0.42** | 0.59 |
| 56– | **0.41** | 0.78 | 1.07 | 0.58 | 0.67 | 0.99 | 0.50 | 0.45 | 0.63 |
| 61– | **0.46** | 0.81 | 1.11 | 0.59 | **0.66** | 1.03 | 0.52 | 0.47 | 0.67 |
| 66– | 0.52 | 0.85 | 1.15 | 0.60 | **0.66** | 1.05 | 0.54 | **0.51** | 0.73 |
| 71– | 0.58 | 0.90 | 1.19 | 0.66 | 0.68 | 1.11 | 0.61 | **0.56** | 0.79 |
| 76– | 0.54 | 0.87 | 1.18 | 0.67 | 0.68 | 1.10 | 0.60 | **0.52** | 0.76 |
| 81– | 0.52 | 0.86 | 1.17 | 0.68 | 0.70 | 1.10 | 0.61 | **0.51** | 0.75 |
| 86– | 0.57 | 0.87 | 1.17 | 0.71 | 0.80 | 1.10 | 0.65 | **0.56** | 0.77 |
| 90+ | 0.59 | 0.88 | 1.19 | 0.71 | 0.78 | 1.11 | 0.64 | **0.57** | 0.79 |

**Bold-face** = minimum distances in rows and columns

In addition to the use of only the between-set distance matrix, we propose to use the minimal row and column distances for clustering. The table of minimal distances is rearranged within rows and within columns (Table 4). Discarded distances are also shown in this table to show their distances to the clusters. Note that we obtain the same clusters as before.

Let us apply this simple clustering procedure to a larger example than the above one. Consider Heuer's shortened version of suicide data (Heuer 1979), in which nine different forms of suicides were tabulated for 18 age groups of each gender. The 36 × 9 contingency table yields eight components. Using all the components,

**Table 6** Suicides and gender-age groups

| Solid materials | Female 10–65 |
| --- | --- |
| Gas at home | Male 36–40 |
| Gas others | Male 26–35 |
| Hanging | Male 10–15, Male 51–90+ |
| Drowning | Female 61–70 |
| Gun | Male 46–50 |
| Knifing | Male 41–60 |
| Jumping | Female 41–90+ |
| Others | Male 16–41 |

the between-set distance matrix was calculated (Table 5). Minimal distances in rows and columns are indicated in bold-face. Eliminating non-minimum distances, we obtain the clusters (Table 6). Note that some clusters are overlapping.

The two procedures proposed here adopted threshold values. The number of clusters depends on the threshold value, and the minimum values of each row and column (the second procedure) lead to the maximum number of clusters. The second method is preferred to the first one for the investigation of multidimensional relations between rows and columns of a two-way table. Although the two procedures are logically simple and robust, we need further explorations of them with respect to the size of data sets and accommodating overlapping clusters.

# 4 Concluding Remarks

Major quantification problems have been worked out, the most significant one being the problem of space discrepancy. Total information analysis combined with cluster analysis is an attempt to deal with the space discrepancy problem. Quantification theory still leaves a number of practical problems (e.g., quantification of combined data such as mixture of ranking and multiple choices; quantification of dominance data by truly ordinal arithmetics). Beyond quantification, we should integrate the space discrepancy problem of principal component analysis into the framework of total information analysis as a tool for identifying row-column relations.

# References

Benzécri JP, et al. (1973) L'Analyse des Données: II. L'Analyse des correspondances. Dunod, Paris

Bock HH (1974) Automatische klassifikation. Vandenhoeck and Ruprecht, Göttingen

Bradley RA, Katti SK, Coons IJ (1962) Optimal scaling for ordered categories. Psychometrika 27:355–374

Cerioli A, Rousseeuw PJ (2006) Robust classification with categorical variables. In: Rizzi A, Vichi M (eds) Proceedings in computational statistics, pp 507–519

Clavel JG, Nishisato S (2008) Joint analysis of within-set and between-set distances. In: Shigemasu K, et al. (eds) New trends in psychometrics. Universal Academy Press, Tokyo, pp 41–50

Clavel JG, Nishisato S (2011) Reduced versus complete space configurations in total information analysis. In: Gaul W, Geyer-Schulz A, Schmidt-Thieme L, Kunze J (eds) Challenges at the interface of data analysis, computer science, and optimization, studies in classification, data analysis, and knowledge organization. Springer, Heidelberg, Berlin

De Leeuw J (1973) Canonical analysis of categorical data. PhD thesis, Leiden University

Eouanzoui KB (2004) On desensitizing data from interval to nominal measurement with minimum information loss. PhD thesis, University of Toronto

Fisher RA (1948) Statistical methods for research workers. Oliver and Boyd, London

Garcia-Escudero LA, Goddaliza A, Matran C, Mayo-Iscar A (2010) A review of robust clustering methods. ADAC 4:89–109

Gifi A (1990) Nonlinear multivariate analysis. Wiley, New York

Gleason HA (1926) The individual concept of the plant association. Bulletin Torrey Botanical Club 53:7–26

Gordon A (1981) Classification. Chapman and Hall, London

Greenacre MJ (1984) Theory and applications of correspondence analysis. Academic Press, London

Heuer J (1979) Selbstmord bei Kindern und jugendlichen. Ernst Klett Verlag, Stuttgart

Hirschfeld HO (1935) A connection between correlation and contingency. Camb Phil Soc Proc 31:520–524

Horst P (1935) Measuring complex attitudes. J Soc Psychol 6:369–374

Kaufman L, Rousseeuw PJ (1987) Finding groups in data. Wiley, New York

Lawrence DR (1985) Dual scaling of multidimensional data structures. An extended comparison of three methods. PhD thesis, University of Toronto

Le Roux B, Rouanet H (2004) Geometric data analysis: From correspondence analysis to structured data. Kluwer, Dordrecht

Lebart L, Morineu A, Tabard N (1977) Techniques de la Description Statistique: Méthodes et Logiciels pour l'Analyse des Grands Tableaux. Dunod, Paris

Lenoble F (1927) A propos des associations végétales. Bull Soc bot Fr 73:873–893

Meulman J (1982) Homogeneity analysis of incomplete data. DSWO Press, Leiden University

Nishisato S (1961) A factor analytic study of anxiety scale. Jpn J Psychol 31:228–236, (in Japanese)

Nishisato S (1966) Minimum entropy clustering of test items. PhD thesis, University of North Carolina at Chapel Hill

Nishisato S (1971) Analysis of variance through optimal scaling. In: Proceedings of the First Canadian Conference on Applied Statistics, Sir George Williams University Press, Montreal, pp 306–316

Nishisato S (1972) Analysis of variance of categorical data through selective scaling. Proceedings of the 20th International Congress of Psychology, Tokio, p 279

Nishisato S (1978) Optimal scaling of paired comparison and rank order data: An alternative to Guttman's formulation. Psychometrika 43:263–271

Nishisato S (1980) Analysis of categorical data: dual scaling and its applications. University of Toronto Press, Toronto

Nishisato S (1984a) Dual scaling by reciprocal medians. Estrato Dagli Atti della XXXII Riunione Scientifica pp 141–147

Nishisato S (1984b) Forced classification: A simple application of a quantification technique. Psychometrika 49:25–36

Nishisato S (1986) Generalized forced classification for quantifying categorical data. In: Diday, et al. (eds) Data analysis and informatics. North-Holland, Amsterdam, pp 351–362

Nishisato S (1987) Robust techniques for quantifying categorical data. In: MacNeil IB, Umphrey GJ (eds) Foundationss of statistical inference, D. Reidel, Dordrecht, pp 209–217

Nishisato S (1991) Standardizing multidimensional space for dual scaling. In: Proceedings of the 20th Annual Meeting of the German Operations Research Society, Hohenheim University, pp 584–591

Nishisato S (1993) On quantifying different types of categorical data. Psychometrika 58:617–629

Nishisato S (1994) Elements of dual scaling: an introduction to practical data analysis. Lawrence Erlbaum Associates, Hilsdale

Nishisato S (1996) Gleaning in the field of dual scaling. Psychometrika 61:559–599

Nishisato S (2003a) Geometric perspectives of dual scaling for assessment of information in data. In: Yanai H, et al. (eds) New developments in psychometrics. Springer, Tokyo, pp 453–462

Nishisato S (2003b) Total information in multivariate data from dual scaling perspectives. Alberta J Educa Res XLIX:244–251

Nishisato S (2005) New framework for multidimensional data analysis. In: Weihs C, Gaul W (eds) Classification - the ubiquitous challenge. Springer, Heidelberg, pp 280–287

Nishisato S (2006) Correlational structure of multiple-choice data as viewed from dual scaling. In: Greenacre M, Blasius J (eds) Multiple correspondence analysis and related methods. Chapman and Hall/CRC, Boca Raton, Chap 6, pp 161–177

Nishisato S (2007) Multidimensional nonlinear descriptive analysis. Chapman and Hall/CRC, Boca Raton

Nishisato S (2010) Data analysis for the behavioral sciences: use of methods appropriate for data types. Baifukan, Tokyo (in Japanese)

Nishisato S, Ahn H (1995) When not to analyze data: Decision making on missing responses in dual scaling. Ann Oper Res 55:361–378

Nishisato S, Arri PS (1975) Nonlinear programming approach to optimal scaling of partially ordered categories. Psychometrika 40:525–548

Nishisato S, Baba Y (1999) On contingency, projection and forced classification of dual scaling. Behaviormetrika 26:207–219

Nishisato S, Clavel JG (2002) A note on between-set distances in dual scaling. Behaviormetrika 30:87–98

Nishisato S, Clavel JG (2008) Interpreting data in reduced space: A case of what is not what in multidimensional data analysis. In: Shigemasu K, et al. (eds) New trends in psychometrics. Universal Academy Press, Tokyo, pp 357–366

Nishisato S, Clavel JG (2010) Total information analysis: Comprehensive dual scaling. Behaviormetrika 37:15–32

Nishisato S, Gaul W (1990) An approach to marketing data analysis: The forced classification procedure of dual scaling. J Market Res 27:354–360

Nishisato S, Lawrence DR (1989) Dual scaling of multiway data matrices: Several variants. In: Coppi R, Bolasco S (eds) Multiway data analysis. Elseveier Science Publishers, Amsterdam, pp 317–326

Nishisato S, Levine R (1975) Optimal scaling of omitted responses. Paper presented at the Annual Meeting of the Psychometric Society

Nishisato S, Sheu WJ (1980) Piecewise method of reciprocal averages for dual scaling of multiple-choice data. Psychometrika 45:467–478

Ramensky LG (1930) Zur Methodik der vergleichenden Bearbeitung und Ordnung von Pflanzen-listen und anderen Objekten, die durch mehrere, verschiedenartig wirkende Faktoren bestimmt werden. Beitr Biol Pft 18:269–304

Richardson M, Kuder GF (1933) Making a rating scale that measures. Person J 12:36–40

Sachs J (1994) Robust dual scaling weights with Tukey's biweight. Appl Psychol Meas 18:301–309

Sokal R, Sneath PHA (1963) Principles of Numerical Taxonomy. W. H. Freeman, London

Stebbins CL (1950) Variations and evolution. Columbus University Press, New York

Takane Y, Shibayama T (1991) Principal component analysis with external information on both subjects and variables. Psychometrika 56:97–120

Takane Y, Yanai H, Mayekawa S (1991) Relationships among several methods of linearly constrained correspondence analysis. Psychometrika 56:667–684

Tanaka Y, Kodake K (1980) Computational aspects of optimal scaling of ordered categories. Behaviormetrika 7:35–46

Tryon RC (1939) Cluster analysis. Edwards Brothers, Ann Arbor

Van Buuren S, Van Rijckevorsel JLA (1992) Imputation for missing categorical data by maximizing internal consistency. Psychometrika 57:567–580

Yanai H, Maeda T (2002) Partial multiple correspondence analysis. In: Nishisato S, Baba Y, Bozdogan H, Kanefuji K (eds) Measurement and multivariate analysis. Springer, Tokyo, pp 57–68

# Part III
# Analysis of m-Mode n-Way and Asymmetric Data

# Modelling Rater Differences in the Analysis of Three-Way Three-Mode Binary Data

Michel Meulders

**Abstract** Using a basic latent class model for the analysis of three-way three-mode data (i.e. raters by objects by attributes) to cluster raters is often problematic because the number of conditional probabilities increases rapidly when extra latent classes are added. To solve this problem, Meulders et al. (J Classification 19:277–302, 2002) proposed a constrained latent class model in which object-attribute associations are explained on the basis of latent features. In addition, qualitative rater differences are introduced by assuming that raters may only take into account a subset of the features. As this model involves a direct link between the number of features $F$ and the number of latent classes (i.e., $2^F$), estimation of the model becomes slow when many latent features are needed to fit the data. In order to solve this problem we propose a new model in which rater differences are modelled by assuming that features can be taken into account with a certain probability which depends on the rater class. An EM algorithm is used to locate the posterior mode of the model and a Gibbs sampling algorithm is developed to compute a sample of the observed posterior of the model. Finally, models with different types of rater differences are applied to marketing data and the performance of the models is compared using posterior predictive checks (see also, Meulders et al. (Psychometrika 68:61–77, 2003)).

## 1 Introduction

When analyzing three-way three-mode binary data (e.g. raters who indicate for each of a set of objects whether they have each of a set of attributes) it is straightforward to use the basic latent class model in order to cluster elements of one mode

M. Meulders (✉)
HUBrussel, Stormstraat 2, 1000 Brussel, Belgium

KUL, Tiensestraat 102, 3000 Leuven
e-mail: michel.meulders@hubrussel.be

W. Gaul et al. (eds.), *Challenges at the Interface of Data Analysis, Computer Science, and Optimization*, Studies in Classification, Data Analysis, and Knowledge Organization, DOI 10.1007/978-3-642-24466-7_13, © Springer-Verlag Berlin Heidelberg 2012

(e.g. raters) on the basis of elements of the other two modes (object-attribute associations). However, as the number of observed variables in such analysis (number of objects × number of attributes) is typically very large, the number of conditional probabilities increases rapidly as more latent classes are added. This may easily result in overparameterized models yielding unreliable parameter estimates.

In order to solve the problem Meulders et al. (2002) proposed a constrained latent class model (further denoted as $M_1$) which explains object-attribute associations on the basis of latent features that may be assigned to objects and attributes with a certain probability. An alternative solution has been proposed by Vermunt (2006). Let $D_{ijk}$ be equal to 1 if object $j$ $(j = 1, \ldots, J)$ has attribute $k$ $(k = 1, \ldots, K)$ according to rater $i$ $(i = 1, \ldots, I)$. Model $M_1$ assumes that observed associations are generated by a specific latent process:

1. When rater $i$ judges whether object $j$ has attribute $k$, she classifies both the object and the attribute with respect to $F$ latent features. More specifically, the binary latent variables $X_{ki}^{jf}$ $(f = 1, \ldots, F)$ equal 1 if feature $f$ is assigned to object $j$ when rater $i$ judges pair $(j, k)$, and 0 otherwise. Likewise, the binary latent variables $Y_{ji}^{kf}$ $(f = 1, \ldots, F)$ equal 1 if feature $f$ is assigned to attribute $k$ when rater $i$ judges pair $(j, k)$, and 0 otherwise. The classification of objects and attributes with respect to the latent features is conceived as a stochastic Bernoulli process:

$$p(x_{ki}^{jf} | \sigma_{jf}) = (\sigma_{jf})^{x_{ki}^{jf}} (1 - \sigma_{jf})^{1 - x_{ki}^{jf}}$$

$$p(y_{ji}^{kf} | \rho_{kf}) = (\rho_{kf})^{y_{ji}^{kf}} (1 - \rho_{kf})^{1 - y_{ji}^{kf}}$$

2. In order to introduce rater differences it is assumed that raters may take into account only a certain subset of the $F$ features. As there are $F$ features, there are $2^F (= Q)$ latent rater classes. Let the latent variable $G_{iq}$ be equal to 1 if rater $i$ belongs to latent class $q$ $(q = 1, \ldots, Q)$ and 0 otherwise. Furthermore let the latent variable $Z_{qf}$ be equal to 1 if feature $f$ is taken into account by raters of class $q$, and 0 otherwise. The probability that a rater belongs to class $q$ equals $\xi_q$ with $\sum_q \xi_q = 1$.

3. It is assumed that the observed judgment $D_{ijk}$ of rater $i$ on object-attribute pair $(j, k)$ is a deterministic function of the latent classifications of the object and attribute with respect to the $F$ features made by rater $i$ and of the latent class the rater belongs to. More specifically it is assumed that the object is associated to the attribute if they have at least one common feature which is also taken into account by the rater, that is, given that rater $i$ belongs to latent class $q$ it holds that:

$$D_{ijk} = 1 \iff \exists f : X_{ki}^{jf} = Y_{ji}^{kf} = Z_{qf} = 1$$

It can be derived from the above assumptions that the probability of an observed association given the latent class membership of the rater equals:

$$P(D_{ijk} = 1 | G_{iq} = 1, \boldsymbol{\sigma}_j, \boldsymbol{\rho}_k) = \pi_{jkq} = 1 - \prod_f (1 - \sigma_{jf} \rho_{kf} z_{qf})$$

Assuming that responses $D_{ijk}$ are independent given the latent class membership of the rater and assuming independent raters the likelihood of the entire data set $\mathbf{d}$ can be written as

$$p(\mathbf{d} | \boldsymbol{\sigma}, \boldsymbol{\rho}, \boldsymbol{\xi}) = \prod_i \sum_q p(\mathbf{d}_i | G_{iq} = 1, \boldsymbol{\sigma}, \boldsymbol{\rho}) p(G_{iq} = 1 | \boldsymbol{\xi})$$

$$= \prod_i \sum_q \left[ \prod_j \prod_k (\pi_{jkq})^{d_{ijk}} (1 - \pi_{jkq})^{1 - d_{ijk}} \right] \xi_q$$

The fact that model $M_1$ includes a direct link between the number of features and the number of latent classes implies that the model includes a large number of latent classes when many features are needed to fit the data. This may be suboptimal for two reasons. First, when using the model for segmentation in an applied setting (e.g. a marketing context) one aims to partition the population into a limited number of interpretable rater classes (or segments) which are sufficiently large so that they can be the subject of a campaign. Obviously, when using model $M_1$ for segmentation one might end up with many small latent classes which cannot be exploited as such. Second, model estimation can become inefficient when in a model with many features, a part of the latent classes included is virtually empty, but these classes are still part of the estimation process.

In order to accommodate for the shortcomings of $M_1$ we will define a new model (further denoted $M_2$) which relaxes the direct link between the number of latent classes and the number of features. The outline of the paper is as follows: First we describe the new model $M_2$. Second, we describe a Gibbs sampling algorithm to compute a sample of the observed posterior distribution of the model. Third, we discuss how to use posterior predictive checks to evaluate the fit of models $M_1$ and $M_2$. Fourth, we apply $M_1$ and $M_2$ to a marketing application on the characteristics of sandwich fillings.

## 2   Extending the Model

In contrast to $M_1$, $M_2$ assumes that features can be taken into account with a certain probability which depends on the latent class the rater belongs to. More specifically, $M_2$ is based on the following assumptions:

1. In the same way as in $M_1$ the judgment of rater $i$ about the association between object $j$ and attribute $k$ is assumed to be based on latent Bernoulli variables $X_{ki}^{jf} \sim \text{Bern}(\sigma_{jk})$, $Y_{ji}^{kf} \sim \text{Bern}(\rho_{kf})$ which indicate the latent feature classification of the object and the attribute, respectively.

2. To introduce rater differences it is assumed that raters belong to one of $Q$ latent classes indicated as $q$ $(q = 1, \ldots, Q)$. The latent variable $G_{iq}$ equals 1 if rater $i$ belongs to latent class $q$, and 0 otherwise. The probability that a rater belongs to class $q$ equals $\xi_q$ with $\sum_q \xi_q = 1$. Given that the rater belongs to class $q$, it is assumed that the judgment about object-attribute pair $(j, k)$ depends on latent Bernoulli variables $Z_{ijk}^{qf} \sim \text{Bern}(\gamma_{qf})$ $(f = 1, \ldots, F)$ which equal 1 if rater $i$ takes into account feature $f$ and 0 otherwise. In other words, $M_2$ introduces rater differences by assuming that raters of distinct latent classes may have different probabilities to take a feature into account.
3. Similar to $M_2$, $M_1$ assumes that the object is associated to the attribute if they have at least one common feature which is also taken into account by the rater, namely:

$$D_{ijk} = 1 \iff \exists f : X_{ki}^{jf} = Y_{ji}^{kf} = Z_{ijk}^{qf} = 1$$

From the above assumptions it can be derived that the conditional probability of an observed association reads as follows:

$$P(D_{ijk} = 1 | G_{iq} = 1, \boldsymbol{\sigma}_j, \boldsymbol{\rho}_k, \boldsymbol{\gamma}_q) = \pi_{jkq} = 1 - \prod_f (1 - \sigma_{jf} \rho_{kf} \gamma_{qf})$$

Assuming independent raters and independent responses given latent class membership, the likelihood of the data $\mathbf{d}$ can be expressed as follows:

$$p(\mathbf{d} | \boldsymbol{\sigma}, \boldsymbol{\rho}, \boldsymbol{\gamma}, \boldsymbol{\xi}) = \prod_i \sum_q p(\mathbf{d}_i | G_{iq} = 1, \boldsymbol{\sigma}, \boldsymbol{\rho}, \boldsymbol{\gamma}) p(G_{iq} = 1 | \boldsymbol{\xi})$$

$$= \prod_i \sum_q \left[ \prod_j \prod_k (\pi_{jkq})^{d_{ijk}} (1 - \pi_{jkq})^{1-d_{ijk}} \right] \xi_q$$

## 3 Parameter Estimation

Meulders et al. (2002) derived an EM-algorithm to locate the posterior mode of $M_1$. In addition, they presented a Gibbs sampling algorithm to compute a sample of the posterior distribution of $M_1$. For parameter estimation of $M_2$ both an EM-algorithm and a Gibbs sampling algorithm were implemented along the same lines as for $M_1$. In this section we will only describe the Gibbs sampling algorithm more in detail. For $M_2$ inference is based on the posterior distribution $p(\boldsymbol{\sigma}, \boldsymbol{\rho}, \boldsymbol{\gamma}, \boldsymbol{\xi} | \mathbf{d}) \propto p(\mathbf{d} | \boldsymbol{\sigma}, \boldsymbol{\rho}, \boldsymbol{\gamma}, \boldsymbol{\xi}) p(\boldsymbol{\sigma}, \boldsymbol{\rho}, \boldsymbol{\gamma}, \boldsymbol{\xi})$. It is convenient to use a conjugate prior distribution:

$$p(\boldsymbol{\sigma}, \boldsymbol{\rho}, \boldsymbol{\gamma}, \boldsymbol{\xi}) = \prod_j \prod_f \text{Beta}(\sigma_{jf} | \alpha_\sigma, \beta_\sigma) \prod_k \prod_f \text{Beta}(\rho_{kf} | \alpha_\rho, \beta_\rho)$$

$$\times \prod_q \prod_f \text{Beta}(\gamma_{qf} | \alpha_\gamma, \beta_\gamma) \text{Dirichlet}(\boldsymbol{\xi} | \delta_1, \ldots, \delta_Q).$$

In order to compute a sample of the posterior distribution a Gibbs sampling algorithm can be used that iterates between the following steps:

1. For each entity $i$ draw the vector $\mathbf{g}_i$ from

$$p(\mathbf{g}_i|\mathbf{d}_i,\boldsymbol{\sigma},\boldsymbol{\rho},\boldsymbol{\gamma},\boldsymbol{\xi}) \propto \prod_j \prod_k p(\mathbf{d}_{ijk}|\mathbf{g}_i,\boldsymbol{\sigma},\boldsymbol{\rho},\boldsymbol{\gamma})p(\mathbf{g}_i|\boldsymbol{\xi})$$

2. For each triple $(i,j,k)$ draw $(\mathbf{x}_{ki}^j,\mathbf{y}_{ji}^k,\mathbf{z}_{ijk})$ from
$p(\mathbf{x}_{ki}^j,\mathbf{y}_{ji}^k,\mathbf{z}_{ijk}|d_{ijk},\mathbf{g}_i,\boldsymbol{\sigma}_j,\boldsymbol{\rho}_k,\boldsymbol{\gamma})$

$$\propto p(d_{ijk}|\mathbf{x}_{ki}^j,\mathbf{y}_{ji}^k,\mathbf{z}_{ijk})p(\mathbf{x}_{ki}^j,\mathbf{y}_{ji}^k,\mathbf{z}_{ijk}|\mathbf{g}_i,\boldsymbol{\sigma}_j,\boldsymbol{\rho}_k,\boldsymbol{\gamma})$$

3. For each pair $(j,f)$ draw $\sigma_{jf}$ from

$$\text{Beta}(\alpha_\sigma + \sum_k \sum_i x_{ki}^{jf}, \beta_\sigma + \sum_k \sum_i (1-x_{ik}^{jf}))$$

4. For each pair $(k,f)$ draw $\rho_{kf}$ from

$$\text{Beta}(\alpha_\rho + \sum_j \sum_i y_{ji}^{kf}, \beta_\rho + \sum_j \sum_i (1-y_{ji}^{kf}))$$

5. For each pair $(q,f)$ draw $\gamma_{qf}$ from

$$\text{Beta}(\alpha_\gamma + \sum_j \sum_k \sum_{i\in q} z_{ijk}^{qf}, \beta_\gamma + \sum_j \sum_k \sum_{i\in q} (1-z_{ijk}^{qf}))$$

6. Draw $\boldsymbol{\xi}$ from Dirichlet$(\delta_1 + \sum_i g_{i1},\dots,\delta_Q + \sum_i g_{iQ})$

For the Gibbs sampler it is known that, under some mild regularity conditions, the simulated sequences converge to the true posterior distribution (Gelfand and Smith 1990). Convergence of the chains will be assessed with the approach suggested by Gelman and Rubin (1992).

## 4   Evaluation of Model Fit

In order to compare the fit of $M_1$ and $M_2$ we will consider two types of measures. First, we will evaluate to what extent the model can reproduce the number of associations between objects and attributes, that is, $F_{+jk} = \sum_i D_{ijk}$. As a measure of descriptive fit we will compute the correlation between observed frequencies and expected frequencies in the marginal objects by attributes table. Expected frequencies are computed as follows:

$$E_{+jk} = \sum_i \sum_q P(D_{ijk} = 1|G_{iq} = 1, \hat{\boldsymbol{\theta}}) P(G_{iq} = 1|\hat{\boldsymbol{\xi}})$$

with $(\hat{\boldsymbol{\theta}}, \hat{\boldsymbol{\xi}})$ an estimate of the posterior mode of the model. Note that for $M_1$, $\boldsymbol{\theta} = (\sigma, \rho)$ and for $M_2$, $\boldsymbol{\theta} = (\sigma, \rho, \gamma)$.

Second, in order to evaluate to what extent the model can capture observed rater differences we will compare observed and expected dependencies between pairs of object-attribute associations by means of a Pearson Chi-square measure defined on the cross-table of all object-attribute pairs $(j, k)$ and $(j', k')$:

$$\chi^2(\mathbf{d}, \boldsymbol{\theta}, \boldsymbol{\xi}) = \sum_l \frac{(O_l - E_l)^2}{E_l}$$

with the index $l$ referring to the four response patterns that can be observed for a particular pair of variables, namely, $(1, 1)$, $(1, 0)$, $(0, 1)$ and $(0, 0)$. The expected frequency of a response pattern is computed as the product of the number of raters $(I)$ and the probability of the response pattern:

$$I \times P(D_{ijk} = u, D_{ij'k'} = v|\boldsymbol{\theta}, \boldsymbol{\xi}) =$$

$$I \sum_q (\pi_{jkq})^u (1 - \pi_{jkq})^{1-u} (\pi_{j'k'q})^v (1 - \pi_{j'k'q})^{1-v} \xi_q$$

To evaluate the significance of the chi-square measures we will use the sample of the posterior distribution to compute posterior predictive check (PPC) $p$-values (see Gelman et al. 1996). In particular, PPC $p$-values are computed as the proportion of replicated datasets in which realized discrepancies $\chi^2(\mathbf{d}^{rep}, \boldsymbol{\theta}, \boldsymbol{\xi})$ exceed or equal observed discrepancies $\chi^2(\mathbf{d}, \boldsymbol{\theta}, \boldsymbol{\xi})$.

## 5  Illustrative Application

In order to compare $M_1$ and $M_2$, both models are applied to binary judgments of 191 first-year psychology students who indicated for 25 types of sandwich fillings and 14 characteristics whether or not a certain sandwich filling has a certain characteristic. In a similar study, Candel and Maris (1997) used a probabilistic feature model without rater differences to fit data on sandwich fillings.

An EM-algorithm is used to locate the posterior mode of model $M_1$ with 1 up to 6 features and to locate the posterior mode of model $M_2$ with 1 up to 6 features and with 1 up to 5 latent classes. For each type of model the EM algorithm is run 10 times using random starting points and the solution with the highest loglikelihood is considered as the mode. As can be seen in Fig. 1, model $M_2$ with 6 features and 5 latent classes (further denoted as $M_2^*$) has a lower BIC value than model $M_1$

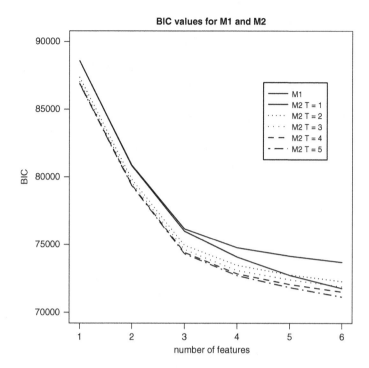

**Fig. 1** BIC values for $M_1$ and $M_2$ as a function of the number features

with 6 features and 63 latent classes (further denoted as $M_1^*$) (i.e. 71,159 versus 71,787, respectively) and hence $M_2^*$ has the best balance between goodness-of-fit and complexity. As a measure of descriptive fit of both models we compute the correlation between the observed and expected frequencies in the marginal objects-by-attributes table. For $M_1^*$ and $M_2^*$ the correlation equals 0.946 and 0.952, respectively, and hence both models explain a high percentage of the variance in the observed frequencies $F_{+jk}$, namely, 89% and 91%, respectively.

In the next step of the analysis, the Gibbs sampling algorithm is used to compute a sample of the posterior distribution of $M_1^*$ and of $M_2^*$. Per model 5 chains are simulated starting from the posterior mode computed by the EM algorithm. This allows us to compute a sample of the posterior mode identified by the EM algorithm. However, during the estimation of $M_2$ we noticed that the Gibbs sampling algorithm often switches between distinct posterior modes in subsequent iterations. This complicates the estimation of the posterior sample in the neighbourhood of the posterior mode. To solve this problem the feature weights $\gamma_{qf}$ are fixed at the posterior mode estimates when applying the Gibbs sampling algorithm. After convergence of the simulated chains is attained with the diagnostic of Gelman and Rubin (1992), a sample of 2,000 draws is constructed using evenly spaced draws of the simulated chains.

In order to evaluate to what extent the models can capture rater differences we compare observed and expected dependencies between pairs of object-attribute associations by means of a Pearson Chi-square measure defined on the cross-table of object-attribute pairs $(j, k)$ and $(j', k')$. For each pair of object-attribute associations a PPC $p$-value is computed based on 2,000 draws of the posterior distribution. The results indicate that for $M_1^*$ 31%, 48% and 63% of the $p$-values are smaller than 0.001, 0.01 and 0.05, respectively. For $M_2^*$ it turns out that 20%, 34% and 50% of the $p$-values are smaller than 0.001, 0.01, and 0.05, respectively. Hence, although both models have a rather poor performance in capturing the dependencies between observed pairs of variables, the new model $M_2^*$ performs considerably better.

## 6 Discussion

As using the basic latent class model for the analysis of three-way three-mode data is not straightforward, Meulders et al. (2002) proposed a constrained latent class model ($M_1$) that explains object-attribute associations on the basis of latent features. Rater differences are introduced by classifying raters on the basis of the subset of the features they take into account. However, as this approach involves a direct link between the number of features and the number of latent classes, the model may not be optimal when many features are needed to fit the data. A new model ($M_2$) was introduced which relaxes the direct relation between the number of features and the number of rater classes by assuming that depending on the latent class, raters take into account a feature with a certain probability. A comparison of both models on associations between 25 sandwich fillings and 14 filling characteristics indicates that models assuming 6 features explain a large proportion of the marginal objects-by-attributes table. However, the more parsimonious model $M_2$ with 5 rater classes yields a better balance between complexity and goodness-of-fit than $M_1$ in terms of BIC. Furthermore, posterior predictive checks on cross tables of pairs of object-attribute variables indicate that $M_2$ can better capture observed dependencies between observed variables. In sum, the more flexible model $M_2$ can clearly be considered an improvement of $M_1$ for deriving rater classes from three-way three-mode binary data. Finally, it would be interesting to compare the results obtained with $M_2$ with mixture extensions of classical dimensional approaches (e.g. probabilistic principal components analysis).

## References

Candel MJJM, Maris E (1997) Perceptual analysis of two-way two-mode frequency data: Probability matrix decomposition and two alternatives. Int J Res Market 14:321–339
Gelfand AE, Smith AFM (1990) Sampling based approaches to calculating marginal densities. J Am Stat Assoc 85:398–409

Gelman A, Rubin DB (1992) Inference from iterative simulation using multiple sequences. Stat Sci 7:457–472

Gelman A, Meng XM, Stern H (1996) Posterior predictive assessment of model fitness via realized discrepancies. Statistica Sinica 4:733–807

Meulders M, De Boeck P, Kuppens P, Van Mechelen I (2002) Constrained latent class analysis of three-way three-mode data. J Classification 19:277–302

Meulders M, De Boeck P, Van Mechelen I (2003) A taxonomy of latent structure assumptions for probability matrix decomposition models. Psychometrika 68:61–77

Vermunt JK (2006) A hierarchical mixture model for clustering three-way data sets. Comput Stat Data Anal 51:5368–5376

# Reconstructing One-Mode Three-way Asymmetric Data for Multidimensional Scaling

Atsuho Nakayama and Akinori Okada

**Abstract** Some models have been proposed to analyze one-mode three-way data [e.g. De Rooij and Gower (J Classification 20:181–220, 2003), De Rooij and Heiser (Br J Math Stat Psychol 53:99–119, 2000)]. These models usually assume triadic symmetric relationships. Therefore, it is general to transform asymmetric data into symmetric proximity data when one-mode three-way asymmetric proximity data are analyzed using multidimensional scaling. However, valuable information among objects is lost by symmetrizing asymmetric proximity data. It is necessary to devise this transformation so that valuable information among objects is not lost. In one-mode two-way asymmetric data, a method that the overall sum of the rows and columns are equal was proposed by Harshman et al. (Market Sci 1:205–242, 1982). Their method is effective to analyze the data that have differences among the overall sum of the rows and columns caused by external factors. Therefore, the present study proposes a method that extends (Harshman et al., Market Sci 1:205–242, 1982) method to one-mode three-way asymmetric proximity data. The proposed method reconstructs one-mode three-way asymmetric data so that the overall sum of the rows, columns and depths is made equal.

A. Nakayama (✉)
Graduate School of Social Sciences, Tokyo Metropolitan University, 1-1 Minami-Ohsawa, Hachioji-shi, Tokyo 192-0397, Japan
e-mail: atsuho@tmu.ac.jp

A. Okada
Graduate School of Management and Information Sciences, Tama University, 4-4-1 Hijirigaoka, Tama-shi Tokyo 206-0022, Japan
e-mail: okada@rikkyo.ac.jp

W. Gaul et al. (eds.), *Challenges at the Interface of Data Analysis, Computer Science, and Optimization*, Studies in Classification, Data Analysis, and Knowledge Organization, DOI 10.1007/978-3-642-24466-7_14, © Springer-Verlag Berlin Heidelberg 2012

# 1 Introduction

In multidimensional scaling (MDS), the prototypical data matrix is square and symmetric one-mode two-way data, which represents a set of empirically obtained similarities or other kinds of proximities between pairs of objects. The relationships among such objects are shown as spatial representation, where pairs of objects perceived to be highly similar locate closely to each other in a multidimensional space of relatively low dimensionality such as two or three dimensions.

However, there is a need for a model that is capable of analyzing asymmetric proximities; the cell entry $\delta_{ij} \neq \delta_{ji}$ for $i, j = 1, \ldots, n; i \neq j$. Asymmetric relationships among objects are common phenomena in marketing research, consumer studies, and so on. For example, asymmetric relationships are the probability of a switching to brand $j$, given that brand $i$ was bought on the last purchase or consumers' preference to brand $j$, given that consumers have already chosen item $i$. There are some approaches to analyzing asymmetric proximities. Some procedures were used (including additive or multiplicative adjustments of rows and columns) to symmetrize asymmetric proximities. The easiest way was to average the conjugate entries, $\delta_{ij}$ and $\delta_{ji}$. In another approach, asymmetric proximities were reconstructed before analysis. For examples, there exist the reconstructing method to remove the extraneous size differences (Harshman et al. 1982) or that on the basis of entropy (Yokoyama and Okada 2006, 2007). Otherwise, rows and columns were treated as separate points by an unfolding-type of MDS analysis or asymmetric proximities were directly analyzed by some asymmetric models such as Okada and Imaizumi (1987), Saburi and Chino (2008), and Zielman and Heiser (1993). In one-mode two-way asymmetric data, Harshman et al. (1982) proposed a method making the overall sum of the rows and columns equal over all objects. Their method reconstructs the data $\delta_{ij}$ so that the sum of each row and column excluding the diagonal element become fixed as follows;

$$\overline{\delta_{i.}} = \frac{1}{2n} \sum (\delta_{ij} + \delta_{ji}) = k. \tag{1}$$

The value of $k$ is usually set at the grand mean of the unadjusted data. An iterative procedure successively adjusts the row and columns of the data until the mean entry for each segment is equal. Their method is effective to analyze the data that have differences among the overall sum of the rows and columns depending on external factors. So, it is possible to show the structural factors of interest clearly. Therefore, the present study proposes a method that extends Harshman et al. (1982)'s method to one-mode three-way asymmetric proximity data. The proposed method reconstructs one-mode three-way asymmetric data so that the overall sum of the rows, columns and depths is equal over all objects.

## 2   The Method

Research interests have been increasing in models with triadic relationships among objects. MDS has often been used to analyze two-way proximity data and to obtain spatial representations for these data defined as the distance between two objects. However, the representations obtained from these analyses have not been sufficiently precise to explain the high-level phenomena underlying the data. Therefore, a need exists for a model that is capable of representing relationships among three or more objects. Some models have been proposed to analyze one-mode three-way proximity data among three objects (e.g. De Rooij and Gower (2003); De Rooij and Heiser (2000)). These models usually assume triadic symmetric relationships except for De Rooij and Heiser (2000). When one-mode three-way asymmetric proximity data $\delta_{ijk}$ are analyzed using MDS, it is usual to transform one-mode three-way asymmetric proximity data $\delta_{ijk}$ among three objects $i$, $j$, and $k$ into symmetric data $\delta'_{ijk}$ as follows;

$$\delta'_{ijk} = (\delta_{ijk} + \delta_{ikj} + \delta_{jik} + \delta_{jki} + \delta_{kij} + \delta_{kji})/6. \tag{2}$$

Symmetrized asymmetric proximities are not able to reflect asymmetric information. Valuable information among objects is lost by symmetrizing asymmetric proximity data. It is necessary to devise this transformation so that valuable information among objects is not lost. The present study proposes a method that extends Harshman et al. (1982)'s method to one-mode three-way asymmetric proximity data. The method reconstructs asymmetric one-mode two-way proximity data so that the sum of each row and column except the diagonal element may become equalized. The method that we propose reconstructs asymmetric one-mode three-way proximity data so that the sum of each row, column, and depth are equalized. The proposed method adds the condition of depth to Harshman et al. (1982)'s method. The method reconstructs asymmetry one-mode three-way proximity data as follows;

$$\overline{\delta_{i..}} = \frac{1}{6n} \sum \sum (\delta_{ijk} + \delta_{ikj} + \delta_{jik} + \delta_{jki} + \delta_{kij} + \delta_{kji}) \tag{3}$$

for each $i$. Diagonal elements are included, so it is thought that they have valuable information among objects. The method that we propose reconstructs one-mode three-way asymmetric proximity data so that the sum of each row, column, and depth is equlized by multiplying row $i$, column $i$, depth $i$ by constant $c_i$. Our proposed method iteratively finds constant $c_i$ by the quasi-Newton method under the following conditions;

$$\sum_i (c_i c_1 c_1 \delta_{i11} + \cdots + c_i c_1 c_n \delta_{i1n} + c_1 c_i c_1 \delta_{1i1} + \ldots$$

$$+ c_n c_i c_1 \delta_{ni1} + c_1 c_1 c_i \delta_{11i} + \cdots + c_1 c_n c_i \delta_{1ni})$$

$$= \sum_i (c_i c_2 c_1 \delta_{i21} + \cdots + c_i c_2 c_n \delta_{i2n} + c_1 c_i c_2 \delta_{1i2} + \cdots$$

$$+ c_n c_i c_2 \delta_{ni2} + c_2 c_1 c_i \delta_{21i} + \cdots + c_2 c_n c_i \delta_{2ni})$$

$$= \sum_i (c_i c_n c_1 \delta_{in1} + \cdots + c_i c_n c_n \delta_{inn} + c_1 c_i c_n \delta_{1in} + \cdots$$

$$+ c_n c_i c_n \delta_{nin} + c_n c_1 c_i \delta_{n1i} + \cdots + c_n c_n c_i \delta_{nni})$$

$$= \frac{1}{n} \sum_i \sum_j (\delta_{ij1} + \cdots + \delta_{ijn} + \delta_{1ij} + \cdots + \delta_{nij} + \delta_{j1i} + \cdots + \delta_{jni}). \tag{4}$$

The difference between the sum of each row, column, and depth which multiplied row $i$, column $i$, and depth $i$ by constant $c_i$ and the grand mean of the unadjusted one-mode three-way asymmetric proximity data may be minimized iteratively to find constant $c_i$ which is able to equalize the sum of each row, column, and depth.

# 3   An Application

We applied the proposed method to consecutive Swedish election data in 1964, 1968, and 1970. As an illustration, we look at the data set obtained from Upton (1978). There are four political parties, Social Democrats (SD), the Center party (C), the People's party (P), and the Conservatives (Con). This ordering is from left-to right-wing parties. The data gives the frequency of 64 possible sequences between these four parties at the three time points. One-mode three-way asymmetric proximity data was calculated from the Swedish election data. The one-mode three-way asymmetric proximity data were reconstructed by our proposed method. For comparison with the results of reconstructed data, we also use the one-mode three-way asymmetric proximity data without reconstructing data. The reconstructed and non-reconstructed asymmetric proximity data were symmetrizied by using (3). These symmetrizing proximity data were analyzed by a generalized Euclidean distance model. In a generalized Euclidean distance model, the triadic distances $d_{ijk}$ are defined as follows:

$$d_{ijk} = (d_{ij}^2 + d_{jk}^2 + d_{ik}^2)^{1/2}, \tag{5}$$

where $d_{ij}$ is Euclidean distance between points $i$ and $j$ representing objects $i$ and $j$. The disparity of the triadic distance $\hat{d}_{ijk}$ satisfies the following conditions;

$$m_{ijk} < m_{rst} \Rightarrow \hat{d}_{ijk} < \hat{d}_{rst} \text{ for all } i < j < k, \ r < s < t, \tag{6}$$

where $m_{ijk}$ shows the reconstructed one-mode three-way symmetrized proximity data. We use non-metric MDS model, so the results are not changed by the magnitude of coefficient $c$.

These analyses were done by using the maximum dimensionalities of nine through five and the minimum dimensionality of one. The smallest stress value in each dimensional space was chosen as the minimized stress value in that dimensional space. The two-dimensional configuration is now discussed in the present analysis. Two-dimensional configuration helps easy understanding of the relationships among the objects. The stress value in two-dimensional space obtained from reconstructed data was 0.226, and the stress value in two-dimensional space obtained from non-reconstructed data was 0.300.

Figure 1 is the two-dimensional configuration obtained from the analysis of reconstructed transition frequency data by the proposed method. Figure 2 is the two-dimensional configuration obtained from the analysis of symmetrized transition frequency data without reconstructing data. In Fig. 1, the positions of parties are based on their characteristics. Left- to right-wing parties locate clockwise. The relationships among parties are able to be expressed pretty well. However, in Fig. 2,

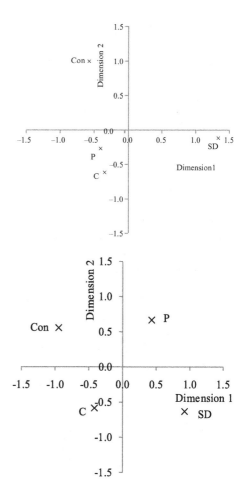

**Fig. 1** The two-dimensional configuration obtained from the analysis of reconstructed transition frequency data by the proposed method

**Fig. 2** The two-dimensional configuration obtained from the analysis of symmetrized transition frequency data without reconstructing data

the positions of each party are not based on the characteristics of them. The relationships among parties are not able to be expressed well. Then, the two-dimensional results of the present method are compared with those of De Rooij and Heiser (2000)'s one-mode three-way asymmetric model. De Rooij and Heiser (2000) analyzed the same transition frequency data of Swedish respondents. The results of De Rooij and Heiser (2000) have the same tendency of the proposed method. The positions of the objects in the result of De Rooij and Heiser (2000) corresponds with those in the proposed method. Left- to right-wing parties locate clockwise. From these results, the proposed method seems to give good results in the present analysis. The analysis of reconstructed one-mode three-way data were able to reveal new relationships among web pages which were not clear in the analysis of the averaged one-mode three-way data. The proposed method may be used successfully to facilitate a clear understanding of the relationships among objects.

# 4  Conclusion and Outlook

The above analysis was carried out to apply the consecutive Swedish election data to our proposed reconstructed method and we compare the results of the reconstructed data with those of non-reconstructed data. As noted above, our reconstructed method seems to have clearly revealed the triadic relationships among objects. However, the effectiveness of our method must be discussed in greater details. Therefore, our proposed reconstructed method were applied to the transition data of web pages. We compare the results of the reconstructed one-mode three-way proximity data based on our method with those of the reconstructed one-mode two-way proximity data based on Harshman et al. (1982)'s method. These analyses were done by using the maximum dimensionalities of nine through five and the minimum dimensionality of one. The smallest stress value in each dimensional space was chosen as the minimized stress value in that dimensional space. The two-dimensional configuration is now discussed in the present analysis. Two-dimensional configuration helps easy understanding of the relationships among web pages. The stress value in two-dimensional space obtained from reconstructed one-mode three-way data was 0.572, and the stress value in two-dimensional space obtained from reconstructed one-mode two-way data was 0.373.

Figure 3 is the two-dimensional configuration of reconstructed one-mode three-way symmetrized transition frequency data among web pages. In Fig. 3, there exist some groups based on relationships with transitions among web pages. One group consists of Pages 2, 3, and 13. Pages 2 and 3 are the pages explaining access analysis and Page 13 illustrates the price of the application. These pages are located near the center and the distance among pages is very short. Therefore, it is thought that these pages are often browsed simultaneously. The second group consists of Pages 12, 14, 15, and 16. These pages show a group relevant to services provided by the company administering the web site such as the pages of flow of the application, introduction example, web site strategic report service, access analysis consulting service. The third group consists of Pages 4, 5, 6, 7, and 8. These pages illustrate a group relevant to the general function provided by the company such as the pages of site analysis, page analysis, and path analysis. The fourth group consists of Pages 9, 10, and 11. These pages explain advertising effectiveness measurement and evaluation of effectiveness of search engine optimization. It is thought that these groups show the tendency of a general browse of the present web site. However these proposals are not generalized from only the results of the present analysis. It is necessary to check the validity of the present analysis in the future. Figure 4 is the two-dimensional configuration of reconstructed one-mode two-way symmetrized transition frequency data among web pages. Figures 3 and 4 show similar tendencies. There exist the same groups in Figs. 3 and 4. However, Fig. 4 partially differs in the tendency from Fig. 3. These differences exist in the pages of the access analysis and the service provided by the company. In Fig. 3, the page of the access analysis such as Pages 2, 3, and 13 are one group. However, only Pages 2 and 3 are one group in Fig. 4. Pages 12, 14, 15, and 16 are one group in Fig. 3, but Pages 12 and 13 are one group and Pages 14, 15, and 16 the other group in Fig. 4. The results of the above analysis provide an important new insight. The analysis of one-mode three-way data revealed new relationships among web pages which were not clear in analysis of one-mode two-way data. Our model clearly identified differences among groups of triadic web pages. These results indicate that our model solution produced more detailed representations among triadic web pages than the two-way distance model.

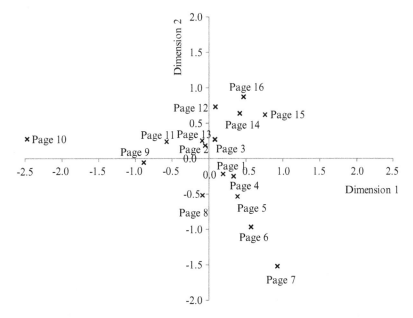

**Fig. 3** The two-dimensional configuration of reconstructed one-mode three-way symmetrized transition frequency data among web pages

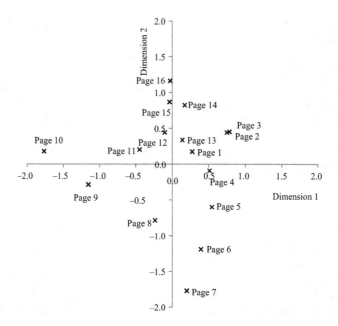

**Fig. 4** The two-dimensional configuration of reconstructed one-mode two-way symmetrized transition frequency data among web pages

In the future study, we would like to establish the validity of the proposed method to apply to various one-mode three-way asymmetric proximity data. The two-dimensional results are discussed in the present analyses because those help easy understanding of the relationships among the objects. So, we would like to consider higher dimensional representations in the future study. We would like to extend the proposed method to the analysis of four-way or more-way data, too.

**Acknowledgements** We would like to express our gratitude to two anonymous referees for their valuable reviews. We wish to thank the Officials of Data Analysis Competition 2006 administered by Joint Association Study Group of Management Science which provided valuable transition data among web pages. We are also greatly indebted to Professor Satoru Yokoyama of Teikyo University for the great support and advice in analyzing data. The major part of the present work done by Nakayama was carried out when he was at Nagasaki University. This work was supported by KAKENHI(20700257).

# References

De Rooij M, Gower JC (2003) The geometry of triadic distances. J Classification 20:181–220
De Rooij M, Heiser WJ (2000) Triadic distances models for the analysis of asymmetric three-way proximity data. Br J Math Stat Psychol 53:99–119
Harshman RA, Green PE, Wind Y, Lundy ME (1982) A model for the analysis of asymmetric data in marketing research. Market Sci 1:205–242

Okada A, Imaizumi T (1987) Nonmetric multidimensional scaling of asymmetric proximities. Behaviormetrika 21:81–96

Saburi S, Chino N (2008) A maximum likelihood method for an asymmetric MDS model. Comput Stat Data Anal 52:4673–4684

Upton GJG (1978) The analysis of cross-tabulated data. Wiley, Chichester

Yokoyama S, Okada A (2006) Rescaling a proximity matrix using entropy in brand-switching data. Jpn J Behaviormetrics 33:159–166

Yokoyama S, Okada A (2007) Rescaling proximity matrix using entropy analyzed by INDSCAL. In: Decker R, Lenz HJ (eds) Advances in data analysis. Springer, Heidelberg, Germany, pp 327–334

Zielman B, Heiser WJ (1993) Analysis of asymmetry by a slide vector. Psychometrika 58:101–114

# Analysis of Car Switching Data by Asymmetric Multidimensional Scaling Based on Singular Value Decomposition

Akinori Okada

**Abstract** Car switching or car trade-in data among car categories were analyzed by
a procedure of asymmetric multidimensional scaling. The procedure, which deals
with one-mode two-way asymmetric similarities, has originally been introduced to
derive the centrality of the asymmetric social network. In the present procedure,
the similarity from a car category to the other car category is represented not by
the distance in a multidimensional space like the conventional multidimensional
scaling, but is represented by the weighted sum of areas with the positive or negative
sign along dimensions of a multidimensional space. The result of the analysis shows
that attributes which already have been revealed in previous studies accounted for
car switching among car categories by a different manner from previous studies, and
can more easily be interpreted than the previous studies.

## 1 Introduction

The car switching or car trade-in data among 16 car categories (Harshman et al.
1982, Table 4) are analyzed in the present study. The car switching among car
categories is inevitably asymmetric. After Harshman et al. (1982) analyzed the
data by their asymmetric multidimensional scaling method, several researchers
analyzed the data (Kiers and Takane 1994; Okada 1988; Okada and Imaizumi 1987;
Zielman and Heiser 1993) by using asymmetric multidimensional scaling based
on different models. In the present study, the data were analyzed by a procedure
of asymmetric multidimensional scaling which is based on the singular value
decomposition. While several procedures of asymmetric multidimensional scaling

A. Okada (✉)
Graduate School of Management and Information Sciences, Tama University, 4-1-1 Hijirigaoka
Tama city Tokyo 206-0022, Japan
e-mail: okada@rikkyo.ac.jp; okada@tama.ac.jp

W. Gaul et al. (eds.), *Challenges at the Interface of Data Analysis, Computer Science,*
*and Optimization*, Studies in Classification, Data Analysis, and Knowledge Organization,
DOI 10.1007/978-3-642-24466-7_15, © Springer-Verlag Berlin Heidelberg 2012

have been introduced (Borg and Groenen 2005, Chap. 23), most of them represent the asymmetric proximity among objects by the distance between points and the geometric concept like a vector or a circle in a multidimensional space to cope with the asymmetry. These models can represent the asymmetry among objects, but the result of the analysis based on these models is not always easy to interpret, because of the geometric concept introduced to cope with the asymmetry.

The procedure utilized in the present study, which originally has been introduced to derive the centrality of the actor of the asymmetric social network (Bonacich 1972; Okada 2010a, in press; Tyler et al. 2003; Wasserman and Faust 1994, pp. 510–511), is reinterpreted as a procedure of asymmetric multidimensional scaling of one-mode two-way similarities among objects. The present procedure does not depend on the distance nor any geometric concept in a multidimensional space. The procedure accounts for the asymmetry among objects by using two concepts; the outward tendency and the inward tendency. The asymmetric similarity from one object to the other object is represented by the sum of areas with the positive or negative sign along dimensions of a multidimensional space. The area with the positive or negative sign is the product of the outward tendency of an object and the inward tendency of the other object along a dimension of a multidimensional space. The product along a dimension can be interpreted as the area of the rectangle with the positive or negative sign in a plane corresponding to the dimension. The manner of interpreting the result is different from the conventional asymmetric multidimensional scaling (cf. Kiers and Takane 1994). The result of the analysis by using the present procedure is easier to interpret than that of the conventional asymmetric multidimensional scaling.

## 2 The Method

Let $\mathbf{A}$ be the matrix of similarities among objects or car categories in the present study. The $(j, k)$ element of $\mathbf{A}$, $a_{jk}$, represents the similarity from objects $j$ to $k$. The $(k, j)$ element of $\mathbf{A}$, $a_{kj}$, represents the similarity from objects $k$ to $j$. The conjugate elements; $a_{jk}$ and $a_{kj}$, are not always equal. Suppose there are $n$ objects, $\mathbf{A}$ is an $n \times n$ asymmetric matrix. The singular value decomposition of $\mathbf{A}$ is

$$\mathbf{A} = \mathbf{XDY'}, \tag{1}$$

where $\mathbf{X}$ is the $n \times n$ matrix of left singular vectors of the unit length, $\mathbf{D}$ is the $n \times n$ diagonal matrix of singular values in the descending order at its diagonal elements, and $\mathbf{Y}$ is the $n \times n$ matrix of right singular vectors of the unit length. The $j$-th element of the $i$-th column of $\mathbf{X}$, $x_{ji}$, represents the outward tendency of object $j$ along dimension $i$. The $k$-th element of the $i$-th column of $\mathbf{Y}$, $y_{ki}$, represents the inward tendency of object $k$ along dimension $i$. The $(j, k)$ element of $\mathbf{A}$, $a_{jk}$, is represented by

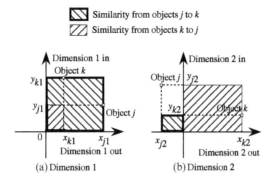

**Fig. 1** (**a**) $x_{j1}y_{k1}$ which corresponds to the similarity from objects $j$ to $k$ is larger than $x_{k1}y_{j1}$ which corresponds to the similarity from objects $k$ to $j$ along Dimension 1. Both $x_{j1}y_{k1}$ and $x_{k1}y_{j1}$ are positive. (**b**) $x_{j2}y_{k2}$ which corresponds to the similarity from objects $j$ to $k$ is negative, and $x_{k2}y_{j2}$ which corresponds to the similarity from objects $k$ to $j$ is positive along Dimension 2

$$a_{jk} = \sum_{i=1}^{n} d_i x_{ji} y_{ki}, \tag{2}$$

where $d_i$ is the $i$-th diagonal element of **D** or the $i$-th largest singular value of **A**. The $i$-th term of the right side of (2), $d_i x_{ji} y_{ki}$, represents the positive or negative component of the similarity from objects $j$ to $k$ along dimension $i$. The term is defined by the product of $x_{ji}$, the outward tendency of object $j$, and $y_{ki}$, the inward tendency of object $k$, along dimension $i$ weighted by the $i$-th singular value $d_i$, and is geometrically represented as the area with the positive or negative sign in a plane corresponding to dimension $i$.

Figure 1a shows $x_{j1}y_{k1}$ and $x_{k1}y_{j1}$ corresponding to the similarity from objects $j$ to $k$ and that from objects $k$ to $j$ along Dimension 1 respectively. The horizontal dimension represents the first left singular vector of **A**, and the vertical dimension represents the first right singular vector of **A**. $x_{j1}y_{k1}$ and $x_{k1}y_{j1}$ are positive, because all outward and inward tendencies are positive. The similarity from objects $j$ to $k$ along Dimension 1 is larger than that from objects $k$ to $j$ ($x_{j1}y_{k1} > x_{k1}y_{j1}$). Figure 1b shows $x_{j2}y_{k2}$ and $x_{k2}y_{j2}$ corresponding to the similarity from objects $j$ to $k$ and that from objects $k$ to $j$ along Dimension 2. The horizontal and the vertical dimensions represent the second left and right singular vectors of **A** respectively. $x_{j2}y_{k2}$ is negative, because $x_{j2}$ is negative and $y_{k2}$ is positive. The similarity from objects $j$ to $k$ is negative and that from objects $k$ to $j$ is positive along Dimension 2.

## 3   The Analysis

The car switching data among 16 car categories can be arrayed in a 16×16 table, whose $(j, k)$ cell consists of the number of cars in car category $j$ which were traded-in to purchase new cars in car category $k$. The $(j, k)$ cell of the table is regarded as the similarity from car categories $j$ to $k$. The 16 car categories are;

(a) Subcompact domestic (SUBD)
(b) Subcompact captive imports (SUBC)
(c) Subcompact imports (SUBI)
(d) Small specialty domestic (SMAD)
(e) Small specialty captive imports (SMAC)
(f) Small specialty imports (SMAI)
(g) Low price compact (COML)
(h) Medium price compact (COMM)
(i) Import compact (COMI)
(j) Midsize domestic (MIDD)
(k) Midsize imports (MIDI)
(l) Midsize specialty (MIDS)
(m) Low price standard (STDL)
(n) Medium price standard (STDM)
(o) Luxury domestic (LUXD)
(p) Luxury import (LUXI).

The abbreviation for each car category in parentheses is used hereafter.

The $(j, k)$ cell of the table is influenced by row $j$ sum (the total number of cars in car category $j$ traded-in) and column $k$ sum (the total number of cars in car category $k$ purchased). The normalization to eliminate the effect of the row sum and the column sum was done, and the normalized car switching data among 16 car categories (Okada 1988, p. 284, Table 2) were analyzed by the present procedure (Okada 2010b). The singular value decomposition of the similarity matrix of the normalized car switching data was done. The five largest singular values were $11,734, 5,259, 3,442, 2,609$, and $1,980 (\times 10^6)$. Three-dimensional result was chosen as the solution. The proportion of the sum of three squared singular values to the sum of all squared singular values or the sum of squares of the data $(1.902 \times 10^{20})$ is 0.93. And the fourth dimension is difficult to interpret. This is the reason why the three-dimensional solution was adopted.

Figure 2 shows the outward and inward tendencies along Dimension 1. The 16 car categories are almost along a line from the upper left to the lower right. The smaller and less expensive car categories are in the upper left area, and have smaller outward and larger inward tendencies. The larger and more expensive car categories are at the lower right area, and have larger outward and smaller inward tendencies. This tells that the larger outward tendency a car category has, the smaller inward tendency it has. All car categories are in the first quadrant, and thus have the positive outward and inward tendencies. The product of the outward and inward tendencies is positive, and the similarity along Dimension 1 is positive for combinations of any two car categories. This suggests that the first set of tendencies or Dimension 1 represents the general dominance or strength of the car category in car switching relationships. The imported car categories do not seem to have any relationship with the outward and inward tendencies along Dimension 1.

The outward and inward tendencies along Dimensions 2 and 3 are not always positive. Figure 3 shows the outward and inward tendencies along Dimension 2.

**Fig. 2** The outward tendency and the inward tendency along Dimension 1. The domestic and captive imported car categories are represented by a solid square, and the imported car category is represented by a gray square

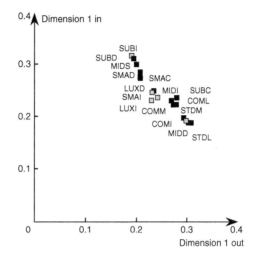

**Fig. 3** The outward tendency and the inward tendency along Dimension 2. As in Fig. 2, the domestic and captive imported car categories are represented by a solid square, and the imported car category is represented by a gray square

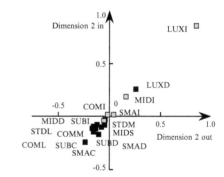

The product of the outward and inward tendencies along Dimension 2 is positive among car categories in the first quadrant, positive among those in the third quadrant, and negative from car categories in the first quadrant to those in the third quadrant, and vice versa. The 16 car categories can be divided almost into two groups; one consists of car categories in the first quadrant (Group 1), and the other consists of those in the third quadrant (Group 2). Group 1 consists of three imported car categories and LUXD. Group 2 includes only one imported car category (SUBI). LUXI, LUXD, and MIDI have larger inward and outward tendencies than most of the car categories in Group 2. This shows that the car switching among these three categories is more frequent than that among car categories in Group 2.

Figure 4 shows the outward and inward tendencies along Dimension 3. There are five domestic car categories (LUXD, STDM, STDL, MIDS, and MIDD) which are larger or more expensive than compact and small specialty car categories in the 16 car categories. Four of these five car categories are in the first quadrant. While MIDD is in the forth quadrant, the other 11 car categories (including all imported car categories) are in the third quadrant. The 16 car categories can be

**Fig. 4** The outward tendency and the inward tendency along Dimension 3. Each of five domestic car categories which are larger or more expensive than compact and small specialty car categories is represented by a gray square. Each of the other 11 car categories is represented by a solid square

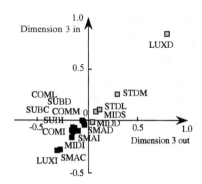

divided almost into two groups; one consists of car categories in the first quadrant (Group 3), and the other consists of those in the third quadrant (Group 4). The car switching among car categories (LUXD, STDM, STDL and MIDS) in Group 3 is more frequent than that among car categories in Group 4. Each of outward and inward tendencies along Dimensions 2 and 3 divided the 16 car categories almost into two groups respectively, where the car switching within a group is frequent and that between two groups is less frequent along these dimensions.

## 4 Discussion

Car switching data among 16 car categories were successfully analyzed by the asymmetric multidimensional scaling based on the singular value decomposition. The present asymmetric multidimensional scaling deals with diagonal elements of the data matrix which would have been ignored by most of the multidimensional scaling, and is also featured by the simpleness of its algorithm. The present model represents not only the similarity by the area of the rectangle, but also the outward and inward tendencies of the object along each dimension. The outward and inward tendencies tell two elements of the asymmetry, suggesting the interpretation of the asymmetry among objects of the present model is easier than that of the other asymmetric multidimensional scaling models.

Similarities along Dimension 1 are positive. Then we regard all 16 car categories constitute Group 0 which represents the general dominance or strength of the car category in car switching suggested by Dimension 1. Five groups defined by Dimensions 1, 2, and 3 are shown in Table 1. They are similar to those derived by ADCLUS (Arabie and Carroll 1980; Arabie et al. 1987; cf. Okada 1988).

The present model bears a resemblance to the generalized GIPSCAL model (Kiers and Takane 1994). Both models do not split the similarity into symmetric and asymmetric (or skew-symmetric) components of the relationship among objects, and represent them geometrically. The present model uses an area of a rectangle with positive or negative sign which corresponds to the component of the similarity along

**Table 1** Five cluster structure

| Car category | Group 0 | Group 1 | Group 2 | Group 3 | Group 4 |
|---|---|---|---|---|---|
| (a) Subcompact domestic (SUBD) | 1 | 0 | 1 | 0 | 1 |
| (b) Subcompact captive imports (SUBC) | 1 | 0 | 1 | 0 | 1 |
| (c) Subcompact imports (SUBI) | 1 | 0 | 1 | 0 | 1 |
| (d) Small specialty domestic (SMAD) | 1 | 0 | 1 | 0 | 1 |
| (e) Small specialty captive imports (SMAC) | 1 | 0 | 1 | 0 | 1 |
| (f) Small specialty imports(SMAI) | 1 | 1 | 0 | 0 | 1 |
| (g) Low price compact (COML) | 1 | 0 | 1 | 0 | 1 |
| (h) Medium price compact (COMM) | 1 | 0 | 1 | 0 | 1 |
| (i) Import compact (COMI) | 1 | 0 | 0 | 0 | 1 |
| (j) Midsize domestic (MIDD) | 1 | 0 | 1 | 0 | 0 |
| (k) Midsize imports (MIDI) | 1 | 1 | 0 | 0 | 1 |
| (l) Midsize specialty (MIDS) | 1 | 0 | 1 | 1 | 0 |
| (m) Low price standard (STDL) | 1 | 0 | 1 | 1 | 0 |
| (n) Medium price standard (STDM) | 1 | 0 | 1 | 1 | 0 |
| (o) Luxury domestic (LUXD) | 1 | 1 | 0 | 1 | 0 |
| (p) Luxury import (LUXI) | 1 | 1 | 0 | 0 | 1 |

each dimension, while the generalized GIPSCAL model uses an inner product which corresponds to the component of the (normalized) similarity along each dimension. The outward and inward tendencies (or the difference of them) of the present model correspond to the radius of the object of Okada and Imaizumi (1987). The effect of the slide vector (Zielman and Heiser 1993) varies according to the angle between the vector and the line connecting two points (which represent two objects). The slide vector represents the asymmetry among objects not based on each object, while the outward and inward tendencies of the present model as well as the radius of the model of Okada and Imaizumi (1987) represent the asymmetry among objects based on each object.

The present data have been analyzed by conventional asymmetric multidimensional scaling. Car switching is accounted for by two dimensions; "size or price" and "domestic/import" (Zielman and Heiser 1993), by those two dimensions and "specialty" dimension (Okada 1988; Okada and Imaizumi 1987). Four dimensions accounted for the car switching in Harshman et al. (1982). The four dimensions consist of "plain large-midsize", "specialty", "fancy large" and "small". The dimensions "plain large-midsize" and "small" seem closely related with "size or price" dimension mentioned in other studies. The generalized GIPSCAL model does not rely on dimensions, but Kiers and Takane (1994) suggested two features; (a) switches from medium or standard cars to small and/or specialty cars, and (b) prominent switches between STDM and LUXD. In the present analysis, three dimensions accounted for the car switching; "general dominance or strength", "domestic/import", and "large or expensive domestic/the others". The "general dominance or strength" dimension seems equivalent to the "size or price" dimension in earlier studies (Okada 1988; Okada and Imaizumi 1987; Zielman and Heiser 1993) and to the first feature suggested by Kiers and Takane (1994). And the "large or expensive domestic/the others" dimension seems equivalent to the "fancy large" dimension mentioned in Harshman et al. (1982), and also corresponds to the second feature suggested by Kiers and Takane (1994). While the manner of representing the asymmetric similarity is different from the conventional asymmetric multidimensional scaling, dimensions already revealed in earlier studies represented the characteristics of the car switching in the present study.

**Acknowledgements** The author would like to express his gratitude to two anonymous referees for their stimulating and helpful comments on the earlier version of the paper.

# References

Arabie P, Carroll JD (1980) MAPCLUS: A mathematical programming approach to fitting the ADCLUS model. Psychometrika 45:211–235

Arabie P, Carroll JD, DeSarbo WS (1987) Three-way scaling and clustering. Sage Publications, Newbury Park, CA

Bonacich P (1972) Factoring and weighting approaches to clique identification. J Math Sociol 2:113–120

Borg I, Groenen PJF (2005) Modern multidimensional scaling: Theory and applications, 2nd edn. Springer, New York

Harshman RA, Green PE, Wind Y, Lundy ME (1982) A model for the analysis of asymmetric data in marketing research. Market Sci 1:205–242

Kiers HAL, Takane Y (1994) A generalization of GIPSCAL for the analysis of nonsymmetric data. J Classification 11:79–99

Okada A (1988) Asymmetric multidimensional scaling of car switching data. In: Gaul W, Schader M (eds) Data, expert knowledge and decisions. Springer, Berlin, Germany, pp 279–290

Okada A (2008) Two-dimensional centrality of a social network. In: Preisach C, Burkhardt L, Schmidt-Thieme L (eds) Data analysis, machine learning and applications. Springer, Heidelberg, Germany, pp 381–388

Okada A (2010a) Analyzing asymmetric relationships of car switching by singular value decomposition. In: Proceedings of the 17th International Symposium on Mathematical Methods Applied to the Sciences, Universidad de Costa Rica, San Jose, Costa Rica, pp 211–212

Okada A (2010b) Two-dimensional centrality of asymmetric social network. In: Palumbo F, Lauro CN, Greenacre MJ (eds) Data analysis and classification. Springer, Heidelberg, Germany, pp 93–100

Okada A (2011) Centrality of asymmetry social network: Singular value decomposition, conjoint measurement, and asymmetric multidimensional scaling. In: Ingrassia S, Rocci R, Vichi M (eds) New perspectives in statistical modeling and data analysis. Springer, Heidelberg, Germany

Okada A, Imaizumi T (1987) Nonmetric multidimensional scaling of asymmetric similarities. Behaviormetrika 21:81–96

Tyler JR, Wilkinson DM, Huberman BA (2003) E-mail as spectroscopy: Automated discovery of community structure within organization. e-print condmat/0303264

Wasserman S, Faust K (1994) Social network analysis: methods and applications. Cambridge University Press, Cambridge, UK

Zielman B, Heiser WJ (1993) Analysis of asymmetry by a slide vector. Psychometrika 50:101–114

# Visualization of Asymmetric Clustering Result with Digraph and Dendrogram

**Yuichi Saito and Hiroshi Yadohisa**

**Abstract** Asymmetric cluster analysis is one of the most useful methods together with asymmetric multidimensional scaling (MDS) to analyze asymmetric (dis)similarity data. In both methods, visualization of the result of the analysis plays an important role in the analysis. Some methods for visualizing the result of the asymmetric clustering and MDS have been proposed (Saito and Yadohisa, Data Analysis of Asymmetric Structures, Marcel Dekker, New York, 2005).

In this paper, we propose a new visualization method for the result of asymmetric agglomerative hierarchical clustering with a digraph and a dendrogram. The visualization can represent asymmetric (dis)similarities between pairs of any clusters, in addition to the information of a traditional dendrogram, which is illustrated by analyzing the symmetric part of asymmetric (dis)similarity data. This visualization enables an intuitive interpretation of the asymmetry in (dis)similarity data.

## 1 Introduction

To analyze an asymmetric (dis)similarity data matrix that describes asymmetric information among objects, asymmetric cluster analysis is an effective choice. However, an ordinary dendrogram that is used for visualizing the result of hierarchical

Y. Saito (✉)
Graduate School of Culture and Information Science, Doshisha University Kyoto 610-0313, Japan
e-mail: dij0024@mail4.doshisha.ac.jp

H. Yadohisa
Department of Culture and Information Science, Doshisha University Kyoto 610-0313, Japan
e-mail: hyadohis@mail.doshisha.ac.jp

W. Gaul et al. (eds.), *Challenges at the Interface of Data Analysis, Computer Science, and Optimization*, Studies in Classification, Data Analysis, and Knowledge Organization, DOI 10.1007/978-3-642-24466-7_16, © Springer-Verlag Berlin Heidelberg 2012

cluster analysis, can not represent information about the asymmetry. To visualize this asymmetry, some extended dendrograms have been proposed.

For example, Okada and Iwamoto (1995) proposed a new dendrogram that represents a ratio of asymmetry between objects. Okada and Iwamoto (1996) proposed a group average algorithm for asymmetric data and a new dendrogram that expresses asymmetry between two clusters as a combining direction. Saito and Yadohisa (2005) extended the updating formula proposed by Lance and Williams (1967) to the asymmetric case and proposed a new dendrogram that represents maximum and minimum distances together with the combined distance.

In this paper, we propose a new visualization method by using cluster analysis and multidimensional scaling (MDS). We decompose an asymmetric (dis)similarity matrix $D$ into the symmetric matrix $S$ and the skew-symmetric matrix $A$. Then we analyze the symmetric part $S$ by symmetric hierarchical cluster analysis and MDS. The dendrogram which represents the result of the clustering is mapped on a two-dimensional space, in which the result of MDS is also represented. Next, we draw the asymmetric part $A$ as arrows between clusters that represent the differences of (dis)similarities between those clusters. We also propose a 3D visualization of the result of asymmetric clustering and MDS, in which clusters are represented as closures and the asymmetry among clusters is represented by arrows.

## 2　Proposal

In this section, we propose an intuitive visualization method for the result of asymmetric agglomerative hierarchical clustering. We use convex hulls to represent clusters and arrows to represent the asymmetric relation.

Here we assume that an asymmetric dissimilarity matrix $D$ is given for analysis. We decompose matrix $D$ into the symmetric matrix $S = (D + D^t)/2$ and skew-symmetric matrix $A = (D - D^t)/2$. We interpret that these two matrices $S$ and $A$, have symmetric and asymmetric information from the data, respectively. We use both the matrices ($S$ and $A$) for visualization. First, we analyze the symmetric part $S$ by hierarchical clustering and MDS. Then, we generate a dendrogram and object coordinates. Next, we analyze the asymmetric part $A$. We consider that matrix $A$ has information that is not expressed in clustering and MDS, and we incrementally display the values of the asymmetric part $A$ in the dendrogram. Details of algorithms to visualize the symmetric part $S$ and the asymmetric part $A$ are described below.

### 2.1　Visualization of the Symmetric Part S

The visualization algorithm of the symmetric part $S$ comprises four steps. This visualization should be performed before the visualization of the asymmetric part $A$.

Step 1: Generate a dendrogram by the hierarchical clustering algorithm for symmetric data.

In step 1, we analyze the symmetric matrix $S$ and generate a dendrogram by applying the hierarchical clustering algorithm. Here, we can use any hierarchical algorithm; for example, the single linkage method, group average method, centroid method and Ward method etc.

Step 2: Generate the object coordinates by MDS.

In step 2, we analyze the symmetric matrix $S$ by MDS and generate two- or three-dimensional coordinates of objects. We propose visualizations of the cluster for two- or three-dimensional cases.

Step 3: Visualize the results of both symmetric clustering and MDS by a dendrogram and object coordinates.

In step 3, we visualize the results of both symmetric clustering and MDS. This step is divided into two cases. When we generate two-dimensional coordinates in step 2, the dendrogram that results from the clustering is mapped on two-dimensional space, in which the result of MDS is also represented. When we generate three-dimensional coordinates in step 2, we only color or shade the objects depending on clusters to which they belong (Fig. 1(a), 1(b)).

Step 4: Surround the clusters with convex hulls and draw a plane that represents the cutting distance.

In step 4, we improve the visualization. For example, we draw convex hulls on clusters and a plane that represents the cutting distance (Fig. 1(c), 1(d)).

## 2.2   Visualization of Asymmetric Part $A$

The visualization algorithm of the asymmetric part $A$ also comprises four processes. In this step, the asymmetric information is appended to a graphic that is generated in the visualization of the symmetric part $S$.

Step 1: Calculate the asymmetry $AS$ between clusters.

In step 1, we calculate asymmetries between clusters. First, we define the skew-symmetric matrix

$$A = (a_{ij}) \quad (i, \ j = 1, \ 2, \ \ldots, \ n). \tag{1}$$

Then, we define an asymmetry from cluster $C_p$ to cluster $C_q$ as

$$AS(C_p, \ C_q) = \frac{1}{\#(C_p)\#(C_q)} \sum_{i \in C_p, \ j \in C_q} a_{ij}. \tag{2}$$

Here, $\#(C_p)$ represents the number of objects in cluster $C_p$.

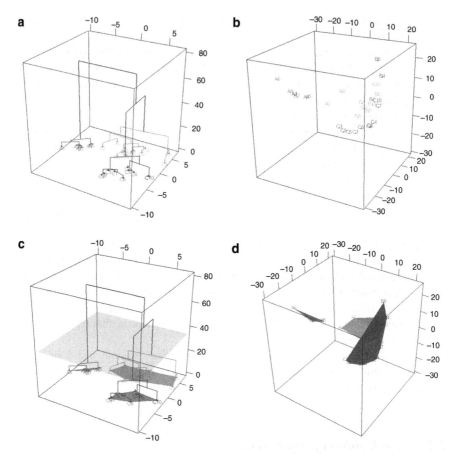

**Fig. 1** (**a**) Two-dimensional visualization in step 3, (**b**) three-dimensional visualization in step 3, (**c**) two-dimensional visualization in step 4 and (**d**) three-dimensional visualization in step 4

We have an equality relation as

$$AS(C_p, C_q) = -AS(C_q, C_p). \tag{3}$$

Step 2: Generate a digraph from $AS$.

In step 2, we generate a digraph from $AS$. Let $\{C_1, C_2, \ldots, C_T\}$ be the set of all clusters. We generate a digraph $(V, E)$ in which a set $V$ is the vertex set and a set $E$ is the arc set defined as

$$V = \{C_1, C_2, \ldots, C_T\}, \ E = \{(C_p, C_q) \mid AS(C_p, C_q) > 0\}. \tag{4}$$

The first element of arc $(C_p, C_q)$ is called the start vertex and the second element of arc $(C_p, C_q)$ is called the end vertex.

Step 3: Determine arrow width from $AS$.

In step 3, we generate an arrow that represents asymmetric information between clusters. This step is divided into two cases. When we generate the two-dimensional coordinates in step 2 of the visualization of $S$, we define coordinates $X = \{(x_i, \ y_i) \mid i = 1, \ 2, \ \ldots, \ n\}$. Then we define the arrow width of arc $(C_p, \ C_q)$ as

$$AW(C_p, \ C_q) = \frac{|AS(C_p, \ C_q)|}{|\max(AS(C_p, \ C_q))|} \left\{ Max - Min \right\} \cdot P, \quad (5)$$

where,

$$Max = \max(\max_i(x_i), \ \max_i(y_i)), \ Min = \min(\min_i(x_i), \ \min_i(y_i)). \quad (6)$$

When the asymmetry $AS$ is large, the arrow width $AW$ becomes large, and vice versa. When we generate the three-dimensional coordinates in step 2 of the visualization of $S$, we define coordinates $X = \{(x_i, \ y_i, \ z_i) \mid i = 1, \ 2, \ \ldots, \ n\}$. Then $AW$ is calculated by the same formula (5), where instead

$$Max = \max(\max_i(x_i), \ \max_i(y_i), \ \max_i(z_i)),$$

$$Min = \min(\min_i(x_i), \ \min_i(y_i), \ \min_i(z_i)). \quad (7)$$

A parameter $P$ is a value that determines the proportion of the arrow width to the display area. In this study, we set $P = 1/16$ due to viewability.

Step 4: Draw arrows between pairs of all centroids of clusters in such a way that the centroids are start or end vertices.

In step 4, we draw arrows between all clusters. We consider all elements of $V$ as centroids of the cluster and draw all elements of $E$ as arrows with arrow width $AW$ between clusters (Fig. 3). Coordinates of each point of an arrow between cluster $C_p$ and $C_q$ are calculated as below. Let $(x_i^{C_p}, \ y_i^{C_p}, \ z_i^{C_p})$ $(i = 1, \ 2, \ \ldots, \ \#(C_p))$ and $(x_j^{C_q}, \ y_j^{C_q}, \ z_j^{C_q})$ $(j = 1, \ 2, \ \ldots, \ \#(C_q))$ be object coordinates that belong to cluster $C_p$ and $C_q$ respectively. When we generate two dimensional coordinates, all $z$-coordinates $z^{C_p}$, $z^{C_q}$ and $z^{C_r}$ are set to zero. Here, parameters $s$, $t$ and $u$ determine the shape of arrows and $\theta_1 and \theta_2$ decide the angle of arrows.

We use the following parametrization

$$s = \frac{4}{5}, \ t = \frac{1}{2} \cdot AW(C_p, \ C_q), \ u = \frac{23}{10} \cdot t$$

**Fig. 2** Each important point of the arrow. Arrow width is shown by $AW$

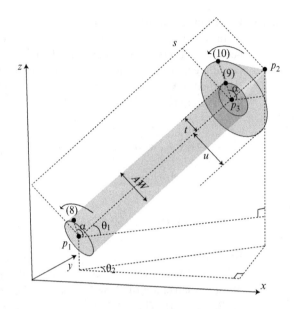

because of viewability. First, we define three vectors that determine important points of the arrow (Fig. 2).

$$\boldsymbol{p}_1 = (x^{C_p}, y^{C_p}, z^{C_p})^t = \left( \frac{1}{\#(C_p)} \sum_{i=1}^{\#(C_p)} x_i^{C_p}, \frac{1}{\#(C_p)} \sum_{i=1}^{\#(C_p)} y_i^{C_p}, \frac{1}{\#(C_p)} \sum_{i=1}^{\#(C_p)} z_i^{C_p} \right)^t,$$

$$\boldsymbol{p}_2 = (x^{C_q}, y^{C_q}, z^{C_q})^t = \left( \frac{1}{\#(C_q)} \sum_{i=1}^{\#(C_q)} x_i^{C_q}, \frac{1}{\#(C_q)} \sum_{i=1}^{\#(C_q)} y_i^{C_q}, \frac{1}{\#(C_q)} \sum_{i=1}^{\#(C_q)} z_i^{C_q} \right)^t,$$

$$\boldsymbol{p}_3 = (x^{C_r}, y^{C_r}, z^{C_r})^t$$
$$= \left( x^{C_p} + (x^{C_q} - x^{C_p}) \cdot s, \; y^{C_p} + (y^{C_q} - y^{C_p}) \cdot s, \; z^{C_p} + (z^{C_q} - z^{C_p}) \cdot s \right)^t.$$

In the three-dimensional case, we define the two angles $\theta_1$ and $\theta_2$ as

$$\theta_1 = \arctan\left( \frac{y^{C_q} - y^{C_p}}{x^{C_q} - x^{C_p}} \right), \; \theta_2 = \arctan\left( \frac{z^{C_q} - z^{C_p}}{((x^{C_q} - x^{C_p})^2 + (y^{C_q} - y^{C_p})^2)^{1/2}} \right),$$

which determine the direction of the arrow. Then, we define three rotation matrices $\boldsymbol{P}_x$, $\boldsymbol{P}_y$ and $\boldsymbol{P}_z$ as

$$\boldsymbol{P}_x = \begin{pmatrix} 1 & 0 & 0 \\ 0 & \cos(-\alpha) & -\sin(-\alpha) \\ 0 & \sin(-\alpha) & \cos(-\alpha) \end{pmatrix}, \; \boldsymbol{P}_y = \begin{pmatrix} \cos(-\theta_2) & 0 & -\sin(-\theta_2) \\ 0 & 1 & 0 \\ \sin(-\theta_2) & 0 & \cos(-\theta_2) \end{pmatrix},$$

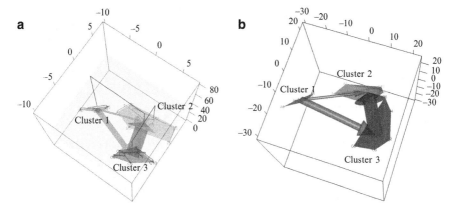

**Fig. 3** Numerical examples of the visualization. (**a**) Overhead view of two-dimensional visualization and (**b**) overhead view of three-dimensional visualization

$$P_z = \begin{pmatrix} \cos(-\theta_1) & -\sin(-\theta_1) & 0 \\ \sin(-\theta_1) & \cos(-\theta_1) & 0 \\ 0 & 0 & 1 \end{pmatrix},$$

where an angle $\alpha$ determines rotation about a line $p_1 - p_2$ (Fig. 2). Furthermore, we define vectors $v_1 = (0, t, 0)^t$, $v_2 = (0, u, 0)^t$ and the arrow is determined by equations that represent three circles as follows:

$$(x_1, y_1, z_1)^t = (P_y(P_z(P_x v_1))) + p_1 \quad (0 \leq \alpha \leq 2\pi), \tag{8}$$

$$(x_2, y_2, z_2)^t = (P_y(P_z(P_x v_1))) + p_3 \quad (0 \leq \alpha \leq 2\pi), \tag{9}$$

$$(x_3, y_3, z_3)^t = (P_y(P_z(P_x v_2))) + p_3 \quad (0 \leq \alpha \leq 2\pi). \tag{10}$$

In the three-dimensional case, we consider that $\alpha$ takes any values in $[0, 2\pi]$. In the two-dimensional case, we consider only $\alpha = \{0, \pi\}$. Figure 2 represents the points of arrows corresponding to each set of coordinates. The labels (8), (9) and (10) in Fig. 2 are related to (8), (9) and (10).

## 3  Numerical Example

In this section, we show a numerical example of our visualization in Fig. 3. We first generate an asymmetric dissimilarity matrix $D$ as follows.

Step 1:  Generate a symmetric matrix $S$ from artificial data that are generated from three-kinds of three-dimensional normal distributions.

Step 2:  Generate a skew-symmetric matrix $A$ from three-kinds of one-dimensional normal distributions.

Step 3:  Finally, we construct the asymmetric dissimilarity matrix $D = S + A$.

We use a $30 \times 30$ asymmetric dissimilarity matrix $D$ that is generated by the above algorithm. We assume that this matrix represents friendliness among 30 people and that these people belong to one of three groups because their data are generated from three different normal distributions.

## 3.1 Interpretation

Figure 3 is the visualization of the result of the analysis for the data generated in the above way. We can interpret it to see that:

- Cluster 1 has (positive) asymmetry to cluster 2 and cluster 3, because cluster 1 has two arrows towards cluster 2 and cluster 3.
- Cluster 3 has the largest (positive) asymmetry to cluster 2, because cluster 3 has the widest arrow towards cluster 2.
- The arrows represent asymmetry that is not used in the clustering process.
- The arrows represent "friendliness" between one cluster and the others.

We can interpret our visualization as representing "some asymmetric information among clusters" (e.g. the friendliness among clusters) and we can see intuitively the asymmetric relationship via arrows. A wide arrow represents a large asymmetry between clusters and vice versa.

## 4 Conclusion

In this paper, we proposed a visualization method that represents the asymmetric relationship between clusters together with the results of clustering and MDS. We combined a dendrogram from cluster analysis and coordinates from MDS and added arrows that represent the asymmetry among clusters. By using convex hulls, we proposed two types of visualization for two- or three-dimensional coordinates that are generated by MDS. This visualization enables us to interpret the asymmetric information among clusters.

In the future, we would like to propose a more intuitive and interactive visualization in which the user could choose the cutting distance interactively and interpret some information more intuitively. This dynamic visualization would enable the user to choose appropriate cutting distances more easily and to interpret the information of asymmetry in the visualization of each cutting distance more efficiently. We would like to improve our visualization method continually.

# References

Lance GN, Williams WT (1967) A general theory of classificatory sorting strategies: 1. Hierarchical systems. Comput J 9:373–380

Okada A, Iwamoto T (1995) An asymmetric cluster analysis study on university enrollment flow among Japanese prefectures. Sociological Theory Methods 10:1–13 (in Japanese)

Okada A, Iwamoto T (1996) University enrollment flow among the Japanese prefectures. Behaviormetrika 23:169–185

Saito T, Yadohisa H (2005) Data analysis of asymmetric structures. Marcel Dekker, New York

# Part IV
# Analysis of Visual, Spatial, and Temporal Data

# Clustering Temporal Population Patterns in Switzerland (1850–2000)

Martin Behnisch and Alfred Ultsch

**Abstract** Spatial planning and quantitative geography face a great challenge to handle the growing amount of geospatial data and new statistics. Techniques of data mining and knowledge discovery are therefore presented to examine by time intervals (=15 decades) the population development of 2,896 Swiss communities. The key questions are how many temporal patterns will occur and what are their characteristics? Relative difference (RelDiff) is proposed as an alternative to relative change calculation. The detection of temporal patterns is based on mixture models and the Bayes' theorem. A procedure of information optimization aims at selecting relevant temporal patterns for clustering. The use of a k-Nearest Neighbor classifier is based on the assumption that similar relevant temporal patterns are a good point of reference for the whole population development. The classification result is explained by significance with already existing classifications (e.g. central-periphery). Spatial visualization leads to the verification in mind of the spatial analyst and provides the process of knowledge conversion.

## 1 Introduction

Urban planners and politicians have several impressions about recent problems of Swiss population losses in peripheral alpine regions as well as about the urban sprawl in the Swiss Plateau (Oswald and Baccini 2003). The long-term development of all Swiss communities is not really quantified and not present in actual planning and decision processes (Perlik et al. 2001). Some temporal classifications (Tappeiner et al. 2001; Bätzing and Dickhörner 2001) have been established for the alpine regions based on hierarchical clustering. It is to emphasize that big

M. Behnisch (✉) · A. Ultsch
Data Bionics Research Group, University of Marburg, United Kingdom
e-mail: Behnisch@informatik.uni-marburg.de; Ultsch@informatik.uni-marburg.de

W. Gaul et al. (eds.), *Challenges at the Interface of Data Analysis, Computer Science, and Optimization*, Studies in Classification, Data Analysis, and Knowledge Organization, DOI 10.1007/978-3-642-24466-7_17, © Springer-Verlag Berlin Heidelberg 2012

and incomparable variances dominate the calculation of Euclidean distances when values are not normally distributed. A deeper explanation of cluster is sometimes missing and such approaches did not take into account knowledge discovery techniques to trigger spatial abstractions. The aim of this contribution is to discover long-term developments using characteristics of 15 decades in Switzerland. Due to the lack of several long-term data dimensions the development of population between 1850 and 2000 is analyzed as a kind of overall indicator for such development of Swiss communities.

## 2 Data Inspection

Fifteen variables are used to measure the change in population by decade (15 decades, 1850–2000). The population data is based on official statistics (Schuler et al. 2002). As an alternative to relative change calculation the authors suggest relative differences. The influence of extreme values is alleviated in view of the fact that the range of relative difference is symmetric and limited. This index is particularly suitable for normalisation and standardization (Ultsch 2008). Nevertheless data inspection of all 15 variables leads to the awareness that there are (wide) edges in the distribution of variables. A similarity measure is necessary, that allows to generate a typology of population dynamics. The idea is to model each of all 15 variables as a mixture of three characteristic developments: losing communities, typical communities (e.g. sum of many unobserved random population is acting independently, CLT theorem), winning communities. The mixture model is a composite of a log-normal, normal, log-normal distribution (see Fig. 1). The expectation maximization (EM-) algorithm is used for parameter estimation. "Good" initial parameters are important as the algorithm only finds a local and not a global optimum. Several re-calculations are necessary to improve the intermediate results. In addition to other mixture models Pareto density estimation (Ultsch 2003) and probability density functions are used to ensure the modeling process. Q-Q-plots confirm the composite distribution. The typical Swiss population change is described by the mean and standard deviation of typical communities.

## 3 Temporal Patterns of Population Change

Posterior probabilities were calculated on basis of the log-normal, normal, log-normal model assumption using Bayes' theorem. This theorem offers advantages through its ability to formally incorporate prior knowledge into model specification via prior distributions and allows considering the variability. A specific dynamic class (e.g. Winner, Typical, Loser) is therefore predominantly observed for a given value of population change in a community (see Fig. 2). Based on the posterior probabilities a scaling procedure was applied. Scaling means the translation of values into the positive or negative range. Abstract numerical values $(-1, 0, +1)$

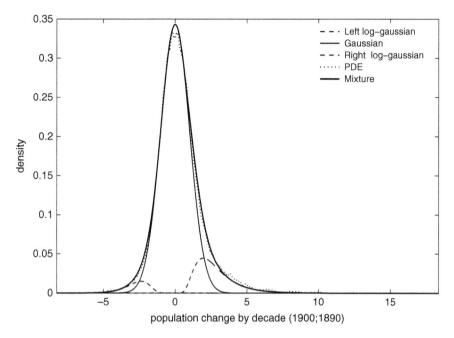

**Fig. 1** Mixture model of population change by decade

are then used to support a precise distinction of dynamic classes by decade (loser, typical, winner). For a better characterization of communities the term temporal pattern is defined to describe specific developments consisting of 15 unique dynamic classes (=15 decades). In view of 15 decades and 3 dynamic classes there are 880 different temporal patterns finally observed. There is one large temporal pattern (observed in 852 communities) clearly representing the "Typical" Swiss population development over time (15 × "Typical"). There are about 775 temporal patterns which are described by only 1 or 2 communities.

## 4   Relevance of Temporal Patterns

The authors aim at selecting relevant temporal patterns for clustering. In view of strategic spatial planning it is a crucial task to measure the relevance of each pattern. Since the Swiss community system is very heterogeneous (e.g. in population, size, etc.) it is by far not sufficient to use the number of communities for the detection of relevant patterns. A pragmatic planning approach is to have a deeper look at the size of population. The impact of population on one temporal pattern is therefore related to the communities and the sum of their average values of population (long-term mean value). A procedure of information optimization provides the selection of temporal patterns based on the Pareto principle (Ultsch 2001). A Lorenz curve presents the association of the number of temporal patterns and the computed

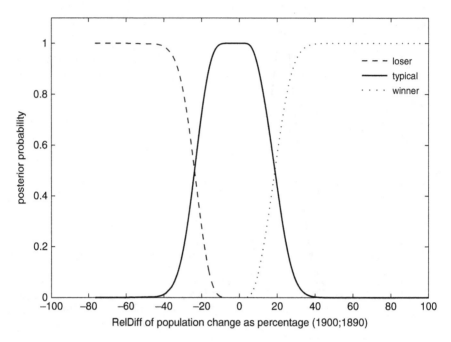

**Fig. 2** Posterior probabilities based on population change and dynamic classes

impact of population. From the ideal point 0% of population impact and 100% of knowledge of the temporal patterns the distance to the real situations on the Lorenz curve is measured. The dashed line in Fig. 3 indicates the shortest one of such distances. From this it is concluded that in order to gain different temporal patterns only about 14% of the temporal patterns, the 14% most relevant ones, should be examined in deep detail. The underlying assumption of the authors is that the minimal value of the population impact on a temporal pattern is within the range of 2,000 to 10,000. This value is assumed and is based on the observed number of communities per temporal pattern and the average mean value of population in Switzerland. The observed value is finally 5,000 per temporal pattern. One hundred and twenty two temporal patterns are selected and declared as relevant. In the presented context, 65% of all 2,896 communities belong to relevant temporal patterns and about 85% of the Swiss population.

## 5  Clustering and Classification of Temporal Patterns

For the purpose of clustering temporal patterns three periodical subdivisions are defined. These periods are extracted based on knowledge about the temporal patterns and their characteristics. The periods are as follows: period 1 (1850–1910,

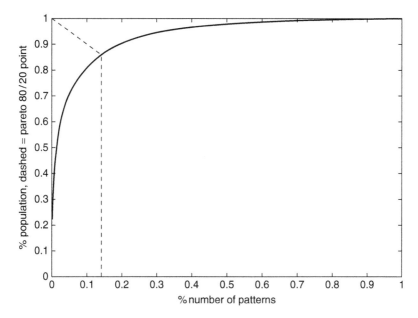

**Fig. 3** Information optimization in view of relevant temporal patterns in Switzerland

industrialization and urban growth), period 2 (1910–1950, World War I, II, subject of separation) and period 3 (1950–2000, urbanization, suburbanization, economic boom). Each period is described by a growth indicator summing up the integer values by decade (see scaling of posterior values in Sect. 3). Large positive values by period indicate a winner, values by zero indicate a typical and large negative values indicate a loser. A precise distinction of temporal patterns is supported by the properties of the growth indicator (limited range, similar scale). Temporal patterns are therefore different from each other by a value of 1. The Euclidean distance is now appropriate and reasonable. A complex variance estimation is not necessary. The Ward algorithm is used as a method for clustering. The information loss refers to the inner and outer cluster differences. It is conceivable when clustering 122 temporal patterns that each cluster will contain about 10 temporal patterns in average. The authors are interested in a clear and compact cluster structure. The authors therefore expect a solution of 6–12 clusters. The interpretation of the dendrogram (see Fig. 4, horizontal dashed line) confirms the expectation and supports an eight cluster solution describing 1,899 communities. The aim is subsequently to allocate all temporal patterns to an observed cluster. This is the basis for a comparison with other existing spatial classifications. A k-Nearest-Neighbor classifier is constructed for this purpose. It is initially trained on the relevant temporal pattern to identify the accuracy of allocation to the given cluster (accuracy = 100%). A typical community is found for each of the 8 clusters based on the average dynamic property by period and the amount of population.

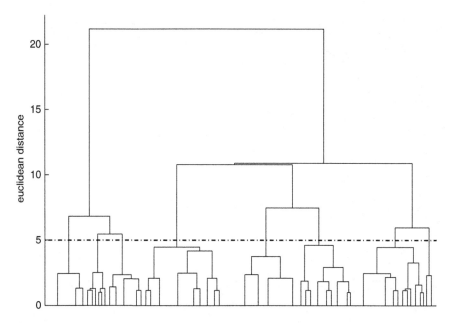

**Fig. 4** Dendrogram of WARD clustering

## 6   Knowledge Discovery

The process of class explanation provides the transition from data to knowledge. The goal is to provide evidence-based insight through a deeper understanding of data (in the mind of the analyst) and to produce results that can be utilized at policy and strategy levels. At first localization of classes is used for spatial verification and spatial reasoning (see Fig. 5). Secondly the detected class partition and other well-known typologies (Schuler et al. 2005) in Switzerland are compared using contingency tables in order to decide whether or not dependencies are significant. Relative differences (Ultsch 2008) are here also appropriate to analyze the deviation of the expected and observed values. Third, structure interpretation and validation in mind of the spatial analyst lead to knowledge about the communities (e.g. discovery of new temporal patterns in data). Such knowledge comprises spatial abstractions of classes and generates hypothesis that might be valuable for further explanations and subsequent analysis. As an example the class "Working Suburbia" is presented as follows: Frequently winning communities (in particular before 1910 and after 1950), workplaces by significance, located in the region Lemanique, in the Midlands and significantly in the Zurich area. The expected values of Out-Commuters are not confirmed. Communities of this class are like "Dietikon" (Typical). Table 1 gives an overview of all classes and their spatial abstractions. Furthermore the specific population dynamic, the amount of communities and the impact of population (see Sect. 4) are summarized. The classes are characterized by the growth indicators of the typical community.

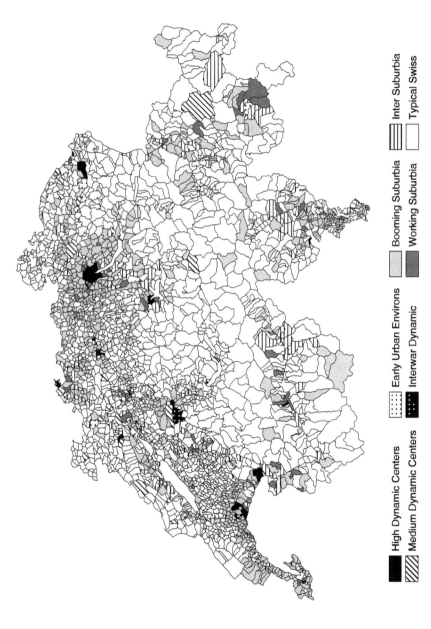

**Fig. 5** Localization of the classification result

**Table 1** Class properties in terms of spatial abstractions and population dynamics by period

| Class label, communities | | Typical | 1850–1910 | 1910–1950 | 1950–2000 | Impact |
|---|---|---|---|---|---|---|
| Early urban environs | 25 | Gossau | Winner, +4 | Typical, 0 | Typical, +1 | 153,773 |
| High dynamic centers | 10 | St. Gallen | Winner, +5 | Typical, 0 | Typical, 0 | 469,374 |
| Medium dynamic centers | 35 | Chaux-de-Fonds | Winner, +3 | Typical, 0 | Typical, 0 | 436,348 |
| Booming Suburbia | 868 | Uster | Typical, 0 | Typical, 0 | Winner, +2 | 907,280 |
| Interwar Dynamic | 8 | Ascona | Typical, 0 | Winner, 0 | Typical, +2 | 40,640 |
| Inter Suburbia | 224 | Zug | Typical, +1 | Typical, 0 | Typical, +1 | 422,423 |
| Working Suburbia | 55 | Dietikon | Winner, +3 | Winner, +1 | Winner, +2 | 197,778 |
| Typical Swiss | 1,671 | Solothurn | Typical, 0 | Typical, 0 | Typical, 0 | 1,684,991 |

# 7 Discussion

Many former demographic studies have been made by using cross-section analysis (=comparison of two time points). In contrast to some former long-term analysis the authors are interested to understand the population data in depth. For example the characterization of the population development was not quantified by 15 decades. Furthermore temporal patterns are defined before clustering. Such procedure forms the basis to take into account the variance of data and the properties of distance measurements. The localization of temporal patterns might be helpful for deeper investigations. The separation of large and small populated communities is observed as expected. One dynamic (15x "Typical") is clearly representing the typical Swiss population development. This dynamic might be helpful as a new spatial semantic in contrast to the well-known semantic alpine regions. Another interesting aspect deals with temporal patterns with one or zero "Non-Typical" (=Winner or Loser). They are characterizing more than 50% of all Swiss communities. This group of observed specific temporal patterns (outliers) is much bigger than expected. Subsets of communities or specific temporal patterns are valuable for detailed investigations using other structural and temporal parameters (e.g. age of population, infrastructure, buildings, etc.). The clustering is based on relevant patterns. In the future other criteria (e.g. number of buildings, percentage of urbanized area per community) should be investigated to confirm the number of relevant patterns. In contrast to the mentioned former approaches the authors suggest contingency tables to explain the cluster with well-known spatial typologies.

# 8 Conclusion

The presented approach (Behnisch 2009) provides the ability to identify 880 temporal patterns within a large amount of Swiss communities. Eight typical Swiss population dynamics are finally extracted. This solution is explained by the population dynamic of three periods and other well-known spatial typologies. Each cluster is finally expressed by one spatial abstraction and a typical community.

**Acknowledgements** Population statistics are from the website of the Federal Office of Statistics (=http:/www.pxweb.bfs.admin.ch, online 30.01.2011). Furthermore, the geometry of communities is from in the repository GEOSTAT (=http:/www.bfs.admin.ch, online 30.01.2011).

# References

Bätzing W, Dickhörner Y (2001) Die Bevölkerungsentwicklung im Alpenraum 1870-1990 aus der Sicht von Längsschnittanalysen aller Alpengemeinden. Revue de Geographie Alpine 89:11–20
Behnisch M (2009) Urban data mining. KIT Scientific Press, Karlsruhe
Oswald F, Baccini P (2003) Netzstadt. Birkhäuser, Basel
Perlik M, Messerli P, Bätzing W (2001) Towns in the alps. Mt Res Dev 21(3):243–252
Schuler M, Ullmann D, Haug W (2002) Eidgenössische Volkszählung 2000 - Bevölkerungsentwicklung der Gemeinden 1850–2000. Tech. rep., Bundesamt für Statistik, Neuchatel, CH
Schuler M, Dessemontet P, Joye D (2005) Eidgenössische Volkszählung 2000 - Die Raumgliederungen der Schweiz. Tech. rep., Bundesamt für Statistik, Neuchatel, CH
Tappeiner U, Tappeiner G, Hilber A, Mattanovich E (2001) The EU agricultural policy and the environment. Blackwell, Berlin
Ultsch A (2001) Eine Begründung der Pareto 80/20 Regel und Grenzwerte für die ABC Analyse. Tech. Rep. 30, Department of Mathematics and Computer Science, University of Marburg, URL http://www.mathematik.uni-marburg.de/databionik/pdf/pubs//ultsch01begruendung
Ultsch A (2003) Pareto density estimation: A density estimation for knowledge discovery. In: Baier D, Wernecke KD (eds) Innovations in classification, data science, and information systems. Springer, Berlin, pp 91–100
Ultsch A (2008) Is log ratio a good value for measuring return in stock investments? In: Fink A, Lausen B, Seidel W, Ultsch A (eds) Advances in data analysis, data handling and business intelligence. Springer, Berlin, pp 203–210

## 8 Conclusion

References

# *p*-adic Methods in Stereo Vision

**Patrick Erik Bradley**

**Abstract** The so-called *essential matrix* relates corresponding points of two images from the same scene in 3D, and allows to solve the relative pose problem for the two cameras up to a global scaling factor, if the camera calibrations are known. We will discuss how *Hensel's lemma* from number theory can be used to find geometric approximations to solutions of the equations describing the essential matrix. Together with recent *p*-adic classification methods, this leads to RanSaC$_p$, a *p*-adic version of the classical RANSAC in stereo vision. This approach is motivated by the observation that using *p*-adic numbers often leads to more efficient algorithms than their real or complex counterparts.

## 1 Introduction

According to (Murtagh 2004), ultrametricity is pervasive in observational data, and this offers computational advantages and a well understood basis for developing data processing tools originating in *p*-adic arithmetic. Consequently, *p*-adic data encoding becomes necessary. In (Bradley 2008) it has been shown that the choice of the prime number $p$ is arbitrary. Hence $p = 2$ can be taken, which is usually the computationally most advantageous prime number. In particular, the *p*-adic Newton iteration method, known in number theory as Hensel's lemma, is most efficient for $p = 2$. We will use this method in order to pave the way for computationally efficient methods for solving the relative pose problem from five corresponding points in stereo vision.

A well known hierarchical image encoding procedure is the (regular) *quadtree*. We will show that it has natural 2-adic encodings which allow to view gray-scale

P. E. Bradley (✉)
Institut für Photogrammetrie und Fernerkundung, Karlsruhe Institut für Technologie (KIT),
Englerstr. 7, 76128 Karlsruhe, Germany
e-mail: bradley@kit.edu

W. Gaul et al. (eds.), *Challenges at the Interface of Data Analysis, Computer Science, and Optimization*, Studies in Classification, Data Analysis, and Knowledge Organization, DOI 10.1007/978-3-642-24466-7_18, © Springer-Verlag Berlin Heidelberg 2012

images as real-valued functions on $p$-adic spaces. This should be understood as an invitation to develop image processing methods originating in $p$-adic functional analysis. In any case, image coordinates are $p$-adic numbers in this situation. Although computationally efficient, the quadtree suffers somewhat from its rigidity when it comes to handling measurement errors. We expect that taking families of 2-adic encodings corresponding to small euclidean perturbations will lead to a dynamic treatment of single images which can overcome this drawback without losing too much computational efficiency.

An Introduction to $p$-adic numbers is e.g. (Gouvêa 1993).

## 2  $p$-Adic Numbers

Kurt Hensel's important contribution to number theory was to view numbers as analytic functions on some imagined "Riemann surface". In this imaginary situation, the "places" are given by the prime numbers $p$ which play the role of a local coordinate[1], and then the number $n$ has "locally" a unique power series expansion

$$n = \sum_{\nu=0}^{\infty} n_\nu p^\nu,$$

which in the case of natural numbers $n$ is in fact a finite expansion with coefficients $n_\nu \in \{0, \ldots, p-1\}$. The $p$-adic metric is given by the length of the common initial part:

$$|n - m|_p = p^{-\nu}, \tag{1}$$

if $m = n_0 + \cdots + n_{\nu-1}p^{\nu-1} + m_\nu p^\nu + \ldots$ and $m_\nu \neq n_\nu$. This is an ultrametric, i.e. the strict triangle inequality

$$|x + y|_p \leq \max \{|x|_p, |y|_p\}$$

holds true. Allowing infinite expansions (1) means completion with respect to the $p$-adic metric, and the completed space $\mathbb{Z}_p$ of $p$-adic integers contains the usual integers $\mathbb{Z}$ as a dense subset. Examples of negative numbers are

$$\sum_{\nu=0}^{\infty} p^\nu = \frac{1}{1-p}, \qquad \sum_{\nu=0}^{\infty} (p-1)p^\nu = -1$$

---

[1]In fact, this dream became true thanks to Grothendieck's concept of *scheme*: The "Riemann surface" is the affine scheme Spec $\mathbb{Z}$, the space whose points are the prime ideals $p\mathbb{Z}$ for $p = 0$ or a prime number. Cf. e.g. (Hartshorne 1993)

The primality of $p$ guarantees that there are no zero-divisors in $\mathbb{Z}_p$, and the field of fractions $\mathbb{Q}_p$ can be formed which densely contains the rational numbers $\mathbb{Q}$. Just like in the function-theoretic case, the $p$-adic numbers thus correspond to the meromorphic functions:

$$\mathbb{Q}_p = \left\{ \sum_{\nu=-N}^{\infty} x_\nu p^\nu \mid x_\nu \in \{0, \dots, p-1\} \right\}$$

and have a "Laurent series" expansion. Observe further that $\mathbb{Z}_p$ is the $p$-adic unit disk:

$$\mathbb{Z}_p = \{x \in \mathbb{Q}_p \mid |x|_p \leq 1\},$$

and we have in $\mathbb{Q}_p$ an ultrametric space on which calculus can be performed.

$p$-adic approximation is given by finite expansions: $x = x_0 + \dots x_{n-1} p^{n-1} +$ higher order terms. That cut-off can be written by a congruence

$$x \equiv x_0 + \dots + x_{n-1} p^{n-1} \mod p^n, \tag{2}$$

from wich it follows that the $p$-adic expansion of $x$ is given by an infinite sequence of congruences (2) with $n = 1, 2, 3, \dots$. And indeed,

$$\left| x - \sum_{\nu=0}^{n-1} x_\nu p^\nu \right|_p \leq p^{-n},$$

we have convergence of these finite expansions to $x$ for $n \to \infty$.

At last we remark that $\mathbb{Q}_p$ is endowed with a Haar measure $dx$ such that $\int_{\mathbb{Z}_p} dx = 1$, i.e. the unit disk has volume 1.

## 3   *p*-Adic Encoding of Images

A 2-adic encoding of square $2^N \times 2^N$-images can be obtained by a hierarchical subdivision as in Fig. 1. Essentially, there are two approaches for the encoding. In the *bottom-up encoding*, the squares at highest resolution are assigned to level $N$,

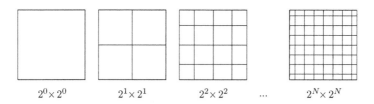

$2^0 \times 2^0$        $2^1 \times 2^1$        $2^2 \times 2^2$        $\dots$        $2^N \times 2^N$

**Fig. 1** Hierarchical subdivision of an image

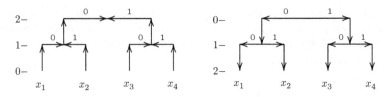

**Fig. 2** Left: bottom-up encoding, right: top-down encoding of a dendrogram

with decreasing level at higher hierarchy, level 0 representing the full image cluster. The encoding scheme for the $x$-coordinate is to traverse a path from bottom to top, and collect a coefficient $a_\nu = 0$ for each right turn, and $a_\nu = 1$ for left turns. This yields the expansion

$$x = \sum_{\nu=0}^{N} a_\nu 2^{-\nu}.$$

Figure 2 (left) exemplifies this with

$$x_1 = 0, \qquad x_2 = 2^{-1}, \qquad x_3 = 2^{-2}, \qquad x_4 = 2^{-2} + 2^{-1}.$$

The intensities (gray values) on the image grid can be viewed as *locally constant* functions $f: \mathbb{Q}_p \to \mathbb{R}$, as vertices in the dendrogram at level $\nu$ can be viewed as $p$-adic disks of radius $p^\nu$. With the bottom-up encoding the functions are constant on all translates of the unit disk $\mathbb{Z}_p$, and methods from $p$-adic functional analysis are ready for application. The functions which are constant on each translate $x + \mathbb{Z}_p$ of the unit disk are in one-one correspondence with functions on the co-set space $\mathbb{Q}_p / \mathbb{Z}_p$.

*Example 1 (p-adic diffusion).* $p$-adic diffusion can be described as a symmetric jump process on $\mathbb{Q}_p / \mathbb{Z}_p$, with the transition probability $P_{xy}$ depending only on their $p$-adic distance:

$$P_{xy} = \rho(|x - y|_p).$$

The equation describing the evolution of probabilities is given by

$$\frac{\mathrm{d}}{\mathrm{d}t} f(x, t) = \int_{\mathbb{Q}_p} (f(y, t) - f(x, t)) \rho(|x - y|_p) \, \mathrm{d}y.$$

By taking as integral kernel the function

$$\rho(|x - y|_p) = \frac{|x - y|_p^{-1-\alpha}}{\Gamma_p(-\alpha)}$$

with $\alpha > 0$, one obtains $p$-adic Brownian motion. Here, the role of the Laplace operator is played by the *Vladimirov operator*

$$D^\alpha f(x,t) = \frac{1}{\Gamma_p(-\alpha)} \int_{\mathbb{Q}_p} \frac{f(x) - f(y)}{|x - y|_p^{1+\alpha}} \, dy,$$

and $\Gamma_p(\alpha) = \frac{1-p^{\alpha-1}}{1-p^{-\alpha}}$ is the *p*-adic gamma function. The diffusion equation

$$\frac{d}{dt} f(x,t) = -D^\alpha f(x,t)$$

can be viewed as the *p*-adic analogue of a scale-space equation, where $t$ plays the role of the scaling parameter. However, a *p*-adic scale-space theory, from which feature detectors and descriptors can be derived, has yet to be developed. More on *p*-adic pseudo-differential equations, functional analysis and mathematical physics can be found in (Vladimirov et al. 1994).

The *top-down encoding* reverses the order of bottom-up, and expansion is in positive powers of 2. This yields 2-adic integers for image coordinates, which turns out useful in the following section. In Fig. 2 (right) one obtains

$$x_1 = 0, \qquad x_2 = 2^1, \qquad x_3 = 2^2, \qquad x_4 = 2^1 + 2^2.$$

## 4   Epipolar Geometry

*Epipolar geometry* is the geometry of two pinhole cameras (Hartley and Zisserman 2008). These are described as projective maps $P_L, P_R : \mathbb{P}^3 \to \mathbb{P}^2$ between projective spaces. Imagine the two cameras viewing the same point $x$ in a 3-dimensional scene. The plane spanned by $x$ and the two camera centers $O_L, O_R$ is called the *epipolar plane*. This plane cuts out in each of the cameras' image planes a line, called the *epipolar line*. To the projection point $x_L = P_L x$ of $x$ onto the left image corresponds the line through $x_R = P_R x$ and $e_R$, where $e_R$ is the intersection of the line $O_L O_R$ with the right image plane. This line is given by

$$(e_R \times x_R)^T x_R = 0.$$

Writing the vector product as multiplication with an anti-symmetric matrix $S_R$ and taking the pseudo-inverse of $P_L$ yields $e_R \times x_R = S_R x_R$, and $x = P_L^+ x_L$. Hence, we obtain

$$0 = (e_R \times x_R)^T x_R = x_R^T (e_R \times x_R) = x_R^T \underbrace{S_R P_R P_L^+}_{=:F} x_L, \tag{3}$$

where $F$ is the *fundamental matrix*. This matrix encodes the relative motion between the two cameras, together with their intrinsic parameters given by the calibration matrices $K_L, K_R$: $F = K_L^{-1} E K_R$. The matrix $E$ is the *essential matrix* and is the fundamental matrix of two *normalised* cameras, which corresponds to the case where both calibrations are known. The essential matrix decomposes into a rotation $R$ and a translation $t$: $E = R[t]_\times$, where $[t]_\times$ is the matrix for the cross product with $t$. This decomposition can be effected, and is unique upto a sign ambiguity. The problem of stereo vision is thus reduced to the estimation of $E$ from two given images of the same scene.

The essential matrix is a projective $3 \times 3$-matrix. Hence, 8 independent equations of the form (3) suffice for determining $E$. The first algorithm starts with

$$x_1^T E x_1' = 0, \quad \dots \quad x_8^T E x_8' = 0,$$

where $(x_i, x_i')$ are 8 generic pairs of corresponding points in the two images.

Although the 8-point algorithm works, its resulting matrix $E$ is not always an essential matrix. This is because further constraints have to be met. For example, from (3) it can be easily seen that an essential matrix must fulfill

$$\det E = 0,$$

which leads to the 7-point algorithm.

The minimal number of point correspondences needed for determining the essential matrix is five. This leads in the generic case to a four-dimensional solution space. Writing the general solution as

$$E = u_1 E_1 + u_2 E_2 + u_3 E_3 + u_4 E_4,$$

one obtains homogeneous linear polynomials in four variables $u_1, \dots, u_4$. The constraints for $E$ are given by the matrix equation

$$E E^T E - \frac{1}{2} \text{Trace}(E E^T) E = 0 \qquad (4)$$

which describes a space $M_E$ containing all essential matrices (Demazure 1988). Hence, one is left with solving a system of homogeneous cubic equations in four variables. Demazure showed that there are up to 10 complex solutions to (4). Nistér's first five-point algorithm reduces the system (4) to a univariate equation of degree 10 which then has to be solved numerically (Nistér 2004).

The importance of having the minimal number of point correspondences in order to find finitely many exact solutions to the problem lies in the fact that outliers generally lead to bad results. Higher stability is obtained by a *Random Sample Consensus* (RanSaC). Here, a random sample of 5 pairs of corresponding points in general position is chosen. The up to 10 real candidate essential matrices are collected, and another sample is taken, etc. After sufficiently many samples, the candidate matrix whose $\epsilon$-neighbourhood contains the highest number of solutions from all samples is declared as the "essential matrix" for the stereo vision problem.

In essence, a classification of the candidates is performed, and the biggest cluster contains the winning candidate.

## 5   Hensel's Lemma and RanSaC$_p$

The projective cameras can also be *p*-adic projective maps $\mathbb{P}^3(\mathbb{Q}_p) \rightarrow \mathbb{P}^2(\mathbb{Q}_p)$. Thus we are lead to estimate *p*-adic essential matrices. The notions of translation and rotation make sense *p*-adically. The latter is defined algebraically: namely a $3 \times 3$-matrix $R$ satisfying

$$R\,R^T = 1, \quad \det R = 1.$$

The top-down encoding from Sect. 3 ensures coordinates from $\mathbb{Z}_p$. In particular, the coefficients of (4) are *p*-adic integers. In this case, Hensel's lifting lemma can be used for *p*-adically approximating the solutions of (4). The point is that a sequence of linear congruences modulo $p^k$ is iteratively solved, which makes the procedure to a *p*-adic version of Newton's method. For convenience, we formulate the lemma in the case of *n* equations in *n* variables in the most familiar version:

**Lemma 5.1 (Hensel)** *Let* $\mathbf{f}(\mathbf{X}) = (f_1(X_1,\ldots,X_n),\ldots,f_n(X_1,\ldots,X_n))$ *be an n-tuple of polynomials in n variables with coefficients from* $\mathbb{Z}_p$. *Let* $\mathbf{a} \in \mathbb{Z}_p^n$ *such that*

$$\mathbf{f}(\mathbf{a}) \equiv 0 \ \mathrm{mod}\ p, \quad and \quad \det \frac{\partial \mathbf{f}}{\partial \mathbf{X}}(\mathbf{a}) \not\equiv 0 \ \mathrm{mod}\ p.$$

*Then there is a unique solution* $\mathbf{a}'$ *of* $\mathbf{f}$ *near* $\mathbf{a}$, *i.e.* $\mathbf{f}(\mathbf{a}') = 0$ *with* $\mathbf{a}' \equiv \mathbf{a} \ \mathrm{mod}\ p$.

A formulation of Lemma 5.1 for the case of possibly more variables than equations is proven in (Bourbaki 1962, Sect. III.4.5, Cor. 2). However, the condition on the Jacobi matrix to be invertible can be relaxed. In fact, a more general multivariate criterion for unique lifting is formulated in (Fisher 1997), where the proof of its correctness is given in the context of rings which are complete with the topology defined by an ideal (polynomial rings with coefficients in $\mathbb{Z}_p$ are a valid example). Unfortunately, it is not always the case that solutions modulo *p* lift to *p*-adic solutions. For example, the equation $x^2 - 5 = 0$ is known to have no solution in $\mathbb{Z}_2$, because any 2-adic *unit* $u$ (i.e. satisfying $u \equiv 1 \ \mathrm{mod}\ 2$) is a square if and only if $u \equiv 1 \ \mathrm{mod}\ 8$ (Gouvêa 1993, Sect. 3.4, Prob. 116). But the equation modulo 2 reads $x^2 + 1 = 0$, and this has the solution $x = 1$ modulo 2.

RanSaC$_p$, the *p*-adic version of Random Sample Consensus (Fischler and Bolles 1981), collects in a set $\mathscr{E}$ all successfully lifted candidate essential matrices obtained from all random samples of five corresponding pairs of image points, and performs a *p*-adic classification. The central elements of the largest cluster determine the choice of solution for the problem.

The classification method proposed in (Bradley 2010) is to find a clustering of $\mathscr{E}$ by minimising, with fixed upper bound $k$ for the number of clusters, the quantity

$$\epsilon_p = \epsilon_p(\mathscr{E}, \mathscr{C}, \mathbf{a}) = \sum_{C \in \mathscr{C}} \sum_{a \in C} \|a - a_C\|_p,$$

where $\mathscr{C} = \{C_1, \ldots, C_\ell\}$, $\ell \leq k$, is a clustering of $\mathscr{E}$, $\mathbf{a} = (a_C)_{C \in \mathscr{C}}$ with $a_C \in C$, and

$$\|(x_1, \ldots, x_n)\|_p = \max\{|x_1|_p, \ldots, |x_n|_p\}$$

is the maximum norm on $\mathbb{Q}_p^n$ (in the present case $n = 9$). The algorithm is a $p$-adic adaptation of the classical split-LBG (Linde, Buzo, Gray) (Linde et al. 1980), a hierarchical version of $k$-means by splitting cluster centers in two and regrouping the cluster around the new centers. The $p$-adic adaptation is called $\mathrm{LBG}_p$ and first splits clusters by replacing vertices in the dendrogram by their children, and afterwards finding centers $a_C$ in cluster $C$ which further minimise $\epsilon_p$ (cf. (Bradley 2009) for details). The result certainly depends on the bound $k$, and if $k$ is too large then all clusters are singletons. In (Bradley 2010), a $p$-adic intra-inter-validity measure is proposed for determining ideal values of $k$.

In the event that there is not a unique biggest cluster, the cluster with highest density can be chosen. This is measured by

$$\delta(C) = \begin{cases} \dfrac{|C| - 1}{\mu(C)}, & |C| > 1 \\ 0, & \text{otherwise} \end{cases}$$

where

$$\mu(C) = \int_{\mathbb{Q}_p^n} 1_{BC} \, dx$$

with $1_{BC}$ the indicator function of the smallest ball in $\mathbb{Q}_p^n$ containing $C$. If two clusters have the same density, the *cluster precision* can be taken as a further tie-breaking rule. It is given by

$$\pi(C) = \frac{1}{\mu(C_c)},$$

where $C_c$ is the intersection of $C$ with the smallest ball in $\mathbb{Q}_p$ containing the centers of $C$.

## 6  Discussion and Outlook

The application of $\mathrm{RanSaC}_p$ for determining the essential matrix in stereo images classifies the $p$-adic lifts of solutions for the equations modulo $p$ obtained by a generalisation of Hensel's lemma 5.1. "Bad" samples are those which do not lead to liftable solutions, and these have to be discarded. However, by (Demazure 1988, Sect. 6), the system (4) describes the intersection points of a union of 4 planes and 3 conics with a linear space inside projective space. A careful analysis of the coefficients yields that the probability of having no liftablity in any of the conic intersections is about 14.8%. This compares well with the real case, where the worst case comes in 12.5% of all cases. Since prime modular arithmetic as a function of the prime number $p$ can be assumed to have an asymptotic time complexity of $O(\log_2 p)$, we can expect the $p$-adic 5-point algorithm to outperform its classical counterpart in terms of efficiency (here we view the real case as arithmetic modulo some large prime $p$). The details of the coefficient analysis will be published elsewhere, and experimental verification is work in progress.

Another issue is that of registration errors. Namely, erroneous point correspondences in the two images lead to erroneous essential matrices. The $p$-adic approach has the drawback that small euclidean inaccuracies can lead to large $p$-adic errors. The research problem is how to overcome this drawback. A promising idea seems for us to study the variation of the 10 points in Demazure's space $M_E$ under the action of the translation group $x \to x + \epsilon$. Namely

$$f(x) \to f(x + \epsilon) \tag{5}$$

amounts to a shift in the division points for the quadtree-like subdivision underlying the encoding. The error $\epsilon$ is controlled by the (inverse) *Monna map* to the real numbers:

$$\epsilon = \sum \epsilon_\nu p^\nu \mapsto M(\epsilon) = \sum \epsilon_\nu p^{-\nu-1},$$

where $M(\epsilon)$ is to be taken small. Here, it is quite tempting to view (5) as part of the Weyl representation of $p$-adic quantum mechanics (Vladimirov and Volovich 1989) (cf. (Takhtajan 2008) for a general introduction), which in particular for processing complex-valued images can lead to exciting new $p$-adic methods.

# 7  Conclusion

Viewing the hierarchical world as ultrametric leads to the consideration of $p$-adic methods for detecting and processing hierarchies. For this, $p$-adic data encoding becomes indispensable. This applied to images yields encodings of special quadtrees, known in image processing. The bottom-up method introduced here opens the way for methods from $p$-adic mathematical physics, whereas the top-down method renders $p$-adic integers as image coordinates. The latter allowed the use of Hensel's lifting lemma to the equations arising in the problem of finding the essential matrix from five point-correspondences in stereo vision.

$p$-adic classification algorithms are known to be more efficient than their classical counterparts. Hence, it is natural to use a recently developed $p$-adic method as part of RanSaC$_p$, a $p$-adic form of the Random Sample Consensus applied to the five-point relative pose problem in order to find the "best" $p$-adic approximation to the essential matrix as the one lying centrally in the biggest cluster.

**Acknowledgements** The author thanks Sven Wursthorn and Boris Jutzi for introducing him to the topic of computer vision, and for valuable discussions. An anonymous referee is thanked for valuable remarks leading to a significant improvement of the exposition.

# References

Bourbaki N (1962) Algbèbre commutative. Hermann, Paris

Bradley PE (2008) Degenerating families of dendrograms. J Classification 25:27–42

Bradley PE (2009) On $p$-adic classification. $p$-Adic Numbers Ultrametric Anal Appl 1:271–285

Bradley PE (2010) A $p$-adic RANSAC algorithm for stereo vision using Hensel lifting. $p$-Adic Numbers Ultrametric Anal Appl 2:55–67

Demazure M (1988) Sur deux probleémes de reconstruction. INRIA Rapports de Recherche 882

Fischler MA, Bolles RC (1981) Random sample consensus: A paradigm for model fitting with applications to image analysis and automated cartography. Comm ACM 24:381–395

Fisher B (1997) A note on Hensel's lemma in several variables. Proc AMS 125:3185–3189

Gouvêa FQ (1993) $p$-adic numbers. An introduction. Springer, Berlin

Hartley R, Zisserman A (2008) Multiple view geometry in computer vision. Cambridge University Press, Cambridge

Hartshorne R (1993) Algebraic geometry. Springer, New York

Linde Y, Buzo A, Gray RM (1980) An algorithm for vector quantizer design. IEEE T Commun 28:84–94

Murtagh F (2004) On ultrametricity, data coding, and computation. J Classification 21:167–184

Nistér D (2004) An efficient solution to the five-point relative pose problem. IEEE T Pattern Anal 26:167–184

Takhtajan LA (2008) Quantum mechanics for mathematicians. AMS, USA

Vladimirov VS, Volovich IV (1989) $p$-adic quantum mechanics. Commun Math Phys 123:659–676

Vladimirov VS, Volovich IV, Zelenov YeI (1994) $p$-Adic analysis and mathematical physics. World Scientific, Singapore

# Individualized Error Estimation for Classification and Regression Models

Krisztian Buza, Alexandros Nanopoulos, and Lars Schmidt-Thieme

**Abstract** Estimating the error of classification and regression models is one of the most crucial tasks in machine learning. While the global error is capable to measure the quality of a model, local error estimates are even more interesting: on the one hand they contribute to better understanding of prediction models (where does and where does not work the model well), on the other hand they may provide powerful means to build successful ensembles that select for each region the most appropriate model(s). In this paper we introduce an extremely localized error estimation, called *individualized error estimation* (IEE), that estimates the error of a prediction model $M$ for each instance $x$ individually. To solve the problem of individualized error estimation, we apply a meta model $M^*$. We systematically investigate various combinations of elementary models $M$ and meta models $M^*$ on publicly available real-world data sets. Further, we illustrate the power of IEE in the context of time series classification: on 35 publicly available real-world time series data sets, we show that IEE is capable to enhance state-of-the art time series classification methods.

## 1 Introduction

Error estimation is one of the most crucial tasks in machine learning. For measuring the overall quality of a model, global error estimations are used, while for the task of analyzing the behavior of a model in different regions of the input space, local error estimations can be performed. This may be interesting on its own, as it contributes to the better understanding of prediction models. Furthermore, by allowing for the selection of the most appropriate model(s) for each region of the input space,

K. Buza (✉) · A. Nanopoulos · L. Schmidt-Thieme
Information Systems and Machine Learning Lab (ISMLL), University of Hildesheim
Marienburger Platz 22, 31141 Hildesheim, Germany
e-mail: buza@ismll.de; nanopoulos@ismll.de; schmidt-thieme@ismll.de

W. Gaul et al. (eds.), *Challenges at the Interface of Data Analysis, Computer Science, and Optimization*, Studies in Classification, Data Analysis, and Knowledge Organization, DOI 10.1007/978-3-642-24466-7_19, © Springer-Verlag Berlin Heidelberg 2012

local error estimation provides powerful means to build ensembles of classifiers or regressors.

In this paper we focus on local error estimation techniques. We introduce the notion of an extremely localized error estimation, called *individualized error estimation* (IEE), that estimates the error of a prediction model $M$ for each instance $x$ individually. To solve the problem of individualized error estimation, we apply a meta model $M^*$. We systematically investigate different combinations of elementary models $M$ and meta models $M^*$ on publicly available real-world data sets.

Furthermore, we show how to exploit IEE's power in the context of time series classification in order to enhance state-of-the art models. We evaluate our approach on 35 publicly available real-world time series data sets. The results show that our approach outperforms state-of-the art time series classification methods.

The paper is organized as follows. In Sect. 2 we introduce individualized error estimation. In Sect. 3 we systematically investigate IEE for various combinations of elementary models and meta models. In Sect. 4 we describe an application of IEE to time series classification, we also present our experimental results. After summarizing related work in Sect. 5, we conclude in Sect. 6.

## 2 Individualized Error Estimation

We illustrate IEE in context of a simple binary classification task of a 2-dimensional data set. Figure 1 depicts a set of labeled instances from two classes that are denoted by triangles and circles. The density in the class of triangles (upper region) is larger than in the class of circles (lower region). We consider two test instances, denoted as "1" and "2", that have to be classified. We also assume that the ground-truth considers test instance "1" as a triangle, and "2" as a circle. Suppose, we use nearest neighbor (NN) models to classify the test instances.

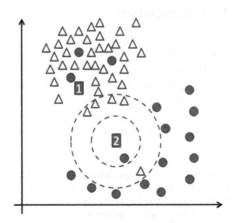

**Fig. 1** A two-dimensional binary classification task. Assuming test instance "1" to be a triangle, and "2" to be a circle (ground-truth), we observe that 1-NN classifies "2" correctly, but misclassifies "1", while 6-NN outputs the appropriate class for "1", and misclassifies "2"

**Table 1** Classification of test instances in the example

| Instance | Ground-truth | Classification with 1-NN | | Classification with 6-NN | |
|---|---|---|---|---|---|
| | | $M_1$ | $M_1^*$ | $M_2$ | $M_2^*$ |
| 1 | Triangle | Circle | Incorrect | Triangle | Correct |
| 2 | Circle | Circle | Correct | Triangle | Incorrect |

Both for "1" and "2", the first nearest neighbor is a circle. Thus, the 1-NN classifies "1" incorrectly, while "2" is classified correctly. However, using e.g. a 6-NN classifier, due to the lower density in the circles' class, we observe "2" to be misclassified (see the large dashed circle around "2"), while "1" is classified correctly.

In this example we are concerned with two models: $M_1$: 1-NN and $M_2$: 6-NN. The perfect meta-model for $M_1$, denoted as $M_1^*$, would output that $M_1$ classifies the first test instance incorrectly, while it classifies the second test instance correctly (see 4th column of Table 1). Similarly, $M_2^*$ (the perfect meta-model for $M_2$) would output that $M_2$ classifies the first test instance correctly, while it classifies the second test instance incorrectly (see the last column of Table 1).

In this simple example, the output of meta-models $M_1^*$ and $M_2^*$ consists of binary decisions whether the classifications by the elementary models $M_1$ and $M_2$ are correct or not. Please note, that one can develop more advanced meta-models, that output e.g. the likelihood (probability) of the correct decision, or, if we use a regression model at the elementary level, the meta-model could output the residuals (difference between predicted and true label).

**Problem formulation** Given a model $M$ and a set of instances $S$ ($M$ predicts the labels of $S$), the task of Individualized Error Estimation (IEE) is to develop a meta-model $M^*$ that is able to estimate the error of $M$ for each instance of $S$ individually.

Various versions of this task can be formulated (both in the context of classification and regression), and this defines the exact meaning of *error* in the above definition. Some of these possible versions include:

1. $M$ is a classification model, and (as in the example above) the meta-model $M^*$ takes binary decisions whether the classification by $M$ is correct or not,
2. $M$ is a classification model, and the meta-model $M^*$ estimates for each instance the likelihood (probability) of the classification being correct,
3. $M$ is a regression model, and $M^*$ estimates the residuals (difference between predicted and true label) for each instance.

This generic definition allows for various classification and regression models, in this paper we are going to explore just a discrete set of models.

**Our approach for IEE** Figure 2 summarizes the training procedure of our approach. Here, we describe the approach in the context of residual estimation, but it can simply be adapted for other versions of the task. The major steps are:

1. Split the labeled training data into two subsets $D_A$ and $D_B$.
2. Train the elementary model $M$ on $D_A$.

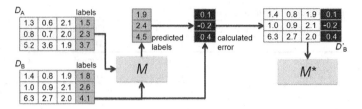

**Fig. 2** Training procedure of our approach

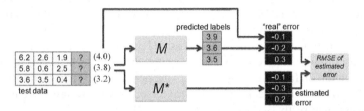

**Fig. 3** Evaluation of IEE

3. Let $M$ predict the labels of $D_B$.
4. As the true labels of $D_B$ are known, we can calculate the error of the predicted labels. In the case of residual estimation, when the labels are continuous, the calculated error $e(i)$ of an instance $i \in D_B$ is: $e(i) = M(i) - l(i)$, where $M(i)$ denotes the prediction of the elementary model $M$ and $l(i)$ is the true label of $i$.
5. Train $M^*$ on $D_B$ using the calculated errors as labels.

When an unlabeled instance $i'$ is processed, the elementary model $M$ predicts its label, while the meta-model $M^*$ estimates the error. In the case of residual estimation, the final predicted label of $i'$ is: $l(i')^{final} = M(i') + M^*(i')$, where $M(i')$ denotes the prediction of the elementary model $M$, and $M^*(i')$ denotes the residual estimated by $M^*$ for instance $i'$.

## 3   IEE with Various Models

Our approach (Sect. 2), is generic as it allows to apply various classification and regression models both at the elementary level (as $M$) and at the meta-level (as $M^*$). Whenever a particular choice of $M$ and $M^*$ is made, we would like to evaluate how successfully $M^*$ could predict the errors of $M$. As this evaluation procedure is non-trivial, we continue by describing our evaluation protocol.

In order to evaluate our approach (see Fig. 3), we first train both $M$, and $M^*$ on training data as described in Sect. 2. Then, using $M^*$, we estimate the error for each instance of the disjoint test data $D_{Test}$. We also predict the labels of $D_{Test}$ using $M$.

Comparing the predicted labels to the ground-truth of the test data, we can calculate the true errors of the predicted labels. Finally, we compare the estimated errors to the true errors. When doing so, we calculate RMSE (root mean squared error) between the vector of true errors and the vector of estimated errors.

In our first experiment, we investigated various combinations of elementary models $M$ and meta models $M^*$ in the residual estimation setting as described above. We used the following publicly available real-world data sets from the UCI repository (Frank and Asuncion 2010): (a) Communities, (b) WineQuality (both red wines and white wines) and (c) Parkinson (both targets: motoric abilities and total abilities). As we observed very similar trends on all data sets, we only report the results on the Communities data set. As a simple baseline we used the meta-model $M^*_{bl}$ that estimates that the prediction of the elementary model $M$ was always correct.

We performed 10-fold-cross-validation: in each round the entire data is divided into 10 splits, out of which 1 serves as test data ($D_{Test}$), the other 9 splits are used as training data. Out of the 9 training splits, 5 splits constituted $D_A$ and the remaining 4 splits belonged to $D_B$ (see Sect. 2 for $D_A$, $D_B$ and the training procedure).

Figure 4 summarizes our results: it shows for all the examined combinations of models and meta-models, in how many folds our approach was better than the baseline. The applied models are listed on the right of the figure, we used the implementations from the WEKA software package (Hall et al. 2009). In the matrix, the horizontal dimension corresponds to the applied meta-model $M^*$, while the elementary models are listed along the vertical dimension. For example, the 4th column position in the 2nd row corresponds to the combination where the model is $k$-NN and the meta-model is LWL. The color of the cell shows how many times (in how many rounds of the 10-fold-cross-validation) our trained meta-model $M^*$ was better than the baseline $M^*_{bl}$. Black cells mean that $M^*$ is better than $M^*_{bl}$ (almost) always (10 or 9 times); dark gray cells indicate that $M^*$ is better than $M^*_{bl}$ 8 or 7 times, while light gray cells denote $M^*$ being better than $M^*_{bl}$ 6 or 5 times.

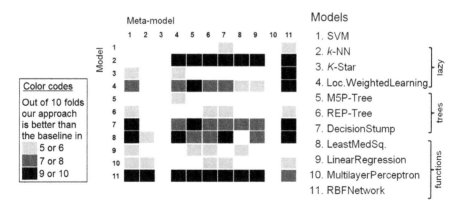

**Fig. 4** Results: IEE with various combinations of models and meta-models (data: Communities). We used the same model types at the elementary and meta levels. A detailed description of the models can be found in the WEKA software package (Hall et al. 2009)

As a general observation, we see that for (almost) all of the elementary models, there are meta-models that are capable to deliver (relatively) good error estimations, although the particular choice of the meta-model is important: some of the meta-models deliver very poor estimations. As further observation, we see, that RBF-Networks and SVMs seem to work generally well as meta-models.

## 4 Time Series Classification Using IEE

In our second experiment, we aimed at improving the accuracy of time series classification with IEE. In the time series domain, according to recent works, e.g. (Xi et al. 2006; Ding et al. 2008), simple nearest neighbor classifiers (using Dynamic Time Warping as distance measure) are generally very difficult to outperform (if it is possible at all). On the other hand, the ideal number of nearest neighbors, $k$, is non-trivial to be chosen and it may vary from region to region, just like in the example in Sect. 2. Therefore, as elementary models $M_i$, we implemented DTW-based nearest neighbor classifiers for time series with odd $k$ values between 1 and 10, i.e. $M_1$: 1-NN-DTW, $M_2$: 3-NN-DTW, ..., $M_5$: 9-NN-DTW. For each of them, we implemented a meta-model $M_i^*$, that aimed at estimating the likelihood of the error of $M_i$, $1 \leq i \leq 5$. This schema is depicted in Fig. 5. We trained the elementary classifiers $M_i$ and the corresponding meta-models $M_i^*$ as described in Sect. 2.

Note, that in contrast to the previous experiment, where residuals were estimated, here, we apply IEE to estimate the error likelihood (see 2nd version of the IEE-problem in Sect. 2). Therefore the training procedure is slightly different: when calculating the individual error of each predicted label in step 4 of our approach, as described in Sect. 2, the error is considered as 0, if the classification produced by $M$ was correct, otherwise the error is considered as 1. These calculated errors serve as meta-level labels. At the meta-level, we use nearest neighbor regression models, that simply average the meta level labels of the nearest neighbors. While the number of nearest neighbors at the elementary level is non-trivial to be selected, we observed that the meta-model for error likelihood estimation is much less sensitive to the

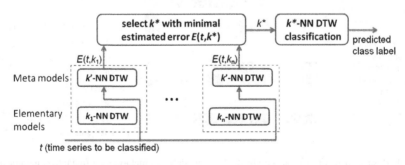

**Fig. 5** IEE for time series classification

**Table 2** Summary of our results for time series classification: number of cases where our approach wins / looses (significantly wins / looses in parenthesis) against the baselines

|                             | $p = 1\%$ | $p = 5\%$ | $p = 10\%$ | total    |
|-----------------------------|-----------|-----------|------------|----------|
| Wins against 1-NN-DTW       | 30 (20)   | 34 (29)   | 34 (31)    | 98 (80)  |
| Looses against 1-NN-DTW     | 5 (1)     | 1 (0)     | 1 (0)      | 7 (1)    |
| Wins against $k$-NN-DTW     | 30 (15)   | 30 (9)    | 28 (14)    | 88 (38)  |
| Looses against $k$-NN-DTW   | 5 (1)     | 5 (1)     | 7 (1)      | 17 (3)   |

number of nearest neighbors, $k'$, and $k' = 5$ (almost) always leads to appropriate error likelihood estimation, therefore we fixed $k' = 5$ for all the meta-models $M_i^*$.

When classifying a test time series $t$, the meta-models estimated the error likelihood for each elementary model $M_1$: 1-NN-DTW, ..., $M_5$: 9-NN-DTW. Then we selected the elementary model with minimal estimated error likelihood and used it to classify $t$. (In this final step of predicting the class label of a test time series $t$, all the training data, $D_A \cup D_B$, is used by the selected nearest neighbor model.)

In our experiments we used 35 publicly available time series data sets from the collection used in (Ding et al. 2008). We omitted 3 data sets due to their tiny sizes: each of them contained less than 100 observations. We performed 10-fold-cross-validation. As baselines we used 1-NN-DTW and $k$-NN-DTW with globally best $k$ selected on a hold-out subset of the training data (after selecting the globally optimal $k$, all the training data is used to classify test time series). We tested statistical significance using t-test at level 0.05. For the original data sets, in the majority of the cases, we did not observe significant differences (often all approaches, ours and both baselines performed very well). As recent research showed, bad hubs[1] are responsible for a surprisingly high portion of the error (Radovanovic et al. 2010). Thus, in order to make the task more challenging, we artificially introduced some bad hubs: we changed the best hubs to bad ones (set the class labels to an artificial "noise" class) by infecting in total $p$ % of the entire data set. Table 2 summarizes our results for 3 different levels of $p$. In the table we report in how many cases our approach wins/looses against the baselines, in parenthesis we report in how many cases the wins/looses are statistically significant.

## 5   Related Work

Error-prediction methods are usually applied globally in order to estimate the overall performance of a classification model (Molinaro et al. 2005; Jain et al. 1987). Closely related to ours is the work of Tsuda et al. (Tsuda et al. 2001),

---

[1]Hubs are time series that appear most frequently as nearest neighbors of other time series. Denote the set of time series for which $t$ is the nearest neighbor as $N_t$. A hub $t$ is a bad hub if its class label is different from the class labels of *many* time series in $N_t$. See also (Radovanovic et al. 2010).

who proposed an individualized approach for predicting the leave-one-out error of vector classification with support vector machines (SVM) and linear programming machines (LPM). Compared to this work, our proposed approach is more generic, as (a) we did not only focus on the overall leave one out error, (b) we did not only focus on vector classification (but also allow for more complex structures like time series), and, most importantly, (c) in the current work we have shown how to exploit IEE to enhance classification.

IEE is also related to boosting, where residuals are estimated in order to enhance classification like in (Duffy and Helmbold 2002). However, IEE is not limited to the estimation of residuals and in this sort of sense IEE is more generic than boosting. Moreover, in the case of boosting a (long) series of (usually weak) models of the same type is used, whereas in IEE we use a pair of models: an elementary model and a meta model, and the models may belong to different (strong) model types.

The presented application of IEE (estimation of the $k$ for time series classification using $k$-NN-DTW) is related to locally adaptive models (Hastie and Tibshirani 1996; Domeniconi and Gunopulos 2001; Domeniconi et al. 2002). In contrast to these works, our approach adapts by selecting the proper value of $k$ and not by determining a localized distance function.

# 6   Conclusion

In this paper we introduced the notion of individualized error estimation (IEE) and defined three versions of the IEE-problem. We systematically investigated IEE in context of various prediction models. We observed RBF-Networks and SVMs to deliver good error estimations compared to the other examined models. We have also shown that IEE is capable to enhance state-of-the art time series classification models. As future work, we aim at exploring new IEE-problem and applications.

# References

Ding H, Trajcevski G, Scheuermann P, Wang X, Keogh E (2008) Querying and mining of time series data: Experimental comparison of representations and distance measures. VLDB Endowment 1(2):1542–1552

Domeniconi C, Gunopulos D (2001) Adaptive nearest neighbor classification using support vector machines. Adv NIPS 14:665–672

Domeniconi C, Peng J, Gunopulos D (2002) Locally adaptive metric nearest-neighbor classification. IEEE Trans Pattern Anal Machine Intell 24(9):1281–1285

Duffy N, Helmbold D (2002) Boosting methods for regression. Mach Learn 47:153–200

Frank A, Asuncion A (2010) UCI machine learning repository. Tech. rep., University of California, School of Information and Computer Sciences, Irvine, URL http://archive.ics.uci.edu/ml

Hall M, Frank E, Holmes G, Pfahringer B, Reutemann P, Witten IH (2009) The WEKA data mining software: An update. SIGKDD Explor 11(1):10–18

Hastie T, Tibshirani R (1996) Discriminant adaptive nearest neighbor classification. IEEE Trans Pattern Anal Mach Intell 18(6):607–616

Jain AK, Dubes RC, Chen CC (1987) Bootstrap techniques for error estimation. IEEE Trans Pattern Anal Mach Intell 5(9):606–633

Molinaro AM, Simon R, Pfeiffer RM (2005) Prediction error estimation: a comparison of resampling methods. Bioinformatics 21(15):3301–3307

Radovanovic M, Nanopoulos A, Ivanovic M (2010) Time-series classification in many intrinsic dimensions. In: Proc. 10th SIAM International Conference on Data Mining, SIAM, pp 677–688

Tsuda K, Rätsch G, Mika S, Müller KR (2001) Learning to predict the leave-one-out error of kernel based classifiers. ICANN 2001, LNCS 2130/2001:331–338

Xi X, Keogh E, Shelton C, Wei L, Ratanamahatana CA (2006) Fast time series classification using numerosity reduction. In: Proc. 23th Int'l. Conf. on Machine Learning, ACM, pp 1033–1040

# Evaluation of Spatial Cluster Detection Algorithms for Crime Locations

Marco Helbich and Michael Leitner

**Abstract** This comparative analysis examines the suitability of commonly applied local cluster detection algorithms. The spatial distribution of an observed spatial crime pattern for Houston, TX, for August 2005 is examined by three different cluster detection methods, including the Geographical Analysis Machine, the Besag and Newell statistic, and Kulldorff's spatial scan statistic. The results suggest that the size and locations of the detected clusters are sensitive to the chosen parameters of each method. Results also vary among the methods. We thus recommend to apply multiple different cluster detection methods to the same data and to look for commonalities between the results. Most confidence will then be given to those spatial clusters that are common to as many methods as possible.

## 1 Introduction

Geographic Information Systems (GIS) and spatial analysis have become valuable and indispensable tools used in day-to-day operations of governmental agencies. This is also true of police departments, which increasingly supplement and enhance their traditional criminological modus operandi with geographical information technologies for tactical and strategic decision-making. To improve the ability to gain knowledge from geospatial data and to understand the spatial processes contributing to the presence or absence of criminal offenses, spatial data mining

M. Helbich (✉)
GIScience, Department of Geography, University of Heidelberg, Berliner Strasse 48, 69120 Heidelberg, Germany
e-mail: marco.helbich@geog.uni-heidelberg.de

M. Leitner
Department of Geography and Anthropology, Louisiana State University, Baton Rouge, LA 70803, USA
e-mail: mleitne@lsu.edu

W. Gaul et al. (eds.), *Challenges at the Interface of Data Analysis, Computer Science, and Optimization*, Studies in Classification, Data Analysis, and Knowledge Organization, DOI 10.1007/978-3-642-24466-7_20, © Springer-Verlag Berlin Heidelberg 2012

tools are essential. The detection of local spatial crime clusters helps to improve the efficiency of strategies for prevention and serves as strategic planning tool for decision-making.

Because it is an important topic in spatial statistics, many different global and local spatial algorithms for different data structures exist to evaluate patterns (e.g. Anselin 1999; Kulldorff 1997; Openshaw et al. 1987). For this reason researchers are called upon to develop easy to understand and easy to use guidelines that practitioners can rely on for choosing appropriate cluster detection methods for different crime distributions. The research presented in this paper focuses on the local – and thus mappable – modeling of spatial clusters and on global trends of a spatial pattern. Following Knox (1989) a spatial cluster is "a geographically bounded group of occurrences of sufficient size and concentration to be unlikely to have occurred by chance." To detect spatial clusters, methods typically screen a study area for evidence of hot spots of points (e.g., offense or disease events) without preconception about their likely locations (Besag and Newell 1991).

Fotheringham and Zhang's research (Fotheringham and Zhan 1996) is one of the few dedicated to the evaluation of the exploratory performance of the Geographical Analysis Machine (GAM, Openshaw et al. (1987) and the Besag and Newell statistic (BNS, Besag and Newell (1991) in conjunction with GIS. Their data set consists of single-family detached dwellings in the U.S. city of Amherst, NY. They concluded that both are suitable within a GIS framework and that BNS is less prone to producing false positive results. Nevertheless, the results of both cluster detection methods are criticized because of perceptual issues of their exploratory maps. Other algorithm comparisons can be found in Kulldorff et al. (2003) and Song and Kulldorff (2003) which emphasize the efficiency of Kulldorff's spatial scan statistic (SSS).

The primary aim of this paper is the evaluation of different local cluster detection algorithms that are frequently used in practice, such as the GAM (Openshaw et al. 1987), the BNS (Besag and Newell 1991), and the SSS (Kulldorff 1997). The main research questions are: Do crime incidences tend to cluster in geographical space? Do the algorithms detect the same spatial clusters? This paper will also discuss the pros and cons of each statistic and it compares each statistic using a real world example of crime locations.

The paper is structured as follows: Sect. 2 presents the study area, the data set, and the necessary data preprocessing steps. Section 3 briefly introduces the methodology. In its subsections the three selected algorithms of local cluster detection are briefly presented. This is followed by a discussion of the results of the empirical analysis. The paper concludes with a summary and some useful recommendation for crime analysts (Sect. 4).


## 2 Study Area and Data

The study area comprises of the U.S. metropolitan area of Houston, TX, which is located inside Harris County. Crime data (e.g., burglaries, robberies, burglaries of motor vehicle, auto thefts) for this study area were received for the entire

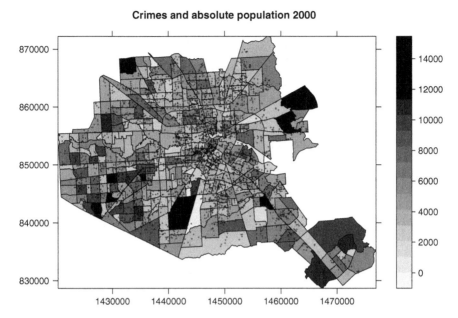

**Fig. 1** Distribution of crime locations (back cross signatures) for August 2005 and the at risk population for the census tracts (shaded polygons). Each point signature represents one crime event

month of August 2005 from the Houston Police Department. The data included the offense date and time, offense type, police beat, and the address of the offense at the street block level. This allowed the geo-coding of crime locations using the TIGER (Topologically Integrated Geographic Encoding and Referencing system) street network data freely available from the U.S. Bureau of Census. Of the total number of crimes (8,528) that occurred during the month of August 2005, 8,057 (94.5%) were successfully geo-coded. As expected, the geo-coding rate varies by police district. Figure 1 shows the spatial distribution of the crime locations as well as the spatial distribution of the population in every census tract. In Figs. 1–5 positions are referenced to the Texas State Mapping System (Lambert Conformal Conic, EPSG Code: 3081), hence the $x$- and $y$-axes denote the distances (in meters) from the Longitude and Latitude of Origin, respectively.

## 3   Methodology and Results

This study compares three different and frequently used local cluster detection algorithms. As shown in Fig. 1 both crime and population patterns are distributed heterogeneously in space and generally there are more offenses in densely populated census tracts. Spearman's rank correlation coefficient confirms this hypothesis and shows some significant positive association ($\rho = 0.449$; $p < 0.001$) between the

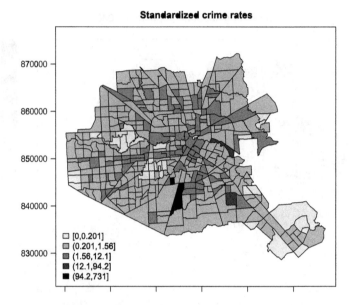

**Fig. 2** Standardized crime rates for August 2005

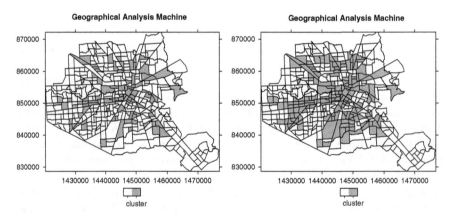

**Fig. 3** Significant clusters (shaded in gray) detected by GAM with different parameter settings. The left map uses a cell width of 500 m and $r = 500$ m. The right map uses a cell width of 500 m and $r = 1,000$ m

number of crimes per census tract and its corresponding heterogeneously distributed population at risk. The latter is based on the 2000 census and assumed to be equal to the residential population in each census tract. We are aware that the estimate of the population at risk is biased in areas such as the central business district, industrial areas, or around big malls. Similarly, in the case of certain types of

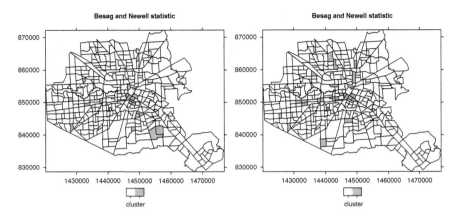

**Fig. 4** Significant ($\alpha = 0.05$) clusters detected by BNS (shaded in gray). The left map uses $k = 20$. The right map uses $k = 30$

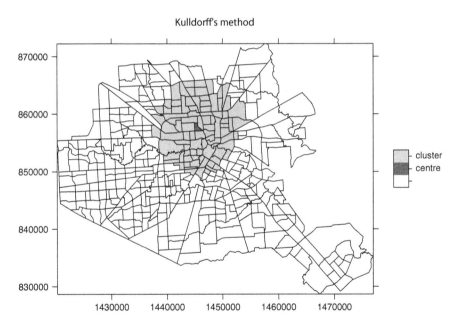

**Fig. 5** Clusters (shaded in gray) detected by SSS with a significance value $\alpha = 0.05$ and a maximum fraction of the total population is 20% of the overall population

property crime (e.g., car theft), human population is only a proxy for the actual (car) population at risk (assuming constant number of cars per capita). Because population data were only available on the administrative level of census tracts, the point locations of crimes were aggregated to the same enumeration units. Thus, our reformulated research question is: Are crime locations more clustered relative to the background population? Nevertheless, this aggregation process is not free of criticism. On the one hand it induces some aggregation bias, because it does not reflect the real-world situation exactly, but on the other hand the consideration of the at risk population leads to more distinguished clusters than purely spatial ones, as stated in Fotheringham and Zhan (1996). However, in the on-going analysis census tracts are represented by their centroids and attributed with total crime incidence and total background population. Because density plots of the number of crime locations and the population size per census tract follow approximately a Poisson distribution, we assumed that this distribution type is appropriate for further analysis. For computation we used the R statistical programming environment, in particular the DCluster package (Gomes-Rubio et al. 2005).

## 3.1 Standardized Crime Rates

A first crude approximation of the spatial crime distribution is by calculating a simple risk estimator like the "raw standardized crime rate" (SCR). SCR is defined as $O_i/E_i$, whereas $O_i$ is the observed number of crimes in census tract $i$ and $E_i$ the expected number. $E_i$ can be estimated as $E_i = POP2000_i * (\sum_{i=1}^{n} O_i / \sum_{i=1}^{n} POP2000_i)$, assuming independence between the crime locations and following a Poisson distribution (Gomes-Rubio et al. 2005). Hence, areas with a SCR above 1 are of interest, because these areas can be considered as high risk areas. Figure 2 visualizes the spatial distribution of the SCR for August 2005 for the study area. In this figure the pattern of high risk areas clearly follows the main traffic axes. Furthermore, the two census tracts with an extremely high SCR value in the south have nearly no residential land use and, therefore, only a marginal number of residential population, leading to more crime locations than inhabitants. It should be mentioned that these findings are also dependent on the observed time-frame. The following Sects. 3.2, 3.3, and 3.4 discuss the results for each algorithm.

## 3.2 Geographical Analysis Machine

The GAM (Openshaw et al. 1987) is the first search method attempting to explore local clusters in point and lattice data sets. The study region is superimposed by a regular grid, and circles with a user specified radius $r$ are drawn around each grid cell. The observed number of crime locations within each circle are compared with its expected number under a Poisson distribution. Cells having a significantly higher than expected number of cases within a circle are saved. Crime clusters are thus shown cell-based on the basis of overlapping circles. This approach has suffered much criticism (Besag and Newell 1991). The major criticism is that GAM does not control for multiple testing, when GAM was first proposed. To counteract this main critique, Openshaw et al. (1987) advocate a significance value of 0.002 to detect apparent clusters, which was used in this study as well when running the GAM. This means that circles with a significance value below 0.002 are declared as significant and therefore saved. Different combinations of parameter settings concerning the grid cell's width (100–1,000 m) and the length of the radii (100–1,000 m) are calculated. Due to a more suitable comparison with the results of the other algorithms (Sect. 4), a polygon-based visualization is applied, whereas each census tract containing at least a significant cell is identified as cluster. As a consequence the resulting clusters can be overestimated in their spatial dimension. Figure 3 shows two examples with different radii.

## 3.3   Besag and Newell Statistic

Another method to explore local clusters is the BNS method (Besag and Newell 1991). This statistic overcomes the drawback of GAM, examining clusters only based on a predefined and constant distance criterion, which is problematic if the population at risk is heterogeneously distributed in space and spatial objects are represented by centroid locations. BNS takes the second issue into account by requiring an a-priori user-defined cluster size $k$. Thus, the scanning circles, centered over each census tract, are expanded until a size of $k$ objects is reached within a cluster. Under the null hypotheses of uniform risk (no clustering), each potential cluster is evaluated by a test statistic $L_i$ concerning its significance. As above, the overall level of simultaneous testing is not controlled for. Researchers (e.g. Fotheringham and Zhan 1996) criticize this approach because of its ad hoc choice of a value for the parameter $k$. In this application the significance level has been set to 0.05. Figure 4 visualizes the detected spatial clusters.

## 3.4   Spatial Scan Statistic

The spatial scan statistic (SSS, Kulldorff (1997)) is a relatively novel methodology. The method can be described as follows: A two dimensional circular scan window moves from census tract $i$ to census tract $j$ throughout the study site and compares the probability of being a case inside the circle given the residential population at risk inside this circle as well as the probability of being a case given the residential population at risk outside (Gomes-Rubio et al. 2005). To detect clusters of different sizes, the scan window continuously increases its size up to a maximum of 20% of the total population falling inside each circle used in SSS. Due to the lack of prior knowledge about the resulting cluster sizes, all potential cluster sizes are being considered, when using this statistic (Kulldorff 1997). For each window size a Poisson-based likelihood function is calculated. The most likely cluster is the one that maximizes this likelihood function. The level of significance is usually set to $\alpha = 0.05$ and a parametric bootstrap of 999 simulation runs based on a

Poisson distribution is used to determine local significance (Gomes-Rubio et al. 2005). Compared to the above mentioned methods, the SSS accounts for multiple hypotheses testing. The result is presented in Fig. 5.

## 4 Discussion of the Results and Conclusions

This paper compares different methods used for exploratory analysis to detect spatial clusters of lattice data. Of the three methods tested, GAM and BNS are very sensitive with regard to the chosen parameters. Table 1 underpins these impressions and compares different aspects of the detected clusters, like the number of census tracts marked as clusters and their size in square miles. For instance, increasing the circle size of GAM leads to larger cluster surfaces. In detail, an increase of the GAM's search radius from 500 m to 1,000 m nearly doubles the detected clusters. Similarly, the setting of different $k$ values of BNS results in different cluster morphologies. Compared to the GAM, BNS finds a lower number of significant clusters, whereas an increase in the parameter $k$ results in more significant clusters. The most likely cluster of the SSS overlaps partially with the results of the GAM but lies spatially in-between the clusters found by the BNS. One advantage of the SSS algorithm is the distinction between a cluster center, having the highest probability of being a cluster, and its assigned parts with a lower but still significant probability. Overall, the number of significant census tracts marked as clusters by the SSS varies considerably, which is of course reflected in the amount of crime affected population, shown in Table 1.

Finally, the correlation between the detected clusters is estimated by Cramer's $V$. This coefficient of contingency measures the strength of the relationship between two nominal scaled variables, whereas 0 means no correlation and 1 a perfect correlation. The results are presented in Table 2. BNS and GAM result in

**Table 1** Number of census tracts including clusters

|  | Number of clusters | Area (sqm.) | Population |
|---|---|---|---|
| GAM ($r = 500$ m) | 84 | 95.9 | 348,654 |
| GAM ($r = 1,000$ m) | 165 | 183.1 | 758,772 |
| BNS ($k = 20$) | 19 | 15.3 | 41,449 |
| BNS ($k = 30$) | 23 | 17.9 | 70,219 |
| SSS | 87 | 77.5 | 345,934 |

**Table 2** Correlations between the detected clusters

|  | SSS | BNS ($k = 20$) | BNS ($k = 30$) | GAM ($r = 500$ m) |
|---|---|---|---|---|
| BNS ($k = 20$) | 0.110 | | | |
| BNS ($k = 30$) | 0.208 | 0.299 | | |
| GAM ($r = 500$ m) | 0.190 | 0.348 | 0.427 | |
| GAM ($r = 1,000$ m) | 0.260 | 0.245 | 0.275 | 0.606 |

roughly similar results, having a higher correlation coefficient. Furthermore, the SSS correlates only slightly with the results of GAM and BNS. An explanation is that the SSS method resulted in one large compact cluster located in the central northern part and literally no clusters found in other parts of the study area.

Overall, the SSS seems most useful, because it has a comprehensive theoretical statistical background, shows the highest flexibility, in terms of availability of data models (e.g., Poisson-based, exponential model) and expandability (e.g., comprise further covariates, space-time analysis), without leaving it up to the user to make subjective decisions about important parameter settings. It is clear that this on-going research is not free of limitations. The main limitation is that a comparison between the cluster results from the different methods was purely based on a visual level and by using a simple correlation measure. Nevertheless, our pragmatic suggestion is to apply different techniques when trying to identify significant clusters and to have the most confidence in those clusters that have been detected by the majority of the cluster algorithms being used. Additional research is required to test and compare alternative cluster methods with each other. Only then can a more complete understanding of an adequate usage of such methods being made.

However, the comparison of these exploratory techniques using a real data set should sharpen the crime mappers awareness concerning the importance of choosing an appropriate method for strategic planning and decision-making because different methods provide different insights into the data.

# References

Anselin L (1999) Local indicators of spatial association. LISA Geogr Anal 27:93–115

Besag J, Newell J (1991) The detection of clusters in rare diseases. J Royal Stat Soc A 154:143–155

Fotheringham A, Zhan F (1996) A comparison of three exploratory methods for cluster detection in spatial point patterns. Geogr Anal 28:200–218

Gomes-Rubio V, Ferrandiz J, Lopez-Quilez A (2005) Detecting clusters of diseases with R. J Geogr Syst 7:189–206

Knox E (1989) Detection of clusters. In: Elliott P (ed) Methodology of enquiries into disease clustering. Small Area Health Statistics Unit, London

Kulldorff M (1997) A spatial scan statistic. Commun Stat Theory Methods 26:1481–1496

Kulldorff M, Tango T, Park P (2003) Power comparisons for disease clustering tests. Comput Stat Data Anal 42:665–684

Openshaw S, Charlton M, Wymer C, Craft A (1987) A mark 1 geographical analysis machine for the automated analysis of point data sets. Int J Geogr Inform Sci 1:335–358

Song C, Kulldorff M (2003) Power evaluation of disease clustering tests. Int J Health Geogr 2(9)

# Checking Serial Independence of Residuals from a Nonlinear Model

Luca Bagnato and Antonio Punzo

**Abstract** In this paper the serial independence tests known as SIS (Serial Independence Simultaneous) and SICS (Serial Independence Chi-Square) are considered. These tests are here contextualized in the model validation phase for nonlinear models in which the underlying assumption of serial independence is tested on the estimated residuals. Simulations are used to explore the performance of the tests, in terms of size and power, once a linear/nonlinear model is fitted on the raw data. Results underline that both tests are powerful against various types of alternatives.

## 1 Introduction

In modelizing observed time series, the most popular approach consists in adopting the class of linear models. In this wide class, the well-known *white noise* process – characterized only by the properties of its first two moments – represents the building block and reflects information that is not directly observable. Nevertheless, the limitations of linear models already appear in Moran (1953). Nowadays, we know that there are nonstandard features, commonly referred to as *nonlinear features* (see Jianqing and Qiwei (2003)) that, by definition, can not be captured in the linear frame. In the attempt to overcome this problem, nonlinear models have to be taken into account (for a discussion on the chief objectives of the

L. Bagnato
Dipartimento di Metodi Quantitativi per le Scienze Economiche e Aziendali, Università di Milano-Bicocca, Milano, Italy
e-mail: luca.bagnato@unimib.it

A. Punzo (✉)
Dipartimento di Impresa, Culture e Società, Università di Catania, Catania, Italy
e-mail: antonio.punzo@unict.it

W. Gaul et al. (eds.), *Challenges at the Interface of Data Analysis, Computer Science, and Optimization*, Studies in Classification, Data Analysis, and Knowledge Organization, DOI 10.1007/978-3-642-24466-7_21, © Springer-Verlag Berlin Heidelberg 2012

introduction of nonlinear models, see e.g., Tjøstheim (1994)). For these kinds of models, as indicated in Jianqing and Qiwei (2003), a white noise process is no longer a pertinent building block as we have to look for measures beyond the second moments to characterize the nonlinear dependence structure. Thus, the white noise has to be replaced by a noise process composed of independent and identically distributed (*i.i.d.*) variables.

As for serial uncorrelation tests in the linear case, tests for validating serial independence are fundamental in the nonlinear approach. Testing serial independence is a preliminary step to model the data generating process (*model selection*). The testing procedure is also a useful statistical tool, once a nonlinear model is fitted to the observed data, to check serial independence of the estimated residuals (*model validation*). This paper focuses on the second type of problem taking into account two tests proposed in Bagnato and Punzo (2010) in the context of model selection. These tests are based on the well-known $\chi^2$-test and, consequently, are asymptotically valid. Both can be applied to those fields of application having a large number of observations such as environmental, financial, and quality control applications.

To summarize, the paper presents in Sect. 2 the problem of checking serial independence in the model validation phase. Here, the contextualization is in the nonlinear frame and the tests proposed in Bagnato and Punzo (2010) are briefly illustrated. In this new context, their performance, in terms of size and power, is studied in Sect. 3 by simulations on some popular models. To find out more in terms of graphical detecting of possible lag(s)-dependences, a simulation is illustrated by the Serial Dependence Plot (SDP; see Bagnato and Punzo (2010)). Finally, in Sect. 4, some conclusions are given.

## 2   Serial Independence Tests Applied to Residuals

Let $\{X_t\}$ be a stationary and ergodic time series. Let $(X_t, \ldots, X_{t-n+1})$ be a single (random) realization of length $n$ of the process at time $t$. In order to describe the process, suppose to use the nonlinear (parametric/nonparametric) model

$$X_t = f\left(\boldsymbol{\xi}_t\right) + \sigma\left(\boldsymbol{\xi}_t\right)\varepsilon_t, \tag{1}$$

where $\boldsymbol{\xi}_t = (X_{t-1}, \ldots, X_{t-d}, \varepsilon_{t-1}, \ldots, \varepsilon_{t-v})$, $f(\cdot)$ and $\sigma(\cdot)$ are known/unknown functions of $d$ past $X_t$'s and $v$ past $\varepsilon_t$'s, with $d < n$ and $v < n$, and where $\{\varepsilon_t\}$ is serially independent with *i.d.* components having zero mean and unitary variance. Model (1) can be viewed, in the nonparametric context, as a generalization of a NARCH model adopted in Jianqing and Qiwei (2003) where $f(\cdot)$ and $\sigma(\cdot)$ depend only on the lagged variables $(X_{t-1}, \ldots, X_{t-d})$. Naturally, a simpler model can be used; here model (1) is considered to highlight the great flexibility in nonlinear model validation provided by the tests proposed in Bagnato and Punzo (2010).

Now, let $p = \max(d, v)$. Once we have estimated model (1), we can obtain the residual vector $\left(\widehat{\varepsilon}_t, \ldots, \widehat{\varepsilon}_{t-n+p+1}\right)$, of length $n - p$, in which the elements are obtained from the simple relations

$$\widehat{\varepsilon}_t = \frac{X_t - \hat{f}\left(\boldsymbol{\xi}_t\right)}{\hat{\sigma}\left(\boldsymbol{\xi}_t\right)}$$

$$\vdots$$

$$\widehat{\varepsilon}_{t-n+p+1} = \frac{X_{t-n+p+1} - \hat{f}\left(\boldsymbol{\xi}_{t-n+p+1}\right)}{\hat{\sigma}\left(\boldsymbol{\xi}_{t-n+p+1}\right)}.$$

Naturally, $\left(\widehat{\varepsilon}_t, \ldots, \widehat{\varepsilon}_{t-n+p+1}\right)$ can be seen as a realization of $\{\varepsilon_t\}$.

Now, suppose we are interested in testing the underlying assumption of serial independence of $\{\varepsilon_t\}$ in model (1). With this aim, we will follow the philosophy presented in Bagnato and Punzo (2010) to check serial independence in the model selection phase, conveniently re-adapted in this new context of model validation. Thus, choose an integer $l$, with $0 < l < n - p$, and consider Table 1 where, for each lag $r$, $r = 1, \ldots, l$, the estimated residuals up to time $t - r$ are contained in the respective column.

Moreover, consider $L = \{1, \ldots, l\}$. Thus, the null hypothesis of interest can be expressed as

$$H_0 : \varepsilon_t \perp\!\!\!\perp \varepsilon_{t-r} \quad \text{for all } r \in L, \tag{2}$$

versus the alternative

$$H_1 : \varepsilon_t \not\!\perp\!\!\!\perp \varepsilon_{t-r} \quad \text{for some } r \in L. \tag{3}$$

Obviously, the number of possible comparisons increases in line with $r$. Before going on, for comprehension sake, consider $r = 1$; accepting $\varepsilon_t \perp\!\!\!\perp \varepsilon_{t-1}$, one implicitly accepts also that $\varepsilon_{t-1} \perp\!\!\!\perp \varepsilon_{t-2}$, $\varepsilon_{t-2} \perp\!\!\!\perp \varepsilon_{t-3}$, and so on. This issue can be easily comprehended considering adjacent columns in Table 1.

**Table 1** Lagged residuals $\varepsilon_{t-r}$, $r = 1, \ldots, l$, $l < n - p$

| | Lagged residuals | | | | |
|---|---|---|---|---|---|
| $\varepsilon_t$ | $\varepsilon_{t-1}$ | $\cdots$ | $\varepsilon_{t-r}$ | $\cdots$ | $\varepsilon_{t-l}$ |
| $\widehat{\varepsilon}_{t-n+p+1}$ | | | | | |
| $\widehat{\varepsilon}_{t-n+p+2}$ | $\widehat{\varepsilon}_{t-n+p+1}$ | | | | |
| $\vdots$ | $\vdots$ | $\ddots$ | | | |
| $\widehat{\varepsilon}_{t-n+p+r+1}$ | $\widehat{\varepsilon}_{t-n+p+r}$ | $\cdots$ | $\widehat{\varepsilon}_{t-n+p+1}$ | | |
| $\vdots$ | $\vdots$ | | $\vdots$ | $\ddots$ | |
| $\widehat{\varepsilon}_{t-n+p+l+1}$ | $\widehat{\varepsilon}_{t-n+p+l}$ | $\cdots$ | $\widehat{\varepsilon}_{t-n+p+l+1-r}$ | $\cdots$ | $\widehat{\varepsilon}_{t-n+p+1}$ |
| $\vdots$ | $\vdots$ | | $\vdots$ | | $\vdots$ |
| $\widehat{\varepsilon}_{t-1}$ | $\widehat{\varepsilon}_{t-2}$ | $\cdots$ | $\widehat{\varepsilon}_{t-r-1}$ | $\cdots$ | $\widehat{\varepsilon}_{t-l-1}$ |
| $\widehat{\varepsilon}_t$ | $\widehat{\varepsilon}_{t-1}$ | $\cdots$ | $\widehat{\varepsilon}_{t-r}$ | $\cdots$ | $\widehat{\varepsilon}_{t-l}$ |

Thus, the "overall" hypotheses in (2) and (3) can be easily re-expressed as follows

$$H_0 : \bigcap_{r=1}^{l} H_0^{(r)} \quad \text{versus} \quad H_1 : \bigcup_{r=1}^{l} H_1^{(r)}, \tag{4}$$

where

$$H_0^{(r)} : \varepsilon_t \perp\!\!\!\perp \varepsilon_{t-r} \tag{5}$$

is the component null hypothesis concerning a sort of independence of lag $r$, while $H_1^{(r)}$ is the negation of (5), $r = 1, \ldots, l$. The superscript "$(r)$" underlines that we are taking into account the comparison of lag $r$, that is, the residual variables $\varepsilon_t$ and $\varepsilon_{t-r}$. In Table 1 the elements considered for this comparison are highlighted. In order to have the largest number of observations for each comparison, we have chosen to fix $\varepsilon_t$ that, from a practical point of view, it is equivalent to fix the first column in Table 1.

In order to test $H_0^{(r)}$, the first step is to create the bivariate joint distribution as shown in Table 2 where, for each variable, $k$ is the number of classes and $a_1^{(r)}, \ldots, a_{k-1}^{(r)}, a = c, d$, are cutoff points defining *equifrequency classes*.

Now, a simple and well-known $\chi^2$-test of independence can be adopted to test the null hypothesis (5) with reference to Table 2. In particular we have denoted the test statistic for $H_0^{(r)}$ by

$$T_r = \sum_{i=1}^{k} \sum_{j=1}^{k} \frac{\left[ n_{ij}^{(r)} - \hat{n}_{ij}^{(r)} \right]^2}{\hat{n}_{ij}^{(r)}}, \tag{6}$$

where $n_{ij}^{(r)}$ are the observed frequencies and $\hat{n}_{ij}^{(r)}$ are the theoretical frequencies under the null hypothesis, $i, j = 1, \ldots, k$. It is well known that under $H_0^{(r)}$ the test statistic (6) is asymptotically distributed as a $\chi^2$ with $(k-1)^2$ degrees of freedom.

As shown in Table 1, when we compare $\varepsilon_t$ with $\varepsilon_{t-r}$, we have $n^{(r)} = n - p - r$ observations (cfr. Table 2). Since we are building equifrequency classes, $k$ is

**Table 2** Contingency table related to $H_0^{(r)}$

| $\varepsilon_t$ \ $\varepsilon_{t-r}$ | $\leq d_1^{(r)}$ | $\cdots$ | $d_{j-1}^{(r)} \dashv d_j^{(r)}$ | $\cdots$ | $> d_{k-1}^{(r)}$ | |
|---|---|---|---|---|---|---|
| $\leq c_1^{(r)}$ | $n_{11}^{(r)}$ | $\cdots$ | $n_{1j}^{(r)}$ | $\cdots$ | $n_{1k}^{(r)}$ | |
| $\vdots$ | $\vdots$ | | $\vdots$ | | $\vdots$ | |
| $c_{i-1}^{(r)} \dashv c_i^{(r)}$ | $n_{i1}^{(r)}$ | $\cdots$ | $n_{ij}^{(r)}$ | $\cdots$ | $n_{ik}^{(r)}$ | |
| $\vdots$ | $\vdots$ | | $\vdots$ | | $\vdots$ | |
| $> c_{k-1}^{(r)}$ | $n_{k1}^{(r)}$ | $\cdots$ | $n_{kj}^{(r)}$ | $\cdots$ | $n_{kk}^{(r)}$ | |
| | | | | | | $n^{(r)}$ |

the only parameter to be chosen. The parameter $k$ may be selected so that the dependence structure between the two variables of interest is maintained. Obviously, in choosing $k$, the number of observations has to be also taken into account. In order to make valid the above asymptotic result, the well-known rule of thumb $\widehat{n}_{ij}^{(r)} \geq 5$ for each $i, j = 1, \ldots, k$ and $r = 1, \ldots, l$, has to hold. This means that the number of observations, in comparison of lag $r$, has to satisfy the inequality $n^{(r)} \geq 5k^2$. Consequently, the number of classes has to be chosen such that $k \leq \sqrt[2]{n^{(r)}/5}$. As highlighted by simulations in Bagnato and Punzo (2010), $k = 5$ reveals to be a good choice when $n \geq 125 + p + l$.

If considering that test statistics $T_u$ and $T_v$ are independent for any $u \neq v$ (as shown in Bagnato and Punzo (2010)), it is possible to summarize the proposed tests in the following way:

**SIS test**: The first, named in Bagnato and Punzo (2010) as Serial Independence Simulataeous (SIS) test, takes advantage of the union-intersection (UI) principle introduced in Roy (1953). In detail, the overall null hypothesis $H_0$ is accepted against the alternative if, and only if, all $H_0^{(r)}$, $r = 1, \ldots, l$, are simultaneously accepted. Taking $p$-values into account, the null hypothesis is accepted if, and only if, the component test with the smallest $p$-value is accepted.

Exploiting the independence between test statistics, the significance levels at which each individual test of independence of lag $r$ has to be performed – in order to preserve the significance level $\alpha$ (chosen in advance) for the overall null hypothesis in (4) – are fixed, using the Šidák inequality (Šidák (1967)), at $1 - (1 - \alpha)^{1/l}$ for any $r = 1, \ldots, l$ (see Shaffer (1995));

**SICS test**: The second, named in Bagnato and Punzo (2010) as Serial Independence Chi-Square (SICS) test, use the (positive) test statistic

$$T = \sum_{r=1}^{l} T_r = \sum_{r=1}^{l} \sum_{i=1}^{k} \sum_{j=1}^{k} \frac{\left[ n_{ij}^{(r)} - \hat{n}_{ij}^{(r)} \right]^2}{\hat{n}_{ij}^{(r)}}, \tag{7}$$

assuming small values under $H_0$. Taking advantage of the independence between test statistics, the asymptotic distribution of $T$ results to be a $\chi^2$ with $l(k-1)^2$ degrees of freedom. Note that the way as the test statistic in (7) is defined is in principle equal to some portmanteau tests, such as both the Box-Pierce test (Box and Pierce (1970)) and the Ljung-Box test (Ljung and Box (1978)), in which the value of $T_r$ is conveniently substituted by the squared autocorrelation coefficient of lag $r$.

# 3 Simulation Study

In order to evaluate the performance of both SIS and SICS tests in the model validation context, their empirical size and power have been analyzed by a simulation study performed using the R environment.

The type of models used to generate the raw series is: nonlinear moving average (D)-(F), linear autoregressive (G), nonlinear autoregressive (H), sign autoregressive (I), bilinear (J), ARCH(1) (K), GARCH(1,1) (L), threshold autoregressive (M), logistic map (N), Hénon map (O) and Hénon map with dynamic noise (P). Details on the adopted models, used also in Diks (2009), are contained in the second column of Table 3.

Note that we have chosen to adopt the same models, and the same model labeling as in Bagnato and Punzo (2010), in order to make comparisons easier. Details on the simulation setting can be easily obtained from the caption of Table 3. The study has also the objective to evaluate the performance of the tests for the lags $l$ between 2 and 5.

Table 3 shows rejection rates, of both tests, applied to raw data and estimated residuals of an AR(1) and of what we denote as ARMA(1,1)-GARCH(1,1). With ARMA(1,1)-GARCH(1,1) we mean that a GARCH(1,1) is fitted on the residuals arising from an ARMA(1,1) estimated on the raw data. As regards the residuals from the AR(1), the observed size of both tests results to be close to the nominal size for model (G), that is an AR(1). In the other cases, the power of both tests drops compared to the same tests based on raw data, which indicates that some of the dependence structure is captured by the AR(1) model. As regards the residuals arising from the ARMA(1,1)-GARCH(1,1), the observed size of both tests results to be close to the nominal size above all for model (G), (K), and (L). In fact, these can be considered as a particular case of an ARMA(1,1)-GARCH(1,1). In the other cases, the power of both tests drops compared to the same tests based on estimated residuals from the AR(1), which indicates that further dependence structure is captured by the ARMA(1,1)-GARCH(1,1) model with respect to the AR(1).

Whenever the null hypothesis of serial independence is rejected on the estimated residuals, it could be useful to investigate the causes in order to improve the adopted model. With this aim, it could be useful to take into account the Serial Dependence Plot (SDP) presented in Bagnato and Punzo (2010). To ascertain this suggestion, we have chosen to generate 10,000 series, of length $n = 1,000$, from an ARCH(2) of equation $X_t = \sqrt{h_t}\varepsilon_t, h_t = 1+0.4X_{t-1}^2+0.3X_{t-2}^2$. On the generated series we have fitted an ARCH(1). The simulated rejection rates, at the nominal level 0.05, of both tests are graphically represented in Fig. 1 by the SDP. Here, the rejection rates are plotted as function of the all possible significant subsets of lags $S \subseteq L$, once fixed $l = 5$. Thus, for example, considering the subset $S = \{2, 4\}$, the simultaneous serial dependence between $\varepsilon_t$ and $\varepsilon_{t-2}$, and between $\varepsilon_t$ and $\varepsilon_{t-4}$, is investigated. Also, considering the subset $S = \{1, 2, 5\}$, the simultaneous serial dependence between $\varepsilon_t$ and $\varepsilon_{t-1}$, between $\varepsilon_t$ and $\varepsilon_{t-2}$, and between $\varepsilon_t$ and $\varepsilon_{t-5}$ is taken into account. Note that a black point on each bar highlights the nominal level. It easy to note as in Fig. 1(a) and 1(b) the higher bar is in correspondence to the set $S = \{2\}$ representing the type of lag-dependence that is not captured from the ARCH(1).

**Table 3** Simulated rejection rates, of SIS and SICS tests, for raw data (see Bagnato and Punzo (2010)) and for estimated residuals of an AR(1) and an ARMA(1,1)-GARCH(1,1); nominal size 0.05 and l = 2,...,5. In the models specification {ε_t} represents a sequence of independent standard normals. Sample size n = 1,000 and number of simulations equal to 1,000 for each independently realized time series from each process

| Model | Specification | Test | Raw data | | | | AR(1) | | | | ARMA(1,1)-GARCH(1,1) | | | |
|---|---|---|---|---|---|---|---|---|---|---|---|---|---|---|
| | | | Lag l | | | | Lag l | | | | Lag l | | | |
| | | | 2 | 3 | 4 | 5 | 2 | 3 | 4 | 5 | 2 | 3 | 4 | 5 |
| (D) | $X_t = \varepsilon_t + 0.8\varepsilon_{t-1}^2$ | SIS | 1.000 | 1.000 | 1.000 | 1.000 | 1.000 | 1.000 | 1.000 | 1.000 | 1.000 | 1.000 | 1.000 | 1.000 |
| | | SICS | 1.000 | 1.000 | 1.000 | 1.000 | 1.000 | 1.000 | 1.000 | 1.000 | 1.000 | 1.000 | 1.000 | 1.000 |
| (E) | $X_t = \varepsilon_t + 0.6\varepsilon_{t-1}^2 + 0.6\varepsilon_{t-2}^2$ | SIS | 1.000 | 1.000 | 1.000 | 1.000 | 1.000 | 1.000 | 1.000 | 1.000 | 1.000 | 1.000 | 1.000 | 1.000 |
| | | SICS | 1.000 | 1.000 | 1.000 | 1.000 | 1.000 | 1.000 | 1.000 | 1.000 | 1.000 | 1.000 | 1.000 | 1.000 |
| (F) | $X_t = \varepsilon_t + 0.8\varepsilon_{t-1}\varepsilon_{t-2}$ | SIS | 0.566 | 0.513 | 0.471 | 0.445 | 0.563 | 0.492 | 0.464 | 0.421 | 0.204 | 0.178 | 0.167 | 0.151 |
| | | SICS | 0.676 | 0.575 | 0.502 | 0.456 | 0.650 | 0.563 | 0.498 | 0.456 | 0.214 | 0.201 | 0.160 | 0.151 |
| (G) | $X_t = 0.3X_{t-1} + \varepsilon_t$ | SIS | 1.000 | 1.000 | 1.000 | 1.000 | 0.061 | 0.047 | 0.046 | 0.054 | 0.032 | 0.037 | 0.038 | 0.039 |
| | | SICS | 1.000 | 1.000 | 1.000 | 0.998 | 0.045 | 0.050 | 0.044 | 0.046 | 0.046 | 0.047 | 0.050 | 0.049 |
| (H) | $X_t = 0.8|X_{t-1}|^{0.5} + \varepsilon_t$ | SIS | 1.000 | 0.998 | 0.994 | 0.993 | 0.702 | 0.650 | 0.615 | 0.586 | 0.665 | 0.621 | 0.584 | 0.561 |
| | | SICS | 1.000 | 1.000 | 1.000 | 1.000 | 0.648 | 0.532 | 0.452 | 0.394 | 0.593 | 0.491 | 0.423 | 0.378 |
| (I) | $X_t = \text{sign}(X_{t-1}) + \varepsilon_t$ | SIS | 1.000 | 1.000 | 1.000 | 1.000 | 0.196 | 0.169 | 0.149 | 0.139 | 0.091 | 0.073 | 0.065 | 0.062 |
| | | SICS | 1.000 | 1.000 | 1.000 | 1.000 | 0.237 | 0.218 | 0.214 | 0.195 | 0.089 | 0.082 | 0.076 | 0.081 |
| (J) | $X_t = 0.6\varepsilon_{t-1}X_{t-2} + \varepsilon_t$ | SIS | 0.784 | 0.751 | 0.726 | 0.702 | 0.768 | 0.736 | 0.712 | 0.686 | 0.267 | 0.259 | 0.228 | 0.213 |
| | | SICS | 0.824 | 0.774 | 0.725 | 0.669 | 0.807 | 0.755 | 0.690 | 0.623 | 0.248 | 0.253 | 0.222 | 0.211 |
| (K) | $X_t = \sqrt{h_t}\,\varepsilon_t,\ \ h_t = 1 + 0.4X_{t-1}^2$ | SIS | 0.885 | 0.817 | 0.799 | 0.781 | 0.842 | 0.804 | 0.784 | 0.759 | 0.048 | 0.045 | 0.044 | 0.037 |
| | | SICS | 0.814 | 0.731 | 0.659 | 0.617 | 0.797 | 0.709 | 0.642 | 0.585 | 0.034 | 0.037 | 0.035 | 0.042 |
| (L) | $X_t = \sqrt{h_t}\,\varepsilon_t,\ \ h_t = 0.01 + 0.8h_{t-1} + 0.15X_{t-1}^2$ | SIS | 0.583 | 0.608 | 0.612 | 0.616 | 0.574 | 0.594 | 0.598 | 0.613 | 0.027 | 0.041 | 0.042 | 0.048 |
| | | SICS | 0.670 | 0.734 | 0.766 | 0.789 | 0.652 | 0.717 | 0.751 | 0.775 | 0.037 | 0.045 | 0.043 | 0.040 |
| (M) | $X_t = \left[-0.5 + 0.9I_{[0,\infty)}(X_{t-1})\right]X_{t-1} + \varepsilon_t$ | SIS | 0.994 | 0.993 | 0.991 | 0.987 | 0.982 | 0.971 | 0.964 | 0.961 | 0.978 | 0.972 | 0.965 | 0.964 |
| | | SICS | 0.986 | 0.957 | 0.939 | 0.919 | 0.957 | 0.912 | 0.881 | 0.839 | 0.958 | 0.900 | 0.853 | 0.807 |
| (N) | $X_t = 4X_{t-1}(1 - X_{t-1}),\ \ 0 < X_t < 1$ | SIS | 1.000 | 1.000 | 1.000 | 1.000 | 1.000 | 1.000 | 1.000 | 1.000 | 1.000 | 1.000 | 1.000 | 1.000 |
| | | SICS | 1.000 | 1.000 | 1.000 | 1.000 | 1.000 | 1.000 | 1.000 | 1.000 | 1.000 | 1.000 | 1.000 | 1.000 |
| (O) | $X_t = 1 + 0.3X_{t-2} - 1.4X_{t-1}^2$ | SIS | 1.000 | 1.000 | 1.000 | 1.000 | 1.000 | 1.000 | 1.000 | 1.000 | 1.000 | 1.000 | 1.000 | 1.000 |
| | | SICS | 1.000 | 1.000 | 1.000 | 1.000 | 1.000 | 1.000 | 1.000 | 1.000 | 1.000 | 1.000 | 1.000 | 1.000 |
| (P) | $X_t = Z_t + \sigma\varepsilon_t,\ \ Z_t = 1 + 0.3Z_{t-2} - 1.4Z_{t-1}^2$ | SIS | 0.989 | 0.993 | 0.992 | 0.993 | 0.974 | 0.985 | 0.984 | 0.985 | 0.174 | 0.244 | 0.226 | 0.201 |
| | | SICS | 0.991 | 0.997 | 0.995 | 0.995 | 0.975 | 0.992 | 0.986 | 0.986 | 0.180 | 0.299 | 0.300 | 0.293 |

**Fig. 1** Rejection rates, at the
nominal level $\alpha = 0.05$,
constructed on 10,000
residual series obtained fitting
an ARCH(1) model on series,
of length $n = 1,000$,
generated from an ARCH(2)

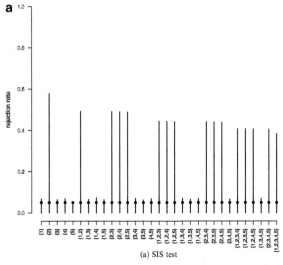

(a) SIS test

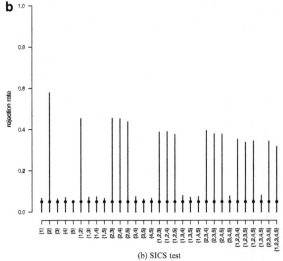

(b) SICS test

## 4    Concluding Remarks

In this paper the two tests for serial independence proposed in Bagnato and Punzo
(2010) are applied in the context of nonlinear model validation. A wide simulation
study has been also performed to investigate their performance. Both tests, which
exploit the well-known $\chi^2$-statistic of independence, show rejection rates, under
null hypothesis, which are very close to the nominal size. They appear also to be
powerful against various types of alternatives.

Furthermore, throughout the Serial Dependence Plot, these tests reveal how
the model could be improved indicating possible lag(s)-dependence to insert in

the model. Further developments will be devoted to check the performance of the tests in the validation of nonparametric structure such as the additive one.

# References

Bagnato L, Punzo A (2010) On the Use of $\chi^2$-Test to Check Serial Independence. Statistica Applicazioni 8(1):57–74

Box GEP, Pierce DA (1970) Distribution of the autocorrelations in autoregressive moving average time series models. J Am Stat Assoc 65(332):1509–1526

Diks C (2009) Nonparametric tests for independence. In: Meyers R (ed) Encyclopedia of complexity and systems science, Springer, Berlin

Jianqing F, Qiwei Y (2003) Nonlinear time series: nonparametric and parametric methods. Springer, Berlin

Ljung GM, Box GEP (1978) On a measure of lack of fit in time series models. Biometrika 65(2):297–303

Moran PAP (1953) The statistical analysis of the Canadian lynx cycle, 1. Structure and prediction. Aust J Zool 1(2):163–173

Roy SN (1953) On a heuristic method of test construction and its use in multivariate analysis. Ann Math Stat 24(2):220–238

Shaffer JP (1995) Multiple hypothesis testing. Ann Rev Psychol 46(1):561–584

Šidák Z (1967) Rectangular confidence regions for the means of multivariate normal distributions. J Am Stat Assoc 62(318):626–633

Tjøstheim D (1994) Non-linear time series: A selective review. Scand J Stat 21(2):97–130

# Part V
# Network Data, Graphs, and Social Relationships

# Analysis of Network Data Based on Probability Neighborhood Cliques

**Andreas Baumgart and Ulrich Müller-Funk**

**Abstract** The authors present the concept of a "probability neighborhood clique" intended to substantiate the idea of a "community", i.e. of a dense subregion within a (simple) network. For that purpose the notion of a clique is generalized in a probabilistic way. The probability neighborhoods employed for that purpose are indexed by one or two tuning parameters to bring out the "degree of denseness" respectively a hierarchy within that community. The paper, moreover, reviews other degree based concepts of communities and addresses algorithmic aspects.

## 1  Communities

### 1.1  Basic Problem

The paper is motivated by the analysis of large and heterogeneous networks. Due to the large size, it is necessary to "regionalize" the analysis in a first step. Only then, i.e. from a mesoscopic point of view, it becomes reasonable to study interactions of individuals and to assign roles (e.g. "drivers", "connectors" etc.). Accordingly, the first step of any such analysis should be the identification of dense subregions and their interrelationship.

The idea is captured by the notion of a *community*. Newman and Girvan (2004) define a "... community structure in networks – [as] natural divisions of network nodes into densely connected subgroups ...". That definition grasps the idea, but is "fuzzy" and not immediately practicable. Accordingly, the concept has to be made precise the one way or the other.

A. Baumgart (✉) · U. Müller-Funk
European Research Center for Information Systems (ERCIS), University of Münster,
Leonardo Campus 3, 48149 Münster, Germany
e-mail: baumgart@ercis.uni-muenster.de; mueller-funk@ercis.uni-muenster.de

W. Gaul et al. (eds.), *Challenges at the Interface of Data Analysis, Computer Science, and Optimization*, Studies in Classification, Data Analysis, and Knowledge Organization, DOI 10.1007/978-3-642-24466-7_22, © Springer-Verlag Berlin Heidelberg 2012

For some authors, "community" is almost synonymous to "cluster", i.e. results from a (disjoint) clustering process – in most cases based on a suitable objective function. Prominent examples include global methods like:

- Trace optimization methods used for optimal cuts or flows, weighted $k$-means or spectral clustering, see Dhillon et al. (2005).
- Modularity optimization methods based on a comparison with random graphs, see Newman (2006).
- Hierarchical clustering methods, divisive or agglomerative; see e.g. Newman and Girvan (2004) resp. Newman (2004).

The focus of these approaches, however, is more on separation (weak ties), less on the identification of dense subregions. For that reason, we shall not pursue these methods. Instead, we shall follow a local approach that works the other way round. First, the notion of a community has to be fixed. Subsequently, an appropriate algorithm has to be developed, which evolves some "seed" (i.e. elementary subgraph) into a community. The last step is repeated – with varying seeds – until the whole network is covered by communities. Unlike clusterings, coverings of that kind allow for overlaps, that help to study the network structure.

In what follows, we restrict attention to the case of simple networks $G = (V, E)$ (i.e. connected graphs without loops) consisting of vertices $v_1, \ldots, v_p \in V$ and undirected, unweighted edges $e_1, \ldots, e_q \in E$. For every (connected) subgraph $H = H_W$ of $G$ induced by $W \subseteq V, |W| = k$, let $\pi_H$ be the *degree probability*, $\pi_H(j)$ expressing the fractions of vertices with degree $j \geq 1$.

## 1.2  Communities Based on Degrees

Whatever way a community is defined, it necessarily must comprise cliques. (A $k$-clique, by definition, is a completely connected set of $k$ vertices, $k \geq 3$ fixed.) Cliques give a "crisp", but rather rigid idea of a community. It seems to be reasonable to look for some weaker notions. Various combinatorial concepts grasping the perception of cliquish subregions can be found in the literature. Some of them are distance-based, e.g. $k$-cliques in the sense of Newman (2010), clans and clubs; see Mokken (1979). There are further concepts, e.g. *dominant sets*; see Pavan and Pelillo (2003), and *pseudo-cliques*; see Haraguchi and Okubo (2006), which, however, do not fit well into the present framework. Here, we shall concentrate on degree-based concepts:

- A $(k, \beta)$-*quasi-clique* $H$ is a connected subgraph with $k$ vertices and at least $\left\lfloor \beta \frac{k(k-1)}{2} \right\rfloor$ edges, $0 < \beta \leq 1$; see Abello et al. (2002), Zeng et al. (2006).
- A $(k, n)$-*core* (or -*plex*) $H$ is a connected subgraph with $k$ vertices, where the degree of each vertex is at least $n$; see Kolaczyk (2009).

For all these concepts there are algorithms providing these quantities for a specific graph. Note, however, that most of these problems are $\mathcal{NP}$-hard (Pardalos and Rebennack 2010).

A more synthetical, but somewhat inflexible, specification of a community is due to Palla et al. (2005), who give a definition based on $k$-cliques:

- A $k$-clique community $H$ is a $k$-clique chain of maximal length, i.e. a union of k-cliques, where adjacent cliques share a $(k-1)$-clique.

A related algorithm is provided by the clique percolation method of Derényi et al. (2005) by which a $k$-clique is "rolled out" through the network: At each step one looks for a $k$-clique not yet included that shares a (1)-clique with one of the $k$-cliques already sampled. Iterate that device as long as possible (See Fig. 1).

A completely different definition is suggested by Lancichinetti et al. (2009). Here, a community is implicitly defined as a maximizer of a Jaccard type fitness function:

- A max-fit community $H$ is a solution of maximizing

$$f_\alpha(H) = \frac{d_{in}(H)}{(d_{in}(H) + d_{out}(H))^\alpha} \tag{1}$$

where $d_{in}(H)$ and $d_{out}(H)$ denote the sums of all internal respectively external degrees of vertices pertaining to $H$ and where $\alpha > 0$ is a constant.

The parameter $\alpha$ controls the size of $H$. A large $\alpha$ corresponds to a small set. Varying $\alpha$, consequently, results in a nested sequence of subgraphs and helps to work out hierarchical structures. According to those authors, the natural choice $\alpha = 1$ is relevant and, in addition, values within the interval 0.5 to 2. That findings seem to originate from computer simulations.

In that paper, an algorithm is proposed as well, which – in a somewhat modified form – will be discussed later on.

It is the primary goal of the present paper to supplement the aforementioned local approaches and to define a flexible, degree-based concept of a community that:

- Unifies notions like quasi-cliques and cores but also comprises clique-chains.
- Is flexible and practicable, grasps the fuzziness of the notion – and does not result in a graph structure that is sensitive to the addition or removal of a few edges.
- Is able to bring out hierarchical, i.e. nested structures among communities as it is the case with the approach by Lancichinetti et al. (2009).

In order to achieve that goal, we start out from a simple characterization of a $k$-clique by means of its degree distribution. Subsequently, that notion is softened or "fuzzified" by specifying suitable neighborhoods – an idea essentially borrowed from robust statistics.

## 2 Probabilistic Neighborhood Approach

### 2.1 *Characterizations via Degree Distribution*

We come back to the combinatorial extensions of a $k$-clique and express them by their degree distributions. (This is possible for other concepts, too, e.g. for $f_\alpha(H)$.)

- $H$ is a $k$-clique iff

$$\pi_H = \delta_{(k-1)} \tag{2}$$

where $\delta_x$ denotes a Dirac measure. In other words, $\pi_H$ has location $(k-1)$ and dispersion 0.

- $H$ is a $(k, \beta)$-quasi-clique iff

$$\mu(\pi_H) \geq \beta(k-1), \quad \mu(\pi_H) \text{ expectation,} \tag{3}$$

$$\text{as} \quad card(E(W)) = \frac{1}{2} \sum_{w \in W} deg(w) = \frac{1}{2} \sum_{j=1}^{k-1} \sum_{\substack{w \in W \\ deg(w)=j}} deg(w)$$

$$= \frac{k}{2} \sum_{j=1}^{k-1} j\pi_H(j) = \frac{k}{2}\mu(\pi_H)$$

- $H$ is a $(k, n)$-core iff

$$\pi_H(0) = \ldots = \pi_H(n-1) = 0. \tag{4}$$

- Consider a $k$-clique chain $H$ obeying a "never visit a vertex twice"-constraint. For the $m$-times rolled out chain, $m \geq k - 1$, we get

$$\pi_H = \frac{2}{m+k-1} \sum_{j=0}^{k-2} \delta_{k+j-1} + \frac{m-k+1}{m+k-1}\delta_{2k-2} \tag{5}$$

$$\approx \delta_{2(k-1)} \tag{6}$$

provided $m$ is sufficiently large. In other words: the $k$-clique chain behaves like a $(2k - 1)$-clique. The approximate identity (6), of course, remains valid if the chain is slightly altered, i.e. if a few edges are added / deleted or a few vertices are revisited.

### *Illustration*

For the graph in Fig. 1, we have $k = 3$ and choose $m = 18$. As a result, we get

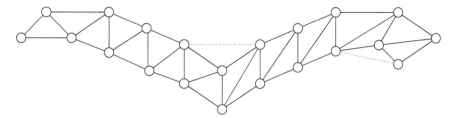

**Fig. 1** Illustration 3-clique chain

$$\pi_H = \frac{1}{10}\delta_2 + \frac{1}{10}\delta_3 + \frac{4}{5}\delta_4 \approx \delta_4, \tag{7}$$

i.e. $\pi_H$ lies close to a $2k - 1 = 5$-clique, even if we include the dashed edges.


## 2.2 Probability Neighborhoods $\mathcal{N}$

Let $\pi(\cdot)$ be a probability on $\{1, \ldots, \text{k-1}\}$. Neighborhoods $\mathcal{N} = \mathcal{N}(\pi)$ can be specified in various ways. Miscellaneous examples, e.g. Hellinger balls, can be found in the literature on robust statistics; see Rieder (1994). In what follows, we are interested in neighborhoods of $\pi = \delta_{(k-1)}$. Examples, corresponding to (2), (3), (4):

- Total variation neighborhoods (i.e. balls in the $l_1$-distance around $\delta_{(k-1)}$, characterized by the $(k - 1)$ dimensional probability vector $(0, \ldots, 0, 1)$):

$$\mathcal{N}_{TV} = \{\pi_H : \sum_{i=1}^{k-2} \pi_H(i) + (1 - \pi_H(k - 1)) < \varepsilon\}, \quad \varepsilon > 0. \tag{8}$$

- Location-dispersion neighborhoods:

$$\mathcal{N}_{LD} = \{\pi_H : \nu(\pi_H) \geq \beta(k - 1), \tau(\pi_H) \leq \varepsilon\}, \quad \varepsilon > 0, \tag{9}$$

where $\nu$ is any measure of location and $\tau$ any measure of dispersion. Examples include the mode, respectively the Gini coefficient.

- n-core neighborhoods:

$$\mathcal{N}_{nC} = \{\pi_H : \sum_{i=1}^{n-1} \pi_H(i) \leq \varepsilon\}, \quad \varepsilon > 0. \tag{10}$$

The choice of the parameters depends on the network and is based on heuristics ("trial and error"). The foregoing considerations motivate the following concepts.

**Definition.** Let $H$ be a connected subgraph of $G$

(a)  $H$ is a $(k, \mathcal{N})$-clique iff $\pi_H \in \mathcal{N} = \mathcal{N}(\delta_{(k-1)})$.

(b)  $H$ is a $(k, \mathcal{N})$-community if it is a $(k, \mathcal{N})$-clique and if it is maximal, i.e. if $\pi_{H'} \notin \mathcal{N}$ for all connected subgraphs $H'$, $H \subset H'$.

# 3   Algorithm

## 3.1   A Greedy Approach

The idea by Lancichinetti et al. (2009) can be modified in order to produce $(k, \mathcal{N})$-cliques.

For this "onion skin" algorithm (see Fig. 2 for the naming), we assume that some $k$-clique ("seed") is already given. (There exist several algorithms, see for example Bron and Kerbosch (1973) for a branch and bound approach, that can be utilized to get cliques within a given graph $G$ as a starting solution. See also Cazals and Karande (2008).

Note that we distinguish between the overall degree $\deg(v)$, where all edges are considered, and the community degree with respect to $H$, say $\deg_H(v)$, where only edges within a certain community are gathered.

In first place, we have to specify the neighborhood $\mathcal{N} = \mathcal{N}(\delta_{(k-1)})$.

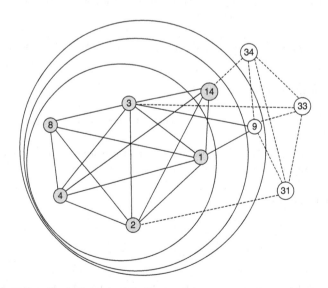

**Fig. 2**  Community with $H_0$ as core clique

1. Find a starting $k$-clique $H_0 = (W_0, E(W_0))$ in $G$

2. Determine all vertices $w$ adjacent to $H_0$ such that

$$\pi_{H(w)} \in \mathcal{N}, \quad \text{where} \quad H(w) = (W_0 \cup \{w\}, E(W_0 \cup \{w\})) \quad \text{"admissibility"} \quad (11)$$

3. Among all adjacent and admissible vertices select the best one, say $w_1$, i.e.

$$\deg_{H_1}(w_1) \text{ is maximal}, \quad H_1 = H(w_1) \tag{12}$$

   (If $w_1$ is not unique, than a plausible tie-breaking rule has to be chosen.)

4. Recursion:

   Case 1: $H_i$ is maximal w. r. to (12). Stop and put $W = W_i$.

   Case 2: Otherwise, go back to step 2 (with $H_i$ in place of $H_0$).

Starting with some clique $H_0' = (W_0', E(W_0'))$, $W_0' \cap W = \emptyset$, the above steps can be repeated with that new seed in order to obtain the next community etc. etc. .

## 3.2 Illustration

For the purpose of illustration, we apply the algorithm to a well-known example – the social network of Zachary's Karate club (Zachary 1977), where friendships among club members are analyzed and different sub-groups can be found. As our starting clique $H_0$, we choose the 5-clique with vertices $\{1, 2, 3, 4, 8\}$. The sub-group membership of the Karate club is denoted by gray and white shading and edges within the community are continuous, others are dashed.

In the first step we choose the $n$-core neighborhood $\mathcal{N}_{nC}$ with $\varepsilon = \frac{1}{7}, n = 4$. Vertex $w_1 = \{14\}$ could be added first because of its highest community degree $\deg_{H_1}(w_1) = 4$, as well as vertex $w_2 = \{9\}$ because of its maximal overall degree $\deg(w_2) = 5$. Notice that vertex $w_2$ belongs to a different sub-group than the starting clique $H_0$. Vertex $w_3 = \{34\}$ could not be added to the community, since the resulting degree distribution $\pi_{H_3} \notin \mathcal{N}_{nC}$ with $\varepsilon = \frac{1}{7}$. Increasing the parameter, e.g. choosing $\varepsilon = \frac{1}{4}$, vertex $w_3$ would be added and a larger community could be identified.

## 4 Outlook

The present approach carries over to digraphs. For any subgraph $H$ the degree distribution is bivariate, i.e. $\pi_H = \kappa(\pi_H^{(i)}, \pi_H^{(o)})$, where $\pi_H^{(i)}$ is the in-degree distribution, $\pi_H^{(o)}$ is the out-degree distribution and $\kappa$ an essentially unique copula describing the dependency between both degrees.

Assume for the moment that $H$ is a clique, i.e. completely connected in the strong sense. In that case, $\pi_H$ equals $\delta_{(k-1,k-1)}$ and the copula $\kappa$ is just the upper Fréchet-bound $\kappa^+(s,t) = \min(s,t)$, $0 \le s,t \le 1$. In that case, it is easy to design neighborhoods to both Dirac measures and the upper Fréchet-bound. A modified analysis will be based on a weakened concept of a directed clique. Details shall be found in a subsequent paper.

# References

Abello J, Resende MGC, Sudarsky S (2002) Massive quasi-clique detection. In: LATIN '02: Proceedings of the 5th Latin American Symposium on Theoretical Informatics, Springer, London, UK, pp 598–612

Bron C, Kerbosch J (1973) Algorithm 457: finding all cliques of an undirected graph. Commun ACM 16(9):575–577, DOI http://doi.acm.org/10.1145/362342.362367

Cazals F, Karande C (2008) A note on the problem of reporting maximal cliques. Theor Comput Sci 407(1–3):564–568, DOI 10.1016/j.tcs.2008.05.010, URL http://dx.doi.org/10.1016/j.tcs.2008.05.010

Derényi I, Palla G, Vicsek T (2005) Clique percolation in random networks. Phys Rev Lett 94(16):160,202, DOI 10.1103/PhysRevLett.94.160202, URL http://dx.doi.org/10.1103/PhysRevLett.94.160202

Dhillon I, Guan Y, Kulis B (2005) A unified view of kernel k-means, spectral clustering and graph cuts. Tech. rep., UTCS Technical Report No. TR-04-25

Haraguchi M, Okubo Y (2006) A method for pinpoint clustering of web pages with pseudo-clique search. In: Jantke K, Lunzer A, Spyratos N, Tanaka Y (eds) Federation over the Web, Lecture Notes in Computer Science, vol 3847. Springer, Berlin/Heidelberg, pp 59–78, DOI 10.1007/11605126\_4, URL http://dx.doi.org/10.1007/11605126_4

Kolaczyk ED (2009) Statistical analysis of network data: methods and models, 1st edn. Springer, New York, DOI 10.1007/978-0-387-88146-1

Lancichinetti A, Fortunato S, Kertész J (2009) Detecting the overlapping and hierarchical community structure in complex networks. New J Phys 11(3):033,015+, DOI 10.1088/1367-2630/11/3/033015, URL http://dx.doi.org/10.1088/1367-2630/11/3/033015

Mokken RJ (1979) Cliques, clubs and clans. Quality Quantity 13:161–173, DOI 10.1007/BF00139635, URL http://dx.doi.org/10.1007/BF00139635

Newman MEJ (2004) Fast algorithm for detecting community structure in networks. Phys Rev E 69(6):066,133, DOI 10.1103/PhysRevE.69.066133, arXiv:cond-mat/0309508

Newman MEJ (2006) Modularity and community structure in networks. Proc Natl Acad Sci USA 103(23):8577–8582, DOI 10.1073/pnas.0601602103

Newman MEJ (2010) Networks: An introduction, 1st edn. Oxford University Press, USA

Newman MEJ, Girvan M (2004) Finding and evaluating community structure in networks. Phys Rev E 69(2):026,113, DOI 10.1103/PhysRevE.69.026113

Palla G, Derényi I, Farkas I, Vicsek T (2005) Uncovering the overlapping community structure of complex networks in nature and society. Nature 435(7043):814–818, DOI 10.1038/nature03607

Pardalos P, Rebennack S (2010) Computational challenges with cliques, quasi-cliques and clique partitions in graphs. In: Festa P (ed) Experimental algorithms, lecture notes in computer science, vol 6049. Springer, Berlin/Heidelberg, pp 13–22, URL http://dx.doi.org/10.1007/978-3-642-13193-6_2

Pavan M, Pelillo M (2003) A new graph-theoretic approach to clustering and segmentation. In: Proc. IEEE Conf. Computer Vision and Pattern Recognition, vol 1, pp 145–152

Rieder H (1994) Robust asymptotic statistics. Springer, Berlin

Zachary WW (1977) An information flow model for conflict and fission in small groups. J Anthropol Res 33:452–473

Zeng Z, Wang J, Zhou L, Karypis G (2006) Coherent closed quasi-clique discovery from large dense graph databases. In: KDD '06: Proceedings of the 12th ACM SIGKDD international conference on Knowledge discovery and data mining, ACM, New York, NY, USA, pp 797–802, DOI 10.1145/1150402.1150506, URL http://doi.acm.org/10.1145/1150402.1150506

# A Comparison of Agglomerative Hierarchical Algorithms for Modularity Clustering

Michael Ovelgönne and Andreas Geyer-Schulz

**Abstract** Modularity is a popular measure for the quality of a cluster partition. Primarily, its popularity originates from its suitability for community identification through maximization. A lot of algorithms to maximize modularity have been proposed in recent years. Especially agglomerative hierarchical algorithms showed to be fast and find clusterings with high modularity. In this paper we present several of these heuristics, discuss their problems and point out why some algorithms perform better than others. In particular, we analyze the influence of search heuristics on the balancedness of the merge process and show why the uneven contraction of a graph due to an unbalanced merge process leads to clusterings with comparable low modularity.

## 1 Introduction

Finding "natural" groups in networks gets increasing attention in recent years as there are various applications for such methods in as different fields as biology, sociology and marketing. As manifold as the application areas are the approaches. Most methods like those based on edge removal (e.g. Girvan and Newman 2002) or label propagation (e.g. Raghavan et al. 2007) work without a measure for the quality of a clustering. With the modularity function, introduced by Newman and Girvan (2004), an explicit formal measure for the quality of a clustering is available. This measure received much attention as it makes algorithms comparable. Modularity opened the way to try all kinds of combinatorical optimization techniques and many

M. Ovelgönne, · A. Geyer-Schulz (✉)
Information Services and Electronic Markets, IISM, Karlsruhe Institute of Technology,
Kaiserstrasse 12, D-76128 Karlsruhe, Germany
e-mail: michael.ovelgoenne@kit.edu; andreas.geyer-schulz@kit.edu

W. Gaul et al. (eds.), *Challenges at the Interface of Data Analysis, Computer Science, and Optimization*, Studies in Classification, Data Analysis, and Knowledge Organization, DOI 10.1007/978-3-642-24466-7_23, © Springer-Verlag Berlin Heidelberg 2012

authors published heuristics for the fast detection of cluster partitions (clusterings) with high modularity.

## 1.1 Modularity

Modularity measures the quality of a cluster partition of an undirected, unweighted graph $G = (V, E)$. The adjacency matrix of $G$ is denoted as $M$. The element of $M$ in the i-th row and j-th column is denoted as $m_{ij}$, where $m_{ij} = m_{ji} = 1$ if $\{v_i, v_j\} \in E$ and otherwise $m_{ij} = m_{ji} = 0$. Let $C = \{C_1, \ldots, C_p\}$ be a non-overlapping clustering of $G$, i.e. a partition of $V$ into groups $C_i$ so that $\forall i, j$ : $i \neq j \Rightarrow C_i \cap C_j = \emptyset$ and $\cup_i C_i = V$. The clustering $C$ of $G$ has a modularity $Q$ given by

$$Q = \sum_i (e_{ii} - a_i^2) \tag{1}$$

with

$$e_{ij} = \frac{\sum_{v_x \in C_i} \sum_{v_y \in C_j} m_{xy}}{\sum_{v_x \in V} \sum_{v_y \in V} m_{xy}} \tag{2}$$

and

$$a_i = \sum_j e_{ij} \tag{3}$$

The term $e_{ij}$ is the fraction of edge endpoints belonging to edges connecting $C_i$ with $C_j$ to the total number of edge endpoints in the graph. The row sums of the matrix $(e_{ij})$ are denoted as $a_i$, which is the fraction of edge endpoints belonging to vertices in $C_i$. The difference $e_{ii} - a_i^2$ compares the fraction of intra-cluster edge endpoints to the expected number when the edges are randomly rewired without changing the degrees of the vertices. In other words, if $\mathcal{G}$ denotes the set of all labeled graphs with the same degree sequence as $G$, then $a_i^2$ is the expected fraction of edge endpoints belonging to edges that connect vertices in $C_i$ for a graph in $\mathcal{G}$.

The higher the modularity of a clustering is, the less random it is. Therefore, we try to maximize modularity to find the best possible clustering of a graph. Let $\Omega$ be the set of all non-overlapping clusterings of G. We try to find a clustering $C^* \in \Omega$ with

$$Q(C^*, G) = max\{Q(C, G) | C \in \Omega\}. \tag{4}$$

## 2 Modularity Maximization Strategies

Finding the clustering with maximal modularity is NP-hard (Brandes et al. 2008) and therefore not feasible in reasonable time for large networks. However, for smaller networks finding optimal solutions might be desirable.

As modularity maximization is a discrete optimization problem, calculating the modularity for all partitionings of a graph would be the straightforward attempt to find the partitioning with maximal modularity. A faster approach is to transform the problem into an integer linear program (IP) (Agarwal and Kempe 2008; Brandes et al. 2008), but solving the IP is NP-hard as well. As exact algorithms fail to cluster graphs with as low as 100 vertices on a standard desktop computer, a lot of scaleable heuristics have been developed in the recent years. We distinguish three types of heuristics: non-hierarchical, divisive hierarchical and agglomerative hierarchical.

Non-hierarchical algorithms start with a (random) initial clustering and iteratively improve the clustering by moving vertices from one cluster to another. Divisive hierarchical algorithms start with a single cluster containing all vertices and recursively cut the clusters in several sub-clusters. Agglomerative hierarchical algorithms work the other way round. They start with singleton clusters containing only one vertex each and merge clusters step by step. The two types of hierarchical algorithms create dendrograms that depict the hierarchical decomposition of a graph rather than a single clustering. With each cut or merge a new clustering is created. We postpone the detailed discussion of agglomerative hierarchical algorithms to Sect. 3.

The non-hierarchical algorithms are predominantly implementations of meta-heuristics. For example, Medus et al. (2005) designed an algorithm based on the simulated annealing optimization scheme. Starting from an initial clustering, one vertex at a time is moved from its current cluster to an empty cluster or a non-empty cluster (which the selected vertex is connected to). The acceptance of a move is determined by the induced modularity difference and the annealing schedule. When more than a preset upper bound of successive move tries are not accepted, the algorithm terminates. Another algorithm with a similar heuristic based on mean field annealing has been proposed by Lehmann and Hansen (2007).

Hierarchical divisive algorithms recursively split a graph into subgraphs. A popular strategy to identify the splits is spectral analysis. Spectral analysis is employed among others by White and Smyth (2005) and Newman (2006). While most spectral algorithms work with bisectioning, the algorithm of Ruan and Zhang (2007) splits a graph into several subgraphs. Further divisive algorithms are the one of Lü and Huang (2009) which employs an iterated tabu search heuristic and the extremal optimization approach of Duch and Arenas (2005).

After the main clustering phase, often a refinement phase follows. Newman (2006) suggested a refinement procedure for clusterings based on the well-known algorithm of Kernighan and Lin (1970). Their algorithm was designed to assign vertices to a fixed number $k$ of clusters, so that the number of edges between clusters is minimal. The algorithm starts with a random split of the vertices in $k$ clusters and then moves vertices between clusters if it decreases the objective function. The adaptation of this idea for refinement is straightforward. In every iteration of the refinement each vertex is selected once and the move of the vertex which increases modularity most is executed. The refinement is done when in one iteration no modularity increasing move has been found. Post-processing algorithms are generally independent of the algorithm used in the main phase. In this paper, we will analyze the core algorithms without additional refinement steps.


## 3   Agglomerative Hierarchical Algorithms

The principle of hierarchical agglomerative clustering in general is to place each vertex in its own cluster and then merge two clusters at a time until all vertices are placed in one cluster. Agglomerative hierarchical modularity algorithms are usually greedy. The decision which clusters should be merged is decided upon the effect of the merge on the increase in $Q(C, G)$. If we consider the clustering $C$ and the clustering $C'$ which results from the merge of the cluster $C_i$ and $C_j$ of $C$ then we get

$$\Delta Q(C_i, C_j) = Q(C, G) - Q(C', G) = e_{ij} + e_{ji} - 2a_i a_j = 2(e_{ij} - a_i a_j) \quad (5)$$

As $\Delta Q(i, j)$ only depends on the clusters $C_i$ and $C_j$ and the connections between them, it is a local measure.

The first algorithm for detecting comunities based on modularity was Newmans's greedy optimization scheme (Newman 2004). In each step of the algorithm those two clusters $C_i$ and $C_j$ get merged that have the highest $\Delta Q(C_i, C_j)$. This proce- dure is very slow, as for each pair of communities that are connected by at least one edge, $\Delta Q$ has to be calculated. Unconnected clusters do not need to be considered for a merge as their $\Delta Q$ is always negative. Clauset et al. (2004) proposed with the Clauset-Newman-Moore (CNM) algorithm an algorithmic improvement that cached the values of $\Delta Q$ so that $\Delta Q(C_i, C_j)$ is only recalculated if this becomes necessary through the merge of one of the clusters i and j with a third cluster. We denote the optimization scheme that is implemented in both algorithms as plain greedy (PG).

Subsequently, many other optimization schemes have been developed that outperformed the first scheme in terms of time and quality. Schuetz and Caflisch (2008) proposed the multistep greedy algorithm (MSG). This algorithm builds classes of joins (= pairs of vertices) with the same $\Delta Q$ and sorts them in descending order. In each step all joins in the top $l$ classes are executed, but every cluster is merged at most once. As CNM does, MSG caches $\Delta Q$ and updates the values when necessary. As the number of classes of joins with the same $\Delta Q$ are large especially in the first steps, MSG achieves a significant speed-up to CNM.

The randomized greedy (RG) scheme of Ovelgönne and Geyer-Schulz (2010) exploits that many joins are equal with regard to their induced $\Delta Q$ and finding a good join does not require to consider all possible joins. RG selects the two clusters to merge from a small random sample of pairs. In each step of the algorithm a set $S$ of $k$ random selected clusters gets created. The pair of clusters with the highest $\Delta Q$ and where one of the clusters is an element of $S$ gets merged. For $k = 1$ this means a random cluster $C_i$ is selected and merged with that neighboring cluster $C_j$

where $\Delta Q(i, j)$ is maximal. RG is very fast due to the very few pairs of clusters that get considered for a merge in any step.

Wakita and Tsurumi (2007) analyzed the merge process of the CNM algorithm after they experienced that the algorithm does not have the reported scalability. They found out that the merge process is not balanced as assumed by Clauset et al. (2004) but highly unbalanced. This unbalancedness causes the slow runtime. In this context, a merging process of an agglomerative hierarchical algorithm is considered as balanced if all (remaining) clusters grow roughly evenly. To receive a balanced growth Wakita and Tsurumi proposed to use a measure they called *consolidation ratio*: $ratio(C_i, C_j) = min(|C_i|/|C_j|, |C_j|/|C_i|)$. For the size of a cluster $|C_i|$ they tested two formulations: the number of vertices the cluster consists of and the number of edges to other clusters. In contrast to the plain greedy scheme where in every step the cluster pair $\{C_i, C_j\}$ with the highest value of $\Delta Q(C_i, C_j)$ is joined, the new algorithm merges the cluster pair with the maximum of $\Delta Q(C_i, C_j) \cdot ratio(C_i, C_j)$. The two algorithm variants speed up the clustering procedure but produce far less modular results than the plain greedy approach. However, Wakita and Tsurumi proposed a third variant (*HE'*) that achieves better quality than PG and is also faster: First determine for every cluster $C_i$ the neighbor $C_x$ where $\Delta Q(C_i, C_x)$ is maximal and than search among those pairs that one, where $\Delta Q(C_i, C_j) \cdot ratio(C_i, C_j)$ is maximal.

## 4 Optimization Scheme Discussion

A critical problem of agglomerative hierarchical clustering algorithms for modularity optimization is the locality problem. The subgraph of Fig. 1 shows a difficult decision situation. From a global view it is obvious that $v_1$ should be assigned to the same clusters as the other vertices of the left clique. Consider completing all dangling edges in Fig. 1 with one vertex. From the narrow local view on the initial singleton clustering the cluster $\{v_1\}$ has 6 links to singleton clusters with equal intensity. The cluster to which vertex $v_1$ will be finally assigned to depends on which of its neighbors have been already merged. If for example $v_2$ and $v_3$ have been

**Fig. 1** Subgraph with dangling edges (Ovelgönne and Geyer-Schulz 2010)

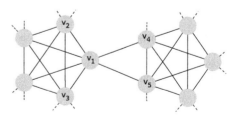

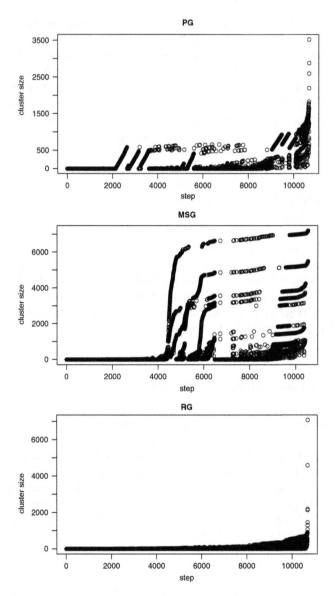

**Fig. 2** Sample merge processes of the PG, the MSG (level parameter $l = 44$), and the RG algorithm on the PGP key signing network from Boguñá et al. (2004). PG achieved a modularity of 0.849, MSG 0.869 and RG 0.874. The depicted cluster size is the size of the larger of both merged clusters. Note the different scales of the y-axis

merged to cluster $\{v_2, v_3\}$, than $\Delta Q(\{v_1\}, \{v_2, v_3\}) = 0.006 > \Delta Q(\{v_1\}, \{v_4\}) = \Delta Q(\{v_1\}, \{v_5\}) = 0.003$. The other way round, if $v_4$ and $v_5$ have been merged to cluster $\{v_4, v_5\}$, than $\Delta Q(\{v_1\}, \{v_4, v_5\}) = 0.006 > \Delta Q(\{v_1\}, \{v_2\}) = \Delta Q(\{v_1\}, \{v_3\}) = 0.003$.

We see that prior mergers in the neighborhood of a cluster influence later merger decisions for this cluster. An unbalanced merge process where some regions of the graph are heavily contracted while other regions are untouched leads to bad clustering results. When, like in the example discussed above, two clusters get merged that have a common neighbor, a bias towards merging the common neighbor $(C_n)$ with the now merged clusters $(C_m)$ arises because the empirical connections between the two clusters $C_m$ and $C_n$ (the $e_{mn}$ in (5)) dominates the whole term $\Delta Q(C_n, C_m)$.

Figure 2 shows the merge processes of PG, MSG and RG. We can see that the PG algorithm steps into the locality pitfall. A single cluster gets merged over and over (up to several hundred times in a row for the test dataset). The self-reinforcing effect that large clusters attract small (especially singleton) clusters lead to poor results in terms of achieved modularity.

The MSG algorithm is able to find clusterings with higher modularity than PG. In the first iteration of MSG a large portion of all mergers is conducted (in our example about one third of all mergers). The number of mergers is significantly lower in the second iteration and decreases further over the rest of the iterations as mergers remove similarities in the graph substructures and thus the size of the equivalence classes of joins decreases. Due to this merge process several clusters grow in parallel. This way, the locality problem has a less negative effect on the merge decisions and clusterings with higher modularity can be found.

In the case of RG, clusters of all sizes grow parallel because the random sampling prevents unbalanced growth. The merge process is even more balanced than that of MSG. PG (192 clusters) and MSG (144 clusters) create clusterings with more clusters than RG (82 clusters) and more very large clusters. The balancedness of RG's merge process prevents the early creation of large clusters which prevent the growth of smaller clusters.

# 5 Summary

In this paper we have analysed the balancedness of the merger pocesses of agglomerative hierarchical algorithms for modularity maximization. Furthermore, we discussed the locality problem that can mislead greedy algorithms. We also showed that unbalanced merger processes are much stronger affected by the locality problem. Altogether, our analysis explains why algorithms for modularity maximization with balanced merger processes find clusterings with superior modularity compared to algorithms with unbalanced merger processes.

**Acknowledgements** The research leading to these results has received funding from the European Community's Seventh Framework Programme FP7/2007-2013 under grant agreement n°215453 - WeKnowIt.

# References

Agarwal G, Kempe D (2008) Modularity-maximizing graph communities via mathematical programming. Eur J Phys B 66:409–418

Boguñá M, Pastor-Satorras R, Díaz-Guilera A, Arenas A (2004) Models of social networks based on social distance attachment. Phys Rev E 70(5):056,122

Brandes U, Delling D, Gaertler M, Görke R, Hoefer M, Nikoloski Z, Wagner D (2008) On modularity clustering. IEEE Trans Knowl Data Eng 20(2):172–188

Clauset A, Newman MEJ, Moore C (2004) Finding community structure in very large networks. Phys Rev E 70

Duch J, Arenas A (2005) Community detection in complex networks using extremal optimization. Phys Rev E 72(2):027,104

Girvan M, Newman MEJ (2002) Community structure in social and biological networks. Proc Natl Acad Sci USA 99(12):7821–7826

Kernighan B, Lin S (1970) An efficient heuristic procedure for partitioning graphs. Bell Syst Technical J 49(1):291–307

Lehmann S, Hansen L (2007) Deterministic modularity optimization. Eur Phys J B 60(1):83–88

Lü Z, Huang W (2009) Iterated tabu search for identifying community structure in complex networks. Phys Rev E 80(2):026,130

Medus A, Acuña G, Dorso C (2005) Detection of community structures in networks via global optimization. Phys A 358(2-4):593–604

Newman MEJ (2004) Fast algorithm for detecting community structure in networks. Phys Rev E 69

Newman MEJ (2006) Modularity and community structure in networks. Proc Natl Acad Sci USA 103(23):8577–8582

Newman MEJ, Girvan M (2004) Finding and evaluating community structure in networks. Phys Rev E 69(2):026,113

Ovelgönne M, Geyer-Schulz A (2010) Cluster cores and modularity maximization. In: ICDMW '10. IEEE International Conference on Data Mining Workshops, pp 1204–1213

Raghavan UN, Albert R, Kumara S (2007) Near linear time algorithm to detect community structures in large-scale networks. Phys Rev E 76(3):036,106

Ruan J, Zhang W (2007) An efficient spectral algorithm for network community discovery and its applications to biological and social networks. In: Seventh IEEE International Conference on Data Mining, 2007, pp 643–648

Schuetz P, Caflisch A (2008) Efficient modularity optimization by multistep greedy algorithm and vertex mover refinement. Phys Rev E 77

Wakita K, Tsurumi T (2007) Finding community structure in mega-scale social networks. CoRR abs/cs/0702048, URL http://arxiv.org/abs/cs/0702048

White S, Smyth P (2005) A spectral clustering approach to finding communities in graphs. In: Proceedings of the Fifth SIAM International Conference on Data Mining, SIAM, pp 274–285

# Network Data as Contiguity Constraints in Modeling Preference Data

Giuseppe Giordano and Germana Scepi

**Abstract** In the last decades the use of regression-like preference models has found widespread application in marketing research. The Conjoint Analysis models have even more been used to analyze consumer preferences and simulate product positioning. The typical data structure of this kind of models can be enriched by the presence of supplementary information observed on respondents. We suppose that relational data observed on pairs of consumers are available. In such a case, the existence of a consumer network is introduced in the Conjoint model as a set of contiguous constraints among the respondents. The proposed approach will allow to bring together the theoretical framework of Social Network Analysis with the explicative power of Conjoint Analysis models. The combined use of relational and choice data could be usefully exploited in the framework of relational and tribal marketing strategies.

## 1 Introduction

Recent studies provided a conceptual framework to model consumer relationships where consumers are both individuals and members of social groups. The consumer, through these relationships, tends to develop its identity/self-concept and to socially identify itself in a group (Brewer and Gardner 1996). In this framework, firms tend to adapt their marketing and communication strategies, finalizing their behavior

G. Giordano (✉)
University of Salerno, Department of Economics and Statistics, Via Ponte Don Melillo,
84084 - Fisciano (Salerno), Italy
e-mail: ggiordan@unisa.it

G. Scepi
University of Napoli Federico II, Department of Mathematics and Statistics, Via Cintia - Monte
S.Angelo, 80126 - Napoli, Italy
e-mail: scepi@unina.it

W. Gaul et al. (eds.), *Challenges at the Interface of Data Analysis, Computer Science,*       233
*and Optimization*, Studies in Classification, Data Analysis, and Knowledge Organization,
DOI 10.1007/978-3-642-24466-7_24, © Springer-Verlag Berlin Heidelberg 2012

to supporting individuals in their social context. The study of the role played by individuals as actors embedded in social relationships is one of the aims of Social Network Analysis (Wasserman and Faust 1994). The network analysis highlights the position of single individuals in the net, describing their peculiar attitude to improve the sharing of resources (knowledge, friendship, confidence, advise, etc.).

The characterization of individuals as potential consumers and social actors, at the same time, leads to a new perspective when analyzing the elicitation preference process. This process has been traditionally analyzed in the context of Conjoint Analysis (Green and Srinivasan 1990). Nowadays, by recognizing the role of the individual as a consumer embedded in a (not only virtual) community, we highlight the relevance to incorporate such knowledge when modeling individual utility functions.

The aim of this paper is to introduce a new model of Conjoint Analysis that makes it possible to take into account: (a) the overall notes of preference revealed on a set of stimuli profiles; (b) the attribute-levels describing the stimuli profiles and (c) the relational data giving information about the set of pairwise ties among respondents and their knowledge exchange in the community. The main goal is to obtain aggregate part-worths estimates considering the neighboring of each individual consumer (ego-network) and comparing the estimated individual utility function with the aggregate utility function associated to the ego-network. Such a comparison might give a major insight about the presence of an ego-network effect in the elicitation preference process.

## 2 The Metric Conjoint Analysis Model: Factorial Approximation and External Information

Conjoint Analysis (Green and Srinivasan 1990) is one of the most popular statistical techniques used in Marketing to estimate preference functions at both individual and aggregate level. In this contribution, we refer to the metric approach to Conjoint Analysis by assuming that a fractional factorial design is realized, the full profile concept evaluation is used in the data collection phase, the part-worth utility function is estimated on a rating scale by means of a metric (OLS) regression. Initially, we consider two sets of variables: the dependent variables in the matrix $Y$ $(N \times J)$ and the explicative ones in the design matrix $X$ $(N \times (K - p))$, where $N$ is the set of Stimuli, $J$ is the number of judges – i.e. the preference responses – and $p$ is the number of experimental factors expanded in $K$ attribute-levels. The Metric Conjoint Analysis model is written as the following multivariate multiple regression model:

$$Y = XB + E_x \qquad (1)$$

Let $B$ be the $(K - p) \times J$ matrix of individual part-worth coefficients and let $E_x$ be the $(N \times J)$ matrix of error terms for the set of $J$ individual regression models. Lauro et al. (1998) proposed a particular factorial decomposition (see (2)) of the estimated coefficient matrix $\hat{B}$ to be read in the within of Principal Component Analysis and related variants such as the Principal component analysis onto a reference subspace (D'Ambra and Lauro 1982) and Redundancy Analysis (Israels 1984). The part-worths decomposition is defined as follows:

$$\hat{B}'X'X\hat{B}u_\alpha = \Lambda u_\alpha \qquad \alpha = 1, \ldots, (K - p) \qquad (2)$$

In the psychometric literature, several authors have contributed to the statistical analysis of data matrices with external information on rows and columns by means of suitable factorial decompositions (Takane and Shibayama 1991). Giordano and Scepi (1999) extended the model (1) to the case of external information observed on the respondents to the Conjoint Analysis task.

Let $Z$ $(J \times (H - q))$ be the matrix of external information on the columns of $B$ holding $(H-q)$ dummy variables related to $q$ categorical variables observed on $J$ judges, then the following multiple multivariate regression model is considered:

$$\hat{B}' = Z\Theta + E_z \qquad (3)$$

where the estimated part-worth coefficients of the Conjoint Analysis are described as a function of the external information on the judges/consumers. The coefficients in the matrix $\Theta$ explain the effect of the individual' characteristic on the stimuli feature, (for instance: "How worth is the *red-painted* car for *women*?"). The matrix $\Theta$ is estimated by:

$$\hat{\Theta} = (Z'Z)^{-1}Z'Y'X(X'X)^{-1} \qquad (4)$$

thus the factorial decomposition proposed by Giordano and Scepi (1999) is obtained by the following Singular Value Decomposition (SVD):

$$\Theta \approx W\Delta V' \qquad with \qquad W'W = V'V = I \qquad (5)$$

where $W$ and $V'$ are the left and right singular vectors with singular values in the diagonal matrix $\Delta$. In the next section the approach will be particularized to the case of relational data as external information on consumers considered as social actors in a network.

## 3 Network Data Matrix as Relational Constraints on Judges

Social Network Analysis deals with ties (relational data) among interacting units to describe patterns of social relationships. In the Economics field, relationships can emerge from different settings: exchange of information, acquaintanceship; trustee; business activity, spatial or geographic contiguity, and so on. The typical data matrix of Social Network Analysis (SNA) is the *Adjacency Matrix*, a squared matrix of order $J$, the number of actors in the net. In this paper, we only refer to the case of binary (presence/absence of ties), symmetric relationships (mutual exchange), so that the adjacency matrix is symmetric and binary valued. Let $A$ $(J \times J)$ be an adjacency matrix, its generic element $a_{ij}$ is 1 if the two corresponding actors $i$ and $j$ are tied, otherwise it is set to 0; $a_{ij} = a_{ji}$ (symmetry); it is assumed that $a_{ii} = 0$ $(i = 1, \ldots, J)$. We particularize now the Giordano and Scepi (1999) approach to the (metric) Conjoint Analysis model with adjacency data as external information on consumers. With the notation introduced in Sect. 2, let $A$ be a squared, binary and symmetric adjacency matrix among $J$ consumers sharing mutual information in a network. The ties can be established by considering an ad-hoc survey on the quality and quantity of exchanged information or by investigating them in a web-based brand forum. Let us consider the model (3) and restate it as:

$$\hat{B}' = A\Theta + E_a \tag{6}$$

its solution leads to:

$$\hat{\Theta} = (AA)^{-1}AY'X(X'X)^{-1} \equiv (AA)^{-1}A\hat{B}' \tag{7}$$

since $A$ is symmetric.

The solution (7) is formed by the matrix multiplication of two quantities, let them be referred as:

$$\hat{\Theta} = (AA)^{-1}A\hat{B}' \equiv \Sigma^{-1}\Upsilon \tag{8}$$

where $\Sigma = (AA)$ and $\Upsilon = A\hat{B}'$. A deeper look at these two quantities shows that:

- $\Sigma$ is the matrix holding the number of pathways of length two from each actor to each other actor. They are a measure of the network connectivity. Its diagonal elements are the actor's degree, i.e. the number of actors belonging to the ego-network of each actor.
- The matrix $\Upsilon$ holds the generic element $\upsilon_{ik}$ which represents the sum of the part-worth coefficients for the attribute $X_k$ estimated for all the actors belonging to the ego-network of actor $i$.
- The generic term of $\hat{\Theta}$, $\hat{\theta}_{i,k}$ is a function of the overall part-worth coefficients of each attribute-level measured on the ego-network of $i$, inversely weighted by the connection strength between each pair of actors/judges.

If we consider only the diagonal elements of $\Sigma$, we define the diagonal matrix $\Delta_\Sigma$, and compute

$$\tilde{\Theta} = \Delta_\Sigma^{-1}\Upsilon \tag{9}$$

The matrix coefficient in (9) has a more straightforward interpretation, since its generic element $\tilde{\theta}_{ik}$ represents the average part-worth coefficient of attribute $X_k$ estimated for all the actors belonging to the ego-network of actor $i$. Let us notice that in the $i$-th row of $\tilde{\Theta}$, the individual utility function (the coefficients in $\hat{B}$) is substituted by the average utility function of judge $i$'s neighbors. The singular value decomposition of $\tilde{\Theta}$ makes it possible to analyze, on a reduced factorial subspace, the proximity between judges sharing similar ego-network preference structures.

**Table 1** Estimated part-worth coefficients for 24 artificial response variables

| J | $X_{11}$ | $X_{12}$ | $X_{21}$ | $X_{22}$ | $X_{31}$ | $X_{32}$ | J | $X_{11}$ | $X_{12}$ | $X_{21}$ | $X_{22}$ | $X_{31}$ | $X_{32}$ |
|---|---|---|---|---|---|---|---|---|---|---|---|---|---|
| J1 | −2.67 | −0.33 | −1.00 | 0.00 | 0.00 | 0.67 | J13 | 0.00 | 3.00 | −0.67 | 0.67 | 0.33 | 0.00 |
| J2 | −3.00 | 0.00 | −0.67 | 0.00 | 0.00 | −0.67 | J14 | 0.00 | 3.00 | −0.67 | 0.00 | 0.00 | 0.33 |
| J3 | −3.00 | 0.00 | −0.33 | 0.33 | −0.33 | 0.00 | J15 | 0.00 | 3.00 | 0.67 | −1.00 | 0.33 | 0.00 |
| J4 | −3.00 | 0.00 | −0.33 | 0.33 | 1.00 | −0.67 | J16 | 0.00 | 3.00 | 0.00 | 0.33 | 0.33 | 0.00 |
| J5 | −3.00 | 0.00 | 0.00 | −0.33 | 0.00 | −0.67 | J17 | 0.00 | 3.00 | 0.33 | −1.00 | −0.33 | 0.00 |
| J6 | −3.00 | 0.00 | 0.67 | 0.00 | 0.33 | 0.00 | J18 | 0.00 | 3.00 | 0.33 | −0.33 | 0.00 | −0.33 |
| J7 | −2.33 | −0.67 | 0.00 | 0.33 | 0.33 | 0.33 | J19 | 3.00 | 0.00 | 0.00 | −0.33 | 0.00 | 0.33 |
| J8 | −1.00 | −2.00 | 1.00 | −0.33 | −0.33 | −1.00 | J20 | 3.00 | 0.00 | −0.33 | 1.00 | 0.00 | −0.33 |
| J9 | 0.00 | −3.00 | 0.00 | −1.00 | 0.00 | 0.00 | J21 | 3.00 | 0.00 | 0.67 | 0.00 | 0.33 | 0.00 |
| J10 | 0.00 | −3.00 | 0.00 | 0.33 | 1.00 | −0.33 | J22 | 3.00 | 0.00 | 0.00 | 0.00 | 0.00 | 0.00 |
| J11 | −0.11 | −2.78 | −0.78 | 0.89 | −0.11 | −0.11 | J23 | 3.00 | 0.00 | −0.67 | 0.33 | 0.33 | −0.67 |
| J12 | 0.11 | −3.22 | 0.11 | −0.22 | 0.11 | −0.22 | J24 | 3.00 | 0.00 | −0.33 | 0.67 | −0.33 | 0.67 |

# 4   Example and Validation Study

The proposed strategy will be illustrated by using an artificial dataset where it is supposed that a set of $J = 24$ judges have been involved in a Conjoint experiment about a generic product described by three factors at three levels each. The estimated part-worths obtained by the metric conjoint model are reported in Table 1. Due to constraints imposed on the coefficients, only six out of nine coefficients have been determined. The 24 judges are arranged according to four different groups of six homogeneous units. Judges 1-6 (first group); Judges 7-13 (second group); Judges 14-18 (third group) and Judges 19-24 (fourth group). So we suppose that each group is a potential market segment and we want to relate these groups with their corresponding relational data.

Let us suppose that relational data were collected on the respondents (for instance, a link has been set if two respondents shared information about the product in a web-forum), the adjacency matrix is in Table 2. Notice that the second and the third group share several ties, whereas the first and the fourth group share fewer ties with the other groups.

The estimated coefficients $\tilde{\Theta}$ stated in (9) are in Table 3.

In order to analyze the network effect on the estimated part-worth coefficients, we start from the idea that if a group of consumers exchanges information to the extent of influencing their own opinions about the strong and weak features characterizing the product, then they should give coherent ratings and thus show similar utility profiles. This produces the consequence that groups of highly connected people in the net should have homogeneous utility profiles. The comparison of the individual part-worth coefficients with the ego-profile (average of the coefficients for the subset of consumers connected with ego) can be realized by looking at the corresponding elements in the matrices $\hat{B}'$ (Table 2) and $\tilde{\Theta}$ (Table 3).

Since the effect can be different for each individual actor in the net, we are interested to compare the coefficients at the individual level. To this aim, we carry out a permutation test based on the following procedure:

**Table 2** The Adjacency matrix **A** for the 24 Judges: higher density of the diagonal blocks simulate the network effect on the four groups of judges

| $J_1$ | $J_2$ | $J_3$ | $J_4$ | $J_5$ | $J_6$ | $J_7$ | $J_8$ | $J_9$ | $J_{10}$ | $J_{11}$ | $J_{12}$ | $J_{13}$ | $J_{14}$ | $J_{15}$ | $J_{16}$ | $J_{17}$ | $J_{18}$ | $J_{19}$ | $J_{20}$ | $J_{21}$ | $J_{22}$ | $J_{23}$ | $J_{24}$ |
|---|---|---|---|---|---|---|---|---|---|---|---|---|---|---|---|---|---|---|---|---|---|---|---|
| 0 | 1 | 1 | 1 | 0 | 0 | 0 | 0 | 0 | 0 | 0 | 0 | 0 | 0 | 0 | 0 | 0 | 0 | 0 | 0 | 0 | 0 | 0 | 0 |
| 1 | 0 | 1 | 0 | 1 | 0 | 0 | 0 | 0 | 0 | 0 | 0 | 0 | 0 | 0 | 0 | 0 | 0 | 0 | 0 | 0 | 0 | 0 | 0 |
| 1 | 1 | 0 | 0 | 1 | 0 | 0 | 0 | 0 | 0 | 0 | 0 | 0 | 0 | 0 | 0 | 0 | 0 | 0 | 0 | 0 | 0 | 0 | 0 |
| 1 | 0 | 0 | 0 | 1 | 0 | 0 | 0 | 0 | 0 | 1 | 1 | 0 | 0 | 0 | 0 | 0 | 0 | 0 | 0 | 0 | 0 | 0 | 0 |
| 0 | 1 | 1 | 1 | 0 | 1 | 1 | 1 | 0 | 0 | 1 | 1 | 1 | 0 | 0 | 0 | 0 | 0 | 0 | 0 | 0 | 0 | 0 | 0 |
| 0 | 0 | 0 | 0 | 1 | 0 | 1 | 0 | 1 | 1 | 0 | 1 | 0 | 1 | 0 | 1 | 0 | 1 | 0 | 0 | 0 | 0 | 0 | 0 |
| 0 | 0 | 0 | 0 | 1 | 1 | 0 | 1 | 1 | 0 | 1 | 1 | 1 | 1 | 0 | 0 | 0 | 0 | 0 | 0 | 0 | 0 | 0 | 0 |
| 0 | 0 | 0 | 0 | 1 | 0 | 1 | 0 | 1 | 1 | 1 | 0 | 1 | 1 | 0 | 1 | 0 | 1 | 0 | 1 | 1 | 0 | 0 | 0 |
| 0 | 0 | 0 | 0 | 0 | 1 | 1 | 1 | 0 | 1 | 1 | 1 | 0 | 1 | 1 | 1 | 0 | 0 | 0 | 0 | 0 | 0 | 0 | 0 |
| 0 | 0 | 0 | 0 | 0 | 1 | 0 | 1 | 1 | 0 | 1 | 1 | 1 | 1 | 0 | 1 | 0 | 1 | 1 | 1 | 1 | 0 | 0 | 0 |
| 0 | 0 | 0 | 1 | 1 | 0 | 1 | 1 | 1 | 1 | 0 | 1 | 1 | 1 | 0 | 0 | 1 | 1 | 0 | 1 | 0 | 0 | 0 | 0 |
| 0 | 0 | 0 | 1 | 1 | 1 | 1 | 0 | 1 | 1 | 1 | 0 | 1 | 1 | 1 | 1 | 1 | 1 | 0 | 0 | 1 | 0 | 1 | 0 |
| 0 | 0 | 0 | 0 | 1 | 0 | 1 | 1 | 0 | 1 | 1 | 1 | 0 | 1 | 1 | 1 | 0 | 0 | 0 | 1 | 1 | 1 | 0 | 0 |
| 0 | 0 | 0 | 0 | 0 | 1 | 1 | 1 | 1 | 1 | 1 | 1 | 1 | 0 | 1 | 1 | 1 | 0 | 0 | 1 | 1 | 1 | 0 | 0 |
| 0 | 0 | 0 | 0 | 0 | 0 | 0 | 0 | 1 | 0 | 0 | 1 | 1 | 1 | 0 | 1 | 0 | 1 | 1 | 1 | 0 | 0 | 0 | 0 |
| 0 | 0 | 0 | 0 | 0 | 1 | 0 | 1 | 1 | 1 | 0 | 1 | 1 | 1 | 1 | 0 | 1 | 0 | 1 | 1 | 1 | 0 | 0 | 0 |
| 0 | 0 | 0 | 0 | 0 | 0 | 0 | 0 | 0 | 0 | 1 | 1 | 0 | 1 | 0 | 1 | 0 | 1 | 1 | 1 | 0 | 0 | 0 | 0 |
| 0 | 0 | 0 | 0 | 0 | 1 | 0 | 1 | 0 | 1 | 1 | 1 | 0 | 0 | 1 | 0 | 1 | 0 | 1 | 1 | 0 | 1 | 1 | 0 |
| 0 | 0 | 0 | 0 | 0 | 0 | 0 | 0 | 0 | 1 | 0 | 0 | 0 | 0 | 1 | 1 | 1 | 1 | 0 | 1 | 1 | 1 | 0 | 0 |
| 0 | 0 | 0 | 0 | 0 | 0 | 0 | 1 | 0 | 1 | 1 | 0 | 1 | 1 | 1 | 1 | 1 | 1 | 1 | 0 | 1 | 0 | 1 | 0 |
| 0 | 0 | 0 | 0 | 0 | 0 | 0 | 1 | 0 | 1 | 0 | 1 | 1 | 1 | 0 | 1 | 0 | 0 | 1 | 1 | 0 | 1 | 1 | 0 |
| 0 | 0 | 0 | 0 | 0 | 0 | 0 | 0 | 0 | 0 | 0 | 0 | 1 | 1 | 0 | 0 | 0 | 1 | 1 | 0 | 1 | 0 | 1 | 0 |
| 0 | 0 | 0 | 0 | 0 | 0 | 0 | 0 | 0 | 0 | 0 | 1 | 0 | 0 | 0 | 0 | 0 | 1 | 0 | 1 | 1 | 1 | 0 | 1 |
| 0 | 0 | 0 | 0 | 0 | 0 | 0 | 0 | 0 | 0 | 0 | 0 | 0 | 0 | 0 | 0 | 0 | 0 | 0 | 0 | 0 | 0 | 1 | 0 |

**Table 3** The matrix $\tilde{\Theta}$: coefficients are the average part-worths of the judge's neighbors

| $J$ | $X_{11}$ | $X_{12}$ | $X_{21}$ | $X_{22}$ | $X_{31}$ | $X_{32}$ | $J$ | $X_{11}$ | $X_{12}$ | $X_{21}$ | $X_{22}$ | $X_{31}$ | $X_{32}$ |
|---|---|---|---|---|---|---|---|---|---|---|---|---|---|
| J1 | −3,00 | 0,00 | −0,44 | 0,22 | 0,22 | −0,44 | J13 | 0,42 | −0,42 | 0,06 | 0,00 | 0,11 | −0,17 |
| J2 | −2,89 | −0,11 | −0,44 | 0,00 | −0,11 | 0,00 | J14 | 0,36 | −0,36 | 0,12 | −0,07 | 0,12 | −0,12 |
| J3 | −2,89 | −0,11 | −0,56 | −0,11 | 0,00 | −0,22 | J15 | 0,76 | 0,72 | −0,15 | 0,01 | 0,10 | −0,03 |
| J4 | −1,42 | −1,58 | −0,42 | 0,08 | 0,00 | −0,08 | J16 | 0,43 | 0,06 | 0,15 | −0,16 | 0,15 | −0,10 |
| J5 | −1,44 | −0,89 | −0,11 | 0,11 | 0,07 | −0,26 | J17 | 0,86 | 0,43 | −0,19 | 0,19 | 0,05 | −0,05 |
| J6 | −0,36 | −0,40 | −0,03 | −0,24 | 0,14 | −0,11 | J18 | 0,73 | −0,45 | 0,09 | −0,03 | 0,12 | −0,21 |
| J7 | −0,88 | −0,63 | −0,04 | −0,04 | 0,04 | −0,21 | J19 | 1,13 | 1,13 | 0,21 | −0,08 | 0,21 | −0,13 |
| J8 | 0,26 | 0,02 | −0,13 | 0,08 | 0,14 | −0,10 | J20 | 0,66 | 0,85 | 0,02 | −0,04 | 0,16 | −0,15 |
| J9 | −0,44 | −0,56 | 0,11 | −0,07 | 0,15 | −0,11 | J21 | 1,11 | 0,08 | −0,12 | 0,18 | 0,18 | −0,19 |
| J10 | 0,42 | 0,08 | 0,03 | 0,06 | 0,08 | −0,11 | J22 | 1,50 | 1,50 | −0,17 | 0,06 | 0,17 | −0,06 |
| J11 | −0,32 | −0,19 | −0,02 | −0,13 | 0,12 | −0,24 | J23 | 2,02 | −0,04 | 0,07 | 0,19 | 0,02 | −0,04 |
| J12 | −0,21 | 0,41 | −0,03 | −0,10 | 0,21 | −0,14 | J24 | 3,00 | 0,00 | −0,67 | 0,33 | 0,33 | −0,67 |

1. For each judge $i = 1, \ldots, J$ compute the mean absolute deviation among the $k=K$-$p$ coefficients (in this example $J = 24$; $K = 9$; $p = 3$; $k = 6$):

$$D_i = k^{-1} \sum_{j=1,k} |\hat{b}_{ij} - \tilde{\theta}_{ij}| \tag{10}$$

2. Produce a random simultaneous permutation of the rows and columns of the observed adjacency matrix, let it be denoted with $\mathbf{A}_{rnd}$.
3. Use $\mathbf{A}_{rnd}$ in computing the coefficients: $\tilde{\Theta}_{rnd}$ according to (9).
4. Get the random individual mean absolute deviations:

$$D_{i,rnd} = k^{-1} \sum_{j=1,k} |\hat{b}_{ij} - \tilde{\theta}_{ij,rnd}| \tag{11}$$

5. Repeat steps from 2 to 4 a very large number of times (we produced 10,000 permutations).
6. Obtain $J$ cumulative distribution functions (*cdf*) of the random individual deviations $D_{i,rnd}$ with $i = 1, \ldots, J$.
7. For each judge, compare the *i-th* observed $D_i$ with the random *cdf* percentiles of the correspondent $D_{i,rnd}$.

In case of a significant effect, the observed $D_i$ should lie in the left tail of the random distribution with a value smaller than a given percentile (say 1%).

The experiment is now conducted on the artificial data set. Having fixed the matrix $\hat{B}$ of the estimated part-worths, we generate 10,000 adjacency matrices by using the rmperm() function of the software package *sna* (ver. 2.1-0; Butts 2010) implemented in the *R* environment (R Development Core Team 2010).

**Table 4** CDF percentiles of the random MAD compared with the observed one (*Obs. MAD*) for 24 judges. The values smaller than the fifth percentile are in bold

| J | 1% | 2% | 5% | 10% | 20% | Obs.MAD | J | 1% | 2% | 5% | 10% | 20% | Obs.MAD |
|---|---|---|---|---|---|---|---|---|---|---|---|---|---|
| J1 | 0,56 | 0,63 | 0,69 | 0,74 | 0,79 | **0.46** | J13 | 0,53 | 0,59 | 0,67 | 0,72 | 0,78 | 0.94 |
| J2 | 0,39 | 0,54 | 0,62 | 0,68 | 0,73 | **0.20** | J14 | 0,46 | 0,54 | 0,61 | 0,67 | 0,71 | 0.86 |
| J3 | 0,44 | 0,52 | 0,60 | 0,65 | 0,70 | **0.24** | J15 | 0,59 | 0,67 | 0,76 | 0,81 | 0,86 | 0.86 |
| J4 | 0,58 | 0,65 | 0,74 | 0,79 | 0,85 | 0.85 | J16 | 0,39 | 0,46 | 0,54 | 0,59 | 0,64 | 0.71 |
| J5 | 0,44 | 0,50 | 0,57 | 0,63 | 0,68 | 0.58 | J17 | 0,55 | 0,63 | 0,72 | 0,78 | 0,84 | 0.93 |
| J6 | 0,49 | 0,56 | 0,62 | 0,67 | 0,71 | 0.71 | J18 | 0,43 | 0,48 | 0,57 | 0,63 | 0,68 | 0.83 |
| J7 | 0,56 | 0,62 | 0,70 | 0,75 | 0,80 | 0.81 | J19 | 0,43 | 0,50 | 0,57 | 0,62 | 0,68 | 0.68 |
| J8 | 0,65 | 0,69 | 0,76 | 0,81 | 0,88 | 1.03 | J20 | 0,48 | 0,58 | 0,68 | 0,73 | 0,79 | 0.82 |
| J9 | 0,54 | 0,59 | 0,65 | 0,70 | 0,75 | 0.70 | J21 | 0,50 | 0,54 | 0,62 | 0,67 | 0,72 | **0.55** |
| J10 | 0,54 | 0,61 | 0,68 | 0,74 | 0,79 | 0.82 | J22 | 0,33 | 0,41 | 0,48 | 0,53 | 0,58 | 0.57 |
| J11 | 0,58 | 0,65 | 0,70 | 0,75 | 0,80 | 0.83 | J23 | 0,50 | 0,59 | 0,69 | 0,75 | 0,81 | **0.48** |
| J12 | 0,46 | 0,50 | 0,56 | 0,61 | 0,66 | 0.73 | J24 | 0,53 | 0,66 | 0,75 | 0,81 | 0,88 | **0.45** |

In Table 4 we have reported the results of the experiment. For each of the 24 judges, we compare the observed mean absolute deviation (MAD) between the two kinds of part-worth coefficients. Small values of the observed MAD indicate the presence of the network effect. In order to assess the significance of the MAD values, we compare them with the lower percentiles of the simulated distributions. The MAD values smaller than the fifth percentile are in bold. Let us notice that three out of six members of group 1 (J1, J2, J3) and three members of group 4 (J21, J23, J24) show significant presence of the network effect. The groups 2 and 3 which share numerous ties (see Table 2) pertain to two different utility profiles (see Table 1) so that for judges labeled from J7 to J18 the network effect is not relevant.

## 5 Concluding Remarks

An extension of the metric approach to the full profile Conjoint Analysis method has been provided. Adjacency data – derived by links in a social network of a brand community – have been used as external information in a multiple regression model with multivariate responses. A validation procedure has been proposed in the framework of random permutation tests. It needs further work to evaluate how to perform the proposed method applied to networks different in size, density and topology.

## References

Benali H, Escofier B (1990) Analyse factorielle lissée et analyse des différences locales. Revue Statist Appl 38(2):55–76

Brewer MB, Gardner W (1996) Who is this "we"? Levels of collective identity and self-representations. J Pers Soc Psychol 71:83–93

Butts CT (2010) sna: Tools for social network analysis. R package version 2.1-0. URL http:// CRAN.R-project.org/package=sna

D'Ambra L, Lauro NC (1982) Analisi in componenti principali in rapporto ad un sottospazio di riferimento. Rivista di Statistica Applicata 15:51–67

Giordano G, Scepi G (1999) Different informative structures for quality design. J Ital Stat Soc 8(2–3):139–149

Giordano G, Vitale MP (2007) Factorial contiguity maps to explore relational data patterns. Statistica Applicata 19(4):297–306

Green PE, Srinivasan V (1990) Conjoint analysis in marketing: New developments with implications for research and practice. J Market 54:3–19

Israels AZ (1984) Redundancy analysis for qualitative variables. Psychometrika 49:331–346

Kempe D, Kleinberg J, Tardos E (2003) Maximizing the spread of influence through a social network. In: KDD 03: Proceedings of the ninth ACM SIGKDD international conference on Knowledge discovery and data mining, ACM, New York, pp 137–146

Kozinets RV (1999) E-tribalized marketing?: The strategic implications of virtual communities of consumption. Eur Manag J 17(3):252–264

Lauro CN, Giordano G, Verde R (1998) A multidimensional approach to conjoint analysis. Appl Stochastic Models Data Anal 14(4):265–274

R Development Core Team (2010) R: A language and environment for statistical computing. R Foundation for Statistical Computing, Vienna, Austria, URL http://www.R-project.org

Takane Y, Shibayama T (1991) Principal component analysis with external information on both subjects and variables. Psychometrika 56:97–120

Wasserman S, Faust K (1994) Social network analysis: methods and applications. Cambridge University Press, Cambridge, New York

# Clustering Coefficients of Random Intersection Graphs

**Erhard Godehardt, Jerzy Jaworski, and Katarzyna Rybarczyk**

**Abstract** Two general random intersection graph models (active and passive) were introduced by Godehardt and Jaworski (Exploratory Data Analysis in Empirical Research, Springer, Berlin, Heidelberg, New York, pp. 68–81, 2002). Recently the models have been shown to have wide real life applications. The two most important ones are: non-metric data analysis and real life network analysis. Within both contexts, the clustering coefficient of the theoretical graph models is studied. Intuitively, the clustering coefficient measures how much the neighborhood of the vertex differs from a clique. The experimental results show that in large complex networks (real life networks such as social networks, internet networks or biological networks) there exists a tendency to connect elements, which have a common neighbor. Therefore it is assumed that in a good theoretical network model the clustering coefficient should be asymptotically constant. In the context of random intersection graphs, the clustering coefficient was first studied by Deijfen and Kets (Eng Inform Sci, 23:661–674, 2009). Here we study a wider class of random intersection graphs than the one considered by them and give the asymptotic value of their clustering coefficient. In particular, we will show how to set parameters – the sizes of the vertex set, of the feature set and of the vertices' feature sets – in such a way that the clustering coefficient is asymptotically constant in the active (respectively, passive) random intersection graph.

E. Godehardt (✉)
Clinic of Cardiovascular Surgery, Heinrich Heine University, 40225 Düsseldorf, Germany
e-mail: godehard@uni-duesseldorf.de

J. Jaworski · K. Rybarczyk
Faculty of Mathematics and Computer Science, Adam Mickiewicz University, Umultowska 87, 61-614 Poznań, Poland
e-mail: jaworski@amu.edu.pl; kryba@amu.edu.pl

W. Gaul et al. (eds.), *Challenges at the Interface of Data Analysis, Computer Science, and Optimization*, Studies in Classification, Data Analysis, and Knowledge Organization, DOI 10.1007/978-3-642-24466-7_25, © Springer-Verlag Berlin Heidelberg 2012

# 1 Introduction

There are several definitions of *clustering coefficients* of a graph. They are, intuitively, graph parameters which describe a local density in the neighborhood of a given vertex or, in the case of a global graph characteristic, an "average" vertex. Strogatz and Watts in (1998) studied the clustering coefficient which measures how much the neighborhood of the vertex $v$ differs from a clique.

**Definition 1.** Let $G$ be a graph with the vertex set $V = V(G)$ and the edge set $E = E(G)$, and let $N(v)$ be the set of all vertices adjacent to a vertex $v \in V$. Denote by $E(G \ [N(v)])$ the set of all edges joining the vertices of $N(v)$. The *clustering coefficient of a vertex* $v \in V(G)$ is defined, for $|N(v)| \geq 2$ by

$$\mathrm{Clus}_\ell(v) = \frac{|E(G \ [N(v)])|}{\binom{|N(v)|}{2}}$$

and 0 otherwise. The *clustering coefficient of a graph* $G$ is defined by

$$\mathrm{Clus}_{g*}(G) = \frac{1}{|V(G)|} \sum_{v \in V(G)} \mathrm{Clus}_\ell(v) .$$

Barrat and Weigt in (2000) and Newman, Strogatz, and Watts in (2001) used yet another definition of a "global" clustering coefficient. Indeed, instead of using the average of the local clustering coefficients as above it is natural to study another average, namely

$$\mathrm{Clus}_g(G) = \frac{\sum_{v \in V} \binom{|N(v)|}{2} \mathrm{Clus}_\ell(v)}{\sum_{v \in V} \binom{|N(v)|}{2}} = \frac{\sum_{v \in V} |E(G \ [N(v)])|}{\sum_{v \in V} \binom{|N(v)|}{2}}$$

$$= \frac{\sum_{v \in V} \{\text{the number of triangles in } G \text{ containing } v\}}{\sum_{v \in V} \{\text{the number of paths of length 2 with the center at } v\}} ,$$

which leads to the following definition:

**Definition 2.** Let $G$ be a graph with at least one vertex of degree greater than 1. The *global clustering coefficient* of a graph $G$ is defined by

$$\mathrm{Clus}_g(G) = \frac{3 \times \{\text{ the number of triangles in } G\}}{\{\text{ the number of paths of length 2}\}} .$$

Finally, in the context of random intersection graphs, Deijfen and Kets in (2009), introduced the clustering coefficient of a vertex $v$ as a probability of an edge $wu$ under the condition that $w$ and $u$ are adjacent to $v$.

**Definition 3.** The *clustering coefficient* $\mathrm{ClusC}(\mathscr{G})$ of a vertex $v$ in a random graph $\mathscr{G}$ is defined by the probability:

$$\mathrm{ClusC}(\mathscr{G}) = \Pr\left\{uw \in E(\mathscr{G}) \mid vw \in E(\mathscr{G}) \text{ and } vu \in E(\mathscr{G})\right\}. \qquad (1)$$

Note that for the homogeneous models of random graphs, where the vertices are indistinguishable, $\mathrm{ClusC}(\mathscr{G})$ does not depend on the particular choice of $v$, $w$ and $u$, and the above clustering coefficient plays the role as a global parameter, as well. Moreover, since for such graphs (1) can be rewritten as

$$\begin{aligned}
\mathrm{ClusC}(\mathscr{G}) &= \frac{3\binom{n}{3}\Pr\{uw \in E(\mathscr{G}) \text{ and } vw \in E(\mathscr{G}) \text{ and } vu \in E(\mathscr{G})\}}{n\binom{n-1}{2}\Pr\{vw \in E(\mathscr{G}) \text{ and } vu \in E(\mathscr{G})\}} \\
&= \frac{3\,\mathbb{E}\big(\text{the number of triangles in } \mathscr{G}\big)}{\mathbb{E}\big(\text{the number of paths of length 2 in } \mathscr{G}\big)},
\end{aligned}$$

it is closely related to the global clustering coefficient given by Definition 2 and it is one of the empirical measures of clustering (see Newman (2003)). Since we will be interested in asymptotical results, namely when the number of vertices of random graph models under consideration grows to infinity, we will also study the limit of the probability given by (1). For simplicity we will use the same notation for the clustering coefficient in the limit case. Moreover all limits as well as notations $o(\cdot)$ and $O(\cdot)$ are as $n \to \infty$. Finally, $(k)_j$ denotes here the falling factorial, i.e., $(k)_j = \binom{k}{j}j! = k(k-1) \times \ldots \times (k-j+1)$.

## 2 Random Intersection Graphs

We will present here results about the clustering coefficients of two random graphs: active and passive random intersection graphs. The models were defined in Godehardt and Jaworski (2002) and they are both generated by a random bipartite graph. Recall that any bipartite graph $\mathscr{BG} = \mathscr{BG}(\mathscr{V} \cup \mathscr{W}, \mathscr{E})$ with the 2-partition $(\mathscr{V}, \mathscr{W})$ of the vertex set and with edge set $\mathscr{E}$ generates two intersection graphs. The first one with the vertex set $\mathscr{V}$ has two vertices joined by an edge if and only if the sets of neighbors of these vertices in $\mathscr{BG}$ have a non-empty intersection. The second intersection graph generated by $\mathscr{BG}$ is defined on the vertex set $\mathscr{W}$, analogously. There are many situations in the study of complex and scale free networks for which the bipartite structure (known or hidden) relating elements of the network to their properties influences the structure of the connections between elements (see Guillaume and Latapy (2004)). In another case, when the vertices of $\mathscr{V}$ represent individuals and the elements of $\mathscr{W}$ correspond to social groups – this defines models for a social network in which two individuals are joined by an edge in the active model if they share at least one group, or we have an edge in the passive case (between two groups) when there exists an individual who chose both groups. Therefore, in both cases there are natural interpretations of the clustering properties. Graph-theoretical concepts are also one of the commonly used tools in data structure analysis. For information on how these concepts can be used in defining cluster models, revealing hidden clusters in a data set, and testing the randomness of such clusters we refer the reader to Bock (1996), Godehardt (1990), Godehardt and Jaworski (1996) for metric data, and Godehardt and Jaworski (2002), Godehardt et al. (2007) for non-metric data. It is natural that in many non-metric data sets relations between objects depend on properties they possess. Therefore, an effective model to analyze the structure of those relations is a random graph, in which edges represent relations based on the objects' properties. The random bipartite graph which will serve to introduce the two models of random intersection graphs is defined as follows.

**Definition 4.** Let $\mathscr{P}_{(m)}$ be a probability distribution with $\mathscr{P}_{(m)} = (P_0, P_1, \ldots, P_m)$, i.e., an $(m+1)$-tuple of non-negative real numbers which satisfy $P_0 + \cdots + P_m = 1$. Denote by $\mathscr{BG}\left(n, m, \mathscr{P}_{(m)}\right)$ a random bipartite graph on a vertex set $\mathscr{V} \cup \mathscr{W}$, where $\mathscr{V} = \{v_1, v_2, \ldots, v_n\}$, $\mathscr{W} = \{w_1, w_2, \ldots, w_m\}$, such that (here and in the following text, $D(z)$ denotes the set of neighbors of a vertex $z$ in the bipartite graph):

- each vertex $v \in \mathscr{V}$ chooses its degree and then its neighbors from $\mathscr{W}$ independently of all other vertices from $\mathscr{V}$;
- the vertex $v$ chooses its degree according to the distribution $\mathscr{P}_{(m)}$,

$$\Pr\{|D(v)| = k\} = P_k, \quad k = 0, 1, \ldots, m;$$

- for every $\mathscr{S} \subseteq \mathscr{W}$, $|\mathscr{S}| = k$, $\Pr\{\mathscr{S} = D(v)\} = P_k / \binom{m}{k}$, i.e., given $|D(v)| = k$, a subset $\mathscr{S}$ of neighbors of a vertex $v$ is chosen uniformly over the class of all $k$-element subsets of $\mathscr{W}$.

We now recall two models of random intersection graphs. The first one, the *active random intersection graph* $\mathscr{G}^{act}\left(\mathscr{BG}\left(n, m, \mathscr{P}_{(m)}\right)\right) = \mathscr{G}^{act}\left(n, m, \mathscr{P}_{(m)}\right)$ is a random graph with the vertex set $\mathscr{V}$, $|\mathscr{V}| = n$, such that any pair $v_i$ and $v_j$ is joined by an edge $v_i v_j$ if and only if the sets of their neighbors (subsets of $\mathscr{W}$ in the original bipartite graph) intersect, i.e., if $D(v_i) \cap D(v_j) \neq \emptyset$. The model can be generalized by replacing the condition above with $|D(v_i) \cap D(v_j)| \geq s$, $s \geq 1$.

Using $\mathscr{BG}\left(n, m, \mathscr{P}_{(m)}\right)$, we can also define the *passive random intersection graph* $\mathscr{G}^{pas}\left(\mathscr{BG}\left(n, m, \mathscr{P}_{(m)}\right)\right) = \mathscr{G}^{pas}\left(m, n, \mathscr{P}_{(m)}\right)$ as a random graph with the vertex set $\mathscr{W}$, $|\mathscr{W}| = m$, such that any pair $w_k$ and $w_l$ of vertices is joined by an edge $w_k w_l$ if and only if the sets of their neighbors (now subsets of $\mathscr{V}$) intersect. Like the previous model, this can also be generalized, by demanding the intersection of the sets of their neighbors to have at least $s$ elements, $s \geq 1$. Note that the random intersection graph considered in Deijfen and Kets (2009) is the special case of both the active and the passive model for which the probability distribution $\mathscr{P}_{(m)}$ is binomial $Bin(m, \mathbf{p})$, where the parameter $\mathbf{p}$ is a random variable itself. Let us mentioned here that it is very interesting to generalize the results presented here for the general case when $s > 1$ (for other results for such general model see Bloznelis et al. (2009), Godehardt and Jaworski (2002), Godehardt et al. (2010), Rybarczyk (2010)).

## 3  Clustering Coefficient of Active Random Intersection Graphs

**Theorem 1.** *Let* $\text{ClusC}_{act}$ *be the local clustering coefficient for the active model* $\mathcal{G}^{act}\left(n, m, \mathcal{P}_{(m)}\right)$ *of a random intersection graph. Let* $Z_n$ *be a random variable with the distribution* $\mathcal{P}_{(m)}$ *and let* $r_n = \sqrt{m/n}$. *If* $r_n = o(\sqrt[3]{m})$, $\Pr\{Z_n \geq 2\} = 1$ *and* $\mathbb{E}\, Z_n = O(r_n)$, *then for any vertex* $v \in \mathcal{V}$

$$\text{ClusC}_{act} = \mathbb{E}\frac{1}{Z_n} + o(1).$$

*Proof.* We will use the following short notation: $\mathcal{G}^{act} = \mathcal{G}^{act}\left(n, m, \mathcal{P}_{(m)}\right)$ and $E(\mathcal{G}^{act}) = E(\mathcal{G}^{act}\left(n, m, \mathcal{P}_{(m)}\right))$. Let $|D(v)| = d_v, |D(w)| = d_w$ and $|D(u)| = d_u$ and let us assume that $2 \leq d_z \leq d_{max} = o(\sqrt[3]{m})$ for $z \in \{u, v, w\}$. For simplicity, let us also use the following notation for the conditional probabilities:

$$\Pr_{vwu}\{\,\cdot\,\} = \Pr\left\{\,\cdot\,\Big|\, |D(v)| = d_v, |D(w)| = d_w, |D(u)| = d_u\right\}$$

and

$$P_{\cap k} = \Pr_{vwu}\left\{|D(v) \cap D(w)| = k\right\},$$

$$P_{1|\cap k} = \Pr_{vwu}\left\{wu, vw, vu \in E(\mathcal{G}^{act}), D(v) \cap D(w) \cap D(u) \neq \emptyset \,\Big|\right.$$
$$\left. |D(v) \cap D(w)| = k\right\},$$

$$P_{2|\cap k} = \Pr_{vwu}\left\{wu, vw, vu \in E(\mathcal{G}^{act}), D(v) \cap D(w) \cap D(u) = \emptyset \,\Big|\right.$$
$$\left. |D(v) \cap D(w)| = k\right\}.$$

Note that

$$\Pr_{vwu}\{wu, vw, vu \in E(\mathcal{G}^{act})\} = \sum_{k=1}^{\min\{d_v, d_w\}} P_{1|\cap k} P_{\cap k} + \sum_{k=1}^{\min\{d_v, d_w\}} P_{2|\cap k} P_{\cap k}. \quad (2)$$

For $d_v + d_w \leq m/2$, $k \geq 1$ and $k \leq \min\{d_v, d_w\}$

$$P_{\cap k+1} = \frac{\binom{d_v}{k+1}\binom{m-d_v}{d_w-k-1}}{\binom{m}{d_w}} = \frac{(d_v)_{k+1}(d_w)_{k+1}}{(k+1)!(m)_{k+1}} \frac{(m-d_v)_{d_w-k-1}}{(m-k-1)_{d_w-k-1}}$$

$$\leq P_{\cap k} \frac{d_v d_w}{m} \frac{2}{k+1} \leq P_{\cap k} \frac{d_v d_w}{m}.$$

Therefore for $n$ large enough

$$P_{\cap 1} \leq \frac{d_v \binom{m-1}{d_w-1}}{\binom{m}{d_w}} = \frac{d_v d_w}{m} \quad \text{and} \quad P_{\cap k} \leq P_{\cap 1} \left(\frac{d_v d_w}{m}\right)^{k-1};$$

$$P_{1|\cap k} \leq \frac{k \binom{m-1}{d_u-1}}{\binom{m}{d_u}} = \frac{k d_u}{m} \leq \frac{d_{max} d_u}{m};$$

$$P_{2|\cap k} \leq \frac{(d_v - k)(d_w - k)\binom{m-2}{d_u-2}}{\binom{m}{d_u}} \leq \frac{d_v d_w d_u(d_u - 1)}{m(m-1)} \leq \frac{d_v d_w}{m} \frac{d_u^2}{m}.$$

Hence for $n$ such that $m/(m - d_v d_w) \leq 2$ we have

$$\sum_{k=2}^{\min\{d_v,d_w\}} P_{1|\cap k} P_{\cap k} \leq P_{\cap 1} \sum_{k=2}^{\min\{d_v,d_w\}} \frac{d_{max} d_u}{m} \left(\frac{d_v d_w}{m}\right)^{k-1} \leq \frac{d_v d_w d_u}{m^2} \frac{2 d_{max}^3}{m},$$

$$\sum_{k=1}^{\min\{d_v,d_w\}} P_{2|\cap k} P_{\cap k} \leq P_{\cap 1} \sum_{k=1}^{\min\{d_v,d_w\}} \frac{d_u^2}{m} \left(\frac{d_v d_w}{m}\right)^{k} \leq \frac{d_v d_w d_u}{m^2} \frac{2 d_{max}^3}{m}$$

and

$$\frac{d_v d_w d_u}{m^2} \left(1 - \frac{3 d_{max}^2}{m}\right) \leq \frac{d_v \binom{m-d_v}{d_w-1} \binom{m-d_v-d_w+1}{d_u-1}}{\binom{m}{d_w} \binom{m}{d_u}} \leq P_{\cap 1} P_{1|\cap 1} \leq \frac{d_v d_w d_u}{m^2}.$$

The above bounds and (2) imply that

$$\frac{d_v d_w d_u}{m^2} \left(1 - \frac{3 d_{max}^2}{m}\right) \leq \Pr_{vwu} \{wu, vw, vu \in E(\mathcal{G}^{act})\} \leq \frac{d_v d_w d_u}{m^2} \left(1 + \frac{4 d_{max}^3}{m}\right).$$

Moreover one can easily show (see the proof of Lemma 3 in Rybarczyk (2011)) that

$$\frac{d_v d_w}{m} \left(1 - \frac{d_{max}^2}{m}\right) \leq \Pr_{vwu}\{vw \in E(\mathcal{G}^{act})\} \leq \frac{d_v d_w}{m},$$

$$\frac{d_v d_u}{m} \left(1 - \frac{d_{max}^2}{m}\right) \leq \Pr_{vwu}\{vu \in E(\mathcal{G}^{act})\} \leq \frac{d_v d_u}{m}.$$

Therefore

$$\Pr_{vwu}\{wu \in E(\mathcal{G}^{act}) | vw, vu \in E(\mathcal{G}^{act})\} = \frac{1}{d_v} \left(1 + O\left(\frac{d_{max}^3}{m}\right)\right),$$

where $O(\cdot)$ is uniformly bounded with respect to $d_v, d_w, d_u \le d_{max}$. Note that if

$$\xi := \Pr\{Z_n > d_{max}\},$$

then $\Pr\{|D(v)| > d_{max} \text{ or } |D(w)| > d_{max} \text{ or } |D(u)| > d_{max}\} \le 3\xi$ and

$$\text{ClusC}_{act} \le \sum_{d_v, d_w, d_u \le d_{max}} \Pr_{vwu}\{wu \in E(\mathscr{G}^{act}) | vw, vu \in E(\mathscr{G}^{act})\}$$

$$\times \Pr\{|D(v)| = d_v, |D(w)| = d_w, |D(u)| = d_u\} + 3\xi$$

$$= \left(1 + O\left(\frac{d_{max}^3}{m}\right)\right) \sum_{d_v \le d_{max}} \frac{1}{d_v} \Pr\{|D(v)| = d_v\}$$

$$\times \sum_{d_w \le d_{max}} \Pr\{|D(w)| = d_w\} \sum_{d_u \le d_{max}} \Pr\{|D(u)| = d_u\} + 3\xi$$

$$\le \left(1 + O\left(\frac{d_{max}^3}{m}\right)\right) \mathbb{E}\frac{1}{Z_n} + 3\xi.$$

Moreover

$$\text{ClusC}_{act} \ge \sum_{d_v, d_w, d_u \le d_{max}} \Pr_{vwu}\{wu \in E(\mathscr{G}^{act}) | vw, vu \in E(\mathscr{G}^{act})\}$$

$$\times \Pr\{|D(v)| = d_v, |D(w)| = d_w, |D(u)| = d_u\}$$

$$= \left(1 + O\left(\frac{d_{max}^3}{m}\right)\right)(1 - \xi)^2 \left(\mathbb{E}\frac{1}{Z_n} - \sum_{d_v > d_{max}} \frac{1}{d_v} \Pr\{|D(v)| = d_v\}\right)$$

$$\ge \left(1 + O\left(\frac{d_{max}^3}{m}\right)\right)(1 - \xi)^2 \left(\mathbb{E}\frac{1}{Z_n} - \frac{\xi}{d_{max}}\right).$$

Let $\omega(n) \to \infty$, $\omega(n) = o(\sqrt[3]{m}/r_n)$ and $d_{max} = r_n \omega(n)$. Then $d_{max}^3 = r_n^3 \omega(n)^3 = o(m)$ and by the Markov inequality

$$\xi = \Pr\{Z_n > d_{max}\} = \Pr\left\{\frac{Z_n}{r_n} > \omega(n)\right\} \le \frac{\mathbb{E}\frac{Z_n}{r_n}}{\omega(n)} = o(1).$$

This implies the thesis immediately. □

*Remark 1.* From the above theorem and the results on the degree distribution of a vertex (see Bloznelis (2008), Rybarczyk (2011)) it follows that the active model of a random intersection graph makes it possible to obtain any arbitrary prescribed value for the clustering coefficient and at the same time have the degrees following the power law distribution. In Rybarczyk (2011) an example of a parameter setting which leads to both properties of complex networks (see Albert and Barabási (2002)) is given.

# 4   Clustering Coefficient of Passive Random Intersection Graphs

**Theorem 2.** *Let* $\mathrm{ClusC}_{pas}$ *be the local clustering coefficient for the passive model* $\mathscr{G}^{pas}\left(n, m, \mathscr{P}_{(m)}\right)$ *of a random intersection graph and let* $Q_2 = \mathbb{E}(Z_n)_2/(m)_2$ *and* $Q_3 = \mathbb{E}(Z_n)_3/(m)_3$*, where* $Z_n$ *is a random variable with the distribution* $\mathscr{P}_{(m)}$*. Then for any vertex* $w \in \mathscr{W}$

$$
\mathrm{ClusC}_{pas} = \frac{1 - 3(1 - Q_2)^n + 3(1 - 2Q_2 + Q_3)^n - (1 - 3Q_2 + 2Q_3)^n}{1 - 2(1 - Q_2)^n + (1 - 2Q_2 + Q_3)^n}.
$$

*Proof.* In order to find the clustering coefficient of the passive random intersection graph we will introduce three events related to a random bipartite graph $\mathscr{BG}\left(n, m, \mathscr{P}_{(m)}\right)$. For given $w, w_1, w_2 \in \mathscr{W}$ let denote by $A_\triangledown = A_\triangledown(w, w_1, w_2)$ the event that there exists a vertex $v \in \mathscr{V}$ such that $\{w, w_1, w_2\} \subset D(v) \subset \mathscr{W}$; by $B_\triangledown = B_\triangledown(w, w_1, w_2)$ the event that there is no vertex $v \in \mathscr{V}$ such that $\{w, w_1, w_2\} \subset D(v)$ but there exist vertices $v_1, v_2, v_3 \in \mathscr{V}$ such that $\{w, w_1\} \subset D(v_1)$, $\{w, w_2\} \subset D(v_2)$ and $\{w_1, w_2\} \subset D(v_3)$ and finally by $C_\vee = C_\vee(w, w_1, w_2)$ the event that there is no vertex $v \in \mathscr{V}$ such that $\{w, w_1, w_2\} \subset D(v)$ but there exist vertices $v_1, v_2, \in \mathscr{V}$ such that $\{w, w_1\} \subset D(v_1)$ and $\{w, w_2\} \subset D(v_2)$. Note that the events $A_\triangledown$ and $B_\triangledown$ are disjoint and both imply that the vertices $w, w_1, w_2$ form a triangle in the passive model $\mathscr{G}^{pas}\left(n, m, \mathscr{P}_{(m)}\right)$. Similarly the events $A_\triangledown$ and $C_\vee$ are disjoint and both imply that the vertices $w, w_1, w_2$ form a path of length 2 centered in $w$ in the passive model $\mathscr{G}^{pas}\left(n, m, \mathscr{P}_{(m)}\right)$. Hence by Definition 3

$$
\mathrm{ClusC}_{pas} = \frac{\Pr\{A_\triangledown\} + \Pr\{B_\triangledown\}}{\Pr\{A_\triangledown\} + \Pr\{C_\vee\}}. \tag{3}
$$

Let us introduce two more events for given $v \in \mathscr{V}$ and $w, w_1, w_2 \in \mathscr{W}$, the first one $A_3(v) = A_3(v; w, w_1, w_2)$ is the event that $\{w, w_1, w_2\} \subset D(v)$, while the second $A_2(v) = A_2(v; w, w_1)$ means that $\{w, w_1\} \subset D(v)$. Note that

$$
\Pr\{A_3(v)\} = \sum_{k=3}^{m} P_k \frac{\binom{m-3}{k-3}}{\binom{m}{k}} = \sum_{k=3}^{m} P_k \frac{(k)_3}{(m)_3} = \frac{\mathbb{E}(Z_n)_3}{(m)_3} = Q_3
$$

and, similarly

$$
\Pr\{A_2(v)\} = \sum_{k=2}^{m} P_k \frac{\binom{m-2}{k-2}}{\binom{m}{k}} = \sum_{k=2}^{m} P_k \frac{(k)_2}{(m)_2} = \frac{\mathbb{E}(Z_n)_2}{(m)_2} = Q_2.
$$

Therefore, since $A_\triangledown = \bigcup_{v \in \mathscr{V}} A_3(v)$ and the events $A_3(v)$, for $v \in \mathscr{V}$ are independent (see Definition 4), we obtain immediately that

$$\Pr\{A_{\triangledown}\} = 1 - \Pr\left\{\left(\bigcup_{v \in \mathscr{V}} A_3(v)\right)^c\right\} = 1 - \prod_{v \in \mathscr{V}} \Pr\{A_3^c(v)\} = 1 - (1 - Q_3)^n. \quad (4)$$

In order to determine the probability of the event $B_{\triangledown}$ let us first notice that it can be described, equivalently, in the following way: there exists an ordered partition of the set $\mathscr{V}$ into four subset, say $(\mathscr{V}_1; \mathscr{V}_2; \mathscr{V}_3; \mathscr{V}_4)$, where the first three are nonempty subsets, such that for any $v_1 \in \mathscr{V}_1$ the event $A_2(v_1; w, w_1) \cap A_3^c(v_1; w, w_1, w_2)$ holds and for any $v_2 \in \mathscr{V}_2$ the event $A_2(v_2; w, w_2) \cap A_3^c(v_2; w, w_1, w_2)$ holds, and for any $v_3 \in \mathscr{V}_3$ the event $A_2(v_3; w_1, w_2) \cap A_3^c(v_3; w, w_1, w_2)$ holds, and $\left|D(v_4) \cap \{w, w_1, w_2\}\right| \le 1$ for any $v_4 \in \mathscr{V}_4$. Moreover all these $n$ events (corresponding to vertices from $\mathscr{V}$) are independent by Definition 4. Therefore using the well know fact that there are $3^k - 3 \cdot 2^k + 3$ ordered partitions of a $k$-element set into three nonempty subsets we have

$$\Pr\{B_{\triangledown}\} = \sum_{k=3}^{n} \binom{n}{k} (3^k - 3 \cdot 2^k + 3)(Q_2 - Q_3)^k (1 - 3(Q_2 - Q_3) - Q_3)^{n-k}$$

$$= (1 - Q_3)^n - 3(1 - Q_2)^n + 3(1 - 2Q_2 + Q_3)^n - (1 - 3Q_2 + 2Q_3)^n. \quad (5)$$

Similarly one can check that

$$\Pr\{C_{\vee}\} = (1 - Q_3)^n - 2(1 - Q_2)^n + (1 - 2Q_2 + Q_3)^n. \quad (6)$$

Now (3), (4), (5) and (6) imply the thesis immediately. □

As a consequence of the above result one can obtain the following two corollaries.

**Corollary 1.** *If $nQ_2 = o(1)$, then for any $w \in \mathscr{W}$*

$$\mathrm{ClusC}_{pas} \sim \frac{n^2 Q_2^3 + Q_3}{nQ_2^2 + Q_3}.$$

*Proof.* Assume that $nQ_2 = o(1)$. It implies that $nQ_3 = o(1)$ as well and therefore

$$\Pr\{A_{\triangledown}\} = 1 - (1 - Q_3)^n = nQ_3 + o(nQ_3). \quad (7)$$

Moreover, since $(3^k - 3 \cdot 2^k + 3)/k! \le 1.5$ for any nonnegative integer $k$, (5) implies

$$\Pr\{B_{\triangledown}\} = (n)_3 (Q_2 - Q_3)^3 (1 + o(1)). \quad (8)$$

Similarly one can check that under our assumption

$$\Pr\{C_{\vee}\} = (n)_2 (Q_2 - Q_3)^2 - (n)_3 (Q_2 - Q_3)^3 (1 + o(1)). \quad (9)$$

Equations (7), (8) and (9) directly imply the thesis. □

**Corollary 2.** *If* $\Pr\{Z_n = d\} = 1$ *for* $d = d(n) \geq 3$ *and* $nd^2 = o(m^2)$, *then*

$$
\text{ClusC}_{pas}(d) \sim \begin{cases} 1 & \text{for } nd/m \to 0 \\ \frac{d-2}{c(d-1)+d-2} & \text{for } nd/m \to c > 0 \\ 0 & \text{for } nd/m \to \infty. \end{cases}
$$

*Proof.* Assume that the distribution $\mathscr{P}_{(m)}$ is degenerated, i.e. $\Pr\{Z_n = d\} = 1$ for some $d = d(n) \geq 3$. Note that in this case $Q_2 = (d)_2/(m)_2 = d(d-1)/(m(m-1))$ and $Q_3 = (d)_3/(m)_3 = d(d-1)(d-2)/(m(m-1)(m-2))$. Therefore, since under our assumptions $d = o(m)$, we have

$$
\text{ClusC}_{pas}(d) \sim \frac{n^2 \left(\frac{d(d-1)}{m(m-1)}\right)^3 + \frac{d(d-1)(d-2)}{m(m-1)(m-2)}}{n \left(\frac{d(d-1)}{m(m-1)}\right)^2 + \frac{d(d-1)(d-2)}{m(m-1)(m-2)}} \sim \frac{\frac{(d-1)^2(m-2)}{(m-1)^2}\left(\frac{nd}{m}\right)^2 + d - 2}{\frac{m-2}{m-1}\frac{nd}{m}(d-1) + d - 2},
$$

which implies the statement.                                                                                    □

*Remark 2.* As for the active model of a random intersection graph, the above corollaries imply that by tuning the parameters of the passive model: $n, m, \mathscr{P}_{(m)}$, we are able to generate a graph with an arbitrary clustering coefficient as well.

**Acknowledgements** J. Jaworski acknowledges the support by the Marie Curie Intra-European Fellowship No. 236845 (RANDOMAPP) within the 7th European Community Framework Programme. This work had been also supported by Ministry of Science and Higher Education, grant N N206 2701 33, 2007–2010.

# References

Albert R, Barabási AL (2002) Statistical mechanics of complex networks. Rev Modern Phys 74:47–97

Barrat A, Weigt M (2000) On the properties of small-world networks. Eur Phys J B 13:547–560

Bloznelis M (2008) Degree distribution of a typical vertex in a general random intersection graph. Lithuanian Math J 48:38–45

Bloznelis M, Jaworski J, Rybarczyk K (2009) Component evolution in a secure wireless sensor network. Networks 53:19–26

Bock HH (1996) Probabilistic models in cluster analysis. Comput Stat Data Anal 23:5–28

Deijfen M, Kets W (2009) Random intersection graphs with tunable degree distribution and clustering probability. Eng Inform Sci 23:661–674

Godehardt E (1990) Graphs as structural models. Vieweg, Braunschweig

Godehardt E, Jaworski J (1996) On the connectivity of a random interval graph. Random Struct Algorithm 9:137–161

Godehardt E, Jaworski J (2002) Two models of random intersection graphs for classification. In: Schwaiger M, Opitz O (eds) Exploratory data analysis in empirical research. Springer, Berlin, Heidelberg, New York, pp 68–81

Godehardt E, Jaworski J, Rybarczyk K (2007) Random intersection graphs and classification. In: Decker R, Lenz HJ (eds) Advances in data analysis. Springer, Berlin, Heidelberg, New York, pp 67–74

Godehardt E, Jaworski J, Rybarczyk K (2010) Isolated vertices in random intersection graphs. In: Fink A, Lausen B, Seidel W, Ultsch A (eds) Advances in data analysis, data handling and business intelligence. Springer, Berlin, Heidelberg, New York, pp 135–145

Guillaume JL, Latapy M (2004) Bipartite structure of all complex networks. Inform Process Lett 90:215–221

Newman MEJ (2003) Properties of highly clustered networks. Phys Rev 68(026121)

Newman MEJ, Strogatz SH, Watts DJ (2001) Random graphs with arbitrary degree distributions and their applications. Phys Rev E 64(026118)

Rybarczyk K (2010) Random intersection graphs. analysis and modeling of networks structure. PhD thesis, Faculty of Mathematics and Computer Science, Adam Mickiewicz University, Poznan, URL http://hdl.handle.net/10593/386

Rybarczyk K (2011) The degree distribution in random intersection graphs. In: Gaul W, Geyer-Schulz A, Schmidt-Thieme L, Kunze J (eds) Challenges at the interface of data analysis, computer science, and optimization, studies in classification, data analysis, and knowledge organization. Springer, Heidelberg, Berlin

Strogatz SH, Watts DJ (1998) Collective dynamics of small-world networks. Nature 393:440–442

# Immersive Dynamic Visualization of Interactions in a Social Network

**Nicolas Greffard, Fabien Picarougne, and Pascale Kuntz**

**Abstract** This paper is focused on the visualization of dynamic social networks, i.e. graphs whose edges model social relationships which evolve during time. In order to overcome the problem of discontinuities of the graphical representations computed by discrete methods, the proposed approach is a continuous one which updates the changes as soon as they happen in the visual restitution. The vast majority of the continuous approaches are restricted to 2D supports which do not optimally match the human perception capabilities. We here present *TempoSpring* which is a new interactive 3D visualization tool of dynamic graphs. This innovative tool relies on a force-directed layout method to span the 3D space along with several immersive setups (active stereoscopic system/visualization in a dome) to offer an efficient user-experience. *TempoSpring* has initially been developed in a particular application context: the analysis of sociability networks in the French medieval peasant society.

## 1 Introduction

Network structures, modeled as graphs, are a key component of how human societies work. There are numerous challenges involved in social network analysis: from the understanding of the social mechanisms implied in the collective structures at different levels to the development of new technologies (e.g. "Social Semantic Web") allowing human beings and companies to collaborate and share knowledge. One of the milestones of social network analysis is the identification of topological structures in the set of relations between the different actors composing the network (Wasserman and Faust 1994). Visualization is well-known to play an important role in this problem by allowing the restitution of the results to domain experts who are

N. Greffard (✉) · F. Picarougne · P. Kuntz
LINA, Polytech'Nantes, rue Christian Pauc BP 50609 F-44306 NANTES Cedex 3, France
e-mail: Nicolas.Greffard@univ-nantes.fr

W. Gaul et al. (eds.), *Challenges at the Interface of Data Analysis, Computer Science, and Optimization*, Studies in Classification, Data Analysis, and Knowledge Organization, DOI 10.1007/978-3-642-24466-7_26, © Springer-Verlag Berlin Heidelberg 2012

not graph specialists, but it also enables the use of the human cognitive system as a powerful pattern recognition tool. Since the pioneering work of (Knuth 1963), the focus of the community shifted from static layouts, the layout of huge graphs, to dynamical layouts thanks to the increasing presence of evolving data. The general problem is the visualization of a graph associated with a system of relationships evolving over time. This evolution usually consists in the addition or suppression of components (nodes, links) or changes in the attributes associated to the nodes.

In this article we focus on such a dynamic visual representation of graphs. The chosen restitution model is the well known node-link model where the vertices are represented by points in the projection space and the edges by lines. We present an innovative approach, called hereafter *TempoSpring*, based on a spring-embedder algorithm (Fruchterman and Reingold 1991) allowing the visualization of the changes appearing in the graph while preserving the mental map of the user. One of the main contributions of our tool relies on the integration of the system in an immersive environment helping the user to easily manipulate the views of the network and to better understand its structure and its evolution.

Our work has initially been developed in a specific application context: the analysis of the evolution of sociability networks in the French Middle-Age farming society. In addition to the identification of the macroscopic structural evolution of the social network, the historians wanted to characterize the communities and their evolution through the identification of cores and peripheral individuals and the detection of individuals linking several communities together ("hubs").

The rest of this article is organised as follows. Section 2 presents a synthetical overview of the node-link visual representations of social networks by focusing on the evolutionary aspect. Section 3 describes the *TempoSpring* algorithm and Sect. 4 briefly introduces the application context of this work. Section 5 concludes on the main perspectives emerging from this work.

## 2   Social Network Visualization

According to a study carried out by (Freeman 2000), the most used visualization models can be classified into two classes. The first paradigm, the closest from the graph theory model, is the node-link one where actors are represented by nodes and social connections by lines. The second paradigm relies on the use of matrices where the rows and columns represent the actors while the entries represent the connections. The node-link model has been the most used by far. This is probably due to its proximity to the combinatorial tools used in the analysis (Fernanda et al. 2004). However, notice that the empirical evaluations of this model against the matrix-based model showed that the last is more efficient in the resolution of low-level tasks (for instance: "Find a common neighbour") (Ghoniem et al. 2005). However, further evaluations carried out by the aforementioned study showed that the identification of some higher-level structures (e.g. articulation points) is easier using the node-link model. Believing that the analysis of a social network cannot be reduced to a succession of low-level tasks and that high-level features are more relevant for practitioners, we here focus on the node-link model.

## 2.1 Node-Link Model

The usual graph layout problems can be dealt with as multi-objective combinatorial optimisation problems (project the vertices in a Euclidean space while minimising the number of crossings and keeping the crossing edge angles shorter than a threshold (Di-Battista et al. 1999)) that ignore the semantics of the manipulated objects. The situation is quite different when some topological structures are explicitly associated to information (Herman et al. 2000). In the context of social networks, a set of spatially close nodes can be interpreted as a community while a social position can be identified by observing a set of actors having their connections to the rest of the network arranged in the same fashion.

We can extract two main approaches from the literature to address this NP-hard problem: methods based on circular layouts and methods based on spring-embedder like algorithms. Whatever the chosen method is, visualization troubles such as occlusions and edge crossings quickly emerge when the actor and link cardinalities increase. Several methods to deal with this problem consist in merging nodes (Archambault et al. 2007) or links (Telea and Ersoy 2010) together, but they inevitably lead to some information loss at the local level. This well-known problem of "Focus & Context" requires additional steps to allow the visualization at different granularity levels, e.g. adding interactive components (for instance, see Sarkar and Brown (1992)).

## 2.2 Dynamic Social Networks

Two distinct methods try to integrate the evolution aspect of the social networks into visualization tools. The "discrete" approach consists in representing an aggregation of the evolutions of the data between different time steps through statistical measure or sub-structure features (Falkowski et al. 2006). Recently, Rosvall and Bergstrom (2010) represented the evolution of the network structure by an alluvial diagram in which the variations of the size, the color and the position of blocks represent the evolution of the communities at each time step. This approach allows the user to efficiently capture the macroscopic evolutions of the network and in particular to easily detect the merging of two communities. However, it requires an additional window to display the dynamic component, and it does not easily allow the user to access finer information that is often mandatory for the functional interpretation. Furthermore, the discrete approaches require the discretization of the time into time steps and produce, at each step $t$, a new representation through a morphing with the previous one. The known drawback of this approach is the discontinuity it introduces in the graphical representation. Indeed, if the changes are numerous, the representation at $t$ can be really different than the one at $t - 1$ and the mental map of the user (Misue et al. 1995) can be perturbed. This induces an additional cognitive cost which can eventually reduce the gain of the process. The "continuous" approach, however, aims at overcoming this issue by displaying each change as soon as it appears. Thus, in order to minimize the re-learning process between two successive iterations, we chose this approach where the visualization is updated as soon as a change appears.

## 3 TempoSpring

### 3.1 Basis

The graph representing the social network is projected into a 3D Euclidean space where the coordinates of the nodes and links are determined by a system of attractive and repulsive forces. More precisely, the nodes are considered to be electrically charged particles of identical polarity and each edge is considered as a spring with an equilibrium length and elasticity specified by the attributes associated with it. On the one hand, the Hooke attraction – describing the elasticity of a spring – is applied between each pair of nodes linked by an edge and tends to maintain them at a defined distance. On the other hand, another force defined by the Coulomb law – measuring the electromagnetic interactions between two electrically charged particles – is applied between all the node pairs. It tends to move each node away from the other nodes. The Hooke attraction $F_H$, applied to a node of coordinates $p_1$ linked to another node of coordinates $p_2$ is specified by (1) where $k$ is the elasticity constant and $R$ the equilibrium length of the edge. The Coulomb repulsion $F_C$, applied between a node of coordinates $p_1$ and another node of coordinates $p_2$ is defined by (2) where $\epsilon_0$ is the electrical constant and $q_1$ and $q_2$ the electrical charges of these particles (nodes).

$$F_H = -k \left( \frac{(|L| - R)L}{|L|} \right) \; with \; L = p_2 - p_1 \tag{1}$$

$$F_C = \frac{1}{4\pi\epsilon_0} \frac{q_1 q_2 (p_1 - p_2)}{|p_1 - p_2|^3} \tag{2}$$

The system tends to minimize the global sum of the forces. In practice, it quickly converges towards a local optimum. Moreover, it produces a physical approximation of the movement of the nodes which can be easily interpreted by a human and

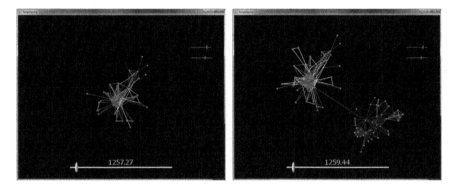

**Fig. 1** Snapshot of a social network at two consecutive time steps. On the right, new nodes have been added without heavily affecting the prior positioning of the graph. The user can control and browse the temporal dimension of the data through the slider located on the bottom part of the application. The sliders located on the top right hand part allow to dynamically control the contribution of the Hooke and Coulomb forces

where the nodes are usually well distributed in the space while preserving the graph symmetries.

In order to allow the user to interpret and understand the changes induced in the representation by the dynamic characteristic of the social network, we have resorted to an adaptation of the incremental version of the spring-embedder algorithm (McCrickard and Kehoe 1997). The apparition of a new node or a new link does not lead to a global reset of the projections. This approach allows the continuous visualization of the evolutions of the whole set of nodes based on the physical motion principles. These principles, inherently present in the environment, seem easily understood by a human being (Fruchterman and Reingold 1991). They significantly preserve the mental map that the user learns between two successive frames (see Fig. 1).

Several extensions have been integrated in order to enhance the resulting visualization. The user can obtain the detailed data held by a node by simply clicking on it. We here only show the main connected component in order to limit the visual load. The user can interactively select other components if he or she wishes to. The gravity center of the selected component is centered at each iteration in the visualization window such that the user can easily and naturally follow the motion of the system. This position shift is done by controlling each node acceleration in order to obtain a fluid motion of the objects between two consecutive frames.

We integrated two filters to highlight the individuals playing a key role in the social network. The *tension* (sum of the absolute values of the forces) applied on a node or link is actually an efficient measure of the importance of the element in the visualization. Consequently, the user can identify the nodes with the greatest tension, and the edges with an important ratio between their equilibrium length and their displayed length.

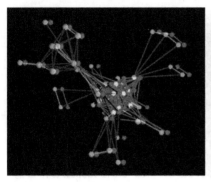 

**Fig. 2** (Left) Anaglyph stereoscopic representation of a social network; (Right) Interactive analysis of a social network projected in a visualization dome

## 3.2 3D and Immersive Restitution

As underlined by C. Ware and P. Mitchell, the first evaluation attempts of 3D visualization were restricted to low resolution displays outmatched by current systems (Ware and Mitchell 2008). However, the technological evolutions of the last few years drove to a brand new generation of displays. They allow a high quality stereoscopic 3D projection and a "physical" immersion of the user within the system. These innovations renewed the interest of the community for immersive displays. Recent works enabled the evaluation of the added value of stereoscopic 3D and other immersive systems (Halpin et al. 2008). In addition to adapted peripherals, these approaches also allow to limit the visual clustering by using projection depth or a wider projection angle. Therefore, along with stereoscopic 3D we propose a projection in a visualization dome (see Fig. 2).

## 4 Application

The application context of our work is the analysis of medieval farming societies. We had access to an astonishing documentation keeping track of notarial documents – mainly agrarian contracts between peasants – on a 10 km$^2$ area in the South-West of France over a period of three centuries. The digitization of this data which are manuscripts transcribed in the XVIII$^{th}$ century required the implementation of a complex and specific form targeted at medieval researchers. The documentation at our disposal holds 6,000 deeds whose digitization is not finished yet.

From these deeds, several social relation networks can be considered depending on the historian's needs. We focused on the relations between the individual names appearing in the deeds: the vertices of the graph thus are the individuals named in the deeds, and there exists an edge between two vertices if the corresponding names were cited during the same transaction.

For the dynamic visualization, the temporal dimension has been reduced to an annual scale which seemed consistent given the historical data we had: the minimal timestep encountered in *TempoSpring* represents a year. Figure 2 shows two views of the social network: a stereoscopic representation and a representation on the dome with a user exploring this network.

These visual restitutions notably allowed the historians to better understand the consequences of the Hundred Years war on the social structure evolution. This war deeply shattered the society, however some similar macroscopic structures were discovered before and after it. Furthermore, historians were able to identify and temporally follow the constitution of the communities of some individuals (lords) and some families.

# 5 Conclusion

This paper presents a new 3D interactive visualization tool of social networks evolving over time. This tool is based on a node-link representation obtained through a spring-embedder algorithm which was specifically adapted to a continuous layout of the evolution of the network. It has been developed and tested within an application context where it allowed to understand, through the identification of articulation points linking different communities, the formation and merging of communities.

However, its actual added value against other approaches has not been statistically appraised. From a methodological point of view, this question is quite sensible in the context of exploratory discovery and is currently widely discussed by the machine learning and information visualization communities (Puolamäki and Bertone 2009). Our experience taught us that low-level based protocols used in the field of Human Computer Interaction are not really adapted to the situations where the user cannot previously formalize its objectives in an accurate fashion and where it is actually the visual support that will guide him through the hypothesis formulation process. Therefore, we made the decision to adopt a behavioural approach by observing the user behaviours through the actions carried out through the effectors and the position of their gaze on the visual restitution by using an eye tracking equipment.

# References

Archambault D, Munzner T, Auber D (2007) Topolayout: Multi-level graph layout by topological features. IEEE Trans Visual Comput Graph 13:2007

Di-Battista G, Eades P, Tamassia R, Tollis IG (1999) Graph drawing - algorithms for the visualization of graphs. Prentice Hall, Upper Saddle River, NJ

Falkowski T, Bartelheimer J, Spiliopoulou M (2006) Mining and visualizing the evolution of subgroups in social networks. In: WI '06: Proceedings of the 2006 IEEE/WIC/ACM International Conference on Web Intelligence, IEEE Computer Society, pp 52–58

Fernanda B, Viégas, Donath J (2004) Social network visualization: Can we go beyond the graph. In: Workshop on Social Networks for Design and Analysis: Using Network Information in CSCW 2004, pp 6–10

Freeman L (2000) Visualizing social networks. J Soc Struct 1:151–161

Fruchterman TMJ, Reingold EM (1991) Graph drawing by force-directed placement. Software Pract Ex 21(11):1129–1164

Ghoniem M, Fekete JD, Castagliola P (2005) On the readability of graphs using node-link and matrix-based representations: a controlled experiment and statistical analysis. Inform Visual 4(2):114–135

Halpin H, Zielinski D, Brady R, Kelly G (2008) Exploring semantic social networks using virtual reality. In: Proceedings of the 7th International Conference on The Semantic Web, Lecture Notes In Computer Science, vol 5318, pp 599–614

Herman I, Melançon G, Marshall S (2000) Graph visualization and navigation in information visualization: a survey. IEEE Trans Visual Comput Graph 6(1):24–43

Knuth D (1963) Computer-drawn flowcharts. Comm ACM 6:555–563

McCrickard DS, Kehoe CM (1997) Visualizing search results using sqwid. In: Sixth International World Wide Web Conference, ACM Press, pp 51–60

Misue K, Eades P, Lai W, Sugiyama K (1995) Layout adjustment and the mental map. J Visual Lang Comput 6(2):183–210

Puolamäki K, Bertone A (2009) Introduction to the special issue on visual analytics and knowledge discovery. SIGKDD Explorations 11(2):3–4

Rosvall M, Bergstrom CT (2010) Mapping change in large networks. PLoS ONE 5(1)

Sarkar M, Brown MH (1992) Graphical fisheye views of graphs. In: CHI '92: Proceedings of the SIGCHI conference on Human factors in computing systems, ACM, pp 83–91

Telea A, Ersoy O (2010) Image-based edge bundles: Simplified visualization of large graphs. Comput Graph Forum 29(3):843–852

Ware C, Mitchell P (2008) Visualizing graphs in three dimensions. ACM Trans Appl Percept 5:2–15

Wasserman S, Faust K (1994) Social network analysis: methods and applications. Cambridge University Press, Cambridge, New York

# Fuzzy Boolean Network Reconstruction

**Martin Hopfensitz, Markus Maucher, and Hans A. Kestler**

**Abstract** Genes interact with each other in complex networks that enable the processing of information inside the cell. For an understanding of the cellular functions, the identification of the gene regulatory networks is essential. We present a novel reverse-engineering method to recover networks from gene expression measurements. Our approach is based on Boolean networks, which require the assignment of the label "expressed" or "not expressed" to an individual gene. However, techniques like microarray analyses provide real-valued expression values, consequently the continuous data have to be binarized. Binarization is often unreliable, since noise on gene expression data and the low number of temporal measurement points frequently lead to an uncertain binarization of some values. Our new approach incorporates this uncertainty in the binarized data for the inference process. We show that this new reconstruction approach is less influenced by noise which is inherent in these biological systems.

## 1 Introduction

Synthetic biology and systems biology are indispensable tools for the understanding of biomolecular systems. Molecular systems biology usually refers to integrated experimental and computational approaches for studying biomolecular networks, such as signal transduction, gene regulation or metabolic systems. At the core of systems biology research lies the identification of these networks from experimental

M. Hopfensitz · M. Maucher
Clinic for Internal Medicine I, University Hospital Ulm, D-89081 Ulm, Germany
e-mail: martin.hopfensitz@uni-ulm.de; markus.maucher@uni-ulm.de

H.A. Kestler (✉)
Clinic for Internal Medicine I, University Hospital Ulm, D-89081 Ulm, Germany and Institute of Neural Information Processing, Ulm University, D-89069 Ulm, Germany
e-mail: hans.kestler@uni-ulm.de

W. Gaul et al. (eds.), *Challenges at the Interface of Data Analysis, Computer Science, and Optimization*, Studies in Classification, Data Analysis, and Knowledge Organization, DOI 10.1007/978-3-642-24466-7_27, © Springer-Verlag Berlin Heidelberg 2012

data via reverse-engineering methods (Markowetz and Spang 2007). We concentrate our efforts on Boolean networks that were first studied by Kauffman (1969, 1993). In recent years inference methods for Boolean networks (Hickman and Hodgman 2009) became popular due to their simplicity and the fact that qualitative predictions of large complicated networks are more manageable (Bornholdt 2005). Liang et al. (1998) presented REVEAL, an algorithm that uses the mutual information between input and output states, i.e. two subsequent measurements of a time series, to infer Boolean dependencies between them. Lähdesmäki et al. (2003) presented an algorithm that can additionally deal with noisy data. To infer a Boolean network solely from quantitative time series data, the continuous data need to be binarized, since the Boolean method assumes only two possible states for each gene. The binarization of the gene expression data is often difficult and unreliable, since the noise and the low number of temporal measurement points can lead to an uncertain binarization. Differences in the binarization result can have strong effects on the resulting Boolean networks, because a state change for a single gene can cause different functions and gene dependencies. In this work, we propose a novel inference method, based on Boolean networks, which determines this uncertainty in the binarization step of the data and incorporates this uncertainty into the reconstruction process. We study whether incorporating the uncertainty in the data for the inference of the Boolean network improves the accuracy of prediction. By incorporating uncertainty we could improve both the state transitions and the network wiring.

## 2  Inference by Using Fuzzy Information

We here provide a detailed description of the algorithm used to reconstruct a Boolean network, which is defined as follows:

**Definition 1.** A Boolean network $G(V, F)$ consists of a set of nodes $V = (x_1, \ldots, x_n)$ and corresponding Boolean functions $F = (f_1, \ldots, f_n)$. Here $f_i : \{0, 1\}^n \to \{0, 1\}$ describes the dependency of node $x_i$. We say that for a function $f$ the input variable $x_j$ is fictitious if for all $x_1, \ldots, x_{j-1}, x_{j+1}, \ldots, x_n$,

$$f(x_1, \ldots, x_{j-1}, 0, x_{j+1}, \ldots, x_n) = f(x_1, \ldots, x_{j-1}, 1, x_{j+1}, \ldots, x_n). \quad (1)$$

If a variable is not fictitious it is called essential. Hence a function $f_i$ depends on $k$ essential variables (input nodes), where the number of input nodes may depend on $i$. The update of the genes is synchronous, depending on the functions corresponding to those genes. For discrete time steps, $x_i$ is derived via the recurrence $x_i(t + 1) = f_i(x_{i_1}(t), \ldots, x_{i_k}(t))$. The list of Boolean functions $F$ contains the rules of regulatory interaction between genes.

Our inference algorithm can be divided into three steps: The fuzzy-c-means clustering algorithm (Bezdek 1981) is used to binarize the given time series, a randomized rounding approach (Raghavan and Thompson 1987) is used to iterate over a large number of possible binarizations of the original time series, and the best-fit approach (Lähdesmäki et al. 2003) is used to construct optimal predictors for binarized time series. We will first give a brief overview over these three techniques and then describe our novel inference method.

Cluster analysis divides a set of objects into groups such that the objects in each cluster are similar to each other, while objects from different clusters should not be similar to each other (Everitt et al. 2001). The fuzzy version of clustering does not necessarily assign each object to a single group, like in conventional (crisp) clustering. Instead, it assigns to each object a membership vector. We interpret these membership coefficients as probabilities that the object belongs to the particular group. Here, we force the coefficients for every object to sum up to 1 (Bezdek 1981). The fuzzy-c-means algorithm computes membership coefficients depending on a weighting exponent $b$ (fuzzifier). This fuzzifier determines the level of "cluster fuzziness", i.e. the tendency to distribute the probabilities evenly among the groups. For binarizing our continuous time series data $D_{cont} \in \mathbb{R}^{n \times m}$, we applied the fuzzy-2-means algorithm on each gene separately. This method resulted in a coefficient matrix $P \in \mathbb{R}^{n \times m}$ where $p_{ij}$ is the probability that the $j$-th element of the time series for gene $i$ belongs to the group of "high expression". The running time of our fuzzy-2-means algorithm grows linearly in the number of iterations as well as in the number of data points. Since only one-dimensional data is clustered, the calculation time of distances can be considered constant.

Randomized rounding is a sampling method used for the optimization of discrete problems (Raghavan and Thompson 1987). It converts an optimal solution of a continuous relaxation of the discrete problem into a vector of probabilities $p \in [0, 1]^l$. Then it repeatedly rounds that vector to a random Boolean vector $r \in \{0, 1\}^l$ with the probabilities $Pr[r_i = 1] = p_i$, $1 \leq i \leq l$. In our inference algorithm, we use this technique in conjunction with the probability vector gained from the fuzzy-c-means algorithm to create slightly different binarized versions of the time series.

To reconstruct the Boolean network from binarized data, we use the best-fit approach (Lähdesmäki et al. 2003). This approach finds the best predictor for each node by computing for every possible combination of $k$ input variables, $0 \leq k \leq k_{max}$, the lowest possible error that can be achieved by a Boolean function. The error $\varepsilon$ of a function $f_i$ is defined as follows:

$$\varepsilon(f_i) = \sum_{t=1}^{m-1} \varepsilon' \text{ with } \varepsilon' = \left\{ \begin{array}{ll} 1, & \text{if } f_i(x_{i_1}(t), ..., x_{i_k}(t)) \neq x_i(t+1). \\ 0, & \text{otherwise} \end{array} \right. \quad (2)$$

Here, we consider combinations of at most $k_{max} = 3$ input variables with a maximum error of $\varepsilon_{max} = 10$. For the reconstruction of the Boolean networks we start with $k = 0$, that is we first test for a constant function without any inputs. Then we examine solutions (combinations of input variables with function of minimal error) with single input variables ($k = 1$), up to $k = k_{max}$. Solutions on any

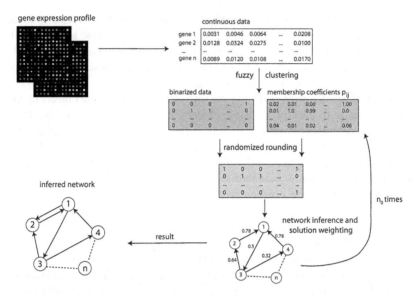

**Fig. 1** Overview of the method. From a time series microarray experiment, the expression levels of all genes are determined. The resulting continuous gene expression data are binarized using fuzzy-clustering. Based on the resulting membership coefficients $p_{ij}$ and the binarized data, a binarized time series is sampled via randomized rounding. Subsequent to the network inference and a weighting of the solutions, another binarized time series is sampled. After $n_0$ runs, the outcome of the algorithm is the highest weighted Boolean network

sets $I_k = (x_{i_1}, \ldots, x_{i_k})$ of variables were only taken into account if their error was strictly less than the error of any solution on a subset of $I_k$.

## 2.1 Algorithm

Our inference algorithm includes the techniques described above in the following way, see Fig. 1. In the first step, the time series gene by gene is discretized with the fuzzy-2-means algorithm, resulting in a probability vector. From this probability vector, we sample a number of discretized time series via randomized rounding. For each of these time series, the best-fit approach is used to find the best solution for every node. Across these repetitions, the scores of found solutions are accumulated. In total the algorithm is run for $n_0$ iterations. This increases the complexity of the network reconstruction by a factor of $n_0$, compared to the best-fit method.

In the end, the best predictors, i.e. the solutions with the highest accumulated score, are returned. In case of two or more equally weighted solutions for a node, all of them are taken into account as predictors. During the inference iterations, the score of a solution depends on its error $\varepsilon$ and the number $k$ of input variables. This reflects the fact that a solution with higher in-degree has a higher chance to result in few errors even if it does not correspond to the underlying model, especially when only short time series are available. We used a scoring function of the form

$$w(\varepsilon, k) = \frac{1}{c_1^{\varepsilon} c_2^{k}} \qquad (3)$$

with constants $c_1, c_2 > 1$. Using this scoring function, a solution attains a high score if it has a low error $\varepsilon$ and indegree $k$. Pseudocode is given in Algorithm 1.

---

**Algorithm 1** Fuzzy Boolean network reconstruction

---

Input: continuous time series data $D_{cont}$
$P \leftarrow$ fuzzy-2-means$(D_{cont})$
**for** $c = 1$ to $n_0$ **do**
    $D_{bin} \leftarrow$ sample according to $P$ via randomized rounding
    sol $\leftarrow$ best-fit$(D_{bin})$
    wsol $\leftarrow$ weight solutions (sol) by the function $w(\varepsilon, k)$
    accumulate the scores of found solutions according to wsol
**end for**
Output: weighted predictors for each node

---

# 3 Results

## 3.1 Artificial Data

We generated artificial time series data, reconstructed possible Boolean functions after adding noise, and analyzed our inference method by comparing it to the best-fit approach. The performance is compared by using the State Transition Error (STE) (Liu et al. 2008), the Positive Predictive Value (PPV) and Sensitivity (Se). The STE represents the fraction of outputs that are incorrectly predicted. It can be defined as follows:

$$\mu = \frac{1}{n} \sum_{i=1}^{n} \frac{1}{2^n} \sum_{x \in \{0,1\}^n} (f_i(x) \oplus f_i^{'}(x)),$$

where $f_i^{'}$ belongs to the reconstructed Boolean functions and $\oplus$ denotes the sum modulo 2. PPV quantifies the percentage of correctly predicted interactions and Sensitivity describes the percentage of true interactions. Based on these three measures, we determine the performance in terms of state transitions and network wiring.

### 3.1.1 Experimental Setup

Random Boolean networks were generated according to Kauffman (1969, 1993), consisting of $n = 5$ nodes and the maximum in-degree $k_{max}$ was set to 3. Based on the original networks, time series consisting of $l$ state transitions were created. For all values of the time series we added Gaussian noise ($\mathcal{N}(0, \sigma)$). Afterwards, the noisy time series data were binarized using fuzzy-c-means on each gene separately. Subsequently we inferred Boolean networks based on the binarized data using the best-fit approach of Lähdesmäki et al. (2003) as well as our new approach, which incorporates resampling (we used $n_0 = 1,000$ as the number of samples) and group membership information in the reconstruction process. If we obtained two or more solutions with the same weight or error size for a node, one of them was chosen randomly as predictor. Since completely determined Boolean functions $f$ are necessary to compute the STE of the network, missing values in consequence of the small number of state transitions $l$, were randomly determined out of $\{0, 1\}$. Based on the original artificial networks, we calculated PPV, sensitivity and STE for the networks reconstructed by the best-fit approach and our new fuzzy method. This was repeated 1,000 times for each combination of $\sigma \in \{0.0, 0.05, \ldots, 0.5\}$ and $l \in \{10, 12, \ldots, 30\}$. We chose the constants $c_1 = 2$ and $c_2 = 1.2$ of our scoring function (determined in prior experiments on networks based on artificial data). The commonly used value for the fuzzifier $b$ is 2, but Dembèlè and Kastner (2003) showed that a lower value for $b$ is often more appropriate for binarizing gene expression data. After experiments we chose the fuzzifier $b = 1.5$.

### 3.1.2 Performance on Artificial Data

The performance of our method was measured by means of PPV, sensitivity, and STE. Figure 2 illustrates that, regarding the PPV, the results of both approaches are very similar up to a noise of $\sigma = 0.2$. For higher noise, the PPV of our new method is higher for each number of state transitions. Our method achieved a PPV improvement by around 15%. Sensitivity (see Fig. 3) is very similar for all numbers of state transitions and noise levels. The results for the STE show (see Fig. 4) that the

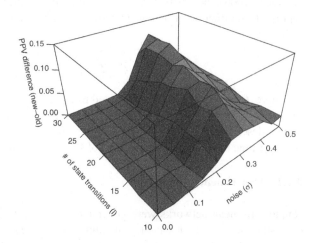

**Fig. 2** Difference between the PPV of our new method and the best-fit method. Values greater than zero indicate an improvement of our new algorithm. For each combination of noise and number of state transitions, the mean difference of 1,000 experiments is presented

**Fig. 3** Difference between
the Sensitivity of our new
method and the best-fit
method. For each
combination of noise and
number of state transitions,
the mean difference of 1,000
experiments is presented

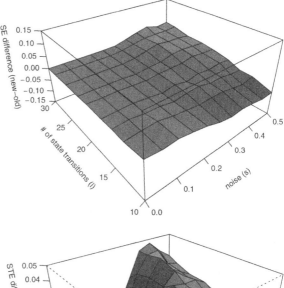

**Fig. 4** Difference between
the STE of the best-fit method
and our new method. Values
greater than zero indicate an
improvement of our new
algorithm. For each
combination of noise and
number of state transitions,
the mean difference of 1,000
experiments is presented

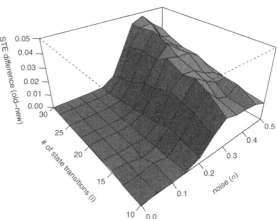

performance of both approaches is again very similar up to a noise level of $\sigma = 0.2$.
But for a noise level higher as $\sigma = 0.2$, the STE of our new method is reduced
by 5%.

## 4   Conclusion

In this work, we present a novel gene network inference method to recover networks
directly from time series gene expression measurements. Boolean network reconstruction requires binary data which is usually obtained by binarizing real-valued
data. Since the real-valued expression values are often noisy and only a low number
of temporal measurements is available, the binarization often leads to an ambiguous
binarization threshold. This can result in values with unreliable binarization. Our

new approach incorporates this uncertainty in the binarized values for the inference process, which can help to compensate for the noise inherent in biological systems while still retaining the good performance of Lähdesmäki's method. We applied our new method to artificial data and observed that incorporating fuzzy information in the inference procedure improved the accuracy both in terms of state transitions and network wiring, compared to the simple inference method without using the group membership information. For both methods the sensitivity is very similar for all numbers of state transitions and noise levels. This means that with both methods we predicted almost the same percentage of true interactions. Together with the result for the PPV, we can conclude that the new approach predicts less incorrect interactions while finding the same number of true interactions for several noise levels directly on the gene expression profiles without prior binarization. Since our algorithm produces more reliable networks, it allows researchers to formulate new hypotheses on gene-regulatory networks with a higher confidence.

# References

Bezdek J (1981) Pattern recognition with fuzzy objective function algorithms. Plenum Press, New York

Bornholdt S (2005) Systems biology. Less is more in modeling large genetic networks. Science 310(5747):449–451

Dembèlè D, Kastner P (2003) Fuzzy C-means method for clustering microarray data. Bioinformatics 19(8):973–980

Everitt B, Landau S, Leese M (2001) Cluster analysis, 4th edn. Oxford University Press, New York

Hickman GJ, Hodgman TC (2009) Inference of gene regulatory networks using Boolean-network inference methods. J Bioinform Comput Biol 7(6):1013–29

Kauffman SA (1969) Metabolic stability and epigensis in randomly constructed genetic nets. J Theor Biol 22(3):437–467

Kauffman SA (1993) The origins of order: self-organization and selection in evolution. Oxford University Press, Oxford

Lähdesmäki H, Shmulevich I, Yli-Harja O (2003) On learning gene regulatory networks under the boolean network model. Mach Learn 52(1-2):147–167

Liang S, Fuhrman S, Somogyi R (1998) Reveal, a general reverse engineering algorithm for inference of genetic network architectures. In: Altman RB, Dunker AK, Hunter L, Klein TED (eds) Proceedings of the Pacific Symposium on Biocomputing, World Scientific, vol 3, pp 18–29

Liu W, Lähdesmäki H, Dougherty ER, Shmulevich I (2008) Inference of boolean networks using sensitivity regularization. EURASIP J Bioinformatics Syst Biol DOI 10.1155/2008/780541

Markowetz F, Spang R (2007) Inferring cellular networks–a review. BMC Bioinformatics 8 Suppl 6:S5

Raghavan P, Thompson CD (1987) Randomized rounding: a technique for provably good algorithms and algorithmic proofs. Combinatorica 7(4):365–374

# GIRAN: A Dynamic Graph Interface to Neighborhoods of Related Articles

**Andreas W. Neumann and Kiryl Batsiukov**

**Abstract** This contribution reports on the development of GIRAN (Graph Interface to Related Article Neighborhoods), a distributed web application featuring a Java applet user front-end for browsing recommended neighborhoods within the network of Wikipedia articles. The calculation of the neighborhood is based on a graph analysis considering articles as nodes and links as edges. The more the link structure of articles is similar to the article of current interest, the more they are considered related and hence recommended to the user. The similarity strength is depicted in the graph view by means of the width of the edges. A Java applet dynamically displays the neighborhood of related articles in a clickable graph centered around the document of interest to the user. The local view moves along the complete article network when the user shows a new preference by clicking on one of the presented nodes. The path of selected articles is stored, can be displayed within the graph, and is accessible by the user; the content of the article of current interest is displayed next to the graph view. The graph of recommended articles is presented in a radial tree layout based on a minimum spanning tree with animated graph transitions featuring interpolations by polar coordinates to avoid crisscrossings. Further graph search tools and filtering techniques like a selectable histogram of Wikipedia categories and a text search are available as well. This contribution portrays the graph analysis methods for thinning out the graph, the dynamic user interface, as well as the service-oriented architecture of the application back-end.

A.W. Neumann (✉)
Institute of Information Systems and Management, Karlsruhe Institute of Technology, Kaiserstr. 12, 76128 Karlsruhe, Germany

msg systems ag, Nauenstr. 67, 4052 Basel, Switzerland
e-mail: andreas.neumann@msg-systems.com

K. Batsiukov
Institute of Information Systems and Management, Karlsruhe Institute of Technology, Kaiserstr. 12, 76128 Karlsruhe, Germany
e-mail: kiryl.batsiukov@gmail.com

W. Gaul et al. (eds.), *Challenges at the Interface of Data Analysis, Computer Science, and Optimization*, Studies in Classification, Data Analysis, and Knowledge Organization, DOI 10.1007/978-3-642-24466-7_28, © Springer-Verlag Berlin Heidelberg 2012

# 1  Wikipedia: A Directed Graph of Articles

Considering today's Internet traffic, the multilingual, web-based, free-content ency-clopedia Wikipedia (www.wikipedia.org) is one of the top most accessed Internet sites of all kinds and is ranked number one within non-commercial websites. Although often solely evaluated by the quality of its content, Wikipedia is more than an alphabetically sorted list of articles. Articles are interconnected by means of links and the link structure itself represents valuable information on the articles. Portrayed as a graph, the articles are the vertices and links between articles are the edges.

Although most users are unaware of the graph structure, they still use features coming from this property. The usual usage pattern of Wikipedia by a unique visitor occurs in the following manner: (a) The first article of current interest is usually found via the Wikipedia search function. (b) Starting from this article most users click on links within this article to go to related material. (c) By repeating step (b) the user explores the neighborhood of the first article by implicitly moving through the graph.

There are three challenges that appear to be critical within this approach:

1. Sometimes the best related articles are not even linked to the current site.
2. Finding the best link to follow within a large article is nearly impossible.
3. The user is completely unaware of the structure of the network of articles.

The Wikipedia community seems to be aware of at least the first challenge. Some articles include a "See also" section made of a manually edited list of links to other articles that are not – or not yet – included in the main article section. Although this approach can be used by a single editor to lessen the first challenge, with every new link the second challenge gets harder.

For each of these challenges a corresponding approach to tackle the problem exists:

1. Determine related articles that are not necessarily currently connected via outbound directed links.
2. Calculate recommended articles including similarity strength of articles.
3. Show a dynamic clickable graph portraying a local view of the neighborhood of related articles centered around the document of current interest while browsing Wikipedia articles.

This contribution reports on the development of GIRAN, a distributed web appli-cation that features solutions to all three challenges by means of the described approaches. The solutions are based on the community created links only, no semantic content analysis is involved.

The data structure and network size of the English Wikipedia used in this work is shown in Fig. 1. The namespace is used to distinguish e.g. encyclopedia content from user pages of editors; all articles belong to the same namespace. Redirects are used to solve the problem of synonyms within the encyclopedia. The article is stored by the name of one of the synonyms, all other synonyms are redirects that bring the

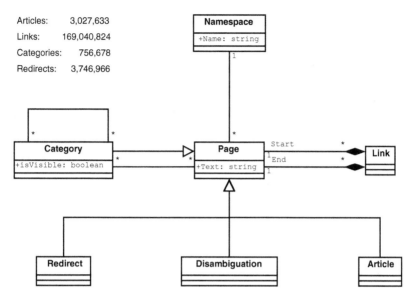

Fig. 1 Data structure and network size of the English Wikipedia (en.wikipedia.org)

user immediately to the article. A special approach to redirects is necessary when analyzing the graph structure for related articles. Disambiguation pages address the problem of homonyms; according to Wikipedia rules, disambiguation pages should not have any direct incoming links, thus no special treatment is necessary. Further on, broken article links exist on a large scale as well since Wikipedia allows to include links to articles that do not exist yet but might be created in the future. These broken links have to be deleted for the graph analysis. The number of links in Fig. 1 only includes links between articles, links by redirects or broken links are not counted.

## 2   Calculation of Neighborhoods of Related Articles

To model the structure of Wikipedia as a graph for the neighborhood analysis the following notation will be used: $G = (V, E)$ is a graph of Wikipedia articles and links. $V = \{v_1, v_2, ..., v_n\}$ is an finite set of vertices of the graph. Each Wikipedia article is represented as a vertex. $E = \{e_1, e_2, ..., e_m\}$ is a finite set of edges. An edge $e \in E = (u, v) \subseteq V \times V$ in a directed graph is an ordered pair having $u$ as a start and $v$ as an end of the edge. If page $u$ has a link which points to page $v$, $v$ exists in Wikipedia, and $v$ is not a redirect then the edge $(u, v)$ is created and included into $E$. If $v$ is a redirect to the page $w$ then an edge $(u, w)$ is included into $E$. As a result of this preprocessing we have a graph representing Wikipedia without broken links and with resolved redirects. Further on, let the input node set of a vertex be defined as $I(u) = \{v \in V : (v, u) \in E\}$ and an output node set as $O(u) = \{v \in V | (u, v) \in E\}$.

In Small (1973) a similar graph model of scientific papers (nodes) and citations (edges) has been analyzed. Small introduced cocitation as a scientometric criterion for finding related papers in such a graph. Two papers are related if there are many other papers which cite both papers together. The cocitations of two papers $u$ and $v$ are defined as follows: $Cocitation(u, v) = I(u) \cap I(v)$. This criterion can be applied to Wikipedia to find related articles. Two articles $u$ and $v$ in Wikipedia are related if there are many other articles, which contain links to $u$ and to $v$ together.

Nevertheless, the amount of cocitations of $u$ and $v$ alone is not sufficient to measure the relatedness of these two articles. Articles with large $I(u)$ tend to have more cocitations in general. Thus a normalization is necessary to find the best related articles and not the most popular ones. Jaccard's index (Jaccard 1901) and Salton's cosine (Salton and McGill 1983) can be both used:

$$J(u, v) = \frac{|I(u) \cap I(v)|}{|I(u) \cup I(v)|}, \quad S(u, v) = \frac{|I(u) \cap I(v)|}{\sqrt{|I(u)| * |I(v)|}}. \tag{1}$$

$J(u, v)$ and $S(u, v)$ both take values in the interval [0,1], are known to be correlated (Hamers et al. 1989), and both reach 1 when the measured nodes are the most similar. It has been stated by Small and Sweeney (1985) that Salton's formula deals more effectively with pairs of highly-cited (large $I(u)$) and low-cited (small $I(u)$) nodes. That is why Salton's cosine was used to normalize cocitation in this work.

The described ideas result in the algorithm for the calculation of the neighborhood of related articles for the article $u$:

1. Build the set of candidates $V_c = I(u) \cup O(u)$
2. Calculate the normalized cocitations $S(u, v) \ \forall v \in V_c$
3. Order $V_c$ by decreasing $S(u, v_c)$
4. Choose the top vertices for the graphical neighborhood representation

## 3  System Architecture and Implementation

To present a smoothly moving dynamic graph to the user the ad hoc computation of the neighborhood after a click needs to be very fast. The algorithm described in the last section can be optimized for the implementation to take care of some rarely existing unusual articles with large corresponding link and cocitation sets. For speed optimization the following adaptations are used. As tests have proven, with the finally chosen thresholds the overall results do not significantly change.

### 3.1  Calculation of the Candidate Set

In case of very large $I(u)$ – and much less often $O(u)$ respectively – random subsets are used. Although this is somehow critical to the results, this only happens very

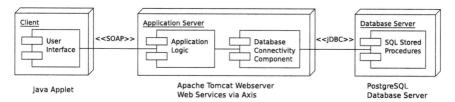

**Fig. 2** GIRAN service-oriented system architecture

rarely with special articles, e. g. the article about the "English language", since the article graph is scale free.

## 3.2  Calculation of the Cocitation

In case of very large input sets random subsets are used. Since cocitation is based on intersections, the correlation between using complete sets compared to not too small subsets is very strong. Thus, this optimization is not critical at all provided the chosen threshold is not too small.

The GIRAN system architecture is portrayed in Fig. 2. The user interface is implemented as a Java applet using the prefuse toolkit (Heer et al. 2005). The underlying scalable service-oriented architecture is relying on a PostgreSQL database and an Apache Tomcat web server with web services implemented in Java via Axis. The preprocessing steps for setting up the system include the graph download via Wikipedia database dump function in MySQL format, the elimination of redirects, the deletion of broken links, the storage of the optimized graph in the local PostgreSQL database, as well as the creation of database indices. Incremental updates of the graph are possible via the MediaWiki API of Wikipedia.

## 4  GIRAN User Interface to Wikipedia

The graphical user interface (see Fig. 3) consists of four parts: the graph display in the center, a panel displaying the up-to-date content of the current article from the graph view on the right, a histogram of Wikipedia categories of the currently visible articles on the left, and a toolbar at the top. GIRAN shows the first and second level of related articles in a radial tree layout based on a minimum spanning tree (see Fig. 4 on the left). Each article from the related neighborhood is represented as a node and the similarity strength between the articles calculated with Salton's cosine is depicted in the graph view by means of the width of the edges. A click on a link in the content panel results in changing the current article in the graph view and vice versa: the graph view and the content panel are automatically

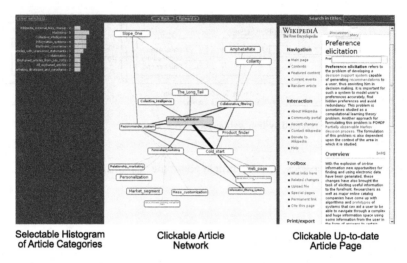

**Selectable Histogram of Article Categories**     **Clickable Article Network**     **Clickable Up-to-date Article Page**

**Fig. 3** The GIRAN interface applet (screenshot of the overall layout)

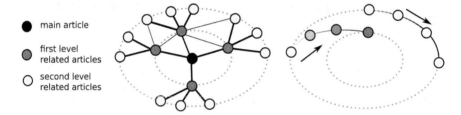

- ● main article
- ◉ first level related articles
- ○ second level related articles

**Fig. 4** Article neighborhood presentation and transition animation

synchronized. When the current article of interest is changed, the related article's neighborhood centered around the new article is calculated on the application server and incrementally loaded. The selected article is then placed in the center of the graph view. The transition is animated to keep it easy to follow. The animation features interpolation by polar coordinates as described in Yee et al. (2001) to avoid nodes crisscrossing each other (see Fig. 4 on the right).

The previous view can be restored using the history function of the applet which allows to navigate back and forth between views using the buttons located in the toolbar (see Fig. 5). Two tools are provided for searching the displayed graph. The panel on the left side shows a histogram which visualizes the distribution of the categories of the currently displayed articles. When the user selects a histogram bar with the mouse, all articles that belong to the selected category are highlighted, while all others are faded into the background; the user can select any set of categories. A second filter, the search box, is located in the upper right of the toolbar. It provides a case-insensitive incremental search in the article titles of all displayed articles. Matches are highlighted as one types to give quick feedback about the issued query.

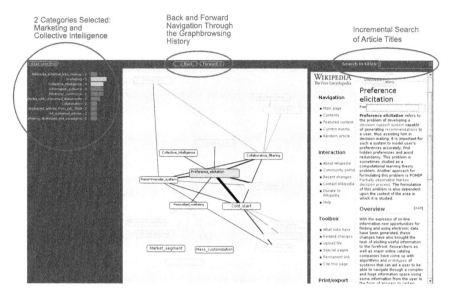

**Fig. 5** Article filtering and history navigation in GIRAN (screenshot)

## 5   Related Work, Evaluation, and Outlook

Table 1 shows three examples of the top places of related articles for three very different Wikipedia entries calculated by GIRAN. Ollivier and Senellart (2007) introduced the Green measure, a similarity measure, which was applied and evaluated in detail for related pages in Wikipedia by means of the same examples. Using Spearman's rank correlation coefficient for top k lists showed that GIRAN results are similar to the well received Green measure, while allowing for incremental graph updates and fast response times of the GIRAN user interface.

The presented work is an advanced version of RecoDiver, a distributed application for browsing behavior-based book recommendation on dynamic graphs on library websites first described in Neumann et al. (2008). RecoDiver relied on a different algorithm for the calculation of the neighborhood and did not show the recommendation (i.e. similarity) strength, but the overall user interface layout and feature set was similar to GIRAN. A usability survey of RecoDiver was conducted and can be found in detail in Neumann (2009). Ninety-three percentage of the users found the graph interface easy to learn, simple to use (93%), and the organization of information clear (90%). In a second comparative user survey most users preferred a graph-based over a list-based interface (like Wikipedia's "See Also" section) for browsing related items. Only for examining the relationship strength the list-based interface was preferred at that time. For GIRAN the feature of showing the similarity strength was introduced to overcome the biggest limitation of RecoDiver. Various tools for static graph visualizations use Wikipedia as a data source, but to the best of our knowledge no other dynamic graph-based interface

**Table 1** Three examples of recommended articles (sorted by similarity strength)

| Star Wars *(large popular article with many links)* | Clique (graph theory) *(short scientific article)* | Pierre de Fermat *(medium sized bibliographic article)* |
|---|---|---|
| George Lucas | Cocoloring | Diophantus |
| Star Wars Episode IV: A New Hope | Complement graph | Leonhard Euler |
| Star Wars Episode VI: Return of the Jedi | Independent set (graph theory) | Fermat's Last Theorem |
| Star Wars Episode V: The Empire Strikes Back | Maximal independent set | Infinite descent |
| Star Wars Expanded Universe | Clique complex | Problem of points |
| List of Star Wars characters | Complete graph | Analytic geometry |
| Star Wars Episode I: The Phantom Menace | Graph coloring | Disquisitiones Arithmeticae |
| Rebel Alliance | Clique problem | Carl Friedrich Gauss |
| Star Wars Episode III: Revenge of the Sith | Bipartite graph | Diophantus II.VIII |
| Coruscant | Subcoloring | Joseph Louis Lagrange |
| Galactic Empire (Star Wars) | Clique-sum | Franciscus Vieta |
| Naboo | Chordal graph | Fermat's principle |
| Star Wars Episode II: Attack of the Clones | Glossary of graph theory | Fermat's theorem on sums of two squares |
| Death Star | Graph (mathematics) | Frans van Schooten |
| Dagobah | Independent set problem | Claude Gaspard Bachet de Méziriac |
| Tatooine | Split graph | Blaise Pascal |
| Star Wars (radio) | Cograph | John Wallis |
| Luke Skywalker | Graph theory | Augustin-Louis Cauchy |
| New Republic (Star Wars) | Turán's theorem | Diophantine equation |

for browsing related articles in Wikipedia can be found in the literature. A larger user survey with GIRAN is planned once the application is released to the general public.

# References

Hamers L, Hemeryck Y, Herweyers G, Janssen M, Keters H, Rousseau R, Vanhoutte A (1989) Similarity measures in scientometric research: The Jaccard index versus Salton's cosine formula. Inform Process Manag 25(3):315–318

Heer J, Card SK, Landay JK (2005) Prefuse: A toolkit for interactive information visualization. In: Proceedings of the SIGCHI conference on Human factors in computing systems, ACM, New York, pp 421–430

Jaccard P (1901) Etude comparative de la distribution florale dans une portion des alpes et du jura. Bulletin de la Societe vaudoise des Sciences Naturelles 37:547–579

Neumann AW (2009) Recommender systems for information providers: designing customer centric paths to information. Physica-Springer, Heidelberg

Neumann AW, Philipp M, Riedel F (2008) RecoDiver: Browsing behavior-based recommendations on dynamic graphs. AI Comm 21(2–3):177–183

Ollivier Y, Senellart P (2007) Finding related pages using Green measures: An illustration with Wikipedia. In: Proceedings of the 22nd National oCnference on Artificial Intelligence, MIT Press, Cambridge, vol 2, pp 1427–1433

Salton G, McGill MJ (1983) Introduction to modern information retrieval. McGraw-Hill, Auckland

Small H (1973) Co-citation in the scientific literature: A new measure of the relationship between two documents. J Am Soc Inform Sci 24(4):265–269

Small H, Sweeney E (1985) Clustering the Science Citation Index using co-citations I. A comparison of methods. Scientometrics 7:391–409

Yee KP, Fisher D, Dhamija R, Hearst M (2001) Animated exploration of dynamic graphs with radial layout. In: InfoVis 2001: IEEE Symposium on Information Visualization, IEEE Computer Society, Washington DC, pp 43–50

# Power Tags as Tools for Social Knowledge Organization Systems

Isabella Peters

**Abstract**  Web services are popular which allow users to collaboratively index and describe web resources with folksonomies. In broad folksonomies tag distributions for every single resource can be observed. Popular tags can be understood as "implicit consensus" where users have a shared understanding of tags as best matching descriptors for the resource. We call these high-frequent tags "power tags". If the collective intelligence of the users becomes visible in tags, we can conclude that power tags obtain the characteristics of community controlled vocabulary which allows the building of a social knowledge organization system (KOS). The paper presents an approach for building a folksonomy-based social KOS and results of a research project in which the relevance of assigned tags for particular URLs in the social bookmarking system delicious has been evaluated. Results show which tags were considered relevant and whether relevant tags can be found among power tags.

## 1  Introduction

In the last years such web services became popular which allow users to collaboratively and intellectually index and describe web resources (e.g. bookmarks) with user-generated keywords, so-called tags. These performed tagging actions result in a folksonomy of this particular collaborative information service (CIS) (Peters 2009). The folksonomy of a CIS (Hotho et al. 2006) $F_{CIS}$ can be defined as a tuple $F_{CIS} := (U, T, R, Y)$ where $U, T, R$ are finite sets of the elements user names $U$, tags $T$, and resource identifiers $R$, and $Y$ is a ternary relation between them, i.e., $Y \subseteq U \times T \times R$ whose elements are called tagging actions. The $F_{CIS}$ is composed of all personomies

I. Peters (✉)
Heinrich-Heine-University Düsseldorf, Universitätsstr. 1, 40225 Düsseldorf, Germany
e-mail: isabella.peters@uni-duesseldorf.de

W. Gaul et al. (eds.), *Challenges at the Interface of Data Analysis, Computer Science, and Optimization*, Studies in Classification, Data Analysis, and Knowledge Organization, DOI 10.1007/978-3-642-24466-7_29, © Springer-Verlag Berlin Heidelberg 2012

$P_1 \ldots P_n$ of a CIS called $P_{CIS}$ and all docsonomies $D_1 \ldots D_n$ of the CIS called $D_{CIS}$. $P_{CIS}$ is defined as a multiset $P_{CIS} := (U, T, X)_b$ where $X \subseteq U \times T$ and $\{(u, t) \in U \times T \mid (u, t) \in X\}, b \in \mathbb{N}^+$. $P_{CIS}$ becomes $P_u$ by substituting $X$ with $X_u$ where $u \in U$. $D_{CIS}$ is defined as a multiset $D_{CIS} := (T, R, Z)_b$ where $Z \subseteq T \times R$ and $\{(t, r) \in T \times R \mid (t, r) \in Z\}, b \in \mathbb{N}^+$. $D_{CIS}$ becomes $D_r$ by substituting $Z$ with $Z_r$ where $r \in R$. It follows that $F_{CIS} \supseteq P_{CIS} \bullet D_{CIS}$ and $P_u \subseteq P_{CIS}$ and $D_r \subseteq D_{CIS}$. The differentiation between $F_{CIS}$ and $D_{CIS}$ (and $D_r$ respectively) and the notion of $D_{CIS}$ as multiset is important for our later discussion of tag distributions.

What is more, folksonomies may allow the multiple assignment of a single tag to a particular resource, so that we can speak of a broad folksonomy in this case. In contrast to this, narrow folksonomies only allow the addition of new tags to the resource (Vander Wal 2005). Typical examples for broad and narrow folksonomies are delicious.com and youtube.com. The major difference between broad and narrow folksonomies is that in broad folksonomies tag frequency distributions on the resource level can be observed. Having this information about folksonomies at hand the paper aims at discussing two research questions (RQ) in detail:

- RQ1: How to build social knowledge organization system (KOS) (automatically) by using folksonomies?
- RQ2: Are Power Tags reflecting collective user intelligence and as such are most relevant for a resource?

## 2 Related Work

Several studies have been concerned with automatic construction of KOS by using folksonomies referring to this idea as "emergent semantics" (Staab et al. 2002). Early work of Schmitz et al. (2006) make use of association rules to detect related tags. Schmitz (2006) discusses how to build an ontology of Flickr tags with a statistical model for subsumption based on tag co-occurrences but does not use broad folksonomies. Marinho et al. (2008) are using frequent itemset mining to enhance ontologies with folksonomic tags. Mika (2007) compares tag clusters either built from the links of tagger and tags or the links of tags and resources. His evaluation shows that tag-resource connections are appropriate for concept mining, whereas the tagger-tag connections can be used for automatic detection of hierarchical relations (e.g. broader-narrower concept) between tags. In our approach we combine the ideas of Mika (2007) for extracting emergent semantics of folksonomies and consider tags which have been tagged for a single resource ($D_r$) and then tags which have been indexed for many resources (tag-resource link; $D_{CIS}$). Differing from Marinho et al. (2008) and Mika (2007) we conduct the first step on the docsonomy level $D_r$ and only respect high-frequent tags of $D_{CIS}$ for processing.

# 3  Using Relevant Power Tags for Constructing Social KOS

In this section we discuss the aforementioned research questions. RQ1 will be answered theoretically whereas RQ2 will be answered by means of empirical data.

## 3.1  RQ1: Folksonomy-Based Social KOS

In this paper we define a "social KOS" as a collaboratively built knowledge representation tool with natural-language terms. Tasks of social KOS are both, finding of appropriate concepts and finding of paradigmatic structures in folksonomies, e.g. hierarchies. The docsonomy reflects via tags the users' collective intelligence (Surowiecki 2005) in giving meaning to this resource. In broad folksonomies the most popular tags for a resource can be determined and it is widely assumed that popular tags are the most important tags for a resource (Weiss 2005) as they reflect an "implicit consensus" of users for describing it. To give credit to the theories of collective intelligence we call these most popular tags "power tags" (Peters and Stock 2010). In order to establish a social KOS we need candidate tags for the finding of concepts and relations. For this purpose we propose power tag pairs.

Let us explain the idea of power tag-based tag pairs with an example. A resource "App Inventor for Android" was saved in the social bookmarking system delicious and was indexed with different tags, e.g. "android", "google", "development" (see Fig. 1). In order to find tag pairs for social KOS we must first define the power tags of this resource reflected in $D_r$. The determination of power tags depends on the tag frequency distribution of a digital resource. The basic assumption is that different distributions of tags may appear in folksonomies: (a) an inverse power law

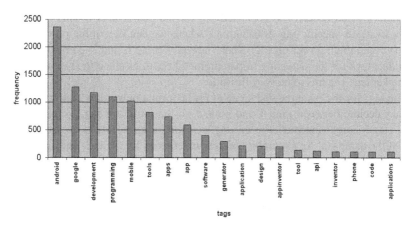

**Fig. 1**  Tag distribution for the delicious-resource "appinventor.googlelabs.com/about"

distribution, a Lotka-like curve, (b) an inverse logistic distribution (Stock 2006), and (c) other distributions. An inverse power law has the form $f(x) = \frac{C}{x^a}$ where $C$ is a constant, $x$ is the rank of the tag relative to the resource, and $a$ is a value ranging normally from about 1 to about 2 (Egghe and Rousseau 1990). If this assumption is true, we see a curve with only few tags at the top of the distribution, and a "long tail" of numerous tags on the lower ranks on the right-hand side of the curve. The discussions about "collective intelligence" are mainly based on this observation: the first $n$ tags of the left hand side of the power law reflect the collective user intelligence.

The inverse logistic distribution shows a lot of relevant tags at the beginning of the curve ("long trunk") and the known "long tail". This distribution follows the formula $f(x) = e^{[-C'(x-1)]^b}$ where $e$ is the Euler number, $x$ is the rank of the tag, $C'$ is a constant and the exponent $b$ is approximately 3 (Stock 2006).

In comparison with the inverse power law the inverse logistic distribution reflects the collective intelligence differently. The curve shows a long trunk on the left and a long tail of tags on the right. Since all tags in the long trunk have been applied with similar (high) frequency, all left-hand tags up to the turning point of the curve should be considered as a reflection of collective intelligence. For the determination of power tags we have to keep in mind both known tag distributions. If the resource-specific distribution of tags follows the inverse power law, the first $n$ tags are considered as "power tags". The value of $n$ must be determined empirically. If the tag distribution forms an inverse logistic distribution, all tags on the left-hand side of the curve (up to the turning point) are marked as power tags.

The concrete processing of power tags works as follows: the first step is to determine power tags for each docsonomy $D_r$ of the CIS. According to the above explanations, two different tag distributions may appear which each identify different numbers of power tags (we call them power tags I). Since these power tags I are important tags in giving meaning to the resource, they have to be processed in the next step. Now, the $n$ numbers of power tags I should each be investigated regarding their relationships to other tags of the whole database $D_{CIS}$ - in other words, a calculation of co-occurrence is carried out for the power tags I. This calculation produces again specific tag distributions, where we can determine power tags as well (we call them power tags II). These new power tag I- and II-pairs are now the candidate tags for the detection of paradigmatic relations since their connection to the power tags I seems to be very fruitful.

The tags for the resource displayed in Fig. 1 form a power law distribution with a sharp decline between rank 1 and rank 2, so that only one tag is considered as power tag I, here: "android". For this tag we examine co-occurrences with all other tags of the database. An exemplary search in delicious for the tag "android" results in the co-occurring tags displayed in Fig. 2 and says that the tag "android" appears together with the tag "mobile" in 70,614 resources. Those co-occurring tags follow an inverse-logistic distribution (see Fig. 2). The first $n$ tags (say $n = 7$) are considered as power tags II. The pairs of power tags I and power tags II are now the source for building a social KOS.

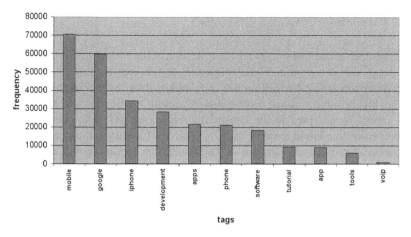

**Fig. 2** Co-occurrence distribution for the tag "android"

**Table 1** Descriptor set for descriptor "android" extracted from tags

| Abbreviation | Descriptor | Name of relation | Type of relation |
| --- | --- | --- | --- |
| Broader term | Phone | Hierarchy, part_of | Meronymy |
| BT | Software | Hierarchy, is_a | Hyponymy |
| Related term | Mobile | Related term, Used_in | Association |
| RT | Google | Related term, produced_by | Association |
| RT | Iphone | Related term, is_competitor | Association |
| RT | Development | Related term | Association |
| RT | Apps | Related term | Association |

Up to this point, the construction of a social KOS can be carried out automatically. But, we state that it is not sufficient to calculate co-occurrences to obtain semantic relations which form the basis for social KOS. Co-occurrences can only work as indicators that some kind of relations may exist between two concepts. Yet, to determine which kind of relation is at hand, we have to add intellectual analyses from people like users or administrators of the CIS.

Figure 2 shows that the tag "android" is frequently combined with the tags "mobile", "google", "iphone", "development", "apps", "phone", and "software". The tag pair "android-mobile" shows an associative relation, when we rely on semantic relations used in thesauri (Aitchison et al. 2000), or a more specific relation like "used_in" , when we consider ontology-like relations. All found relations are shown in Table 1.

The given assumptions result in an entry of controlled terms, e.g. descriptors, with which we now can build a social knowledge organization system. Here we are able to answer research question one. Yes, it is possible to construct a social KOS by using the users' tags. But: no, it is not possible to automate this process to the full extent. We are always in need of some intellectual endeavors to identify the type of a relation between tags.

## 3.2   RQ2: Relevance of Power Tags

If we presuppose that power tags are the reflection of the collective user intelligence and along with it the most relevant tags for the resource we have to prove that this assumption is true by making use of empirical data. As tagging behavior and tags are highly subjective and error-prone (regarding spelling etc.) as well as such facts cannot be accepted in the construction of social KOS, the question arises whether collective intelligence of users is capable of ironing out too personal and erroneous tags so that all users are satisfied with high-frequent power tags as best matching keywords for the resource. That is why we have conducted a survey with 20 information science students at the HHU Düsseldorf as assessors to determine the relevance of power tags for resources.

We downloaded 30 resources from delicious in February 2010 which fulfilled two preconditions:

1. Resources must be tagged with the tag "folksonomy" to guarantee that students have the technical knowledge to judge the relevance of the tags and can as such serve as independent domain experts.
2. Resources must be tagged from at least 100 users to guarantee that broad tag distributions with recognizable characteristics are downloaded.

Each assessor received each resource and the tags assigned to it, but the assessors did not know which rank that tag gained from the delicious-users. To be able to judge the relevance of the tags the assessors had access to the resources. Following the common procedure in information retrieval system evaluation the students indicated each and every tag of the resource with 1 for a relevant tag and with 0 for a non-relevant tag. In this context "relevant" means that a particular tag describes the content of the resource in an appropriate manner. Please note that we worked with binary relevance judgments, similar to the now "classic" methods of Cranfield and TREC (Voorhees 2001). Moreover, the relevance judgments of the assessors were considered objective and neutral as assessors are not regarded as system users (and along with it not as tagging users) in Cranfield-like retrieval system test settings (Kamps et al. 2009).

Receiving 20 assessors' relevance judgments we constructed tag relevance distributions for every resource. After this we defined which ratio of assessors' judgments determines the relevance of a resource's tag, which was 50%, and we extracted the Top 10-tags for every resource inclusive tag frequency from delicious. The process for an example is displayed in Fig. 3 where column "relevance frequency" reflects how many assessors considered this tag relevant and where column "tag frequency" shows how many delicious-users used this tag for indexing. Pearson correlation is also given for the two columns. The average value of Pearson correlation for 30 resources was 0.49. The summarization of all assessor judgments can be found in Fig. 4. It presents in how many cases a tag on rank 1...10 was judged relevant by the majority of the assessors. The evaluation shows a clear result: in most cases only the first two tags, which are the power tags, have been considered relevant from the assessors.

| delicious-rank | delicious-tags | tag frequency | relevance frequency | relevant? | Pearson |
|---|---|---|---|---|---|
| 1 | folksonomy | 122 | 13 | 1 | 0,75 |
| 2 | tagging | 95 | 11 | 1 | |
| 3 | authority | 88 | 3 | 0 | |
| 4 | wikipedia | 66 | 4 | 0 | |
| 5 | web2.0 | 50 | 9 | 0 | |
| 6 | tags | 45 | 6 | 0 | |
| 7 | search | 30 | 1 | 0 | |
| 8 | findability | 28 | 5 | 0 | |
| 9 | ia | 22 | 3 | 0 | |
| 10 | article | 21 | 0 | 0 | |

**Fig. 3** Are power tags the most relevant tags for a resource? Example

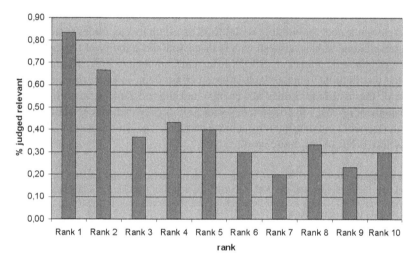

**Fig. 4** Assessors' relevance judgments for Top 10-tags of 30 delicious-resources

During the analysis of the assessors' relevance judgments it was noticeable that they show a slight bias to German tags as the mother tongue of most students was German and, therefore, German tags were more relevant for them. Students also preferred the original spelling of proper nouns, e.g. the tag "LibraryThing" was considered more relevant than "librarything".

## 4   Discussion and Outlook

We presented an approach to construct social KOS on the basis of broad folk-sonomies. It could be shown that tag co-occurrence analyses of power tags I and power tags II produce fruitful tag pairs which may work as candidate tags for further semantic enrichment of folksonomies via defined relations between tags. Although the extraction of power tag pairs can be carried out automatically, the need for their intellectual processing to assess the the type of tag relation cannot be denied. So, folksonomies deliver concept candidates for social KOS which make use of intellectually built tag structures.

We have also shown that power tags do reflect the collective user intelligence in giving meaning to the resources as assessors judged the top 2-delicious-tags relevant in most cases. As such, power tags work as community-proved controlled vocabulary which can be utilized for social KOS. The method could be optimized with a combination of morphologically - not semantically - similar tags (e.g. "folksonomy"and "folksonomies"), the combination of phrase tags (e.g. "information" and "architecture" to "information architecture"), the unification of multi-term tags (e.g. "social_classification", "socialclassification", and "SocialClassification") and the joining of cross-language synonyms (e.g. "classification" and "Klassifikation").

A relevant aspect for building social KOS based on folksonomies is still open for research although it has yet been widely discussed: the stability of tag distributions (amongst others: Halpin et al. (2007), Kipp and Campbell (2006)). Research shows that tag distributions remain stable after a certain point in time, meaning that the shape of the distribution does not change and that no rank permutations in tags appear anymore. The construction of social KOS is only then able to start, when the resource-specific tag distributions achieve stability. The determination of the point of stability is left to our future work.

**Acknowledgements** The author would like to thank her colleagues and students from the HHU Düsseldorf, the participants of the 34th Annual Conference of the German Classification Society and the reviewers for their helpful comments. The presented work is funded by the German research fund DFG (STO 764/4-1).

# References

Aitchison J, Gilchrist A, Bawden D (2000) Thesaurus construction and use. Europa Publishing, London, New York

Egghe L, Rousseau R (1990) Introduction to informetrics. Elsevier, Amsterdam

Halpin H, Robu V, Shepherd H (2007) The complex dynamics of collaborative tagging. In: Williamson CL, Zurko ME, Patel-Schneider PF, et al. (eds) Proc. of the 16th Int. WWW Conf. Banff, Alberta, Canada, ACM, New York, pp 211–220

Hotho A, Jaeschke R, Schmitz C, et al. (2006) Information retrieval in folksonomies: Search and ranking. Lect Notes Comput Sci 4011:411–426

Kamps J, Lalmas M, Larsen B (2009) Evaluation in context. Lect Notes Comput Sci 5714:339–351

Kipp M, Campbell D (2006) Patterns and inconsistencies in collaborative tagging systems: An examination of tagging practices. In: Grove A (ed) Proc. of the 17th Annual Meet. of the Am. Soc. for Inf. Sci. and Tech, Austin, Texas, USA (CD-Rom)

Marinho LB, Buza K, Schmidt-Thieme L (2008) Folksonomy-based collabulary learning. Lect Notes Comput Sci 5318:261–276

Mika P (2007) Ontologies are us: A unified model of social networks and semantics. J Web Semant 5 (1):5–15

Peters I (2009) Folksonomies: Indexing and retrieval in Web 2.0. De Gruyter Saur, Berlin

Peters I, Stock WG (2010) Power tags in information retrieval. Libr Hi Tech 28(1):81–93

Schmitz C, Hotho A, Jaeschke R, et al. (2006) Mining association rules in folksonomies. In: Batagelj V, Bock H, Ferligoj A, et al. (eds) Data science and classification. Springer, Berlin, Heidelberg, pp 261–270

Schmitz P (2006) Inducing ontology from Flickr tags. In: Carr L, De Roure D, Iyengar A, et al. (eds) Proc. of the 15th Int. WWW Conf., Edinburgh, Scotland, ACM, New York

Staab S, Santini S, Nack F, et al. (2002) Emergent semantics. Intell Syst 17(1):78–86

Stock WG (2006) On relevance distributions. J Am Soc Inf Sci Tech 57(8):1126–1129

Surowiecki J (2005) The wisdom of crowds. Anchor Books, New York

Vander Wal T (2005) Explaining and showing broad and narrow folksonomies. URL http://www.vanderwal.net/random/entrysel.php?blog=1635, cited 13 Aug 2010

Voorhees EM (2001) The philosophy of information retrieval evaluation. Lect Notes Comput Sci 2406:355–370

Weiss A (2005) The power of collective intelligence. NetWorker 9(3):16–23

# The Degree Distribution in Random Intersection Graphs

Katarzyna Rybarczyk

**Abstract** We study the degree distribution in a general random intersection graph introduced by Godehardt and Jaworski (Exploratory Data Analysis in Empirical Research, pp. 68–81, Springer, Berlin, 2002). The model has shown to be useful in many applications, in particular in the analysis of the structure of data sets. Recently Bloznelis (Lithuanian Math J 48:38–45, 2008) and independently Deijfen and Kets (Eng Inform Sci 23:661–674, 2009) proved that in many cases the degree distribution in the model follows a power law. We present an enhancement of the result proved by Bloznelis. We are able to strengthen the result by omitting the assumption on the size of the feature set. The new result is of considerable importance, since it shows that a random intersection graph can be used not only as a model of scale free networks, but also as a model of a more important class of networks – complex networks.

## 1 Introduction

Graph-theoretical concepts belong to the commonly used tools in data structure analysis, for example in revealing hidden clusters in a data set, testing the randomness of such clusters (see for example Bock (1996), Godehardt (1990) for metric data, and Godehardt and Jaworski (2002), Godehardt et al. (2007) for non-metric data) or the structure analysis of large networks (Albert and Barabási 2002). It is natural that in many non-metric data sets relations between objects depend on properties they possess, therefore an effective model to analyse the structure of those relations is a random graph, in which edges represent relations based on the objects' properties. Let a set of vertices $\mathcal{V}$ represent the set of objects and an auxiliary

K. Rybarczyk (✉)
Faculty of Mathematics and Computer Science, Adam Mickiewicz University, Umultowska 87, 61-614 Poznań, Poland
e-mail: kryba@amu.edu.pl

W. Gaul et al. (eds.), *Challenges at the Interface of Data Analysis, Computer Science, and Optimization*, Studies in Classification, Data Analysis, and Knowledge Organization, DOI 10.1007/978-3-642-24466-7_30, © Springer-Verlag Berlin Heidelberg 2012

set of features $\mathcal{W}$ represent the set of properties. We consider a random bipartite graph in which edges join vertices from $\mathcal{V}$ (representing objects) with features from $\mathcal{W}$ (representing properties) according to some random procedure. In the related random intersection graph two vertices are joined by an edge if and only if they represent those objects having at least one common property.

In this paper we concentrate on finding a good graph theoretical model representing data sets, which are included in a wide family of so-called scale free networks (i.e. the degree distribution follows the power law) and its subfamily of complex networks (they have also a large clustering coefficient). For precise definitions and examples of scale free and complex networks see the paper of Albert and Barabási (2002). As stated in Albert and Barabási (2002) none of the traditionally used graph models has the power law degree distribution and asymptotically constant clustering coefficient at the same time, so they are not good models of complex networks. New results show that in many complex and scale free networks the bipartite structure (known or hidden) relating elements of the network to their properties influences the structure of the connections between elements (see Guillaume and Latapy (2004)). Moreover, new results on random intersection graphs (see for example Bloznelis (2008), Deijfen and Kets (2009)) suggest that the structure analysis of random intersection graphs is the right direction for further research of the theoretical model. In this paper we study the degree distribution in a random intersection graph. The obtained result combined with the one presented in (Godehardt et al. 2011) shows how to set parameters of a random intersection graph to get a good model of complex networks.

## 2 Previous Work and Results

**Definition 1.** Let $s$ be a positive integer, $\mathcal{V} = \{v_1, \ldots, v_n\}$ and $\mathcal{W} = \{w_1, \ldots, w_{m(n)}\}$ be disjoint sets and $\mathscr{P}_{(m)} = (P_0, P_1, \ldots, P_m)$ be a probability distribution. Moreover let $D(v_1), \ldots, D(v_n)$ be a family of random subsets of $\mathcal{W}$ generated according to the following procedure. Independently for all $1 \leq i \leq n$:

1. First $Z_i$, the cardinality of a set of properties $D(v_i)$, is assigned to $v_i$ according to the probability distribution $\mathscr{P}_{(m)}$ (i.e., $\Pr\{Z_i = d\} = P_d$ for all $0 \leq d \leq m$).
2. Then given $Z_i = d$, a set of $d$ properties is assigned to $v_i$ uniformly over the class of all $d$-element subsets of $\mathcal{W}$, i.e., for a given $d$-element subset $A \subseteq \mathcal{W}$

$$\Pr\{D(v_i) = A \mid Z_i = d\} = \binom{m}{d}^{-1}.$$

A random intersection graph $\mathscr{G}_s(n, m, \mathscr{P}_{(m)})$ is a graph with vertex set $\mathcal{V}$ and edge set $E = \{\{v_i, v_j\} : |D(v_i) \cap D(v_j)| \geq s\}$.

For the remainder of this paper we consider only the case $\mathscr{G}_1(n, m, \mathscr{P}_{(m)})$, which we denote by $\mathscr{G}(n, m, \mathscr{P}_{(m)})$. Moreover all limits as well as the notation $o(\cdot)$ are as $n \to \infty$.

The main theorem of this paper establishes the asymptotic distribution of the vertex degree in $\mathscr{G}(n, m, \mathscr{P}_{(m)})$ for a wide class of distributions $\mathscr{P}_{(m)}$. The first exact formula describing vertex degrees in a random intersection graph was given by Godehardt and Jaworski (2002). In the same paper they gave the asymptotic result in the case $P_d = 1$ for some $d = \sqrt{cm/n}$. The next asymptotic result on the topic was given by Stark (2004), where a special case of the model, introduced in (Karoński et al. 1999), was considered.

Further research, on the one hand concerned finding sufficient and necessary conditions under which the vertex degree is asymptotically Poisson distributed (see the results of Jaworski and Stark (2008), and of Jaworski et al. (2006)) and on the other hand concentrated on finding conditions on $\mathscr{P}_{(m)}$ imposing vertex degree distribution in $\mathscr{G}(n, m, \mathscr{P}_{(m)})$ which differ much from the Poisson (i.e. also the power law distribution). Deijfen and Kets (2009) considered $\mathscr{G}(n, m, \mathscr{P}_{(m)})$ in which $\mathscr{P}_{(m)}$ is binomially distributed $\mathrm{Bin}(n, p)$ with $p$ being a random variable. They gave an example of $\mathscr{P}_{(m)}$ for which the degree distribution in $\mathscr{G}(n, m, \mathscr{P}_{(m)})$ follows the power law and $\mathscr{G}(n, m, \mathscr{P}_{(m)})$ has an asymptotically constant clustering coefficient. Independently Bloznelis (2008) showed a similar result concerning vertex degrees in $\mathscr{G}(n, m, \mathscr{P}_{(m)})$ with another distribution $\mathscr{P}_{(m)}$. The theorem introduced in this paper is an enhancement of the result of Bloznelis. This improvement is important due to the fact that the theorem presented in this paper shows the parameters for which $\mathscr{G}(n, m, \mathscr{P}_{(m)})$ has a power law degree distribution and asymptotically constant clustering coefficient (see Remark 1).

Let us introduce some additional notation. Let $T_\lambda$ be a random variable with Poisson distribution with expected value $\lambda$ $(\mathrm{Po}(\lambda))$ and let $\mathbb{T}_N$ be a random variable with the mixture of Poisson distributions, i.e. a random variable with Poisson distribution $\mathrm{Po}(N)$, where $N$ is a random variable with real non negative values. Therefore, for a random variable $Y$ with real positive values and any integer $k \geq 0$

$$\Pr\{\mathbb{T}_{Y\mathbb{E}Y} = k\} = \mathbb{E}\left(\frac{(Y\mathbb{E}Y)^k e^{-Y\mathbb{E}Y}}{k!}\right). \tag{1}$$

**Theorem 1.** *Let* $X_n = X_n(v) = \deg(v)$ *be a random variable equal to the degree of a vertex* $v \in \mathscr{V}$ *in* $\mathscr{G}(n, m, \mathscr{P}_{(m)})$, $Z_n$ *be a random variable with distribution* $\mathscr{P}_{(m)}$ *and* $Y_n = \frac{Z_n}{r_n}$, *where* $r_n = \sqrt{\frac{m}{n}}$. *If*

*(i)* $Y_n \xrightarrow{D} Y$, *i.e.* $Y_n$ *tends in distribution to a random variable* $Y$;

*(ii)* $\mathbb{E}Y_n \to \mathbb{E}Y$;

*(iii)* $\mathbb{E}Y_n < \infty$ *and* $\mathbb{E}Y < \infty$,

*then* $X_n$ *tends in distribution to a random variable with the mixture of Poisson distributions* $\mathbb{T}_{Y\mathbb{E}Y}$.

*Remark 1.* Bloznelis (2008) noticed that $Y$ may be chosen such that the random variable $\mathbb{T}_{Y\mathbb{E}Y}$ follows a power law (i.e. $\mathscr{G}(n, m, \mathscr{P}_{(m)})$ is a good model of scale free networks). However in (Bloznelis 2008) an additional assumption $n = o(m)$ is made. Under the assumptions of the theorem if $m = o(n^3)$ and $n = o(m)$ by a theorem from Godehardt et al. (2011) the clustering coefficient equals

$$\text{Clus}(\mathscr{G}(n, m, \mathscr{P}_{(m)})) = \mathbb{E}\frac{1}{Z_n} + o(1) = \sqrt{\frac{n}{m}} \mathbb{E}\frac{1}{Y_n} + o(1) = o(1).$$

However if $m = cn$ for some constant $c > 0$, and $\Pr\{Z_n \geq 2\} = 1$ by the same theorem

$$\text{Clus}(\mathscr{G}(n, m, \mathscr{P}_{(m)})) = \sqrt{c}\, \mathbb{E}\frac{1}{Y_n} + o(1) \to \sqrt{c}\, \mathbb{E}\frac{1}{Y}$$

(the convergence is true since $Y_n = Z_n/\sqrt{c} \geq 2/\sqrt{c}$ with probability 1). Therefore, under the assumptions of the theorem the clustering coefficient does not tend to 0, only in the case when $n$ and $m$ are of the same order of magnitude. Thus only in this case $\mathscr{G}(n, m, \mathscr{P}_{(m)})$ may be a good model of complex networks (see Albert and Barabási 2002).

## 3 Proof

We start with a definition and two lemmas useful for Poisson approximation. Let $X$ and $Y$ be random variables with values in the same countable set $P$. The total variation distance between $X$ and $Y$ is given by

$$d_{TV}(X, Y) = \sup_{A \subseteq P} |\Pr\{X \in A\} - \Pr\{Y \in A\}| = \frac{1}{2}\sum_{a \in P} |\Pr\{X = a\} - \Pr\{Y = a\}|.$$

The following lemma is shown in Barbour et al. (1992) (see equation (1.23) page 8).

**Lemma 1.** *Let $X$ and $Y$ be random variables with binomial distribution $\text{Bin}(N, p)$ and Poisson distribution $\text{Po}(Np)$, respectively. Then $d_{TV}(X, Y) \leq p$.*

**Lemma 2.** *Let $X$ and $Y$ be random variables with Poisson distributions $\text{Po}(\lambda)$ and $\text{Po}(\lambda + \varepsilon)$, respectively. Then $d_{TV}(X, Y) \leq \varepsilon$.*

*Proof.* By (1.9) from Barbour et al. (1992) for any bounded function $g : \mathbb{N} \to \mathbb{R}$:

$$\mathbb{E}((\lambda + \varepsilon)g(Y + 1) - Yg(Y)) = 0. \tag{2}$$

For any $A \subseteq \mathbb{N}$, let $g_A : \mathbb{N} \to \mathbb{R}$ be defined as in (1.10) in Barbour et al. (1992). Therefore $g_A(0) = 0$ and $\lambda g_A(j + 1) - jg_A(j) = I_A(j) - \Pr\{X \in A\}$ for $j \geq 0$, where $I_A$ is the characteristic function of $A$. Then by Lemma 1.1.1 from Barbour et al. (1992) the function $g_A$ is bounded. Moreover $\sup_{j \in \mathbb{N}} \sup_{A \subseteq \mathbb{N}} |g_A(j)| \leq 1$ and thus

$$
\begin{aligned}
|\Pr\{Y \in A\} - \Pr\{X \in A\}| &= |\mathbb{E}I_A(Y) - \Pr\{X \in A\}| \\
&= |\mathbb{E}(I_A(Y) - \Pr\{X \in A\})| = |\mathbb{E}(\lambda g_A(Y+1) - Y g_A(Y))| \\
&= |\mathbb{E}((\lambda + \varepsilon)g_A(Y+1) - Y g_A(Y)) - \mathbb{E}(\varepsilon g_A(Y+1))| \\
&\overset{(2)}{=} \varepsilon |\mathbb{E}(g_A(Y+1))| \leq \varepsilon.
\end{aligned}
$$
$\square$

Now we give estimates on the edge probability in a random intersection graph. Denote

$$
\begin{aligned}
P(d, d') &= \Pr\{vw \in E(\mathscr{G}(n, m, \mathscr{P}_{(m)})) \mid |D(v)| = d, |D(w)| = d'\}; \\
P(d) &= \Pr\{vw \in E(\mathscr{G}(n, m, \mathscr{P}_{(m)})) \mid |D(v)| = d\}.
\end{aligned} \tag{3}
$$

Obviously $P(d) = 0$ for $d = 0$ and $P(d, d') = 0$ for $d = 0$ or $d' = 0$.

**Lemma 3.** *Let $\Delta$ and $d$ be any positive integers. Then*

$$
\frac{d \, \mathbb{E}(Z_n)}{m} - \frac{d \, \mathbb{E}(Z_n \mathbb{I}_{Z_n > \lfloor \Delta/d \rfloor})}{m} - \frac{d^2 \mathbb{E}(Z_n^2 \mathbb{I}_{Z_n \leq \lfloor \Delta/d \rfloor})}{m^2} \leq P(d) \leq \frac{d \, \mathbb{E}(Z_n)}{m},
$$

*where $\mathbb{I}_{\mathscr{A}}$ is an indicator random variable of the event $\mathscr{A}$.*

*Proof.* For any integers $d, d' \geq 0$

$$
\Pr\{D(v) \cap D(w) \neq \emptyset \mid |D(v)| = d, |D(w)| = d'\} \leq \frac{d \binom{m-1}{d'-1}}{\binom{m}{d'}} = \frac{dd'}{m};
$$

$$
\Pr\{D(v) \cap D(w) = \emptyset \mid |D(v)| = d, |D(w)| = d'\} = \frac{\binom{m-d}{d'}}{\binom{m}{d'}} = \prod_{i=0}^{d'-1}\left(1 - \frac{d}{m-i}\right)
$$

$$
\leq \left(1 - \frac{d}{m}\right)^{d'} \leq 1 - \frac{dd'}{m} + \binom{d'}{2}\left(\frac{d}{m}\right)^2 \leq 1 - \frac{dd'}{m} + \left(\frac{dd'}{m}\right)^2.
$$

Therefore for any integers $d, d' \geq 0$

$$
\frac{dd'}{m} - \frac{(dd')^2}{m^2} \leq P(d, d') \leq \frac{dd'}{m} \quad \text{and} \quad P(d) = \sum_{d'=0}^{m} P(d, d') \Pr\{|D(w)| = d'\}.
$$

Combining the above estimates we get for $d \geq 1$

$$P(d) \leq \sum_{d'=0}^{m} \frac{dd'}{m} \Pr\{|D(w)| = d'\} = \frac{d\mathbb{E}Z_n}{m} \quad \text{and}$$

$$P(d) \geq \sum_{d'=0}^{\lfloor \Delta/d \rfloor} P(d, d') \Pr\{|D(w)| = d'\}$$

$$\geq \sum_{d'=0}^{\lfloor \Delta/d \rfloor} \frac{dd'}{m} \Pr\{|D(w)| = d'\} - \sum_{d'=0}^{\lfloor \Delta/d \rfloor} \frac{(dd')^2}{m^2} \Pr\{|D(w)| = d'\}$$

$$= \frac{d\,\mathbb{E}(Z_n)}{m} - \frac{d\,\mathbb{E}(Z_n \mathbb{I}_{Z_n > \lfloor \Delta/d \rfloor})}{m} - \frac{d^2 \mathbb{E}(Z_n^2 \mathbb{I}_{Z_n \leq \lfloor \Delta/d \rfloor})}{m^2}.$$

$\square$

Now we proceed with the proof of the main theorem.

*Proof (Theorem 1).* Let

$$\Delta = \Delta_n = o(m) \quad \text{and} \quad \frac{m}{n} = o(\Delta_n). \tag{4}$$

Moreover, let $X_n^d = X_n^d(v)$ be a random variable representing the degree of a vertex $v \in \mathcal{V}$ in $\mathcal{G}(n, m, \mathscr{P}_{(m)})$ under the condition $|D_n(v)| = d$. By independence of choices of sets $D(\cdot)$, $X_n^d$ has binomial distribution $\text{Bin}(n - 1, P(d))$ (where $P(d)$ is given by (3)). If we put $\lambda_d = d\mathbb{E}Z_n/r_n^2$, then by Lemma 3 for $d \geq 1$

$$\lambda_d - \frac{d\mathbb{E}(Z_n)}{m} - \frac{nd\mathbb{E}(Z_n \mathbb{I}_{Z_n > \lfloor \Delta/d \rfloor})}{m} - \frac{nd^2 \mathbb{E}(Z_n^2 \mathbb{I}_{Z_n \leq \lfloor \Delta/d \rfloor})}{m^2} \leq (n - 1)P(d) \leq \lambda_d.$$

Therefore by Lemma 1 and Lemma 2 for $d \geq 1$

$$d_{TV}(X_n^d, T_{(n-1)P(d)}) \leq P(d) \leq \frac{d\mathbb{E}(Z_n)}{m},$$

$$d_{TV}(T_{(n-1)P(d)}, T_{\lambda_d}) \leq \frac{d\mathbb{E}(Z_n)}{m} + \frac{nd\mathbb{E}(Z_n \mathbb{I}_{Z_n > \lfloor \Delta/d \rfloor})}{m} + \frac{nd^2 \mathbb{E}(Z_n^2 \mathbb{I}_{Z_n \leq \lfloor \Delta/d \rfloor})}{m^2},$$

since $|\lambda_d - (n - 1)P(d)| \leq \frac{d\mathbb{E}(Z_n)}{m} + \frac{nd\mathbb{E}(Z_n \mathbb{I}_{Z_n > \lfloor \Delta/d \rfloor})}{m} + \frac{nd^2 \mathbb{E}(Z_n^2 \mathbb{I}_{Z_n \leq \lfloor \Delta/d \rfloor})}{m^2}$. The total variation distance fulfils the triangle inequality, therefore we obtain

$$d_{TV}(X_n^d, T_{\lambda_d}) \leq \varepsilon(n, d, \Delta),$$

where $\varepsilon(n, d, \Delta) = \frac{2d\mathbb{E}(Z_n)}{m} + \frac{nd\mathbb{E}(Z_n \mathbb{I}_{Z_n > \lfloor \Delta/d \rfloor})}{m} + \frac{nd^2 \mathbb{E}(Z_n^2 \mathbb{I}_{Z_n \leq \lfloor \Delta/d \rfloor})}{m^2}$. Thus for any $A \subseteq \mathbb{N}$ and $d \geq 1$

$$\Pr\{T_{\lambda_d} \in A\} - \varepsilon(n, d, \Delta) \leq \Pr\{X_n^d \in A\} \leq \Pr\{T_{\lambda_d} \in A\} + \varepsilon(n, d, \Delta). \quad (5)$$

Here it should be added that $(n-1)P(d) = \lambda_d = 0$ for $d = 0$, and thus

$$\Pr\{T_{\lambda_d} \in A\} = \Pr\{X_n^d \in A\} \text{ for } d = 0. \quad (6)$$

Notice that

$$\sum_{d=0}^{m} \Pr\{X_n^d \in A\} \Pr\{Z_n = d\} = \Pr\{X_n \in A\} \quad \text{and}$$

$$\sum_{d=0}^{m} \Pr\{T_{\lambda_d} \in A\} \Pr\{Z_n = d\} = \Pr\{T_{Z_n \mathbb{E} Z_n / r_n^2} \in A\} = \Pr\{T_{Y_n \mathbb{E} Y_n} \in A\}.$$

Therefore by (5) and (6)

$$\Pr\{T_{Y_n \mathbb{E} Y_n} \in A\} - \varepsilon(n) \leq \Pr\{X_n \in A\} \leq \Pr\{T_{Y_n \mathbb{E} Y_n} \in A\} + \varepsilon(n), \quad (7)$$

where $\quad \varepsilon(n) \quad = \quad \sum_{d=1}^{m} \left( \frac{2d\mathbb{E}(Z_n)}{m} + \frac{nd\mathbb{E}(Z_n \mathbb{I}_{Z_n > \lfloor \Delta/d \rfloor})}{m} + \frac{nd^2 \mathbb{E}(Z_n^2 \mathbb{I}_{Z_n \leq \lfloor \Delta/d \rfloor})}{m^2} \right)$
$\Pr\{Z_n = d\}$.

Now we show that $\varepsilon(n) = o(1)$. First by the assumptions (ii) and (iii) of the theorem

$$\sum_{d=1}^{m} \frac{2d\mathbb{E}Z_n}{m} \Pr\{Z_n = d\} = 2\frac{(\mathbb{E}Z_n)^2}{m} = 2\frac{(\mathbb{E}Y_n)^2}{n} = o(1)$$

and by (4) and again by the assumptions (ii) and (iii)

$$\sum_{d=1}^{m} \frac{nd^2 \mathbb{E}(Z_n^2 \mathbb{I}_{Z_n \leq \lfloor \Delta/d \rfloor})}{m^2} \Pr\{Z_n = d\}$$

$$= \frac{n}{m^2} \sum_{d=1}^{m} \sum_{d'=1}^{\lfloor \Delta/d \rfloor} (dd')^2 \Pr\{Z_n = d\} \Pr\{Z_n = d'\}$$

$$\leq \frac{n}{m^2} \Delta \sum_{d=1}^{m} \sum_{d'=1}^{\lfloor \Delta/d \rfloor} dd' \Pr\{Z_n = d\} \Pr\{Z_n = d'\} = \frac{\Delta(\mathbb{E}Y_n)^2}{m} = o(1).$$

It remains to show that $\sum_{d=1}^{m} \frac{nd\mathbb{E}(Z_n \mathbb{I}_{Z_n > \lfloor \Delta/d \rfloor})}{m} \Pr\{Z_n = d\} = o(1)$. Notice that

$$\sum_{d=1}^{m} \frac{nd\, \mathbb{E}\left(Z_n \mathbb{I}_{Z_n > \lfloor \Delta/d \rfloor}\right)}{m} \Pr\{Z_n = d\}$$

$$\leq \frac{n}{m} \sum_{d=1}^{m} \sum_{d'=\lfloor \Delta/d \rfloor}^{m} dd' \Pr\{Z_n = d\} \Pr\{Z_n = d'\}$$

$$\leq \frac{n}{m} \sum_{d=1}^{\lfloor \sqrt{\Delta} \rfloor} \sum_{d'=\lfloor \sqrt{\Delta} \rfloor}^{m} dd' \Pr\{Z_n = d\} \Pr\{Z_n = d'\}$$

$$+ \frac{n}{m} \sum_{d=\lfloor \sqrt{\Delta} \rfloor}^{m} \sum_{d'=1}^{m} dd' \Pr\{Z_n = d\} \Pr\{Z_n = d'\}$$

$$= 2\mathbb{E}(Y_n)\mathbb{E}(Y_n \mathbb{I}_{Y_n \geq \sqrt{n/m} \lfloor \sqrt{\Delta} \rfloor})$$

By (ii) in order to prove that $\varepsilon(n)$ tends to 0 we are left with showing that $\mathbb{E}(Y_n \mathbb{I}_{Y_n \geq \sqrt{n/m} \lfloor \sqrt{\Delta} \rfloor}) = o(1)$.

Let $\delta > 0$ and $C$ be such that $\Pr\{Y = C\} = 0$ and $\mathbb{E}Y \mathbb{I}_{Y \geq C} \leq \delta/3$. (Since $\mathbb{E}Y < \infty$, there exists such $C$). Moreover by convergence in distribution $\mathbb{E}Y \mathbb{I}_{Y < C} - \mathbb{E}Y_n \mathbb{I}_{Y_n < C} = o(1)$ and thus there exists such $N$ that for any $n > N$

$$\mathbb{E}Y \mathbb{I}_{Y < C} - \mathbb{E}Y_n \mathbb{I}_{Y_n < C} < \delta/3, \quad \mathbb{E}Y_n - \mathbb{E}Y < \delta/3 \quad \text{by (ii)}$$

$$\text{and} \quad \sqrt{n/m} \left\lfloor \sqrt{\Delta_n} \right\rfloor > C \quad \text{by (4)}.$$

Therefore for any $\delta > 0$ there exists such $N$ that for any $n > N$

$$\mathbb{E}Y_n \mathbb{I}_{Y_n \geq \sqrt{n/m} \lfloor \sqrt{\Delta} \rfloor} \leq \mathbb{E}Y_n \mathbb{I}_{Y_n \geq C}$$

$$= (\mathbb{E}Y_n - \mathbb{E}Y) + (\mathbb{E}Y \mathbb{I}_{Y < C} - \mathbb{E}Y_n \mathbb{I}_{Y_n < C}) + \mathbb{E}Y \mathbb{I}_{Y \geq C} < \delta.$$

Thus $\mathbb{E}Y_n \mathbb{I}_{Y_n \geq \sqrt{n/m} \lfloor \sqrt{\Delta} \rfloor} = o(1)$. This finishes the proof of the fact that $\varepsilon(n) = o(1)$.

Therefore by (7) $d_{TV}(X_n, \mathbb{T}_{Y_n \mathbb{E}Y_n}) = o(1)$. Thus for any $k \in \mathbb{N}$

$$|\Pr\{X_n = k\} - \Pr\{\mathbb{T}_{Y_n \mathbb{E}Y_n} = k\}| = o(1). \tag{8}$$

We will use a standard property of the convergence in distribution of random variables, that under assumptions (i) and (ii) we have $Y_n \mathbb{E}Y_n \xrightarrow{D} Y\mathbb{E}Y$. Substitute $f_k(x) = \Pr\{T_x = k\} = \frac{x^k}{k!}e^{-x}$. The function $f_k(x)$ is bounded and continuous, therefore by the assumptions of the theorem and convergence in distribution

$$\mathbb{E}f_k(Y_n\mathbb{E}Y_n) \to \mathbb{E}f_k(Y\mathbb{E}Y).$$

This, by definition of the mixture of Poisson distributions (see also (1)), implies

$$\Pr\{\mathbb{T}_{Y_n\mathbb{E}Y_n} = k\} \to \Pr\{\mathbb{T}_{Y\mathbb{E}Y} = k\}. \tag{9}$$

Thus by (8) and (9) for any $k \in \mathbb{N}$ we have $\Pr\{X_n = k\} \to \Pr\{\mathbb{T}_{Y\mathbb{E}Y} = k\}$, which for random variables with values in $\mathbb{N}$ is equivalent to $X_n \xrightarrow{D} \mathbb{T}_{Y\mathbb{E}Y}$. $\qquad\square$

**Acknowledgements** This work has been partially supported by the Ministry of Science and Higher Education, grant N N206 2701 33, 2007–2010.

# References

Albert R, Barabási AL (2002) Statistical mechanics of complex networks. Rev Modern Phys 74:47–97

Barbour AD, Holst L, Janson S (1992) Poisson approximation. Oxford University Press, Oxford

Bloznelis M (2008) Degree distribution of a typical vertex in a general random intersection graph. Lith Math J 48:38–45

Bock HH (1996) Probabilistic models in cluster analysis. Comput Stat Data Anal 23:5–28

Deijfen M, Kets W (2009) Random intersection graphs with tunable degree distribution and clustering probability. Eng Inform Sci 23:661–674

Godehardt E (1990) Graphs as structural models. Vieweg, Braunschweig

Godehardt E, Jaworski J (2002) Two models of random intersection graphs for classification. In: Schwaiger M, Opitz O (eds) Exploratory data analysis in empirical research. Springer, Berlin – Heidelberg – New York, pp 68–81

Godehardt E, Jaworski J, Rybarczyk K (2007) Random intersection graphs and classification. In: Decker R, Lenz HJ (eds) Advances in data analysis. Springer, Berlin – Heidelberg – New York, pp 67–74

Godehardt E, Jaworski J, Rybarczyk K (2011) Clustering coefficients of random intersection graphs. In: Gaul W, Geyer-Schulz A, Schmidt-Thieme L, Kunze J (eds) Challenges at the interface of data analysis, computer science, and optimization, studies in classification, data analysis, and knowledge organization. Springer, Heidelberg, Berlin

Guillaume JL, Latapy M (2004) Bipartite structure of all complex networks. Inform Process Lett 90:215–221

Jaworski J, Stark D (2008) The vertex degree distribution of passive random intersection graph models. Combinator Probab Comput 17:549–558

Jaworski J, Karoński M, Stark D (2006) The degree of a typical vertex in generalized random intersection graph models. Discrete Math 306:2152–2165

Karoński M, Scheinerman ER, Singer-Cohen KB (1999) On random intersection graphs: The subgraph problem. Combinator Probab Comput 8:131–159

Stark D (2004) The vertex degree distribution of random intersection graphs. Random Struct Algorithm 24:249–258

# Application of a Community Membership Life Cycle Model on Tag-Based Communities in Twitter

Andreas C. Sonnenbichler and Christopher Bazant

**Abstract** Social networks are the backbone of Web 2.0. More than 500 million users are part of social networks like Twitter, Facebook, discussion boards or other virtual online communities. In this work we report on a first empirical study of the conceptional community membership life-cycle model of (Sonnenbichler, A Community Membership Life Cycle Model, Sunbelt XIX International Social Network Conference, University of California, San Diego, USA, 2009) applied on message data from the micro-blogging service Twitter. Based on hash tags we analyze ad-hoc communities of Twitter and we operationalize the roles of the conceptional model with the help of activity-levels and the local interaction structure of community members. We analyze the development of roles over the life-time of the community. Our explorative analysis supports the existence of the roles of the conceptional model and is a first step towards the empirical validation of the model and its operationalization. The knowledge of a community's life-cycle model is of high importance for community service providers, as it allows to influence the group structure: Stage transitions can be supported or harmed, e.g. to strengthen the binding of a user to a site and keep communities alive.

## 1 Introduction

Web 2.0 is transforming the internet. Internet users become more information producers rather than only information consumers. In the early 1980s Toffler introduced the term "prosumers" (Toffler 1984). In social networks like Facebook (http://www.facebook.com), LinkedIn (http://www.linkedin.com) or Twitter (http://twitter.com) users create the content for other users. In such networks there is high

A.C. Sonnenbichler (✉) · C. Bazant,
Information Services and Electronic Markets, IISM, KIT, Kaiserstrasse 12, D-76128 Karlsruhe, Germany,
e-mail: andreas.sonnenbichler@kit.edu; christopher.bazant@student.kit.edu

W. Gaul et al. (eds.), *Challenges at the Interface of Data Analysis, Computer Science, and Optimization*, Studies in Classification, Data Analysis, and Knowledge Organization, DOI 10.1007/978-3-642-24466-7_31, © Springer-Verlag Berlin Heidelberg 2012

information exchange between the users and those networks are living from user generated content.

Twitter is an online messaging service. It allows users to transmit short messages up to 140 characters. Only characters can be sent, binary data like pictures or videos can be transmitted indirectly by links (URLs). Every Twitter user may subscribe to messages of other users. Twitter calls this "following someone". By this, a user may have followers and follow others. The first ones are informed about all messages he sends out. From the latter ones the user receives all messages.

Our research objectives are two-fold. First, we are interested if communities get established in Twitter simply by posting messages upon the same topic (hash-tag). Of course, this must not necessarily be the case: It is one thing to post messages and a second to establish friendship networks specialized on certain topics. Please note that we do not focus on friendship relations in general, but topic-based relations. Second, we are interested in the development of such a community over time. How does a community – if it exists – evolve? Can we identify opinion leaders? Who are wrappers, who are trend-setters? Is our methodology able to detect such roles?

The benefit of knowing the roles in a community are obvious: For a service provider (like Twitter, message forums, discussion boards etc.) of a community network we expect it to be of high interest to keep the community alive. Otherwise the service would not be offered. To be able to influence the healthiness it is important to know, how the inner structure of a community looks like and how it can be influenced. The return of the efforts is much higher if the influence spreads through the network rather than to trickle away. To identify social roles in a network, we make use of the Community Membership Life Cycle Model (Sonnenbichler 2009).

We will give an overview of the underlaying community membership life-cycle model and related work in Sect. 2. In Sect. 3 we will discuss the methodology and operationalize it. As we analyze communities in Twitter, we will discuss in this chapter how Twitter works and how we define a community. In Sect. 4 we will describe very brief the technical implementation. Section 5 discusses the results. We end this paper in Sect. 6 with the conclusion and future work plans.

## 2  The Community Membership Life-Cycle Model and Related Work

Already in the early 1990s Rheingold published his experience in virtual online communities (Rheingold 1994, 2000). Communities form as well in an online context as in a real world context. This has been shown by several scientists: Carolin Haythornthwaite reported about virtual online student classes at the university of Illinois at Urbana-Champaign (Haythornthwaite et al. 2000). Her focus was on group development over time based on the well known small group stages observed and defined by Tuckman and Jensen (1964; 1977). Palloff and Pratt (1999) did

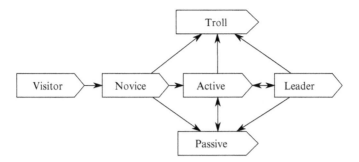

**Fig. 1** The model of a Community Membership Life Cycle Model following Sonnenbichler (2009). Each community role has defined transitions to successor roles

research in online learning communities. They describe the inner mechanisms in learning communities, especially in discussions about new content.

A community membership life cycle model is suggested by Sonnenbichler (2009). The model defines the social roles a community member can hold at a given point of time. It describes how a user's role can evolve over time and how the role of a user can be identified making use of social network analysis. The model distinguishes six different roles (see Fig. 1).

- The first contact someone may have with a community is in the role of a visitor. Visitors can not be identified by an account as the decision to sign up for one has not yet been made.
- Novices have signed-up or can be identified reliably. They are very new in the community and learn the rules and issues.
- Actives are the backbone of a community. They participate in discussions and the community and represent the average. They are Toffler's prosumers.
- Passives often form the silent majority of a group. They mainly consume content and are not active all the time.
- The leaders are opinion leaders, issue makers and form the soul and heart of the community. They set trends. Content provided by them influences many others.
- Trolls are a collection of negative roles. They want to disturb the community.

## 3 Operationalization of Community Roles

We collect data sets via the Twitter API (2010). We used key words to identify communities. Keywords can be declared explicitly by users posting messages through Twitter by putting a hash sign in front of the keyword(s). Twitter calls these terms hash-tags.

We use hash-tags to identify people twittering on a specific topic. Periodically we analyzed all messages recently posted including the selected hash-tag. For every

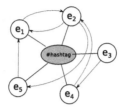

**Fig. 2** Example of one fixed period t. The hash-tag is used to identify users $e_{1..5}$ posting on this topic. The social network is established in a second step by linking followers to the users they follow with a directed edge

person identified we retrieve all followers. We interpret the result as a directed graph $\mathbf{G}_t = (\mathbf{V}, \mathbf{E})$ in period $t$. The nodes $v_{it} \in \mathbf{V}_t$ represent the twitter user having posted using the hash-tag in period $t$. An edge $e_{ijt} \in \mathbf{E}_t$ is drawn from node $v_{it}$ to $v_{jt}$ if user $i$ is following $j$ in period $t$. Figure 2 illustrates an example.

Data is collected in the time periods $t = t_1, t_2, .., t_n$. The superset of nodes $\mathscr{V}_{t_i}$ in period $t_i$ is defined as union of all users having posted since analysis period $t_1$.

$$\mathscr{V}_{t_i} = \cup_{t_1}^{t_n} \mathbf{V}_t \tag{1}$$

$$\mathscr{E}_{t_i} = \mathbf{E}_{t_i} \tag{2}$$

The superset of edges $\mathscr{E}_{t_i}$ in period $t_i$ nevertheless is always the latest available network of follower links and thereby not cumulative. This means, that nodes are never deleted during all observation periods. An observed poster of a tweet will be in $\mathscr{E}_{t_i}$ ever after his first post in $t_i$.

For the classification of the community members according to the community membership life cycle model the following rules have been defined:

- An active member must post at least one message within two weeks. Additionally his in- $d^+$ and out-degree $d^-$ meet the interval

$$(\mu - \sigma, \mu + \sigma) \tag{3}$$

  with $\mu$ being the mean value and $\sigma$ the standard deviation of the in- respectively out-degree of all users at this point of time. Both values are calculated separately for the in-degree $d^+$ and out-degree $d^-$.

  The in-degree calculates the number of links toward a node, the out-degree the number of out-going connections. For further details about measures in the social network analysis we refer to Wasserman and Faust (1994).
- Members are classified as passive if no posts were observed during the last 2 weeks.
- Leaders are identified by having an in-degree $d^+$ above the interval $(\mu - \sigma, \mu + \sigma)$. In addition, their proximity value, betweenness value or PageRank value must be above average, thus the same interval for the respective values.
- A user is classified as novice if his sign-up in Twitter has taken place since the last week. As users may be a novice in a hash-tag induced community but not new

to Twitter, we need more criteria. Therefore, a user is also classified as novice if his in- or out-degree is below the defined interval. Nevertheless, a user who has been classified as active, troll, or passive will not be able to be classified as novice (again).

- Visitors can not be detected due to the experimental setup.
- Trolls are identified by a sudden activity peak while having no or only a small change in in- and out-degree.

# 4 Data Collection from Twitter and Implementation

The data collection was done on a Linux system running an Apache web server, Postgres Database and PHP.

1. In a first step the Twitter API (2010) is queried for all tweets including the hash-tag since the last retrieval date. The API offers RESTful services. For an introduction and definition of REST we refer to Fielding (2000). The XML response is transformed through XPath and data stored in a PostgreSQL database.
2. In the second step we retrieve all followers of the found Twitter users in step 1 and store them by the same technology.
3. The third step comprises of the calculation of the network $G_t$ by interpreting all users found since period $t_1$ as vertices and all current follower relationships in $t_i$ as edges. The network is visualized using UCINet (Borgatti et al. 2002).
4. Then all users are classified by using the community membership life cycle model described in Chap. 3.

# 5 Explorative Analysis and Results

For our analysis we used three hash-tags Australien (Australia), Motorrad (motor-bike) and Tierschutz (animal protection). We want to present results for the hash-tag Australien here. All three hash tags led to comparable results so that the choice for Australien does not represent a special case.

The analysis for Australien was done at 12 points in time between the 15th April 2010 and 27th May 2010. Three to five days passed between each analysis point. The time period was chosen so that all tweets (messages) in this period could be retrieved and no post was lost. During the experiment 175 different community members have been observed posting messages including the hash tag. For each date the followers' network has been calculated as described in Sect. 3. Figure 3 depicts the network at date 1, 6 and 12.

The directed graph $G_t$ is not completely connected. Nevertheless we can easily identify one large component, which is loosely connected. This result supports the assumption of a community. In case we would not find at least *one* large component the assumption of a community based on hash-tags within Twitter would

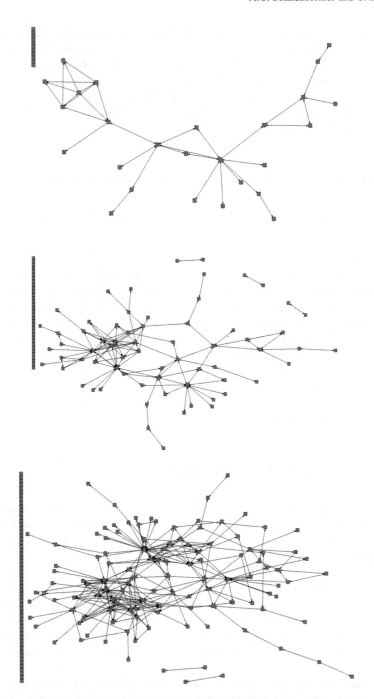

**Fig. 3** Graph $G_1$, $G_6$, $G_{12}$ of hash-tag Australien (Australia). It represents all community members detected at this point of time and their follower structure which we interpret as community structure. The number close to the nodes are the twitter ids

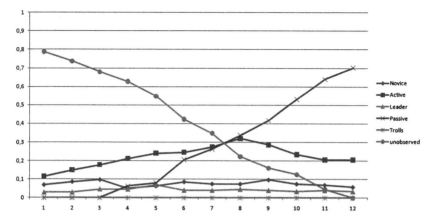

**Fig. 4** Development of role identification in the Twitter community Australien over all observation periods (1..12)

be compromised. The results retrieved in our experiment show that at any given time at least one large loosely connected component can be found in every hash-tag induced network. This is not limited to the tag Australien, but also can be observed in Tierschutz and Motorrad. This is an argument in favor of hash-tag induced communities in Twitter. On the other hand, we can identify several islands in the graph which are totally unconnected.

The application of the community membership life cycle model (see Sects. 2 and 3) is depicted in Fig. 4.

The number of the leaders fluctuates between 4% and 7% of all community members. Interestingly enough the identification of the leaders succeeds very early in the observation process. It is also stable over time. Visitors could not be detected as the design of the experiment excluded their observation: A visitor is defined being a user before signing up to the service. Without registering in Twitter, one is not able to post messages which prevents being observed in our experiment. Another finding was that we did not observe any Trolls. We identified some activity peaks for users. Nevertheless manual data analysis showed no offensive content. The identification of active users was not as stable as the recognition of leaders. We observed between 11% and 32% actives. The lower stability in detecting actives seems reasonable as an active user might not post during every period in time and so stay undetected. Furthermore, actives and passives can be distinguished after some observation time.

As expected from the definition the identification of passive community members took the longest time: As passive members rarely post messages themselves but are defined mainly as content consumers their observation with our experiment takes time (or is not possible, if a passive does not post in the observation period). The finding that the unobserved community members (until period 12) mainly are classified as passive members seems conclusive. The passive members form up to 70% of the community. We want to emphasize that of course we do not know about all community members, so the percentage of unobserved members can not be interpreted as a global value.

In order to validate the results we identified the leaders in our community. To our surprise, we found several of the members classified as leaders being related to travel communities or website owners about Australia. This finding is very conclusive and supports the correct classification as leaders: We expect a travel company and website owner to have high interests in influencing a community for commercial benefits.

# 6 Conclusion

Our work aims to operationalize the community membership life-cycle model suggested in Sonnenbichler (2009). The knowledge of a community's life cycle, its current stages, respectively the stage distributions, is of high value for community service providers as it allows to influence the group structure. Role distributions can be supported or harmed, e.g. to strengthen the binding of a user to a site and keep communities alive.

To provide first results, we collected hash tag-based Twitter data and carried out an explorative analysis of the data by applying the community membership life-cycle model on it.

In our results we found strong arguments that such hash-tag induced social networks in Twitter can be interpreted as communities. At any given point in time we found one large loosely connected component. This finding is true for all analyzed hash-tags. We argue that this result supports the assumption that there are communities in Twitter. Of course, this work can only be seen as a first result. Further work has to be done in investigating the structure and details of such networks.

Furthermore, we applied the community membership life cycle model on a social network. The identification of the role leader has shown to be stable after a very short period of observation. The correct identification of actives required some longer observation. Passives have shown to be the hardest role for identification as it needed the longest observation time. Future work is planned to cross-validate also the classification of actives and passives. The application of the community membership life cycle model on other communities is planned.

**Acknowledgements** The research leading to these results has received funding from the European Community's 7th Framework Program FP7/2007-2013 under grant agreement n°215453 – WeKnowIt.

# References

Borgatti SP, Everett MG, Freeman LC (2002) UCINet - Software for social network analysis. Analytic Technologies, Havard, Massachusetts

Fielding RT (2000) Architectural styles and the design of network-based software architectures. PhD thesis, University of California, Irvine, CA, URL http://www.ics.uci.edu/~fielding/pubs/dissertation/fielding_dissertation.pdf

Haythornthwaite C, Kazmer MM, Robins J, Shoemaker S (2000) Community development among distance learners: Temporal and technological dimensions. J Comput Mediat Comm 6(1), http://jcmc.indiana.edu/vol6/issue1/haythornthwaite.html, accessed at 2009-03-28

Palloff RM, Pratt K (1999) Building learning communities in cyberspace. Jossey-Bass, San Francisco, CA

Rheingold H (1994) A slice of life in my virtual community. In: Harasim LM (ed) Global networks: Computers and international communication. MIT Press, Cambridge, pp 57–80

Rheingold H (2000) The virtual community: homesteading on the electronic frontier, revised edition. MIT Press, Cambridge, New York

Sonnenbichler AC (2009) A Community Membership Life Cycle Model. In: Sunbelt XIX International Social Network Conference, University of California, San Diego, USA, avaiable online at http://arxiv.org/abs/1006.4271

Toffler A (1984) The third wave. Bantam Books, USA

Tuckman BW (1964) Developmental sequence in small groups. Psychol Bull 63(6):384–399

Tuckman BW, Jensen MA (1977) Stages of small group development revisited. Group Org Studies 2:419–427

Twitter (2010) Twitter application programming interface. Http://apiwiki.twitter.com/, accessed at 2010-5-26

Wasserman S, Faust K (1994) Social network analysis: methods and applications. Cambridge University Press, Cambridge, New York

# Measuring the Influence of Network Structures on Social Interaction over Time

**Christoph Stadtfeld**

**Abstract** Communication decisions in networks can be described as a two-level decision process. The second decision about event receivers is a multinomial logistic regression model with an unknown vector of parameters. These parameters evaluate network structures that enforce or weaken the probability for choosing certain actors. However, in many cases those parameters may change over time. In this paper a sliding window approach is introduced, that can be used to understand whether there is evolution of behavior in an observed data set. For future work, it is proposed to develop a statistical test on normalized decision statistics.

## 1 Introduction

The dynamics of social networks are often measured on longitudinal data of single data snap shots (Snijders 2005) or on static networks (Monge and Contractor 2003; Robins et al. 2007). Communication data, however, is often not collected as static networks, but can be measured by logging the computer mediated communication data instead. This consists of a set of dyadic, directed, time-stamped and sometimes weighted *events*. In Stadtfeld (2010) a stochastic model was proposed that allows to explain how new events are determined and how structures in the communication network, other networks, or user attributes influence these decisions. Using a maximum likelihood optimization, the relevance of these structures can be estimated for a given data stream. However, it may happen that structural influences on communication decisions change over time. Instead of analyzing the event stream as a whole, a sliding window approach can be used to reveal these changes.

C. Stadtfeld (✉)
Information Services and Electronic Markets, Karlsruhe Institute of Technology,
Karlsruhe, Germany
e-mail: christoph.stadtfeld@kit.edu

W. Gaul et al. (eds.), *Challenges at the Interface of Data Analysis, Computer Science, and Optimization*, Studies in Classification, Data Analysis, and Knowledge Organization, DOI 10.1007/978-3-642-24466-7_32, © Springer-Verlag Berlin Heidelberg 2012

There are a lot of reasons, why structural effects of communication may not be stable over time. Actor strategies can vary at different times in one data set. These variations may be slow and continuous processes, e.g. when in a new organizational team informal communication structures evolve. In contrast, changes may sometimes be sudden with a high absolute change of communication behaviour due to some structural breaks or external changes.

For example, when communication on a social media web site is observed, technical changes may have a direct and significant effect on how people communicate. In Stadtfeld and Geyer-Schulz (2011) actors in a web community started to communicate in different structures. This was probably caused by some technical change, like the inclusion of direct links to a communication service within the community.

This paper presents an idea about how those structural changes can be measured using a sliding window approach. In Sect. 2 the basic idea of the abstract interaction (communication) model is briefly explained. Section 3 introduces a concrete dyadic communication model, which can be used to simulate artificial event streams. This is illustrated in Sect. 4. In Sect. 5 two event streams that include structural changes are simulated, estimated using a sliding window and plotted. In Sect. 6 the results are briefly summarized and some ideas about future work are proposed.

## 2   Basic Ideas of the Abstract Model

Computer mediated communication data is often dyadic, directed and time-stamped (an ordered list of so called *communication events*). In this section, we briefly introduce a stochastic model, that describes how these events emerge, based on a communication network represented by the graph

$$X = (x_{ij}) \tag{1}$$

that expresses the previous directed communication intensity from actors $a_i$ to $a_j$ in the set of actors $A$. This set of actors is

$$A = \{a_1, \ldots, a_n\}. \tag{2}$$

As explained in Stadtfeld et al. (2010), the abstract model is described as a two-level decision process:

1. Each actor decides whether he starts an interaction at all with an individual Poisson rate (activity).
2. Given that an actor gets active, he decides about recipients depending on the underlying network structures.

The activity of each actor $a_i$ is given by a Poisson process with parameter $\rho_i$. This means, that for each individual, the time between two consecutive events is exponentially distributed with an average length of $\frac{1}{\rho_i}$ time units. The probability density function for a time span $\delta_i$ between two events of actor $a_i$ is therefore

$$f(\delta_i; \rho_i) = 1 - e^{-\rho_i \delta_i}. \tag{3}$$

The time span $\delta$ between two consecutive events in the event stream is distributed with the probability density function

$$f(\delta; \rho_1, \ldots, \rho_n) = 1 - e^{-(\sum_{a_i \in A} \rho_i)\delta}. \tag{4}$$

Given that an actor starts an interaction (this is determined by the Poisson process with exponentially distributed inter-event times), he decides about the communication recipient based on $m$ network structures in the set $Q$.

$$Q = \{q_1, \ldots, q_m\} \tag{5}$$

The set $Q$ defines a concrete model based on the abstract one. The choice of this set depends on the context in which the event stream data was collected.

Structures of communication networks that typically influence future communication are the re-using of existing communication paths (people repeatedly communicate with the same people), reciprocal (bi-directional communication patterns) and transitive communication. Dependent on the context, the structures that describe communication well, vary. Some examples are given in Stadtfeld (2010). In a professor-student communication network or in the network of hierarchically structured organisations, reciprocity in communication is most likely weaker than in a network of peers.

Given that an actor starts an interaction, he makes the second decision based on the structures included in the concrete model. Structures are measured and represented by a $(n - 1) \times m$ matrix $S(X, i)$ with $a_i$ determined to be the *sending* actor of the event. $n - 1$ is the number of all possible recipients (all $n$ actors in the dataset except for $a_i$) and $m$ is the number of network structures that describe communication choices in the concrete model.

$s(X, i)_{kl}$ is therefore the value (the statistic) of the $l$-th structure, given that actor $a_k$ was chosen as a recipient.

The probabilities for choosing a certain recipient $a_j$ (the second decision) is given by the probability $p(j; S(X, i), \beta)$:

$$p(a_j; S(X, i), \beta) = \frac{1}{\sum_{a_k \in A} \exp\left(\sum_{q_l \in Q} s(X, i)_{kl} \beta_l\right)} \exp\left(\sum_{q_l \in Q} s(X, i)_{jl} \beta_l\right) \tag{6}$$

$\beta$ is a vector of length $m$ of real values, that assigns weight to each structure $q_l$ in $Q$. High positive values $\beta_l$, for example, indicate that the corresponding structure increases the probability for choosing a recipient with a rather strong structure compared to a random decision. For negative values the contrary effect holds.

If the denominator of the expression is interpreted as a constant (it is invariate for all possible recipients of one decision), this probability distribution belongs to the exponential family of probability distributions (Robins et al. 2007; Young and Smith 2005). The values in the matrix $S(X, i)$ are then sufficient statistics. The probability function can be interpreted as a multinomial logistic regression model describing an actor choice (see McFadden 1974). More details about the model framework and its estimation can be found in Stadtfeld et al. (2010).

## 3  Exemplary Model

Imagine a communication network in which the choice of recipients is very well explained by two network structures:

- *Re-using of existing ties ($q_1$):* Expresses how far people like to re-use a communication path they have used before.
- *Reciprocal communication ($q_2$):* Expresses how far people like to reciprocate incoming communication events.

Both structures are dyadic, which means they do not include third actors, like in a transitive structure, for example. In most communication networks, both structures should be expected to have positive weights $\beta_1, \beta_2 > 0$. People often have a restricted set of communication partners, so they rather prefer to re-use communication paths (see Monge and Contractor (2003)). Also, communication is bidirectional – most people communicate in both directions instead of one-way. A significantly positive parameter would indicate, that people chose those structures significantly more often than in random choices.

In Fig. 1 two different second decisions (about the event recipient) are shown. The sender of the message is marked with an ellipse. In case of Fig. 1(a) it is actor B, in case of Fig. 1(b) it is actor A. The two other actors in the subfigures are potential

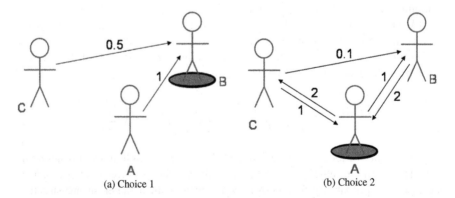

(a) Choice 1                                        (b) Choice 2

**Fig. 1** Two different cases in which a given actor (B and A) decides about the event recipient

recipients. Dependent on the weight of the corresponding parameters $\beta_1$ and $\beta_2$, the probabilities for the two decisions differ. If $\beta_1$ (weighting re-using of ties) had a (high) positive weight, actor B in the first example would be indifferent, as he has no outgoing ties, yet. In the second example actor $A$ would prefer actor C over actor B, as this outgoing tie is stronger. If $\beta_1 = 1$ and $\beta_2 = 0$, the probability for choosing actor C in the example of Fig. 1(b) is

$$
p\left(C; \begin{pmatrix} 1 & 2 \\ 2 & 1 \end{pmatrix}, \begin{pmatrix} 1 \\ 0 \end{pmatrix}\right) = \frac{\exp(2)}{\exp(2) + \exp(1)} = 72.1\%.
\tag{7}
$$

As explained in (6), the probability depends on a structural matrix $S(X, i)$, which includes all possible recipients in set $A\backslash i$. The first row represents actor $B$, the first column the statistic of the first structure $q_1$ *Re-using of existing ties*. The statistics of the second structure are measured by taking the minimum value of the in- and outgoing ties *after* including a new tie after the event. A more detailed example is given in Stadtfeld et al. (2010).

If specific communication could theoretically very well be explained with these two effects, then it is also possible to simulate artificial communication streams that are similar to real-world communication. So, if we had a model that reduces unexplained variance very well, the model can also be simulated. Such a simulation can be applied to test in how far real-world changes are reflected in the estimators. Note that simulating an estimated model is not the same as predicting future communication behavior.

## 4  How to Simulate

As the stochastic process is described as a two-level decision process, the simulation is also processed in two steps. First, the simulation processor determines the actor who gets active next and after what time span this happens. Then the time dependent decay is applied on the communication network. Second, the recipient is determined based on the introduced probability distribution.

Every actor $i$ in $A$ has a Poisson process activity rate. All actors are assumed to get active completely independently from all other actors. The next active actor can therefore be determined by randomly drawing one time span for each of the actors. The shortest simulated time span is chosen and the corresponding actor is set active, after the time decay for the resulting time span has been applied.

The second decision, depending on that the sending actor of the $v$-th event has randomly been identified as $\hat{i}_v$ is expressed by a discrete probability distribution over all possible recipients in $A\backslash\hat{i}_v$. The updated network at the new time is $X_v$. Then the decision matrix $S_v(X, \hat{i}_v)$ can be calculated. Using the predefined vector $\beta$, a random choice following the multinomial model can be drawn.

Both the continuous probability distribution of the first decision and the discrete probability distribution of the actor's choice are simulated using uniformly distributed random numbers. Details about how this works can be found in Waldmann and Stocker (2004).

## 5 Exemplary Simulation

Two different event streams are simulated using the introduced simple, dyadic model. Afterwards, a sliding window is used to visualize the estimates over time.

For both simulations holds the following: The first decision about actor activity is similarly distributed for all actors. This means that every actor $a_i$ has the same activity with a parameter of $\rho_i = 1$. Each actor is therefore expected to start one communication event per time unit.

The second decision about event recipients only depends on the two introduced dyadic structures. The model explains whether people tend to re-use existing communication paths ($q_1$) and whether they communicate reciprocally ($q_2$).

The data stream consists of a set of 30 actors. 2,000 events are simulated. After 1,000 events, the actor communication strategies change. The simulated parameters $\beta_1$ and $\beta_2$ for both halves of the event stream are given in Table 1 for the first of two simulations and in Table 2 for the second one.

After simulating the event streams, a sliding window approach is applied to visualize the change of parameter estimates. The sliding window has a width of 200 events and is shifted by 50 events per step. The resulting plots for simulation I and II are shown in Figs. 2 and 3.

The darker line shows estimates of structure $q_1$, the lighter one of $q_2$. The x-axis indicates the number of the sliding window. Window 1, for example, starts at the first event and ends at the 200th event. The y-axis gives the estimators of the $\beta$-values. Two bars in the middle show those sliding windows that include parts of both actor strategies that changed after half of the time.

Basically, both plots reflect the general tendency of the estimates. As there are only 200 second decisions in each window with 29 possible recipients, the accuracy is limited, but the estimators reflect the general tendency very well. It can be seen

**Table 1** Simulation parameters of simulation I

| Events: | 0–1,000 | 1,001–2,000 |
|---------|---------|-------------|
| $\beta_1$ | 1 | −1 |
| $\beta_2$ | −1 | 1 |

**Table 2** Simulation parameters of simulation II

| Events: | 0–1,000 | 1,001–2,000 |
|---------|---------|-------------|
| $\beta_1$ | 1 | 1 |
| $\beta_2$ | −1 | 1 |

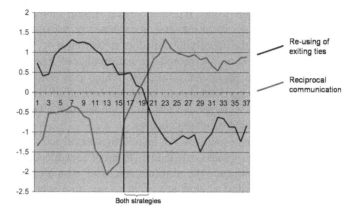

**Fig. 2** Parameter changes of simulation I. The transition takes place between the vertical bars

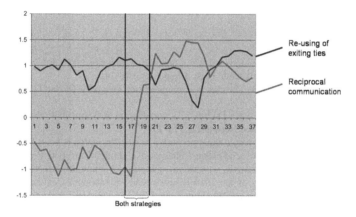

**Fig. 3** Parameter changes of simulation II. The transition takes place between the vertical bars

that the change in communication structures can be identified very soon after the first events are part of the analyzed windows.

Of course, these are only two exemplary simulations. The results of this sliding window approach may give hints about what is happening in the data set, but do not provide any statistical test, yet. Ideas about further research are briefly discussed in the concluding section.

## 6  Conclusions

This article introduced a way to detect structural changes of communication patterns over time. A sliding window approach was used to visualize those changes. Two artificial communication events based on a simple, dyadic model were simulated to demonstrate how this procedure works. It turned out that visualization can be a good exploratory method to find out about the general tendency of network estimators, especially if the tendency in the underlying dynamics is very clear.

However, this approach does not provide a statistical test. There is always a certain randomness in the parameter estimation, which may lead to the false conclusion, that the underlying dynamic structures have changed. Therefore, one focus of future work will be to develop a test to decide whether changes are just random or caused by underlying structural changes.

Also, the estimates are not normalized. An increasing general activity in the network or a decreasing number of nodes may lead to different network characteristics. Those characteristics also influence the absolute values of estimated parameters. Therefore, the introduced sliding window test is only a very first approach towards the goal of automatically assessing behavioral changes in networks.

# References

McFadden D (1974) Conditional logit analysis of qualitative choice behavior. In: Zarembka P (ed) Frontiers in econometrics. Academic Press, Chap 4, pp 105–142

Monge PR, Contractor NS (2003) Theories of communication networks. Oxford University Press, New York, USA

Robins G, Pattison P, Kalish Y, Lusher D (2007) An introduction to exponential random graph (p*) models for social networks. Soc Netw 29(2):173–191

Snijders TAB (2005) Models for longitudinal network data. In: Carrington P, Scott J, Wasserman S (eds) Models and methods in social network analysis. Cambridge University Press, New York

Stadtfeld C (2010) Who communicates with whom? measuring communication choices on social media sites. In: Proceedings of the 2010 IEEE Second International Conference on Social Computing (SocialCom), Minneapolis, Minnesota, USA, pp 564–569

Stadtfeld C, Geyer-Schulz A (2011) Analysing event stream dynamics in two mode networks: An exploratory analysis of private communication in a question and answer community. Soc Netw (under revision)

Stadtfeld C, Geyer-Schulz A, Waldmann KH (2010) Estimating event-based exponential random graph models. In: Dreier T, Krämer J, Studer R, Weinhardt C (eds) Information management and market engineering, studies on eorganisation and market engineering, vol 2. KIT Scientific Publishing, pp 79–94

Waldmann KH, Stocker U (2004) Stochastische Modelle: Eine Anwendungsorientierte Einführung, Springer, Berlin

Young GA, Smith RL (2005) Essentials of statistical inference. Cambridge Series in Statistical and Probabilistic Mathematics, Cambridge University Press, Cambridge, UK

# Identifying Artificial Actors in E-Dating: A Probabilistic Segmentation Based on Interactional Pattern Analysis

Andreas Schmitz, Olga Yanenko, and Marcel Hebing

**Abstract** We propose different behaviour and interaction related indicators of artificial actors (bots) and show how they can be separated from natural users in a virtual dating market. A finite mixture classification model is applied on the different behavioural and interactional information to classify users into bot vs. non-bot-categories. Finally the validity of the classification model and the impact of bots on sociodemographic distributions and scientific analysis is discussed.

## 1 Introduction

Social networking services have extensively proliferated in the last years and thereby, they increasingly enable users to relocate various practices into the Internet. For social sciences this development is a possibility to acquire a new kind of data on social processes by directly observing agency, interactions and social networks. One example is the new approach of studying issues of mate choice processes constituted by recording and analysing interactions within online mate markets (Schmitz et al. 2009). Since those behavioural records are not originally arranged for scientific intentions, there are particular methodological aspects to be taken into account before analysing the data against a theoretical background. A considerable problem

A. Schmitz,
Chair of Sociology I, University of Bamberg, Wilhelmsplatz 3, 96045 Bamberg, Germany
e-mail: andreas.schmitz@uni-bamberg.de

O. Yanenko, (✉)
Chair for Computing in the Cultural Sciences, University of Bamberg, Feldkirchenstrasse 21, 96045 Bamberg, Germany
e-mail: olga.yanenko@uni-bamberg.de

M. Hebing
Socio-Economic Panel Study (SOEP), DIW Berlin, Mohrenstrasse 58, 10117 Berlin, Germany
e-mail: mhebing@diw.de

W. Gaul et al. (eds.), *Challenges at the Interface of Data Analysis, Computer Science, and Optimization*, Studies in Classification, Data Analysis, and Knowledge Organization, DOI 10.1007/978-3-642-24466-7_33, © Springer-Verlag Berlin Heidelberg 2012

with analysing and interpreting this data arises with the presence of artificial actors or so-called *bots*. Bots are automated third-party programs trying to make users engage into contact and eventually into an over-priced and useless external product. Beyond a diminished benefit of the user and reputational losses of the particular dating-provider, data quality becomes an important issue from a scientific point of view. In this paper we develop behaviour and interaction related indicators of bot-presence and separate artificial users from humans by applying a finite mixture model on this informational vectors. The classification results suggest an amount of approximately 3.72% artificial users. Contrasting our latent class prediction with a manual screening for bot-presence yields a high validity. We demonstrate that those entities generate a disproportionately high amount of first contacts (30.5%) and significantly distort socio-demographic distributions, such as gender and age. We conclude with an outlook and a discussion about practical implications.

## 2  Bots in the Social Web

The term bot is the abbreviation for robot and implies different automated programs that have the ability to act autonomously to some extent (Gianvecchio et al. 2008). There is a vast variety of bots which were mostly implemented for performing tasks in different areas of the World Wide Web. The field of bot activities varies from collecting information about websites for search engine ranking generation to text editing as accomplished by Wikipedia[1] bots for instance (Fink and Liboschik 2010). These bots are created to support their operators in performing boring or recurring jobs. But there are also numerous bots, which were deployed for financial purposes. Most of these bots have a further-reaching negative impact for Internet users they are interacting with. The aims of such bots vary from copyright fraud (Poggi et al. 2007) or identity theft to gross marketing tricks. The most common operating mode for bots is to establish a direct contact with other people, usually by writing spam emails. These spam mails differ from conventional advertisement since they are formulated as personal messages and are therefore often not recognised as malicious. Human spammers exist as well, but since the manual distribution of emails is very time-consuming, it is a common practice to automate this process by implementing a bot. The progression of today's spam filters forces bot developers to strike new paths. To equip bots with some kind of personality is therefore a major strategy in creating new malicious programs. For this reason, bots are opening up new areas of social acting on the web by creating their own accounts on different Web 2.0 sites, such as online-dating platforms, and masquerade as human actors. In the social web context the spam bot problem is also referred to as Spam 2.0 (Hayati et al. 2010). Usually a Spam 2.0 bot contacts a person only once with a nice sounding text at some point telling the receiver to follow

---

[1]http://en.wikipedia.org/

a link to some website, to call some costly phone number (Dvorak et al. 2003) or just to change the communication channel to private email in order to collect email adresses for subsequent spamming. Particularly on dating platforms people are in search of (communication) partners and therefore prone to contacts with fake profiles.

Since bots are a real threat for other users of chat rooms and social network communities, there has been a big interest in identifying and banning them. Because of the wide range of bot types and usage scenarios many different approaches were introduced in the past. Spam bots are usually recognised by means of identifying suspect text patterns or sender addresses since no other information is available. Behavioural bots in contrast can be determined by considering the social structure of their actions. The major fact for differentiating between humans and bots is that human behaviour is not easily computable while bots always show some recurring behavioural patterns.

## 2.1 Bot Detection

Since spam bots are the main problem in the area of social web communities, content analysis is used for their identification. Therefore lists with keywords that are known to be used by bots are compared with messages. This method is not very reliable as bots regularly update their vocabulary depending on the blacklists. Another point is that bots often add random characters to the keywords that look like typing errors and on the one hand let them seem more human and on the other hand avoid matching a word from the blacklist (Gianvecchio et al. 2008). In some cases there is no possibility to analyse the text contents, for example due to privacy reasons.

The detection of bots by analysing their behavioural patterns is a common practice for identifying bots in online games like World of Warcraft[2] (Chen et al. 2006, 2009). Similar techniques can be used in social network scenarios as well. Gianvecchio et al. (2008) proposed a method for detecting bots in Internet chat rooms. They collected interactional data on the Yahoo! chat system[3] which then was manually labelled as human, bot or ambiguous. Additionally, two indicators, namely inter-message delay and message size, were defined to help identifying the bots. The classification was made by a machine learning approach, where half of the available data was used for training the classifiers and the other half for testing them. However, bot detection methods introduced in the past rely either on pure content analysis or behavioural patterns. Thus, we argue that interactional patterns should be taken into account as well and conjecture that they provide better evidence as they include further information.

---

[2]http://eu.battle.net/wow/

[3]http://messenger.yahoo.com/

## 2.2 Bot-Induced Problems with Data Quality and Theory-Testing

But what kind of problems can bots cause in our field of application? From a commercial perspective the presence of artificial actors represents a risk of lowering the quality of the product and thereby the customer retention. But if there is a difference between natural and artificial actors some analytical problems arise as well. First, the total amount of the sample might be overrated to the extent of the presence of bots. Second, the distribution of relevant characteristics or marker parameters as for example sex (female), age (young), mating preferences (short term), and outer appearance (very attractive) might be biased. This can be expected as programmers rely on their stereotypes when coding a bot with an expected maximum of success probability. The same holds true for manual spammers. Third, the behavioural and interactional patterns, being the core advantage of process-generated data (Schmitz et al. 2009), might be biased. As for example young attractive women are usually selective with regard to their contact and answer behaviour, whereas artificial actors with similar profile information do probably not reveal this behaviour, as it would be irrational to be confined to only a few male victims of fraud. This problem of misrepresentative interactional patterns becomes worse if we take into account that a bot usually contacts a lot of users, leading to an aggregate pattern where usually choosy persons show up as extreme outgoing ones. Thus, results of previous research on e-dating using web-generated process data have to be interpreted with caution.

## 3 Method

### 3.1 Sample

The raw data is provided as anonymised dump files from an operative SQL database of the cooperative maintaining company. It consists of real time click streams, which first have to be edited and then converted to flat file formats readable for common statistical packages. This process-generated interaction data (who sends whom a message and at what time) is integrated with profile information (what do users tell other users about themselves on their profile pages) via a unique user-id. The sample used in this paper consists of 32,365 active users who generated 683,312 messages from January 1st 2009 to April 14th 2010, forming 362,067 dyads.

### 3.2 Behavioural and Interactional Indicators

Based on the information of the sender's and his contact's user-ids in combination with the time stamp of the message, we identify interactional dyads. In a next step, we deduce information about their behavioural and interactional patterns. We combine this information with indicators about the content of the messages to identify messages containing email or web addresses and to count the number of characters in the messages. This information is aggregated on the level of the individual user. We define two types of indicators: behaviour and interaction related indicators. Behaviour related indicators measure how users send messages. We expect artificial actors to differ from humans in some key properties:

- *First Contact Sending Rate (log)* [FCSR]: Bots need to produce a huge amount of messages in order to enhance the intended reaction rate. As the total number of first contacts depends on the duration of membership on the platform, we calculate a logarithmised time-dependent rate.
- *High-Speed Sending-Ratio* [HSSR]: Bots are able to send messages in a faster sequence than humans are. We calculate a ratio, comparing the amount of messages sent in less then 20 seconds divided by the total amount of messages sent.
- *Message Length Standard Deviation (log)* [MLSD]: Most bots send standardised messages. Even if they are able to exchange the name of their contact, messages should not differ in length. We use the standard deviation of characters in the messages of one user as an indicator for the variation of message content.
- *Inspection vs. Contacted-Ratio (log)* [IVCR]: There is no reason for bots not to contact a user after visiting his profile. As it is not necessary to visit a user's profile before contacting him, artificial actors should initiate more contacts then visiting profiles.
- *Web-* and *Email-Address-Ratio* [WAR/EAR]: Bots are programmed to initiate a contact outside the platform, so they send further contact information. The challenge is to identify this contact information as the programmers try to codify them to avoid spam filter. For example, you will find (at) instead of an @-sign. This indicator is defined as the amount of messages containing such expressions, divided by the total amount of messages.

In addition, we define interaction related indicators, which take into account that the human reaction towards artificial actors differs from the reaction towards human actors and thereby reveals a lot about the character of the contacting user.

- *Mean Conversation Length (log)* [MCL]: If two users hold up a conversation for a longer time, it is an indication for reciprocal human interaction. We calculate the mean of the number of interactions in the conversations initiated by the users.
- *Response-Received-Ratio* [RR]: In contrast, bots will less often receive answers on their messages as humans identify bots in most cases by common sense. This indicator measures the ratio of contacts initiated by one user receiving an answer in relation to the total number of initiated contacts by the same user.
- *Blocked User* [BU]: Users sometimes complain about artificial and commercial actors, which eventually leads to a block of the particular profile.

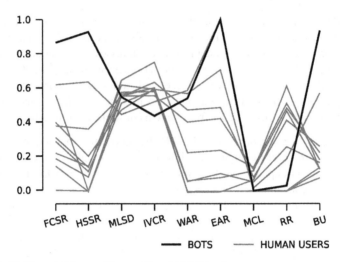

**Fig. 1** Profile-Plot of 13 Latent Classes (Class 9 highlighted)

## 3.3 A Finite Mixture Analysis of Bot Indicators

Usually different information on behaviour is used univariately in bot-detection approaches. In order to distinguish between natural and artificial actors, we apply a finite mixture model with mixed indicators (Vermunt and Magidson 2002) on the described behavioural and interactional indicators. Since the finite mixture model is independent from the scaling (e.g. categorical, ordinal, continuous, Poisson, etc.) of the variables used it is an appropriate choice for categorising online-daters based on the indicators described in the previous section. Thus, the mixed indicators model is a general latent variable approach that specifies a model with manifest variables of different scaling and C categorical latent variables. Those C unobserved categorical variables can be identified by maximum likelihood estimation under the assumption of multivariate normal distributions given the latent profiles. The joint density of the observed indicators, $f(y)$, is a function of the prior probability of belonging to latent class $\pi_k$ with class-specific mean vectors and covariance matrices $f_k(y_{ij}|\theta_{jk})$, given the model parameters $\theta$.

$$f(y_i|\theta) = \sum_{k=1}^{K} \pi_k \prod_{j=1}^{J} f_k(y_{ij}|\theta_{jk})$$

The observed indicators in our model, the different information derived from the click stream data, are simultaneously modelled by this technique. The models are estimated with the statistic software Latent Gold 4.5.

According to BIC and CAIC-information criteria the optimal solution yields 13 latent classes. Figure 1 shows the profile plot of the optimal solution. Consider class 9 which covers 3.72% of all users in the sample. This class strongly differs from the other classes by means of its high First Contact Sending Rate and its considerable High-Speed Sending-Ratio. Furthermore, the Inspection vs. Contacted-Ratio shows that actors belonging to this class are less interested in the profile information of contacted users. In addition, this class has a very high amount of messages containing email addresses and hyperlinks. The average length of interaction with other users is very short and they can be characterised by a low Response-Received-Ratio. Finally, this class consists of users who are with a very high probability (.94) blocked by the provider. Thus, class 9 differs clearly from classes 1–8 by means of their behavioural and interactional patterns. But class 10, containing 3.22% of all actors, shows a distinctive pattern as well. Users in this class reveal the second highest First Contact Sending Rate and the second fastest High-Speed Sending-Ratio. They have the lowest standard deviation of their text messages and thereby use the same text several times in their communication. According to the second lowest Inspection vs. Contacted-Ratio, they contact most of the profiles they visited but not all of them, pointing out to some degree of selectiveness. Similar to class 9, the probability of sending an email or web address is very high in class 10. On average their interaction courses are very short and their Response-Received-Ratio is adverse. Finally, users of class 10 have the second highest probability of being blocked by the provider (.57). We interpret actors belonging to class 10 as spammer, people trying to scam users in respect to financial purposes. Class 10 shows similar behavioural patterns as bots do, though due to their manual restrictions they cannot contact as many users as fast as bots can.

In a next step, we validated our bot analysis by contrasting our prediction with a direct visual check of the text messages for bot presence[4]. Table 1 shows that 98.23% of latent class 9 revealed bot behaviour in their text messages. In total 54.95% of all bots are present in class 9[5]. Table 2 compares bots and human users

**Table 1** Manual check for bots

|  | Humans | Bots |  |
|---|---|---|---|
| Class 9 | 1.77% | 98.23% | 100.00% |
| Classes 1–8, 10–13 | 95.16% | 4.84% | 100.00% |

**Table 2** Impact of bot presence

|  | Women | Age (mean) | BMI (mean) | First contacts |
|---|---|---|---|---|
| Humans | 48.38% | 37.23 | 23.89 | 69.55% |
| Bots | 99.50% | 28.74 | 20.58 | 30.45% |
| Total | 53.97% | 36.19 | 23.67 | 100.00% |

---

[4]Due to privacy reasons the visual message check was done by the online-dating provider.
[5]A chi-square test on predicted vs. actual bots yielded a Cramer's V of 0.7156 ($\chi^2 = 1,000$, $p = 0.00$).

by means of their socio-demographic characteristics. Bots are nearly always female, about 10 years younger than average and show a favourable body-mass-index. This class of artificial actors generates 30.5% of all first contact attempts.

# 4 Conclusion

When analysing social web data, it is necessary to realise that there are several new issues and challenges of data quality that come into play. We discussed the presence of artificial actors called bots foiling the potential of research designs utilising web-generated process data of human interactions. Previous research on online mating behaviour for example did not consider that bots can dramatically reduce the quality of observed interactional records by suggesting a larger sample and by inducing a huge amount of artificial contact patterns. Based on assumptions of bot behaviour, we were able to construct behaviour and interaction related indicators based on web-based process-generated data. We applied a finite mixture model on the different indicators in order to encircle the artificial actors from different directions. The model identified a group that conformed to our theoretic conditions. A group of 3.72% of the actors showed a pattern where all variables indicated bot behaviour. Another class revealed similar behaviour and was interpreted as human spammers. A comparison of predicted and visually identified bots showed a high validity of our model. Bots differ clearly from human actors by means of their socio-demographic profiles. This is considerably problematic as bots produce about one third of all contact attempts. Further research is needed to improve the detection model and to analyse the impact of bots on substantial models such as regression statistics on human choosing behaviour. Furthermore, the proposed classification approach can be useful whenever spam behaviour of bots or human spammers occurs and log-file data on behaviour and interaction is recorded. For example social web platforms as well as instant messenger systems could be improved by removing unwanted bots. Future research has to deal with the distinction between such unwanted and requested bots on the one hand and the differentiation between human spammers and artificial behaviour on the other hand. The latent variable approach was useful to get first insights into the nature of artifical web behaviour so that future research can rest on these findings. Since the goal of our research is the assignment of actors to a fixed number of classes (actually bot vs. non-bot) future work will focus on the improvement of our results by applying different machine learning techniques, such as classification trees.

# References

Chen K, Jiang J, Huang P, Chu H, Lei C, Chen W (2006) Identifying MMORPG bots: A traffic analysis approach. In: Proceedings of the ACM SIGCHI International Conference on Advances in Computer Entertainment Technology, ACM, New York, vol 266:4

Chen K, Liao A, Pao HK, Chu H (2009) Game bot detection based on avatar trajectory. In: Stevens SM, Saldamarco SJ (eds) Proceedings of the 7th international Conference on Entertainment Computing, Springer, Berlin, Heidelberg, Lecture Notes In Computer Science, vol 5309, pp 94–105

Dvorak JC, Pirillo C, Taylor W (2003) Online! The Book. Prentice Hall, Upper Saddle River, NJ

Fink RD, Liboschik T (2010) Bots–Nicht-menschliche Mitglieder der Wikipedia-Gemeinschaft. Working Paper. Online: http://www.wiso.tu-dortmund.de/wiso/ts/Medienpool/AP-28-Fink-Liboschik-Wikipedia-Bots.pdf (accessed: 13. February 2011).

Gianvecchio S, Xie M, Wu Z, Wang H (2008) Measurement and Classification of Humans and Bots in Internet Chat. USENIX Security Symposium pp 155–170

Hayati P, Potdar V, Talevski A, Firoozeh N, Sarenche S, Yeganeh EA (2010) Definition of spam 2.0: New spamming boom. In: IEEE Digital Ecosystem and Technologies, Dubai, UAE

Poggi N, Berral JL, Moreno T, Gavalda R, Torres J (2007) Automatic detection and banning of content stealing bots for e-commerce. In: NIPS Workshop on Machine Learning in Adversarial Environments for Computer Security., Whistler, Canada, pp 7–8

Schmitz A, Skopek J, Schulz F, Klein D, Blossfeld HP (2009) Indicating mate preferences by mixing survey and process-generated data. The case of attitudes and behaviour in online mate search. Historical Social Research 34(1):77–93

Vermunt J, Magidson J (2002) Latent class cluster analysis. In: Hagenaars J, McCutcheon (eds) Applied latent class analysis. Cambridge University Press, Cambridge, UK, pp 89–106

# Part VI
# Text Mining

# Calculating a Distributional Similarity Kernel using the Nyström Extension

**Markus Arndt and Ulrich Arndt**

**Abstract** The analysis of distributional similarities induced by word co-occurrences is an established tool for extracting semantically related words from a large text corpus. Based on the co-occurrence matrix $C$ the basic kernel matrix $K = CC^T$ reflects word–word similarities. In order to considerably improve the results, a similarity kernel matrix is expressed as $G = U_k U_k^T$, where $U_k$ are the first $k$ eigenvectors of the eigendecomposition $K = U\Sigma U^T$. Clearly, the bottleneck of this technique is the high computational demand for calculating the eigendecomposition. In our study we speed up the calculation of the low-rank similarity kernel by means of the Nyström extension. We address in detail the inherent challenge of the Nyström method, namely selecting appropriate kernel matrix columns in such a way that the fast approximation process yields satisfactory results. To illustrate the effectiveness of our method, we have built a thesaurus containing 32,000 entries based on 0.5 billion corpus words (nouns, verbs, adjectives and adverbs) extracted from the Project Gutenberg text collection.

## 1 Introduction

The generation of thesauri based on a vector space representation of words for which word distribution data is derived from large text corpora is analyzed, and by applying latent semantic analysis (LSA) a remarkably good representation of semantic information is obtained (Rapp 2008). The underlying algorithm can be

M. Arndt (✉)
European Patent Office, Erhardt Str. 27, 80649 Munich, Germany
e-mail: marndt@epo.org

U. Arndt
data2knowledge GmbH, Fahrenheitstr. 1, 28359 Bremen, Germany
e-mail: ulrich.arndt@data2knowledge.de

W. Gaul et al. (eds.), *Challenges at the Interface of Data Analysis, Computer Science, and Optimization*, Studies in Classification, Data Analysis, and Knowledge Organization, DOI 10.1007/978-3-642-24466-7_34, © Springer-Verlag Berlin Heidelberg 2012

summarized as follows: pre-process the text corpus, map the corpus words to a vector space, perform dimension reduction by matrix decomposition, calculate word–word similarities and present for each target word the most semantically related words (Turney and Pantel 2010). In this study, we examine a variant of the original latent semantic analysis algorithm (Landauer and Dumais 1997), whereby two methods for calculating the complex and expensive matrix decomposition are presented, namely Nyström extension and singular value decomposition (SVD). The Nyström method is used to approximately calculate the eigenvalues and eigenvectors of a Gram matrix. The method is particularly attractive when dealing with large matrices as it only involves matrix multiplications and the eigenvalue decomposition of a small submatrix. However, when applying the Nyström extension, one must provide a sampling procedure which selects a subset of columns of the original matrix (Drineas and Mahoney 2005; Kumar et al. 2009). We consider three different ways of sampling. In contrast to the approximation method, the singular value decomposition provides exact results. It can be efficiently calculated for sparse matrices.

Both methods for performing matrix decomposition are evaluated by comparing the calculated word associations with WordNet (Fellbaum 1998). By this procedure, we are able to empirically benchmark the Nyström extension and the sampling methods involved based on the semantic information provided by latent semantic analysis.

## 2   Vector Space Representation of Semantic Information

### 2.1   Corpus Pre-processing

English texts of the Gutenberg Project[1] form the text corpus for this study. These texts mainly concern fictional literature. But the corpus also contains a wide variety of scientific texts and other non-literature texts such that the collection can be regarded as a source of richly featured language. The corpus contains 24,885 documents and has a size of approximately 9 GB of raw text. The complete collection is part-of-speech tagged and all words are mapped to their lemma. Adjectives, adverbs, verbs and nouns are extracted excluding proper nouns and words from a stopword list. In addition, noun sequences are concatenated as a simple measure to deal with compound words. The pre-processing of the corpus data results in a sequence of about 500 million words. By comparison, the British National Corpus (BNC) comprises about 100 million words. Applying the presented pre-processing steps results in a list of approximately 30 million words for this corpus.

---

[1] www.gutenberg.org

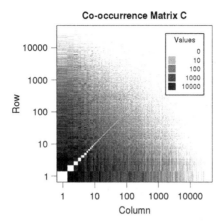

**Fig. 1** Visualization of the co-occurrence matrix $C$, the grey intensity reflecting the content of the matrix. Each row and column which are logarithmically scaled correspond to a word whereby the rows and columns are arranged in decreasing word frequency order, a row vector thus represents the word contexts as determined by counting co-occurrences. The diagonal of the matrix contains zero values since self co-occurrences were not counted

## 2.2 Vector Space Model

The vector space model (VSM) is used to represent word semantics (Turney and Pantel 2010). To this end, the top $n = 32{,}000$ frequent words of the corpus are mapped to a vector space $V^n$. Within the vector space, each dimension corresponds to one of the $n$ selected words. Practically, using the pre-processed corpus, direct consecutive word–word co-occurrences are counted. The resulting word pair counts are stored in the symmetric co-occurrence matrix $C \in \mathbb{R}^{n \times n}$. Hence, each row vector of $C$ is an element of $V^n$ and represents a word whereby the row values describe in which contexts the word appears in the corpus. The sparse matrix $C$ has a density of 7.5%, a visualization of the matrix is shown in Fig. 1. An association measure was applied on the entries of $C$ to dampen the influence of highly frequent words, the resulting matrix is called association matrix $A$ with elements $a_{ij} = \sqrt{c_{ij}}$.

## 2.3 Low Rank Kernel Matrix Reflecting Word–Word Similarities

For building the kernel matrix reflecting word–word similarities, we start by calculating a normalized Gram matrix, which is identical to calculating the cosine between any two row vectors of the matrix A. Hence, let $K = AA^T$ and $K_N$ : $k_{Nij} = \frac{k_{ij}}{\sqrt{k_{ii} k_{jj}}}$. Moreover, let $K_N = U \Sigma U^T$ be the eigenvalue decomposition of $K_N$. Using the top $k$ eigenvectors of $U$, the low rank matrix $G_k = U_k U_k^T$ is calculated. The normalized kernel matrix $G_{kN}$ describes the similarity between any two words $t_i$ and $t_j$. The larger a value within the matrix, the stronger the relatedness of the associated words. The similarity kernel is only based on the eigenvectors as suggested in (Shawe-Taylor and Cristianini 2004). Clearly, this method does not scale. Directly performing an eigenvalue decomposition on a 32,000 × 32,000 matrix is simply too expensive. However, the Nyström extension provides a means to approximately calculate the eigenvectors. Formally, let

$$K_N = \begin{bmatrix} X & Y^T \\ Y & Z \end{bmatrix} \succeq 0 \tag{1}$$

be a block matrix partition of $K_N$ with $X \succeq 0, X \in \mathbb{R}^{s \times s}, Y \in \mathbb{R}^{n-s \times s}$ and $Z \in \mathbb{R}^{n-s \times n-s}$. Moreover, let $X = Q \, S \, Q^T$ ($Q, \, S \in \mathbb{R}^{s \times s}$) be the eigendecomposition of $X$. Then, the matrix

$$\tilde{U}_k = \begin{bmatrix} Q_k \\ YQ_k S_k^{-1} \end{bmatrix} \quad \tilde{U}_k \in \mathbb{R}^{n \times k}; \; Q_k, \, S_k \in \mathbb{R}^{s \times k}; \; k \leq s$$

represents an approximation of the $k$ eigenvectors of $U$ corresponding to the $k$ largest eigenvalues of $S$, whereby $k$ is to be selected such that the $k$ eigenvalues are greater than zero. For details on the Nyström-extension, we refer to the literature (Drineas and Mahoney 2005; Kumar et al. 2009).

For calculating the word–word similarities $U_k$ is used to create the similarity matrix $\tilde{G}_k = \tilde{U}_k \tilde{U}_k^T$. This matrix is again normalized and the approximated similarity matrix $\tilde{G}_{Nk}$ is obtained.

The values for $s$ and $k$ are determined empirically. The question remains of how one should choose the partition of (1), that is, how to sample appropriate columns from $K_N$ to build $\begin{bmatrix} X \\ Y \end{bmatrix} \in \mathbb{R}^{n \times s}$. We analyse three methods:

- Randomly sampling uniformly $s$ columns without replacement.
- Sampling according to word frequency (1): Choose the columns of $K_N$ which correspond to the $s$ most frequent words
- Sampling according to word frequency (2): Pick from the ordered list of most frequent words every $n/s$ word and choose the columns of $K_N$ accordingly.

Random sampling can be regarded as the baseline sampling method for the Nyström extension (Kumar et al. 2009). The second method which is called "top frequent sampling", is motivated by the hope that the dense contexts of the top frequent words are a good selection for the extrapolation process of the Nyström extension. On the other hand, for the third method ("constant step sampling"), the samples are equally distributed over the full frequency spectrum of words in a well-defined and structured way. Finally, the approximation involving the sampling methods is compared with results obtained by calculating the exact eigenvectors by a singular value decomposition which does not involve any sampling.

Calculating the similarity kernel by means of the singular value decomposition is based on the observation that the kernel matrix $K_N$ can also be expressed in matrix notation involving an appropriately defined matrix $D$. To this end, let $K_N = DKD$ where $D$ is a diagonal matrix with elements $d_{ii} = \frac{1}{\sqrt{k_{ii}}}$ and $k_{ii}$ is the $i$th diagonal element of $K_N$. It follows that $K_N = DA \, (DA)^T$. Calculating the SVD for $DA = U\Lambda V$ yields $K_N = U\Lambda^2 U^T$ which shows a basic property of linear algebra, namely that the left singular vectors of a matrix are equal to the eigenvectors of an eigenvalue decomposition of a corresponding Gram matrix. As the matrix product $DA$ is a sparse matrix, the SVD can be efficiently calculated for some $k$ singular vectors. The normalized similarity kernel is then $G_{kN}$ based on $G_k = U_k U_k^T$.

## 3    Evaluation

As we want to retrieve as many highly related words as possible for a given target word, a precision and a recall value are defined in analogy to precision and recall in information retrieval. As a gold standard, **WordNet** (Fellbaum 1998) is used, which is a manually produced lexical database for the English language. Therefore, in order to objectively evaluate the information content of the similarity matrix, precision $P$ and recall $R$ are defined by

$$P = \frac{\sum_i^n |S_{Ai} \cap S_{Wi}|}{\sum_i^n |S_{Ai}|} \, , \quad R = \frac{\sum_i^n |S_{Ai} \cap S_{Wi}|}{\sum_i^n |S_{Wi}|}$$

where $S_{Ai}$ is the set of the 32 nearest neighbors of the target word $t_i$; the neighborhood is determined by the similarity matrices $G_{kN}$ and $\tilde{G}_{kN}$, respectively; only words which are of the same word type as the target word are added to $S_{Ai}$; and where $S_{Wi}$ is the set of related WordNet words for the target word $t_i$. If $t_i$ is of the type noun, then the WordNet synonyms, hyponyms, hypernyms and antonyms are considered; for a verb, the synonyms, troponyms, hypernyms and antonyms; and for an adjective or adverb the synonyms and antonyms. Within these settings, $\sum_i^n |S_{Ai}|$ and $\sum_i^n |S_{Wi}|$ assume constant values. Therefore, we simply use $\text{argmax}_{s,k \in \mathbb{N}} \, P$ for determining the values for $s$ and $k$ as well as for comparing the sampling methods. An example for calculating precision $P$ is presented in Fig. 2. There exist other methods for evaluating an automatically created thesaurus. One example is the widely used **TOEFL** test (Landauer and Dumais 1997), which basically provides synonym pairs for evaluation. However, this method and its variants only examine a small set of words, whereas our evaluation method uses all the 32,000 words contained in the thesaurus.

## 4    Results

The experiments were carried out on a server comprising 8 CPUs and 16 GB RAM. Except for performing SVD, the algorithms were implemented in Java; the software packages used are listed in Fig. 3; multi-threaded parallel matrix multiplication is

Term  t  =  abbey

$S_{Wt}$  =  {church, convent, monastery, church building, abbey}

$S_{At}$  =  {monk, convent, mosque, ruin, priory, chapel, hermitage, chantry, friar,
minster, chapterhouse, west front, manor house, abbey, abbot, parish-church,
church, cathedral church, cloister, abbess, nunnery, diocese, manor, chateau,
tower, refectory, nun, monastery, castle, cathedral, palace, bishopric}

$S_{At} \cap S_{Wt}$  =  {church, convent, monastery, abbey}

$P_t$  =  $\frac{|S_{At} \cap S_{Wt}|}{|S_{At}|} = 4/32 = 12.5\%$

**Fig. 2** Example evaluation and calculation of presicion $P$

| | |
|---|---|
| SVD: SVDLIBC | tedlab.mit.edu/~dr/SVDLIBC/ |
| Linear Algebra packages: Colt and jlapack | acs.lbl.gov/software/colt/ |
| | www.netlib.org/java/f2j/ |
| Java API for WordNet Searching (JAWS) | lyle.smu.edu/~tspell/jaws/ |
| Stanford Log-linear Part-Of-Speech Tagger | nlp.stanford.edu/software/tagger.shtml |

**Fig. 3** Software packages used to implement the algorithms

based on our own development. In a first series of experiments, precision $P$ and the Frobenius norms $\| K_N - K_{Nk} \|_F$ and $\| K_N - \tilde{K}_{Nk} \|_F$ are determined where $K_{Nk} = U_k \Lambda_k^2 U_k^T$ is the SVD based low rank approximation of $K_N$ and $\tilde{K}_{Nk} = \tilde{U}_k S_k \tilde{U}_k^T$ is the Nyström low rank approximation. The sample size is fixed to $s = 4,000$ corresponding to 12.5% of the columns of $K_N$ and the number of eigenvectors ranges from $k = 100$ to 4,000 corresponding to 0.3% ... 12.5% of the columns. Figure 4a and b show the results of these experiments. The precision curve based on SVD shows a sharp increase and then reaches a plateau within the examined range of singular vectors. The maximum precision value is obtained for about 10% singular vectors. This corresponds to an average agreement $|S_{Ai} \cap S_{Wi}|$ of 2.7 (standard deviation $\sigma = 1.5$). As expected, the Frobenius norm, which provides an indication of the accuracy of the low rank approximation, decreases continuously as more singular vectors are used. The results for the sampling method "top frequent" are almost identical to the SVD results when just a few eigenvectors are used. However, the results deviate significantly when using more and more eigenvectors. The sampling method "constant step" closely follows the SVD results. The maximum value is also obtained for about 10% eigenvectors. A similar curve is obtained when applying random sampling. However, the results are constantly slightly below "constant step". Looking at the Frobenius plot, we also observe that the overall distance of a sampling curve from the SVD curve is an indication of the performance of the sampling method. For the second batch of experiments, the number of eigenvectors is fixed while the sample size varies (see Fig. 5). The results are similar to the results presented in connection with a fixed sample size, whereby the "constant step" sampling method performs best.

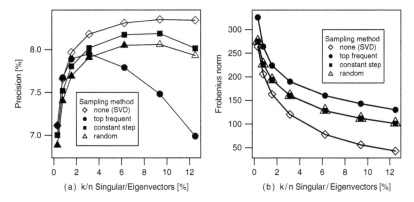

**Fig. 4** In both plots the sample size for the Nyström extension is fixed to $s = 4,000$. Concerning the random sampling method, a solid triangel indicates that the precision value is the average of 16 runs (standard deviation $\sigma = 0.01...0.02$), while the value for a non-solid triangle is the result of a single run. (**a**) Precision $P$ is plotted versus the relative number $k/n$ ($n = 32,000$) of singular/eigenvectors used to calculate the similarity kernel. (**b**) Frobenius norms $\| K_N - K_{Nk} \|_F$ and $\| K_N - \tilde{K}_{Nk} \|_F$ versus the relative number of singular/eigenvectors

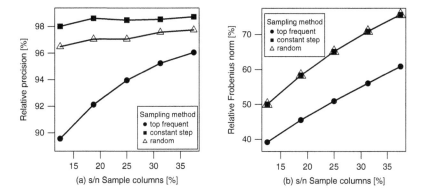

**Fig. 5** In both plots the number of singular/eigenvectors used to calculate the similarity kernel is fixed to $k = 3,000$. (**a**) Relative precision $P_{\text{Nyström}}/P_{\text{SVD}}$ is plotted versus the relative number of sample columns $s/n$ ($n = 32,000$) selected for the Nyström method. (**b**) Relative Frobenius norm $\| K_N - K_{Nk} \|_F / \| K_N - \tilde{K}_{Nk} \|_F$ versus the relative number of sample columns

## 5 Conclusion

Naturally, the SVD method returns the best precision results. Alternatively, the Nyström method results in a good approximation, whereby the quality of the results depends on the sampling method. In particular, the results from the "constant step" sampling method come close to the SVD results based on a relatively small-sized sample. The highly parallel calculation of the Nyström method in comparison to the single threaded SVD calculation results in up to a three times faster calculation and gives the possibility to scale even into larger matrix sizes. Comparing "constant step" and random sampling, the former always performs better in the experiments, which leads to the question whether this observation can be generalized and theoretically justified. In this connection, it is important to note that arranging the columns in the co-occurrence matrix in decreasing word frequency order is approximately the same as arranging the columns in decreasing column norm order which can be derived from Fig. 1.

Our experiments have not been able to validate the assumption in previous works on latent semantic analysis (Rapp 2008; Landauer and Dumais 1997), that some hundred singular-/eigenvectors are sufficient to produce best results. Within the context of the presented LSA method and evaluation procedure, which takes the semantic information of the complete corpus into account, we saw that some thousands of singular-/eigenvectors were necessary to maximize precision. Hence, further investigation is needed to clarify the interaction and dependencies between the corpus, the feature mapping, the dimension reduction method, the similarity calculation and the evaluation procedure.

In future work, we intend to make use of the information content provided by the similarity kernel. A first idea is to determine characteristic features of a text corpus by ranking words of a word type according to their distance from the respective word type centroid. Applying this idea to our corpus, we get the list of nouns: *man, thing, people, boy, person, house, woman, girl, place, friend, time, fellow*, which appears to be a fairly good representation for a fictional literature corpus. A further application would be to perform word sense induction by grouping the nearest neighbors of a target word by a local cluster algorithm. In first experiments, we obtained promising results. For example clustering the nearest noun neighbours of *bass* resulted in *bass, bass voice, fife – baritone, contralto, alto, soprano, treble, tenor, falsetto – pickerel, trout, shad, minnow, catfish, cod, salmon, sturgeon – cello, bassoon, oboe – solo, accompaniment, duet, recitative*.

## Disclaimer

The article represents the views and opinions of the authors alone.

## References

Drineas P, Mahoney MW (2005) On the Nyström method for approximating a Gram matrix for improved kernel-based learning. J Mach Learn Res 6:2153–2175
Fellbaum C (1998) WordNet: An Electronic Lexical Database. Cambridge, MA: MIT Press.
Kumar S, Mohri M, Talwalkar A (2009) Sampling techniques for the Nyström method. In: Twelfth International Conference on Artificial Intelligence and Statistics (AISTATS 2009), pp 304–311
Landauer TK, Dumais ST (1997) A solution to Plato's problem: The latent semantic analysis theory of the acquisition, induction, and representation of knowledge. Psychol Rev 104:211–240

Rapp R (2008) The automatic generation of thesauri of related words for English, French, German, and Russian. Int J Speech Technol 11:147–156

Shawe-Taylor J, Cristianini N (2004) Kernel methods for pattern analysis. Cambridge University Press, Cambridge

Turney PD, Pantel P (2010) From frequency to meaning: Vector space models of semantics. J Artif Intell Res 37:141–188

# Text Categorization in R: A Reduced N-Gram Approach

Wilhelm M. Geiger, Johannes Rauch, Patrick Mair, and Kurt Hornik

**Abstract** For the majority of Natural Language Processing methods, identifying the language of the processed text is one of the key tasks. Corresponding Natural Language Processing techniques often have language specific conditions, i.e., selecting the correct stop word list or the correct set of rules for stemming. Among various different approaches for language identification or more generally, text categorization, a rather large proportion is based on the word N-gram approach pioneered by Cavnar and Trenkle. In this contribution we will show how to produce language and document profiles using a reduced version of Cavnar and Trenkle's original algorithm. In addition, performance for N-gram based text classification employing both the original and the reduced approach, is compared. For this purpose, two groups of language profiles were used. One is composed of heterogeneous text data and the other one is solely based on articles from Wikipedia. Within this context we present the R package `textcat`. It enables the user to generate language profile databases as well as document profiles and allows to perform text classifications according to both the original and the reduced N-gram approach.

## 1 Introduction

Working with written text usually requires knowledge about the language used. For example, a text mining process requires the language of the document to be analyzed, such that the correct stop-word list and the correct set of rules for stemming can be selected. Therefore, modern text processing tools heavily rely on highly effective algorithms to complete the language identification.

W.M. Geiger (✉) · J. Rauch · P. Mair · K. Hornik
Institute for Statistics and Mathematics, WU Vienna University of Economics and Business,
Vienna, Austria
e-mail: wilhelm.geiger@gmail.com

W. Gaul et al. (eds.), *Challenges at the Interface of Data Analysis, Computer Science,
and Optimization*, Studies in Classification, Data Analysis, and Knowledge Organization,
DOI 10.1007/978-3-642-24466-7_35, © Springer-Verlag Berlin Heidelberg 2012

**Table 1** Traditional method
for generating N-grams

| $N = 1$ | $N = 2$ | $N = 3$ | $N = 4$ | $N = 5$ |
|---|---|---|---|---|
| -  | _g  | _gf   | _gfk  | _gfkl   |
| g  | gf  | gfk   | gfkl  | gfkl_   |
| f  | fk  | fkl   | fkl_  | fkl_ _  |
| k  | kl  | kl_   | kl_ _ | kl_ _ _ |
| l  | l_  | l_ _  | l_ _ _ | l_ _ _ _ |

The method presented in this paper will concentrate on the Cavnar-Trenkle approach (Cavnar and Trenkle 1994) and, in addition, we will also elaborate a reduced version of the N-gram approach. Both methodologies are implemented in R (R Development Core Team 1994) by means of the package textcat (Hornik 2010).

## 2  The Original N-Gram Approach by Cavnar & Trenkle

An N-gram, as defined by Cavnar and Trenkle (1994, p. 163) is a contiguous N-character slice of a longer string. For instance, an N-gram of length one is called "unigram", length two is a "bigram", length three is a "trigram", and length four or more is simply called an "N-gram". Obviously, these strings can vary in length and also overlap each other. Typically, underscores are used to identify, whether an N-gram includes the first letter (e.g, _fi), letters from the middle (e.g., dd, no underscore here), or the very end (e.g., nd_) of a word. Hence, a word of length $k$ will always result in $k + 1$ N-grams. As an example, Table 1 shows an N-gram decomposition of gfkl.

Basically N-gram based language identification is a two-step approach. First, it starts with the generation of different *language profiles* which are essentially frequency distributions of the N-grams. A sample corpus of different documents, all written in the same known language, has to be compiled. Subsequently, N-grams are generated from each word of each document within this corpus. All those N-grams are then sorted and ranked. Finally, the 300[1] most frequent N-grams are stored as the language profile for the chosen language. This small amount of N-grams suffices, as the N-gram distribution follows Zipf's law (Zipf 1935) [see also Cavnar and Trenkle (1994, p. 162)].

"The *n*th most common word in a human language text occurs with a frequency inverse proportional to *n*."

In a second step, a *document profile* must be generated that has to be classified in terms of language. Again, this document profile is a frequency distribution of the N-grams. The N-grams are generated for a certain amount of words from the

---

[1]This number was chosen by Cavnar and Trenkle (1994).

document, sorted, ranked, and the 300 most frequent ones are stored. Finally, one compares the profile of the document to be classified to each of the language profiles to see, which one fits best. This is judged by a distance measure on the frequency distributions. Typically, these N-gram frequency profiles follow a Zipf distribution (Zipf 1935, 1949). The algorithm classifies the document as belonging to the category having the smallest distance. Cavnar and Trenkle (1994, p. 167f) used the *Out-Of-Place Measure*.

# 3   The Reduced N-Gram Approach

Let us start this section with a closer look at the N-gram decomposition by means of the term gfkl already used above. The counting results of the classical N-gram approach are given in the second column of Table 2. In order to find ways for improvement and to remove redundancy, let us discuss various properties and key issues of the N-gram approach:

**Table 2** N-Gram frequency table of the abbreviation gfkl

| N-gram | Traditional method | Reduced approach |
| --- | --- | --- |
| _ | 1 | 0 |
| _g | 1 | 1 |
| _gf | 1 | 1 |
| _gfk | 1 | 1 |
| _gfkl | 1 | 1 |
| g | 1 | 0 |
| gf | 1 | 0 |
| gfk | 1 | 0 |
| gfkl | 1 | 0 |
| gfkl_ | 1 | 0 |
| f | 1 | 1 |
| fk | 1 | 1 |
| fkl | 1 | 0 |
| fkl_ | 1 | 1 |
| fkl_ _ | 1 | 0 |
| k | 1 | 1 |
| kl | 1 | 0 |
| kl_ | 1 | 1 |
| kl_ _ | 1 | 0 |
| kl_ _ _ | 1 | 0 |
| l | 1 | 0 |
| l_ | 1 | 1 |
| l_ _ | 1 | 0 |
| l_ _ _ | 1 | 0 |
| l_ _ _ _ | 1 | 0 |

## 3.1   Blanks

The blank is the most frequent N-gram in all languages. Counting them will not give any advantage. If a fixed number of words are taken to generate N-grams, this will lead to precisely the same number of blanks for each document.

## 3.2   Position

According to Cavnar and Trenkle (1994), all N-grams should be counted, including those with and without the extra information about their position, i.e., the underscores. In our example, one can find both gf and _gf. In fact, counting gf as done in the classical approach, would mean that the N-gram gf can be found in the middle of the abbreviation gfkl. This is not the case, of course. Only those N-grams containing the first or last letter of a word have be counted, if the additional information about their position is a part of the N-gram.

## 3.3   Redundancy

Including kl_ as well as kl__ is redundant. The information that there is a kl at the end of the abbreviation gfkl is of importance only.

## 3.4   Short Words

Using the traditional method short words can not be represented correctly. Let us look at the following simple sentence: *This is a good example.*. Using Cavnar and Trenkle's method, the word is would wrongly be represented as _is and is_. This is not true, of course, there is no is at the beginning of a word, nor at the end. Hence, only _i, _is_ and s_ are the correct representations. The same applies to the word a.

To solve the shortcomings addressed above, we will add a set of rules to the original method:

- Blanks are not counted.
- N-grams containing the first letter of a word, without the additional information about this position, i.e. without underscores, are not counted.
- N-grams containing the last letter of a word, without the additional information about this position, i.e. without underscores, are not counted.
- All N-grams containing more than one underscore at the end are not counted.

- Given the number of characters $k$, if $k \leq n$ then only N-grams may be generated for which the constraint $n \leq k$ holds true.
- For each word with $k < (n - 2)$, N-grams with $n \leq k$ must be generated as well as an N-gram of the type _xy_.
- A word with $k = 1$ will result in an N-gram of the type _x_.

The results of applying those rules to the process of generating N-grams for the abbreviation gfkl can be seen in the third column of Table 2.

This rule set reduces the redundancy within the language profile drastically. As already mentioned, only the 300 most frequent N-grams were used by (Cavnar and Trenkle 1994) to form a language profile. As an example, let us assume that we have many words ending with s. In this case 4 out of 300 places within the language profile will be blocked just to supply this one information (i.e., s_, s__, s___ and s____). Hence, quite a large proportion of the entire information supplied by the N-gram frequency distribution of a language profile is redundant.

The same thought applies to N-grams containing the first or last character of a string, without the additional information about its position. For instance, if we would count _gfk as well as gfk, we waste one of the 300 places without getting any helpful, additional information. Furthermore, we do not store wrong information, since gfk, as a character slice from the middle of the word, actually does not exist. Third, it now is possible to correctly store short words.

# 4 Data Sources, Preparation, and Classification Results

First of all language profiles have to be trained using the R (R Development Core Team 1994) package textcat (Hornik 2010).

```
> install.packages("textcat")
> require(textcat)
```

To train language profiles, a certain amount of training data has to be supplied. We used the following datasets in order to explore the performance of the classical and the reduced N-gram approach.

First, training data for *WIKI* profiles were created using Wikipedia content for 10 key words in 25 languages. The keywords were philosophy, mathematics, statistics, France, USA, religion, wikipedia, internet, medicine and rice. Due to the different extent of Wikipedia articles in different languages the total length of the training data sets varies form 53,000 characters for Latin to 530,000 characters for German.

The second dataset are the classical ECI benchmark dataset (European Language Resource Association 2008; Project Gutenberg 2008) and private sources. The training data for the *ECI* profiles range from 145,000 for Lithuanian to 1,060,000 for English. For the classification, documents in Dutch, English, French, German, Italian and Spanish were used that were randomly composed of the respective documents of the *ECI* training data set. Each document had an approximate length of 30,000 characters.

The number of words per text unit to be classified (*numwd*) was varied from 100 to 10, 5 and 1. Five different language profiles were used to classify the different text units. *ECI_new* and *ECI_old*, *WIKI_new* and *WIKI_old* and *ECI_PKG* which is the currently implemented standard in textcat. The sub tag *_new* signifies profiles generated with the reduced N-gram approach, *_old* profiles generated with the original approach. *ECI_PKG*, as it is the current standard for textcat, is of course generated using the reduced approach. To actually classify the text units the *Out-Of-Place Measure* (Cavnar and Trenkle 1994, p. 167f) was used.

Figures 1–3 show the classification results for *numwd* = 1, 5, 10. Note that using 100 words per text unit to be classified 100% correct classifications were achieved regardless the source of training data and the N-gram method used.

A clear tendency is that the less words are used for one classification unit, the more the positive classification rate decreases and, correspondingly, the range between the single classification results becomes bigger. Moreover, there is no clear tendency neither for the N-gram approach applied nor for the training data used. So far unknown effects of the training data are for example the categorization results for *en* using the *ECI_new* language profile. While its relative performance for *numwd* = 10 and 5 were the poorest of all five language profiles, it performed second best for *numwd* = 1. Similar results are also to be found for example using the *WIKI_new* language profile categorizing *es* or for *WIKI_old* categorizing *en*.

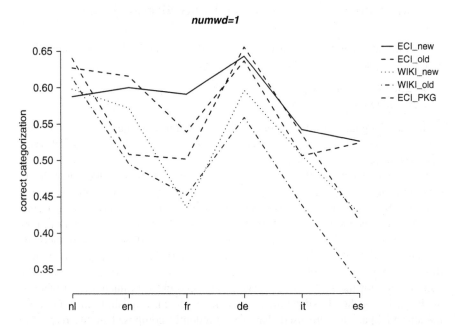

**Fig. 1** Classification results for *numwd* = 1

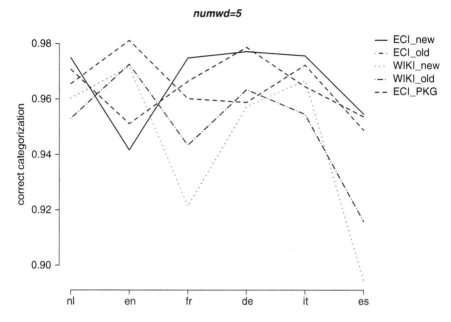

**Fig. 2** Classification Results for *numwd* = 5

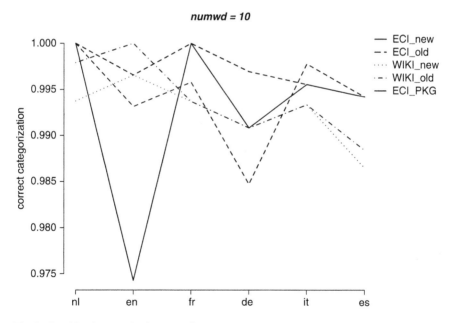

**Fig. 3** Classification Results for *numwd* = 10

**Table 3** Number of best/worst performances by approach new, old, new & old

|              | New | Old | New & Old |
|--------------|-----|-----|-----------|
| $numwd = 1$  | 2/4 | 3/2 | 1/0       |
| $numwd = 5$  | 1/2 | 4/4 | 1/0       |
| $numwd = 10$ | 2/0 | 1/4 | 3/2       |

Moreover Table 3 shows the number of best/worst performances for categorizations using either the new or the old approach. Again, there is no clear tendency to be found in the results but a clear advise for further research efforts.

# 5 Conclusion

In this article we presented the N-gram approach for text classification and introduced a reduced version. Both algorithms are implemented in the textcat package.

The reduced N-gram approach is mainly characterised by eliminating redundancies enclosed in the Cavnar-Trenkle approach, which is neither improving nor worsening the performance. However with respect to the eliminated redundancies the efficiency has improved from a computational perspective.

Taking into account that assembling the heterogeneous *ECI* training corpus requires much more effort than assembling a similar corpus from Wikipedia content relieves the process of creating individual language profiles of similar quality. As we have seen in Figs. 1–3 in the previous section although there is quite some variation in the classification results, this variation is caused by hitherto unknown characteristics but not by the chosen N-gram approach.

This study has shown, that Wikipedia content is as adequate for creating language profiles for text categorization as heterogeneous text sources including professionally composed language collections (*ECI*). In terms of future research, on the one hand we aim for an explanation of the possible causes for the variation of the categorization results and develop appropriate solutions. On the other hand, we will explore ways of improving the categorization performance for very short text units ($numwd = 1, \ldots, 5$).

# References

Cavnar WB, Trenkle JM (1994) N-gram-based text categorization. In: Proceedings of SDAIR-94. 3rd Annual Symposium on Document Analysis and Information Retrieval, pp 162–175

European Language Resource Association (2008) Catalog reference: W0004. URL http://catalog.elra.info

Hornik K (2010) textcat: N-Gram Based Text Categorization. R Package Manual, Cited 25 Aug 2010

Project Gutenberg (2008) About project gutenberg. URL http://www.gutenberg.org

R Development Core Team (1994) A Language Environment for Statistical Computing. R Funda-
tion for Statistical Computing, Vienna, Austria, URL http://www.R-project.org, cited 25 Aug
2010

Zipf GK (1935) The psychobiology of language. Hougthon-Mifflin, Bosten, MA

Zipf GK (1949) Human behavior and the principle of least effort: An introduction to human
ecology. Addison-Wesley, Reading, MA

# Part VII
# Dimension Reduction

# HOMALS for Dimension Reduction in Information Retrieval

Kay F. Hildebrand and Ulrich Müller-Funk

**Abstract** The usual data base for multiple correspondence analysis/homogeneity analysis consists of objects, characterised by categorical attributes. Its aims and ends are visualisation, dimension reduction and, to some extent, factor analysis using alternating least squares. As for dimension reduction, there are strong parallels between vector-based methods in Information Retrieval (IR) like the Vector Space Model (VSM) or Latent Semantic Analysis (LSA). The latter uses singular value decomposition (SVD) to discard a number of the smallest singular values and that way generates a lower-dimensional retrieval space. In this paper, the HOMALS technique is exploited for use in IR by categorising metric term frequencies in term-document matrices. In this context, dimension reduction is achieved by minimising the difference in distances between objects in the dimensionally reduced space compared to the full-dimensional space. An exemplary set of documents will be submitted to the process and later used for retrieval.

## 1 Introduction

The *vector space model (VSM)* is an instrument of *Information Retrieval (IR)* that does not rely on string pattern matching, but represents both documents and queries as *term vectors*. They may be of the following form:

$$d_i = (d_{i1}, d_{i2}, \ldots, d_{it}) \tag{1}$$

denotes a document vector and

$$q_i = (q_{i1}, q_{i2}, \ldots, q_{it}) \tag{2}$$

K.F. Hildebrand (✉) · U. Müller-Funk
European Research Center for Information Systems (ERCIS), University of Münster, Germany
e-mail: hildebrand@ercis.uni-muenster.de; funk@ercis.uni-muenster.de

W. Gaul et al. (eds.), *Challenges at the Interface of Data Analysis, Computer Science, and Optimization*, Studies in Classification, Data Analysis, and Knowledge Organization, DOI 10.1007/978-3-642-24466-7_36, © Springer-Verlag Berlin Heidelberg 2012

would be a query vector. Here, $d_{ik}$ and $q_{ik}$ indicate the value of the $k$th term in $d_i$ or $q_i$, respectively (Salton 1988). In their most basic forms, these values will be $d_{ik} = 1$, if term $k$ appears in $d_i$ and $d_{ik} = 0$ if not. The same is true for queries analogously.

To convey the central notion of this paper, in Sect. 2 LSA is introduced as a concept for IR. Section 3 deals with HOMALS as an instrument for dimension reduction and visualisation. In the central Sect. 4, the algorithm of HOMALS shall then be transferred to the context of IR. Eventually, findings will be summarized and an outlook is given in Sect. 5.

# 2 LSA

LSA is a variant of the VSM. In the VSM terms form the dimensions of a space that is used to project queries into and measure similarities between their vectors and those of documents. LSA follows the same approach, but the dimensions of the indexing space are not made from terms, but derived through truncation of an SVD of the term-document input matrix (Dumais 2007). Furthermore, LSA – by means of truncation – maps a high-dimensional space onto a low-dimensional one. In doing that, it tries to create a mapping that provides the best results for the purpose of dimension reduction – in this case IR. Consequently the axes of the reduced space are meant to cover the greatest amount of variation in that dimension. The first dimension is supposed to cover a maximum of all existing variation, the second dimension would then deal with the remaining variation in data. Thus, LSA is closely related to PCA (Manning and Schütze 1999).

## 2.1 Conducting LSA

In order to show the complete conduct of an LSA, it is helpful to use an example that is employed throughout the process. This process has been described by Dumais as consisting of four stages:

1. Term-document matrix.
2. Transformed term-document matrix.
3. Dimension reduction.
4. Retrieval in reduced space (Dumais 2007).

During the first step, all documents and terms are represented in a term-document matrix $A$, just like in the VSM. The second step contains the weighting of raw term frequencies. The third step features the main difference to the VSM: Dimensions are reduced via SVD. Retrieval happens in step four. The query is converted into a vector and compared with the existing document vectors.

Every non-zero entry of $A$ is weighted. First, the frequency contained in each cell increased by 1 and converted to its log (Landauer et al. 1998). Afterwards the entropy measure $1 + \sum_j^M \frac{p_{ij} \log_2(p_{ij})}{\log_2 M}$, with $p_{ij} = \frac{\text{tf}_{ij}}{\text{cf}_i}$ is applied to each non-zero cell of $A$ (Martin and Berry 2007). As a consequence, the effect of the transformation is

> "to weight each word-type occurrence directly by an estimate of its importance in the passage and inversely with the degree to which knowing that a word occurs provides information about which passage it appeared in." (Landauer et al. 1998)

The term-by-document matrices, like $A$, are considered sparse. In practice they only posses about one per cent of non-zero entries. This is due to the fact that documents only use a small subset of the existing vocabulary.

## 2.2 Dimension Reduction

Once an input matrix has been prepared, the key computations of LSA can be carried out. The goal is to exploit the truncation of vectors when transforming them into a term and document vector space via *orthogonal decomposition* (Martin and Berry 2007). The instrument used for this is called SVD. SVD has formerly been used to solve unconstrained linear least squares problems, estimate matrix ranks or for canonical correlation analysis (Berry et al. 1994). In case of LSA, its purpose is to uncover the underlying (i.e. latent) semantic structure of word usage (Dumais et al. 1988).

The orthogonal decomposition that the SVD provides, preserves some properties of the input matrix: The norms and distances of row and column vectors remain the same. Also, the $L_2$ norm of $A$ is kept. The result is a set of orthogonal matrices.

There are other means for decomposing term-by-document matrices like $A$ into orthogonal components. One method is QR factorisation, another would be SDD (Berry et al. 1999, pp. 8) as well as (Kolda and O'Leary 1998). Nevertheless, SVD remains the first choice in the context of LSA for various reasons. Martin and Berry (2007) state, that:

- SVD decomposes $A$ into orthogonal factors that represent both terms and documents.
- The vector representations for terms and documents are achieved simultaneously.
- SVD allows for adjustments in the number of dimensions that the semantic vector space is composed of.
- SVD is computationally manageable even for large datasets.

Input to the procedure of SVD is a weighted term-document matrix $A$. The general structure of an SVD can be written as in (3) (Dumais 1991).

$$A = U \Sigma V^{\mathrm{T}} \tag{3}$$

Here, $U$ and $V$ are orthogonal matrices (i. e. $U^T U = I$ and $V^T V = I$), while $\Sigma$ is diagonal (i.e. $\Sigma$ has the values $(\sigma_1, \sigma_2, \ldots, \sigma_n)$ on its main diagonal and zeros elsewhere). $A$ shall be of rank $r$.

There are $r$ orthonormal eigenvectors in the first $r$ columns of $U$. Those eigenvectors are associated with the $r$ nonzero eigenvalues of $AA^T$. Accordingly, the first $r$ columns of $V$ consist of orthonormal eigenvectors, associated with the $r$ nonzero eigenvalues of $A^T A$. At the same time, the first $r$ entries of the main diagonal of $\Sigma$ are the nonnegative square roots of the nonzero eigenvalues $(\lambda_1, \ldots, \lambda_r)$ of $A^T A$ and $AA^T$ (Berry et al. 1994). Finally, the rows of $U$ are called left singular vectors. They represent the term vectors. The rows of $V$ on the other hand are the right singular vectors and represent the documents. The elements on the diagonal of $\Sigma$ are called singular values.

In SVD dimensions of the resulting vector space are to be reduced. This is done by setting all but the $k$ largest singular values in $\Sigma$ to zero. This leads to the matrices $\Sigma_k, U_k$ as well as $V_k^T$. Since $\Sigma$ is structured in a way that the largest singular value appears in the top-left corner and they decrease towards the bottom-right corner, the dropped $r - k$ singular values are also the smallest ones. If $U_k, \Sigma_k$ and $V_k^T$ are multiplied, they yield a matrix

$$A_k = U_k \Sigma_k V_k^T \tag{4}$$

that is the closest rank-$k$ approximation to the original matrix $A$. This means that the distance between $A$ and $A_k$ is minimised (Berry and Browne 1999). This process is called truncation and it reduces the dimensionality of the vector space from $r$ to $k$. The remaining $k$ dimensions cover the essential usage of terms, the dropped dimensions are referred to as *noise* or variability in using terms to describe semantic concepts in texts. Furthermore, terms can appear close to each other (i. e. similar) in $k$-dimensional space even if they never appeared in the same document (Martin and Berry 2007).

## 2.3 Retrieval in Reduced Space

A query in $k$-dimensional space is treated like a document and referred to as a *pseudo document* (Deerwester et al. 1990). In order to produce a set of retrieval results, the query has to be compared with all documents in the reduced-rank space. The ones that produce the highest cosine measure (i.e. the nearest vectors) are seen as best results.

A query is created by weighting the sum of its term vectors scaled by the inverse of the singular values. This leads to an individual weight for each dimension in the $k$-dimensional term-document space. Thus, a query may be written in the following form:

$$\hat{q} = q^T U_k \Sigma_k^{-1}, \tag{5}$$

where $\hat{q}$ is the resulting truncated vector and $q^T$ contains the weighted term frequencies and otherwise zeros. This vector is generated through the same process of removing high-frequency words, stemming and weighting as described above.

Generally, there are three different comparisons to be made in the $k$-dimensional space. Terms can be compared to each other, as can documents, and terms may be compared to documents. The comparison between terms $t_i$ and $t_j$ requires a view on the dot product between the row vectors of $A_k$. Let $A_k = U_k \Sigma_k V^{T_k}$, then the dot product of any two term vectors reflects the similarity between those vectors. It would be convenient to create a square similarity matrix for term similarities. This can be achieved in the following way: If

$$A_k A_k^T = U_k \Sigma_k V_k^T (U_k \Sigma_k V_k^T)^T = U_k \Sigma_k V_k^T V_k \Sigma_k U_k^T = U_k \Sigma_k^2 U_k^T \qquad (6)$$

is this similarity matrix, then the dot products for any two terms is the dot product of rows $i$ and $j$ of $U_k \Sigma_k$. Consequently, in order to compute a similarity in $k$-dimensional space, the term vectors scaled by the singular values are utilised, no matter which similarity measure is actually used. This reasoning can be applied to document comparisons analogously. The resulting computations can be seen in (7). Now, the matrix $V_k \Sigma_k$ yields all dot products of interest. If its rows $i$ and $j$ are compared in a dot product, this equals the comparison of two documents.

$$A_k^T A_k = (U_k \Sigma_k V_k^T)^T U_k \Sigma_k V_k^T = V_k \Sigma_k U_k^T U_k \Sigma_k V_k^T = V_k \Sigma_k^2 V_k^T \qquad (7)$$

Comparison of terms and documents has to be handled differently. Again, if $A_k = U_k \Sigma_k V^{T_k}$, then the cell $a_{ij}$ of $A_k$ contains the important information. It is obtained by computing the dot product of the $i$th row of $U_k \sqrt{\Sigma_k}$ and the $j$th row of $V_k \sqrt{\Sigma_k}$. Important here is the consequence of matrices $U_k \Sigma_k$ and $V_k \Sigma_k$ used for comparisons between entities of the same kind, and $U_k \sqrt{\Sigma_k}$ as well as $V_k \sqrt{\Sigma_k}$ for comparisons between entities of different kind. Because of the different scaling ($\Sigma_k$ vs. $\sqrt{\Sigma_k}$), between and within comparisons cannot be made in the same configuration of points. Axes will always be stretched or shrunk by a factor of $\sqrt{\Sigma_k}$, precisely (Deerwester et al. 1990; Martin and Berry 2007).

## 3   HOMALS

Homogeneity analysis using alternating least squares (HOMALS) can be motivated from different angles. Here, its ability to reduce dimensions of categorical data is in focus. Generally, this reduction is carried out as far as reaching two dimensions which enables an easily accessible plot of data objects (Michailidis and Leeuw 2005). In detail, HOMALS assumes $N$ objects with $J$ categorical variables assigned $l_j$ possible values (i. e. categories) for variables $j \in \mathbf{J} = 1, 2, \ldots, J$. Following a graph approach, it is fair to interpret the $N$ objects and $\sum_{j \in \mathbf{J}} l_j$ categories of

$J$ variables as a bipartite graph. Each object is connected to the respective category nodes of its variables. Consequently, object nodes will have degree $J$ and the degree of category nodes equals the number of objects that belong to this category.

Given this setup, the goal is to provide a mapping to low-dimensional Euclidean space and preserve as much information from the original positioning in high-dimensional space. $X$ is the $N \times k$ matrix that contained the coordinates of $N$ objects in $R^k$. Furthermore, we need $j \in \mathbf{J}$ matrices $Y_j$ of $l_j \times k$ for the coordinates of the $l_j$ category vertices of variable $j$. While $X$ is the matrix of object scores, $Y_j$ denotes the category quantification matrix. $G_j$ are indicator matrices in which an entry $G_j(i, t) = 1$ if an object $i$ belongs to a category $t$ and 0 if not. One can construct an adjacency matrix $G = (G_1, \ldots, G_J)$ of the corresponding graph.

In these settings, Michailidis and Leeuw denote the average squared edge length over all variable as

$$
\begin{aligned}
\sigma(X; Y_1, \ldots, Y_j) &= J^{-1} \sum_{j=1}^{J} \mathrm{SSQ}(X - G_j Y_j) \\
&= J^{-1} \sum_{j=1}^{J} \mathrm{tr}(X - G_j Y_j)'(X - G_j Y_j).
\end{aligned}
\tag{8}
$$

SSQ being the sum of squares of the elements of a matrix. Equation 8 will be minimized over $X$ and the $Y_j$s. There are two constraints in order to prevent trivial solutions:

$$
X'X = N I_k,
\tag{9}
$$

$$
u'_N X = 0.
\tag{10}
$$

Equation (9) requires the columns of $X$ to be orthogonal in multidimensional solutions. It also normalises the squared length of object scores to be $N$.

In order to solve the problem given in (8), an alternating least squares algorithm is used. In a first step, $X$ is fixed and (8) is minimized w.r.t. $Y_j$. The solution looks as follows:

$$
\hat{Y}_j = D_j^{-1} G'_j X, \ j \in \mathbf{J},
\tag{11}
$$

with $D_j = G'_j G_j$ being the $l_j \times l_j$ diagonal matrix containing relative category frequencies of variable $j$. Once this step has been carried out, X is determined with fixed $Y_j$s. $\hat{X}$ is given by

$$
\hat{X} = J^{-1} \sum_{j=1}^{J} G_j Y_j.
\tag{12}
$$

In a third step, object scores are column centered and orthonormalised in order to satisfy (9) and (10). The loss function can be written as

$$
Np - J^{-1} \sum_{j=1}^{J} \mathrm{tr}(\hat{Y}'_j D_j \hat{Y}_j),
\tag{13}
$$

with $\sum_{j=1}^{J} \mathrm{tr}(\hat{Y}'_j D_j \hat{Y}_j)$ being the fit of the solution (Gifi 1992).

## 4 HOMALS in IR

In order to apply HOMALS to the concept of latent semantic information retrieval, it is necessary to reinterpret some of HOMALS' features. In the IR context, objects of HOMALS will be documents as they would appear in a corpus, i.e. collections of words that may have been submitted to a stop word[1] or stemming[2] procedure. Consequently, we perceive $N$ as the number of documents and $X$ becomes an $N \times k$ matrix of document vertices.

Furthermore, for every word in the corpus, there will be a variable that each object possesses. This means that $J$ equals the number of distinct words in the corpus and $\mathbf{J}$ denotes the set of variables, one for each word. An adjustment that requires further inspection is that of categories $l_j$: In LSA, each word-document-pair receives a frequency which is metric. Now, variables are categorical and the term frequencies have to be categorised. A number of categories per variable (i.e. word) has to be determined. A first approach may be to use highest and lowest frequency, decide upon a number of categories and give each one equivalent width: $w(j) = \frac{1}{l_j} * \max(t) - \min(u), t, u \in T$, with $w$ being the width of all categories, regardless the variable they belong to and $T$ being the set of term-frequency values that have been assigned.

Another way may be to put term frequencies in ascending order and build categories that have the same number of elements in them. Thus, $t_p \leq t_{p+1} \forall t_p \in T$ and $w(j) = |C_i| = |C_{i+1}| = \lfloor \frac{|T|}{l_j} \rfloor$, with $C_i \subset T \forall i \in 1, \ldots, l_{j-2}$ and $|C_{l_j}| = (\frac{|T|}{l_j} - \lfloor \frac{|T|}{l_j} \rfloor) l_j$. This means that the first $l_j - 1$ categories have equal cardinality, while the last one carries the largest term frequencies. These are just the remainder that could not build a category of their own, because the number of frequencies per category w(j) had to be $\in N$. As a result, the elements of a category can be noted as

$$C_i = \bigcup_{h=1}^{w(j)} t_{(h+(i-1)w(j))}. \tag{14}$$

In the second case it is worth thinking about zero entries. Since the input matrices are sparse, there will be many term frequencies equal to zero. This again leads to the first categories mainly consisting of zeros. It might be worthwhile to put all zero entries into $C_1$ and apply the formula to $C_2$ up to $C_{l_j}$.

The indicator matrices $G_j$ do not change essentially. Only now, the $(i, t)$-th entry of $G_j$ corresponds to a membership of an object to part of the term-frequency spectrum of word $j$. The following minimisation remains the same, too.

---

[1] Here, words that cannot discriminate between documents and do not carry any content like *a* or *and* are removed.

[2] In stemming, certain endings are removed or merged in order to map words with identical stems to the same item.

The result has some interesting properties that can be interpreted in terms of IR: (1) The configuration represents documents and term-frequency categories in the same $k$-dimensional space. Also, (2) documents will be close to each other, if their term-frequency quantifications are similar. Thus, in contrast to LSA, not angles between document vectors but distances between document representations. Identical documents will receive identical scores. (3) The frequency categories are in the centre of objects belonging to that category. (4) Terms that have better ability to discriminate have category scores further apart from each other. (5) Documents and frequency categories coincide if the document is the only one possessing the respective term in that frequency. (6) Unique documents are further away from the origin that average ones. (7) Solutions are nested: A $k$-space contains the first $k$ dimensions of a $k + 1$-space. (8) The solutions for subsequent dimensions are ordered in terms of maximum eigenvalues.

# 5 Conclusion

We have introduced LSA as an instrument in IR. Homogeneity Analysis using Alternating Least Squares has been introduced as a means to visualise categorical data in a convenient and feature-preserving way. Furthermore, we have shown that it is indeed possible to use the concepts of HOMALS for IR. There have to be adjustments in input data and some reinterpretations of results. The algorithm itself, however, does not change. Thus, future work consists of using the approach on test data as well as elaborating on how specific properties of a HOMALS solution could be facilitated in the context of IR.

# References

Berry MW, Browne M (1999) Understanding search engines: mathematical modeling and text retrieval. Society for industrial and applied mathematics. Philadelphia, PA, USA

Berry MW, Dumais ST, O'Brien GW (1994) Using linear algebra for intelligent information retrieval. Tech. Rep. UT-CS-94-270, University of Tennessee

Berry MW, Drmac Z, Jessup ER (1999) Matrices, vector spaces, and information retrieval. SIAM Rev 41:335–362

Deerwester SC, Dumais ST, Landauer TK, Furnas GW, Harshman RA (1990) Indexing by latent semantic analysis. J Am Soc Inform Sci 41(6):391–407

Dumais ST (1991) Improving the retrieval of information from external sources. Behav Res Meth Instrum Comput 23(2):229–236

Dumais ST (2007) LSA and Information retrieval: Getting back to basics, Lawrence Erlbaum associates. Mahwah, NJ, Chap. 16, pp 293–321

Dumais ST, Furnas GW, Landauer TK, Deerwester SC, Harshman RA (1988) Using latent semantic analysis to improve access to textual information. In: CHI '88: Proceedings of the SIGCHI conference on Human factors in computing systems. ACM Press, New York, NY, pp 281–285

Gifi A (1992) Nonlinear multivariate analysis. Comput Stat Data Anal 14(4):548–544, URL http://econpapers.repec.org/RePEc:eee:csdana:v:14:y:1992:i:4:p:548–544

Kolda TG, O'Leary DP (1998) A semidiscrete matrix decomposition for latent semantic indexing information retrieval. ACM Trans Inf Syst 16(4):322–346

Landauer TK, Foltz PW, Laham D (1998) Introduction to latent semantic analysis. Discourse Process 25:259–284

Manning CD, Schütze H (1999) Foundations of statistical natural language processing. MIT Press, Cambridge, MA

Martin DI, Berry MW (2007) Mathematical foundations behind latent semantic analysis, Lawrence Erlbaum associates. Mahwah, NJ, Chap. 2, pp 35–55

Michailidis G, Leeuw JD (2005) Homogeneity analysis using absolute deviations. Comput Stat Data Anal 48(3):587–603

Salton G (1988) Automatic text processing: The transformation analysis and retrieval of information by computer. Addison-Wesley

# Feature Reduction and Nearest Neighbours

**Ludwig Lausser, Christoph Müssel, Markus Maucher, and Hans A. Kestler**

**Abstract** Feature reduction is a major preprocessing step in the analysis of high-dimensional data, particularly from biomolecular high-throughput technologies. Reduction techniques are expected to preserve the relevant characteristics of the data, such as neighbourhood relations. We investigate the neighbourhood preservation properties of feature reduction empirically and theoretically. Our results indicate that nearest and farthest neighbours are more reliably preserved than other neighbours in a reduced feature set.

## 1 Introduction

Feature reduction constitutes a major step in the processing chain of high-dimensional biological data, such as microarray data (Guyon and Elisseeff 2003). To design specialized diagnostic chips and assays, it is often necessary to select a small set of candidate markers from large-scale screening experiments. Measuring such small sets of markers is often more accurate and less expensive than a whole-genome microarray experiment, and the results are easier to interpret.

From a computational point of view, a reduced feature set can facilitate subsequent machine learning tasks, such as classification and clustering. For example,

L. Lausser · M. Maucher
Internal Medicine 1, University Hospital Ulm, 89081 Ulm, Germany
e-mail: ludwig.lausser@uni-ulm.de; markus.maucher@uni-ulm.de

C. Müssel
Institute of Neural Information Processing, Ulm University, 89069 Ulm, Germany
e-mail: christoph.muessel@uni-ulm.de

H.A. Kestler (✉)
Institute of Neural Information Processing, Ulm University, 89069 Ulm and Internal Medicine 1, University Hospital Ulm, 89081 Ulm, Germany
e-mail: hans.kestler@uni-ulm.de

W. Gaul et al. (eds.), *Challenges at the Interface of Data Analysis, Computer Science, and Optimization*, Studies in Classification, Data Analysis, and Knowledge Organization, DOI 10.1007/978-3-642-24466-7_37, © Springer-Verlag Berlin Heidelberg 2012

a smaller number of weights has to be adapted for perceptron-based neural networks when using a reduced feature set. This reduces the parameter search space significantly.

Reduction techniques are expected to preserve the relevant characteristics of the datasets. In particular, they should preserve distance relations among the samples in the datasets: many machine learning algorithms are distance-based, e.g. $k$-means clustering (MacQueen 1967), $k$-Nearest Neighbour (Ripley 1996), LVQ (Kohonen 1989), support vector machines (Vapnik 1998), or neural networks. Distance preservation has previously been studied theoretically. Johnson and Lindenstrauss state that distances are preserved up to a factor of $1 \pm \epsilon$ after certain projections with random matrices (Johnson and Lindenstrauss 1984). Beyer et al. (1999) and Hinneburg et al. (2000) describe that the distance to the nearest data point approaches the distance to the farthest data point with increasing dimensionality.

For many distance-based approaches, it is sufficient to preserve neighbourhood relations. That is, the order of the distances between the data points should remain unchanged, but the distances themselves or their proportions need not be preserved. Beyer et al. define stability of nearest neighbours by the number of data points lying in an $\epsilon$ range around the nearest neighbour. Radovanović et al. analyse the distribution of the $k$ nearest neighbours in high-dimensional data and observe that hub points emerge as the dimensionality increases (Radovanović et al. 2009). Aggarwal et al. (2001) show that stability is influenced by the employed distance measure and conclude that the Manhattan distance is more preferable than the Euclidean distance for high-dimensional data.

In this paper, we study the stability of neighbourhoods across feature reductions. In particular, we investigate the question whether there are differences in the stability of the close neighbours and the neighbours that are farther away. This investigation explores the preservation of distances when features are removed from classification data in order to reduce that data from a high-dimensional space to a low-dimensional subspace. For certain distance based classifiers, this allows for using a low-dimensional feature space for classification.

In the first part, we empirically examine the neighbourhood structure of artificial and real-world data sets. In the second part, we characterize the stability of neighbour relations theoretically.

## 2  Definitions

The following introduces some central terms to describe neighbourhood relations. We define neighbourhoods according to the Manhattan distance of two $d$-dimensional points $p_1$ and $p_2$:

$$\|p_1 - p_2\| = \sum_{i=1}^{d} |p_1^{(i)} - p_2^{(i)}|$$

The *neighbourhood index* of a point $p_i$ in a set of points $P = \{p_1, \ldots, p_n\}$ with respect to another point $p$ is

$$k_p(p_i) = \text{rank}_D(d_i)$$

where $D = \{d_1, \ldots, d_n\} = \{\|p - p_1\|, \ldots, \|p - p_n\|\}$ is the set of distances of point $p$ to the points in $P$. $\text{rank}_D(d_i)$ is the rank of distance $d_i$ in $D$, with $D$ sorted in increasing order. Ties are broken by taking the average rank. A *k-th neighbour* of a point $p$ is a point $p_i$ with $k_p(p_i) = k$.

## 3  Empirical Evaluation of Neighbourhood Stability

We first assess the stability of neighbourhoods in microarray data. We employ the breast cancer data set by West et al. (2001) comprising $d = 7,129$ genes and $n = 49$ samples in two classes (25 tumors positive to estrogen receptor, 24 tumors negative to estrogen receptor).

The experimental design is as follows: For each pair of points $(p, q)$, we determine the neighbourhood index $k_p(q)$ of $q$ with respect to $p$. We now reduce the data set by $\Delta = 1, \ldots$ features using a simple correlation-based ranking: For each feature, the Pearson correlation to the class label is measured. The $\Delta$ features with the lowest correlation are removed. In the reduced data set, we again determine the neighbourhood indices $k_{p'}(q')$ of all pairs of points $(p', q')$. By comparing the values of $k_p(q)$ with the corresponding values of $k_{p'}(q')$, we can analyse how the order of the neighbours changes in the reduced data sets.

The results are shown in Fig. 1. In this figure, the original neighbourhood index $k_p(q)$ is plotted on the $x$ axis, and the neighbourhood index in the reduced data set $k_{p'}(q')$ is plotted on the $y$ axis. The $z$ axis indicates how many of the pairs of data points with an original neighbourhood index $k_p(q)$ have a neighbourhood index of $k_{p'}(q')$ after feature reduction. Hence, a high $z$ value on the diagonal indicates that the neighbourhood relations are stable.

Panel A shows a data set reduced by 500 features. Apparently, data point pairs with a small or large value of $k_p(q)$ are more likely to have the same neighbourhood relation $k_{p'}(q')$ in the reduced data set than data points with an intermediate value of $k_p(q)$: in the middle of the plot, the neighbourhood positions deviate stronger from the diagonal. In other words, the nearest and the farthest neighbours are more stable than other neighbours. For Panel B – with 1,000 eliminated features – this effect is even stronger: the intermediate neighbours change their neighbourhood indices in the majority of cases, whereas the nearest and farthest neighbours mostly remain unchanged. Panels C and D show that neighbourhood relations are more and more disrupted when the majority of features (around 70% for Panel C and around 90% for Panel D) is removed. Still, the nearest and the farthest neighbours are most stable.

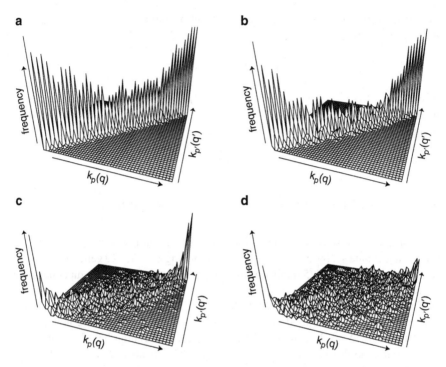

**Fig. 1** Changes of the neighbourhoods in the West dataset reduced by 500 features (Panel A), 1,000 features (Panel B), 5,000 features (Panel C), and 6,500 features (Panel D). The original neighbourhood index $k_p(q)$ is plotted on the $x$ axis, and the neighbourhood index in the reduced data set $k_{p'}(q')$ is plotted on the $y$ axis. The $z$ axis indicates how many of the pairs of data points with an original neighbourhood index $k_p(q)$ have a neighbourhood index of $k_{p'}(q')$ after feature reduction

To further investigate this phenomenon, we visualize the distance distributions of the data points. Figure 2 shows the distributions of the standardized distances from each of the data points to the other data points. The small local maxima at the right side of the plot belong to three outliers that have very high distances to all other points in the data set (data not shown here). The distributions of the remaining distances are similar to a normal distribution. This observation possibly explains the higher stability of the nearest and farthest neighbours: The distances of nearest and farthest neighbours are less frequent than intermediate distances, and therefore they cannot easily change their rank in the neighbourhood.

To study the described behaviour in a controlled generic scenario, we generate an artificial data set with $d = 1,000$ features and $m = 100$ samples uniformly at random. Instead of selecting features according to a specific feature reduction technique, we randomly eliminate a single feature. Figure 3 illustrates the results: Again, the nearest and farthest neighbours are preserved much better than intermediate neighbours. The closer the neighbourhood indices are to the extremes, the higher is the frequency of unchanged neighbourhood relations (i.e., values on the diagonal).

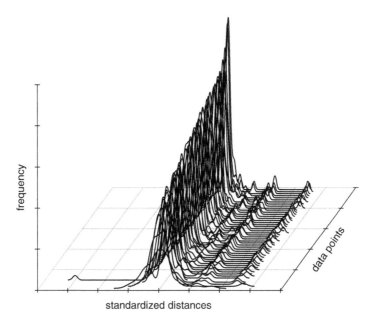

frequency

standardized distances

**Fig. 2** Histograms of the distance distributions of the data points in the West data set estimated with a Gaussian kernel. Each curve on the $y$ axis depicts the distances of one data point to all other data points. The distances are standardized by subtracting the mean distance and dividing by the standard deviation to be comparable. The $z$ axis corresponds to the frequency of the observed distances

## 4 Theoretical Assessment

Based on the previous controlled scenario, we now prove the observed fact that nearest and farthest neighbours are preserved better in a feature reduction than intermediate neighbours. First, the following lemma shows that it is less likely to find two points with similar distances to a third point if these distances are extreme.

**Lemma 1.** *Consider a $d$-dimensional feature space $S := [0, 1]^d$. Fix a point $p \in S$, and choose $n$ points $P = \{p_1, \ldots, p_n\}$ in $S$ i.i.d. at random. Let $d_1, \ldots, d_n$ be the Manhattan distances $\|p - p_i\|, i = 1, \ldots, n$. Then, finding two points $q, r \in P$ with $|\|p - q\| - \|p - r\||$ less than a fixed $\epsilon$ is less likely if these distances are extremely small or extremely large compared to the mean distance.*

*Proof.* Since the features are i.i.d. and have limited variance, the distances $d_i$ are approximately normally distributed with mean $\mu$ and standard deviation $\sigma$ according to the Central Limit Theorem. Choose a second point $q \in P$, and let $d_q$ be the distance between $p$ and $q$.

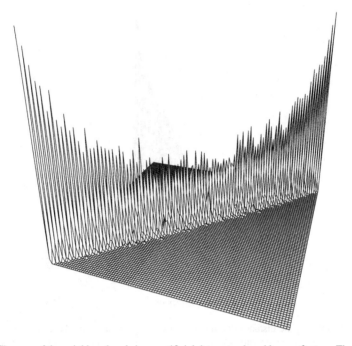

**Fig. 3** Changes of the neighbourhoods in an artificial data set reduced by one feature. The original neighbourhood index $k_p(q)$ is plotted on the $x$ axis, and the neighbourhood index in the reduced data set $k_{p'}(q')$ is plotted on the $y$ axis. The $z$ axis indicates how many of the pairs of data points with an original neighbourhood index $k_p(q)$ have a neighbourhood index of $k_{p'}(q')$ after feature reduction

Then the probability that a point $r \in P$ has a similar distance $d_r \in [d_q - \epsilon, d_q + \epsilon]$ is

$$Pr(|d_r - d_q| < \epsilon) \approx \frac{1}{\sigma\sqrt{2\pi}} \int\limits_{d_q - \epsilon}^{d_q + \epsilon} e^{-\frac{1}{2}\left(\frac{x-\mu}{\sigma}\right)^2} dx$$

This probability is monotonically decreasing in $|d_q - \mu|$. The probability that *at least one* of the points in $P$ has a similar distance $d_i \in [d_q - \epsilon, d_q + \epsilon]$ is

$$Pr(\exists p_i : |d_i - d_q| < \epsilon) = 1 - \prod_{i=1}^{n}\left(1 - Pr(|d_i - d_q| < \epsilon)\right)$$

Since $Pr(|d_i - d_q| < \epsilon)$ is monotonically decreasing in $|d_q - \mu|$, the total probability $Pr(\exists p_i : |d_i - d_q| < \epsilon)$ is monotonically decreasing in $|d_q - \mu|$ as well. That is, finding two points that have a similar distance to a third point is less likely if these distances are extremely small or extremely large.

Using the lemma, we can prove our empirical results:

**Theorem 1.** *Fix three points* $p, q, r \in S$, *and let* $d_q$ *and* $d_r$ *be the distances between* $p$ *and* $q$ *or* $r$ *respectively. Let* $d_r - d_q = \delta > 0$, *i.e.* $k_p(r) > k_p(q)$. *We apply a feature reduction* $\mathcal{F}_j : S \to [0,1]^{d-1}$ *that eliminates a random feature* $j$. *If* $d_q$ *is extremely small or extremely large, the probability that* $k_p(r) < k_p(q)$ *after eliminating one feature is smaller than in the case where* $d_q$ *has an average value.*

*Proof.* The distances in the reduced space are

$$\tilde{d}_q = \sum_{i \neq j} |p^{(i)} - q^{(i)}| = d_q - d_q^{(j)} \text{ and}$$

$$\tilde{d}_r = \sum_{i \neq j} |p^{(i)} - r^{(i)}| = d_r - d_r^{(j)}$$

The probability that the feature reduction leads to $\tilde{d}_q > \tilde{d}_r$ is

$$Pr(\tilde{d}_q > \tilde{d}_r) = Pr(d_q^{(j)} - d_r^{(j)} > \delta)$$

This probability decreases with increasing $\delta$, independent of the distribution of $q$ and $r$. Combining this with Lemma 1, we can conclude that for a point $q$ with an extremely small or large distance to $p$, it is less likely that any other point $r$ with a higher distance to $p$ gets closer than $q$ after eliminating one feature.

# 5 Discussion

In this paper, we examined the neighbourhood preservation properties of feature reduction empirically and theoretically. Although the assumption of uniformly distributed features in our artificial data may be an oversimplification, we observed that the derived assumption of normally distributed distances is often valid for real-world microarray data sets. Even for non-uniformly distributed features, the Central Limit Theorem still holds as long as the features are i.i.d. If features are highly correlated, Burghouts et al. (2008) show that distances are Weibull-distributed which approaches the normal distribution for certain values of the shape parameter $\gamma$.

In our theoretical assessment, we consider the case of eliminating a single feature. As can be seen from our experiments with real data (Fig. 1), removing multiple features shows a similar distribution of neighbourhood changes, with the probability of a change increasing with the number of eliminated features.

Our experiments are based on the simplifying assumption that all features are relevant. In real data sets, usually not all features are relevant. As a consequence, it is not always possible to define which are the "true" neighbourhood relations: The neighbourhood relations in the full data set may be influenced by irrelevant and noisy features. In this context, preserving neighbourhoods must be defined according to the unknown set of relevant features.

Our empirical and theoretical results indicate that nearest and farthest neighbours are more reliably preserved than other neighbours in a reduced feature set. It may be helpful to consider this fact when designing machine learning systems. For example, this may justify preferring the 1-Nearest Neighbour classifier over $k$-Nearest Neighbour classifiers with $k > 1$ in certain contexts.

# References

Aggarwal CC, Hinneburg A, Keim DA (2001) On the surprising behavior of distance metrics in high dimensional spaces. In: Proceedings of the 8th International Conference on Database Theory, Springer, London, UK, pp 420–434

Beyer K, Goldstein J, Ramakrishnan R, Shaft U (1999) When is "nearest neighbor" meaningful? In: Proceedings of the 7th International Conference on Database Theory. Springer, London, UK, pp 217–235

Burghouts G, Smeulders A, Geusebroek JM (2008) The distribution family of similarity distances. In: Platt JC, Koller D, Singer Y, Roweis S (eds) Advances in neural information processing systems 20, MIT Press, Cambridge, MA, USA, pp 201–208

Guyon I, Elisseeff A (2003) An introduction to variable and feature selection. J Mach Learn Res 3:1157–1182

Hinneburg A, Aggarwal CC, Keim DA (2000) What is the nearest neighbor in high dimensional spaces? In: Proceedings of 26th International Conference on Very Large Data Bases. Morgan Kaufmann, San Francisco, CA, USA, pp 506–515

Johnson WB, Lindenstrauss J (1984) Extensions of Lipshitz mapping into Hilbert space. Contemp Math 26:189–206

Kohonen T (1989) Self-organization and associative memory, 3rd edn. Springer, Berlin, Germany

MacQueen JB (1967) Some methods for classification and analysis of multivariate observations. In: Proceedings of the 5th Berkeley Symposium on Mathematical Statistics and Probability, University of California Press, Berkeley, CA, USA, vol 1, pp 281–297

Radovanović M, Nanopoulos A, Ivanović M (2009) Nearest neighbors in high-dimensional data: The emergence and influence of hubs. In: Proceedings of the 26th International Conference on Machine Learning, Omnipress, Madison, WI, USA, pp 865–872

Ripley BD (1996) Pattern recognition and neural networks. Cambridge University Press, Cambridge, UK

Vapnik V (1998) Statistical learning theory. Wiley, Chichester, GB

West M, Blanchette C, Dressman H, Huang E, Ishida S, Spang R, Zuzan H, Olson JA, Marks JR, Nevins JR (2001) Predicting the clinical status of human breast cancer by using gene expression profiles. PNAS 98(20):11462–11467

# Part VIII
# Statistical Musicology

# Musical Instrument Recognition by High-Level Features

**Markus Eichhoff and Claus Weihs**

**Abstract** In this work different high-level features and MFCC are taken into account to classify single piano and guitar tones. The features are called high-level because they try to reflect the physical structure of a musical instrument on temporal and spectral levels. Three spectral features and one temporal feature are used for the classification task. The spectral features characterize the distribution of overtones and the temporal feature the energy of a tone. After calculating the features for each tone classification by statistical methods is carried out. Variable selection is used and an interpretation of the selected variables is presented.

## 1 Introduction

What characterizes the sound of a musical instrument? Because pitch and loudness are not specific enough to discriminate between single tones of different instruments it is important to look at timbre represented by the distribution of overtones in periodograms. This distribution depends on the physical structure of the musical instrument (Fletcher 2008; Hall 2001). On the other side the energy of every single tone has got a temporal envelope that differs from one musical instrument to the other and is therefore considered in this work. In total four groups of features are taken into account: A spectral envelope derived from linear predictor coefficients (LPC), Mel-frequency Cepstral Coefficients (MFCC), Pitchless Periodogram (short: Chroma) and an Absolute Amplitude Envelope. MFCCs have already shown to be useful for classification tasks in speech processing (Rabiner and Juang 1993) as well as in musical instrument recognition (Krey and Ligges 2009; Wold et al. 1999; Brown 1999).

M. Eichhoff (✉) · C. Weihs,
Chair of Computational Statistics, TU Dortmund, Germany
e-mail: eichhoff@statistik.tu-dortmund.de; weihs@statistik.tu-dortmund.de

W. Gaul et al. (eds.), *Challenges at the Interface of Data Analysis, Computer Science, and Optimization*, Studies in Classification, Data Analysis, and Knowledge Organization, DOI 10.1007/978-3-642-24466-7_38, © Springer-Verlag Berlin Heidelberg 2012

**Fig. 1** ADSR-curve of a
musical signal

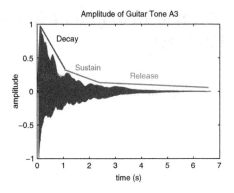

**Fig. 2** Blocks of a tone

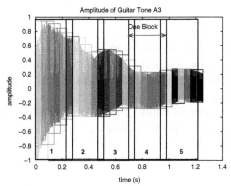

Each piano or guitar tone is windowed by half-overlapping segments $w_s$, $s \in \{1, \ldots, 25\}$ of a size of 4,096 samples. It can be divided into four phases: Attack, decay, sustain and release (see Fig. 1). Correspondingly, blocks of five consecutive segments each are fixed (see Fig. 2) and the block-features are calculated on each block.

## 2 Groups of Features

Each tone consists of an audio signal $x[n]$, $n \in \{1, \ldots 52,920\}$, the first 1.2 s are used for the calculation of the feature vectors. The signal has a sampling rate of $sr = 44,100$ Hz. If not denoted separately, the calculation of the features is carried out for each of the five consecutive time windows by calculating the median of the latters component-by-component.

### 2.1 Absolute Amplitude Envelope

Taking the upper and lower shape of the energy envelope of a tone into account the absolute values $|x[n]|$ define the so-called *Absolute Amplitude Envelope* $e \in IR^{1 \times 132}$

**Fig. 3** Musical signal and its envelope

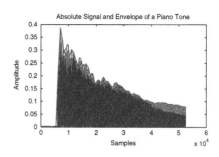

as follows:

$$e = \left( \max_{1 \le i \le 400} \{|x[i]|\}, \max_{401 \le i \le 800} \{|x[i]|\}, \ldots, \max_{l-399 \le i \le l} \{|x[i]|\} \right),$$

with $l = \lfloor \frac{N}{400} \rfloor \cdot 400$.

Calculation was done by using non-overlapping frames of size 400. A visualization of the Absolute Amplitude Envelope of one guitar tone A3 you see in Fig. 3.

## 2.2 Pitchless Periodogram

The Pitchless Periodogram (see Weihs and Ligges 2003) describes the distribution of overtones of a tone. It is based on the *periodogram*. For a given sequence of samples $X \equiv x[j]$, $j = 1, \ldots, M$ the *periodogram* $P_X$ is defined as

$$P_X(k) = \frac{1}{N} \left| \sum_{j=0}^{N-1} x[j] e^{-2\pi i \frac{k}{N} j} \right|^2,$$

$k = 1, \ldots, N/2$. $N$ is the number of time samples and $k/N$ are called Fourier frequencies.

At first the ground-frequencies $\hat{f}_0^{w_s}$ are estimated per window $w_s$, $s \in \{1, \ldots, 25\}$ and as described at the beginning of the second section the estimated block-ground-frequencies $\hat{f}_0^{b \lceil \frac{s}{5} \rceil}$ and block-overtones $\hat{f}_i^{b \lceil \frac{s}{5} \rceil}$ are calculated as

$$\hat{f}_0^{b \lceil \frac{s}{5} \rceil} = \text{median} \left( \hat{f}_0^{w_r}, \hat{f}_0^{w_{r+1}}, \ldots, \hat{f}_0^{w_{r+4}} \right) \text{ and } \hat{f}_i^{b \lceil \frac{s}{5} \rceil} = (i+1) \hat{f}_0^{b \lceil \frac{s}{5} \rceil},$$

$r \in \{1, 6, 11, 16, 21\}, i \in \{0, 1, \ldots, 13\}$.

After calculating the thirteen block-overtones and block ground-frequencies, the *pitchless periodogram* $p \in I\!R^{70}$ can be calculated. It is defined as

**Fig. 4** Pitchless periodogram

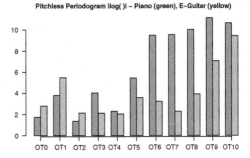

$$p = \left( p_1^{k_0}, p_1^{k_1}, \ldots, p_1^{k_{13}}, p_2^{k_0}, \ldots, p_2^{k_{13}}, \ldots, p_5^{k_0}, \ldots, p_5^{k_{13}} \right),$$

$$p_{\lceil \frac{r}{5} \rceil}^{k_i} := \operatorname{median} \left( P_{x_{w_r}}(k_i), P_{x_{w_{r+1}}}(k_i), \ldots, P_{x_{w_{r+4}}}(k_i) \right),$$

with $k_i$ defined by

$$\left| \hat{f}_i^{b \lceil \frac{r}{5} \rceil} - k_i / 4096 \cdot sr \right| = \min_{1 \le j \le 2048} \left| \hat{f}_i^{b \lceil \frac{r}{5} \rceil} - j / 4096 \cdot sr \right|,$$

$i \in \{0, 1 \ldots, 13\}, r \in \{1, 6, 11, 16, 21\}$.

The name "pitchless" is chosen because the pitch does not appear in the figure anymore and the distances between the peak events of the periodograms have vanished.

In Fig. 4 you see the Pitchless Periodogram of one piano and one guitar tone (1st block) with nine overtones. A log-transformation $|log(p)|$ of the original pitchless periodogram feature vector is carried out to get a better visualization.

## 2.3 Mel-Frequency Cepstral Coefficients

At first a preemphasis

$$y[n] = x[n] - 0.97 \cdot [n-1]$$

is applied on the signal $x$ to boost the higher harmonics of the sound. Then a windowed Fourier Transformation of $y[n]$ is calculated and the power spectrum $P(f)$ is derived. $f$ are the so-called Fourier Frequencies.

Each Fourier Frequency is transformed by a Mel-scale transformation

$$M(f) = 2595 \log_{10}(1 + f/700)$$

that leads to the power spectrum $P(M)$ depending on the Mel-frequencies $M(f)$. The warped power spectrum $P(M)$ is then convolved with the triangular band-pass filter to $\theta(M)$ with $X_k = \ln(\Theta(M_k)), k = 1, \ldots, K$. The MFCC are the result of the application of the Discrete Cosine Transformation

$$MFCC(d) = \sum_{k=1}^{K} X_k \cos[d(k - 0.5)\pi/K], \ d = 1, \ldots, D.$$

Calculation is done by using the MIRToolbox (Lartillot and Toiviainen 2007). See Fig. 6 for an example of 16 MFCCs of the guitar tone A3.

## 2.4 LPC Simplified Spectral Envelope

For each time window the coefficients of a $p$th-order linear predictor (FIR filter) (see Jackson 1989, pp. 255–257) are calculated with $p = \lfloor 2 + sr/1000 \rfloor = 46$ (rule of thumb for formant estimation). So the current value of the signal $x[n]$ in segment $k$ can be estimated by the past samples:

$$\hat{x}^k(n) = -a_2^k x^k(n - 1) - a_3^k x^k(n - 2) - \cdots - a_{p+1}^k x^k(n - p).$$

The 512-points complex frequency response vector $H$ of the filter can be interpreted as the transfer function evaluated at $z = e^{i\omega}$:

$$H^k(e^{i\omega}) = \left( \sum_{l=1}^{p+1} a_l^k e^{-i\omega l} \right)^{-1}, \ k \in \{1, \ldots, 25\}, \ a_1^k = 1$$

where $a_l^k$ are the linear predictor coefficients. This frequency response is calculated for each time window $k$ and so yields a matrix $K \in IR^{512 \times 25}$, with $K_{\cdot,j} = 20 \log_{10} |H^j|, \ j \in \{1, \ldots, 25\}$. With $r \in \{1, 6, 11, 16, 21\}$ define

$$v^r := \text{median} \left( K_{\cdot,r}, K_{\cdot,r+1}, K_{\cdot,r+2}, K_{\cdot,r+3}, K_{\cdot,r+4} \right).$$

This yields $V = \left( v^1, v^6, v^{11}, v^{16}, v^{21} \right) \in IR^{512 \times 5}$. The *Simplified LPC Spectral Envelope* $s \in IR^{1 \times 125}$ is then the maximum of each subsequent 20 rows of $V$:

$$s = \left( \max_{1 \le j \le 20} \{V_{j,1}\}, \max_{21 \le j \le 40} \{V_{j,1}\}, \ldots, \max_{501 \le j \le 512} \{V_{j,1}\}, \right.$$

$$\vdots$$

$$\left. \max_{1 \le j \le 20} \{V_{j,21}\}, \max_{21 \le j \le 40} \{V_{j,21}\}, \ldots, \max_{481 \le j \le 501} \{V_{j,21}\} \right).$$

See Fig. 5 for an example of the Simplified LPC Spectral Envelope of the guitar tone A3.

**Fig. 5** LPC simplified
spectral envelope

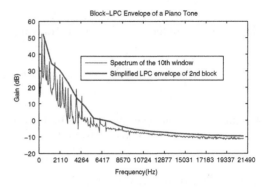

**Fig. 6** MFCCs

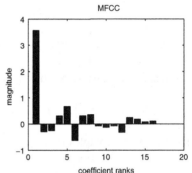

## 3   Classification and Evaluation

Classification was carried out using 270 guitar and 275 piano tones to train the
models and 4,309 guitar and 1,345 piano tones for evaluation. The training was
done by using a 10-fold cross-validation.

In some cases variable selection is used. For each iteration of the cross-validation
a logistic model is calculated and by stepwise forward selection those variables are
selected that minimize the (AIC)-criterion (Akaike 1974). This procedure is done
for three randomly built 10-fold cross-validations. ($V \geq 5$) means: Those variables
are used that have been chosen at least five of ten times.

The tones were taken from the McGill University Master Samples (McGill 2010),
RWC database (Goto et al. 2003) and music instrument samples of Electronic Music
Studios, Iowa (University of Iowa 2010).

In Tables 1 and 2 the evaluation results (mean misclassification error) of the
classification with the eight different statistical classification methods like linear or
multivariate discriminant analysis (LDA, MDA), logistic regression (LOGREG),
decision trees (RPART), support vector machines (SVM), boosting, $k$NN and
random forests are shown. Only the last five methods have hyperparameters. The
mean misclassification error is then related to the best performance, that means
the hyperparameter combination that leads to the lowest mean misclassification

**Table 1** Evaluation results (mmce) in % with or without variable filtering

| Dimensions: | 70 | 276 | 212 | 6 | 6 |
|---|---|---|---|---|---|
| Methods | P | PLM | PLM (Mnw) | PLM ($V \geq 9$) | PLM (Mnw, $V \geq 8$) |
| LDA | 41.26 | 13.06 | 12.25 | 7.05 | 4.83 |
| MDA | 35.08 | 14.17 | 12.98 | 8.14 | 6.46 |
| LOGREG | 25.80 | 17.01 | 19.69 | 7.12 | 4.87 |
| RPART | 79.06 | 12.25 | 10.80 | 12.25 | 11.24 |
| SVM | 79.06 | 59.25 | 50.05 | 6.44 | **3.78** |
| ADABOOST | **16.66** | **3.50** | **1.72** | **6.3** | 4.60 |
| KNN | 26.70 | 16.03 | 18.63 | 8.57 | 6.41 |
| RANDOM FOREST | 17.85 | 4.82 | 4.15 | 6.05 | 5.87 |

P = Pitchless Periodogram, L = LPC (downsampled to 22,050 Hz), M = MFCC, Mnw = MFCCs not windowed

**Table 2** Classification results by using sequential forward selection

| Methods | PLM | Dim. | PLM (Mnw) | Dim. |
|---|---|---|---|---|
| LDA (sfs) | **6.75** | 7 | 6.40 | 6 |
| LOGREG (sfs) | 7.56 | 7 | 5.05 | 7 |
| ADABOOST (sfs) | 7.12 | 6 | **3.35** | 10 |

sfs = sequential forward selection

error of the 10-fold cross-validation. The computational calculation was done with MATLAB (Lartillot and Toiviainen 2007) and the R-packages tuneR and mlr (Ligges 2010; Bischl 2010).

As one can see in Table 1 the results get better by using additional features to the Pitchless Periodogram. In the non-windowed version the combination PLM shows the best result with the Adaboost method. It can also be seen clearly that the variable filtering reduces the error rate by ca. 50 % in case of some methods like LDA or LOGREG. The error rates of more than 20% can be explained by the fact that the number of training observations (545) is too small for this method. Table 2 shows the results when using sequential forward selection, i.e. when including feature selection in the chosen classification method.

# 4 Interpretation of Selected Features

In the following you see four examples of the selected features (name_blocknumber_featurenumber):

- PLM (Mnw, $V \geq 8$): lpc_block5_6, lpc_block5_8, mfcc_1, mfcc_2, chroma_block1_1stOT
- PLM ($V \geq 9$): lpc_block5_8, mfcc_block1_1, mfcc_block1_4, mfcc_block5_1, mfcc_block5_2

- PLM (Mnw), ADABOOST (sfs): mfcc_2, mfcc_1, chroma_block1_6thOT, lpc_block4_8, lpc_block5_6, lpc_block1_10, lpc_block5_11, chroma_block2_7thOT, lpc_block1_9, lpc_block4_3
- PLM, LDA (sfs): mfcc_block5_2, mfcc_block1_1, mfcc_block5_1, chroma_block4_1stOT, lpc_block5_8, lpc_block1_6, lpc_block4_24

Which conclusion can be drawn from the selected list of features?

- *Generally:* The 1st and 5th block are most important. The first block matches with the first 0.279s and contains the attack phase of the tone that is rather specific for musical instruments and therefore seems to be selected here. The fifth block starts from 1,117s. This fact shows that also the declining part of the energy (sustain or decay-phase) is useful to discriminate between piano and guitar tones.
- *MFCC:* 1st and 2nd coefficient play an important role (timbre).
- *LPC:* Frequencies $\geq$ 4,737 are not important enough for classification of piano and guitar tones of this downsampled (22,050 Hz) signal.
- *Pitchless Periodogram:* 1st, 6th and 7th overtone are often chosen.

## 5    Conclusions

As the classification of piano and guitar tones shows good results – especially in case of variable selection – it may be assumed to get good results by classifying other musical instruments like strings, piano and guitar or even additionally reed instruments. It is, however, not possible to draw a conclusion which methods may be the best ones for a given set of feature groups.

## References

Akaike H (1974) A new look at the statistical model identification. IEEE Trans Automat Contr 19(6):716–723, doi:10.1109/TAC.1974.1100705

Bischl B (2010) Machine learning in R. In: R-Package, TU Dortmund, URL http://r-forge.r-project.org/projects/mlr/

Brown JC (1999) Computer identification of musical instruments using pattern recognition with cepstral coefficients as features. J Acoust Soc Am 105(3):1933–1941

Fletcher NH (2008) The physics of musical instruments. Springer, New York

Goto M, Hashiguchi H, Nishimura T, Oka R (2003) RWC music database: Music genre database and musical instrument sound database. In: ISMIR 2003 Proceedings, pp 229–230

Hall DE (2001) Musical acoustics, 3rd edn. Brooks Cole

Jackson LB (1989) Digital filters and signal processing. 2nd edn. Kluwer Academic Publishers, Norwell, MA

Krey S, Ligges U (2009) SVM based instrument and timbre classification. In: Locarek-Junge H, Weihs C (eds) Classification as a tool for research. Springer, Berlin-Heidelberg-New York, pp 759–766

Lartillot O, Toiviainen P (2007) MIR in Matlab (II): A toolbox for musical feature extraction from audio. In: Proceedings of the 8th International Conference on Music Information Retrieval, ISMIR, pp 127–130

Ligges U (2010) tuneR: Analysis of music. http://r-forge.r-project.org/projects/tuner

McGill (2010) Mcgill master samples collection on DVD. http://www.music.mcgill.ca/resources/mums/html

Rabiner L, Juang BH (1993) Fundamentals of speech recognition. Prentice-Hall PTR, Upper Saddle River, NJ

University of Iowa (2010) Electronic music studios. Musical instrument samples. http://theremin.music.uiowa.edu

Weihs C, Ligges U (2003) Voice prints as a tool for automatic classification of vocal performance. In: Kopiez R, Lehmann AC, Wolther I, Wolf C (eds) Proceedings of the 5th Triennial ESCOM Conference, Hanover University of Music and Drama, Germany, pp 332–335

Wold E, Blum T, Keislar D, Wheaton J (1999) Classification, search and retrieval of audio. In: Furht B (ed) Handbook of multimedia computing. CRC Press, Boca Raton, FL, pp 207–226

# The Recognition of Consonance Is not Impaired by Intonation Deviations: A Revised Theory

Jobst Peter Fricke

**Abstract** The recognition of musical intervals is investigated comparing neuro-biological and theoretical models (Tramo et al., The Biological Foundations of Music, Annals of the New York Academy of Sciences, 930, pp. 92–116, 2001; Ebeling, Verschmelzung und neuronale Autokorrelation als Grundlage einer Konso-nanztheorie, Lang, Peter, Frankfurt/Main, 2007). The actual analyses focus on pitch tolerances of consonance identification. The mechanisms are different in models and the neurobiological process (pulse width and time latency) that the listener tolerates the deviation of the exact ratio of frequencies in the recognition of consonance. The neurobiological process is characterized by the spontaneous neural activity which is described by a Poisson distribution. Event related activities may be displayed in interspike-interval (ISI) and peri-event-time-histograms (PETH). Consonant musical intervals are characterized by periodicity in all-order ISI-histograms. This result is explained by the frequency ratio of the interval of pitches. These ISI-histograms also display subharmonics which are explainable as artifacts because of methodical issues. In contrast, the peridocity indicates the frequency of the residue. In order to adapt the model to reality, the width of the statistical distribution of the neural impulses should be considered. The spike-analysis for the recognition of periodicity is investigated on the basis of the statistical distribution and compared with the statistical results of the listener's assessment of muscial intervals. The experimental data were taken from a study dealing with the assessment of intervals in a musical context (Fricke, Classification: The Ubiquitous Challenge, pp. 585–592, Berlin, Springer, 2005).

J. P. Fricke (✉)

Department of Systematic Musicology, Institute for Musicology, University of Cologne, Germany

e-mail: alm01@uni-koeln.de

W. Gaul et al. (eds.), *Challenges at the Interface of Data Analysis, Computer Science, and Optimization*, Studies in Classification, Data Analysis, and Knowledge Organization, DOI 10.1007/978-3-642-24466-7_39, © Springer-Verlag Berlin Heidelberg 2012

# 1 Introduction

The presented theory is a revised version of the concept proposed by Licklider (1951). It is now adapted to new insights into neurobiological information processing. Licklider's model suffers from its own precision. It is based on autocorrelation analysis to provide pitch detection. Neuronal processes of coding sound information and its transfer to the cortex are characterized by variance at each level of information processing. Thus, the actual considerations on statistics in the analyses of neuronal information coding processes are implemented into the model. Autocorrelation analysis is also an appropriate tool for detecting musical consonance because these musical intervals are characterized by the periodicity of their sound signals.

Langner and Schreiner could prove in 1988 that a periodicity analysis in the auditory organ exists and that the periodicity pitch is neurally represented in the Inferior Colliculus and Cortex. Their dimension is independent of the tonotope representation and is about orthogonal to the other dimension. The periodicity of sound signals of consonant intervals could be neuronally proven by means of a periodicity analysis too (Tramo et al. 2001). The periodicity of the acoustic signals however is imperfect in music performance. Intonation deviations which are a disturbance of the periodicity are tolerated in the hearing process to a considerable extent. This can be seen particularly in the judgement of consonant intervals. Depending on the musical context, standard deviations of 13 cents (1/100 of a tempered semitone) for the optimal intonation were measured for the fourth as well as for the fifth. In total the variation was even at 70 cents. For in this range of experimentally determined hearing tolerances, the statistical process of neural coding and processing, in particular the neural integration for the autocorrelation, seems to be responsible.

# 2 Analysis of Stochastic and Periodic Neuronal Activity

Spontaneous neuronal activity of a unit is best described by a Poisson-distribution (Koch 1999, p. 353, equation(15), 2). This is based on the understanding of coding as digital (all-or-none response), yet our model also needs to take into consideration the refractory period, which limits the minimum time distance between two action potentials (spikes) to about 1 to 2 msec. For this reason, the random Poisson process including the refractory period has been computed. Representative spike train, interspike interval (ISI) histogram and power spectrum from a extracellularly recorded cell showed good concordance compared to a modeled "point process" (Koch 1999, p. 355). PETH (peri event time histograms) and ISI-histograms are useful tools to recognize the coding of events in the statistical distribution of spikes (Abeles 1994, p. 131). Beyond that, all order ISI-histograms are also suited to detect periodicities in neuronal coding (Keidel 1989, p. 180, Fig. 114), as they are comparable to an autocorrelation:

$$a_t(\tau) = \int_{-\infty}^{\infty} f(t) \cdot f(t + \tau) dt$$

If the delay lag period - as an independent variable - is varied, the result obtained is the distance between identical events, i.e. the time distances between which events repeat themselves. The result also indicates the multiples of these time intervals, because it includes the distances to the second and the third event as well. The work by Tramo et al. (2001) shows all-order ISI-histograms that were recorded while a cat was exposed to consonant and dissonant musical intervals. For that purpose, action potentials were derived at the auditory nerve of a cat from a bundle of approximately 90 neurons. The periodicity of oscillatory progress of the sound event – a criterion for consonant intervals – can clearly be recognised in the registration of spike volleys. This proves that periodically repeated stimuli patterns that correspond to the perception of a periodic pitch activate – independent of the frequency information depicted tonotopically – certain areas in the cortex corresponding to the respective repetitive frequency (periodotopic organization Langner and Schreiner 1988). Since this leads to the assumption that periodic timing functions can be distinguished from aperiodic ones, one can conclude that cats can distinguish gradual levels of consonance and dissonance.

## 3   Integration Capability of a Nerve Cell

Generally, the threshold of excitement for discharging an impulse in a nerve cell is not reached by a single impulse from a pre-synaptic cell. It takes several input impulses within the integration period to raise the electric potential of a cell to a value that is required for triggering. The integration period relates to the refractory period, which in turn can be described by the decay time of the action potential. The decay time must be taken as approximately 5 msec. Therefore, the result of the summation of the input potential is higher, the smaller the distance between the input impulses. An integration period of 1 msec can be considered as guide value for successful summation. A nerve cell has several thousand inlets and it is connected – via synapses – to just as many other cells. Every neuron in a brain contacts approximately 20,000 others and receives input signals from just as many still (Singer 2002, p. 64). Synapses are located preferentially on the cell body and at the start of the axon. In essence then, summation of synaptic inputs is spatially and temporally organized. Accordingly, two types of summation (that are not entirely independent of each other) need to be distinguished (Dudel 1990, p. 52).

Because of the summation within the integration period, the cell provides a constant coincidence check of input signals. In conjunction with run times that delay signals in various ways on different paths, time differences occur that are required for autocorrelation (see formula for autocorrelation above). If, for periodic signals, the time difference (due to delay) is identical to the duration of a period, the cell is stimulated because of the coincidental input signals and will send out a signal when the threshold has been crossed. In this case, one must take into consideration that the cell here is being stimulated (nearly) simultaneously via many synapses. Consequently, this creates the necessity for stochastic treatment of neural processes.

## 4 Precision and Ambiguity for Coding Periodic Signals

The discharge of impulses – apart from being a spontaneous activity – is based on the triggering of ganglia cells and subsequent neurons, within the beat of oscillations of the basilar membrane. The development of volleys occurs through triggering populations of neighbouring neurons. Characteristic of these volleys is that not all of the impulses are discharged simultaneously. This scattering of impulses within the volleys is the first source of ambiguity for the evaluation of periodicity (B1 in Fig. 1). Here, the distance between two successive volleys represents the period duration T of the sound signal. Instead, the distance between two impulses in two successive volleys is decisive for coincidence detection by an integration neuron (B2 in Fig. 1). The determination of coincidental impulses in two successive volleys is based on the impulses themselves. Therefore, scattering within volleys adds up during determination of period duration. The third source of ambiguity in the evaluation of periodicity lies in the coincidence control as such. As shown above, based on the temporal summation, a neuron can register synchronicity only within one time interval. It would certainly not be justified to simply add up the statistical distributions. The combination of each spike of a volley with the spike of the successive volley (superimposed on the time interval of summation) would result in greater ambiguity than has been shown in the experiment by Tramo et al. (2001, p. 94, Fig. 1 I-L). Instead, it should be assumed that the information in the neuron populations is weighted. Within these volleys, spikes from neurons that have been tuned optimally to the characteristic frequency (CF) (Langner and Schreiner 1988, pp. 1802, 1804) have more weight than those in populations along the marginal zones, which fire less frequently.

In contrast to the formula for autocorrelation in which the $\tau$-variable is located, the auditory system possesses a bank of parallel arranged correlators. The correlators, located along the cochlear duct (Fig. 2), all show a fixed lag period which

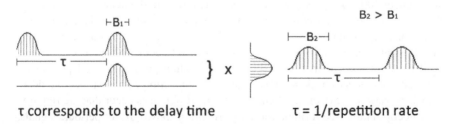

$\tau$ corresponds to the delay time                    $\tau = 1/$repetition rate

**Fig. 1** The correlation of spike volleys (containing statistically distributed spikes) with a fuzzy coincidence pattern results in an even greater variance of periodotopic representation

generates a spatially organised autocorrelation. This has significant consequences for interpreting the events of the autocorrelation. When an autocorrelator with a fixed lag time is supplied with a variable frequency, it responds to the fundamental frequency as well as to the harmonic overtones of the fundamental frequency.

$$\tau = T_1 = 2T_2 = 3T_3 = \ldots = nT_n$$

$$with \quad f_1 = f_2/2 = f_3/3 = \ldots = f_n/n,$$

$$consequently \quad f_n = nf_1, \quad n = 1, 2, 3, \ldots$$

If an autocorrelator with fixed delay time has to analyse a frequency mixture, it will be ambiguous concerning the harmonics. This ambiguity is limited in the auditory system by low-pass filtration in the cochlear duct. This is because the succession of serially-coupled low-pass filters with decreasing upper edge frequency (Fig. 2) continuously removes the upper frequency components from the traveling wave moving along the basilar membrane until only infra-sound remains at the helicotrema. Conversely, however, the periodicity of lower frequencies can be diverted at registration sites of higher frequency.

In the work of (Tramo et al. 2001, p. 101) the plot is called an all-order ISI histogram, which is equivalent to the autocorrelation of the spike train. In both of these cases, the delay lag period acts as independent variable. With a fixed input frequency, this histogram shows – just like the autocorrelation formula – the series of sub-harmonics on the axis of the variable lag period. The integral of the product will

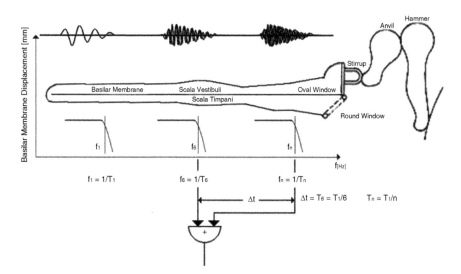

**Fig. 2** Spatially organized autocorrelation. A filterbank of parallely organized correlators is arranged along the cochlear duct. Each correlator has a fixed time delay. A sequence of low pass filters prevents the higher frequencies from arriving in the higher cochlea helix where lower frequencies are processed

then produce a maximum when the delay amounts to one period or multiples thereof. Consequently, if in the integral as well as in the ISI-histogram the lag period is the variable, the frequencies of the subtone series occur as multiples of that period. This is an artifact for which there is no analogy in the neural network. Neural connections capable of realizing a variable delay have so far not been found in the neural network of the brain. No neural connections where the run time can be varied continuously are known to exist, and it is impossible to imagine how such connection might work in any case. However, there are connections between receptors and neurons and connections among neurons along which spikes travel for different lengths of time over different pathways. These are comparable to a bank of parallel autocorrelators each of which shows fixed lag periods. Arrangements of parallel delay lines used to depict time differences with the aid of different running speeds, can be found along the cochlear duct, for instance. There, axons originating from inner hair cells are linked to each other via synapses. On the basilar membrane, traveling waves move along with a comparatively low speed of 1–3 m/sec, whereas the velocity of neural fibers must be taken as approx. 28 m/sec. The velocities of the traveling wave (place- and frequency dependent: in the beginning faster, later more slowly) can be derived from illustrations published by v. Békésy (1947) and Eldredge (1974). The different speed of the mechanical wave and the continuation of the action potential triggered by the wave that is passed on in parallel to the wave but at higher speed creates the time differences required for autocorrelation.

## 5   Experimental Setting and Results: Evidence of Statistical Distribution of Judgements

Pairs of successive chords played by either harpsichord or a group of violins were presented to music experts (choir directors, music theory teachers and conservatory students). In most cases, a dissonant chord (containing a diminished or augmented fourth, an augmented fifth and a minor sixth, respectively) resolves into a consonant chord following the rules of classical harmony. When hearing these successions of chords in which one note was varied in its frequency, the subjects were asked to decide between (a) optimal, perfectly in-tune intonation and (b) tolerated, "still barely acceptable" divergence from optimal intonation. (For more on the experimental setting see Fricke 2005, p. 587).

Consonant intervals (fifth, fourth and thirds) included in both the dissonant and the consonant chords were to be judged. These intervals were evaluated separately according to their musical context. Some examples of fifths (Figs. 3, 4) and fourths (Fig. 5) are presented here to demonstrate the extent of variation and statistical distribution.

The statistical distribution of judgements on tone intervals is a result of the (at least threefold) statistical distribution in the information processing of the neuronal network. If judgements are divided according to different musical contexts the evaluation shows two different distributions with two different means. This

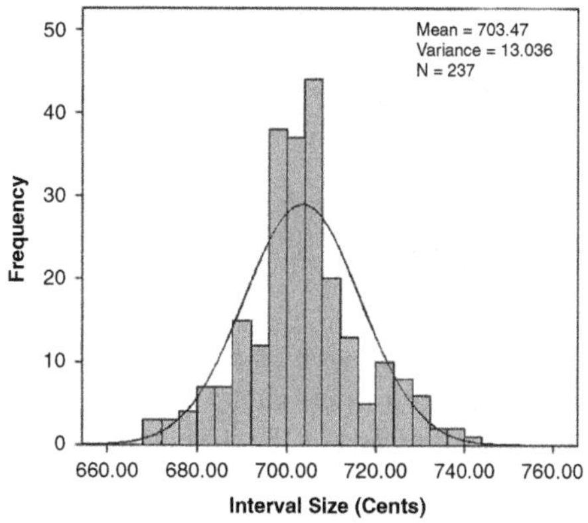

**Fig. 3** Optimal, perfectly in-tune intonation of the fifth in different musical contexts

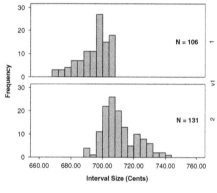

**Fig. 4** Perfectly in-tune intonation of the fifth in two different musical contexts in comparison

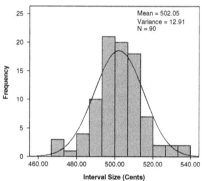

**Fig. 5** The statistical distribution of judgements of optimal, perfectly in-tune intonation of the fourth in different musical contexts

should not be possible, as the fifth, defined by the frequency relation 3/2, has the value 702 cents independently from the context.

The variance of judgements illustrated in the histograms is much larger than the hearing tolerance for pitch measured as DL's (difference limens). Furthermore, the pictures show different results for different contexts. This would not be possible if harmonic partials of the acoustic material (e.g. by producing beats and roughness as postulated in common theories of consonance) were of any importance. Instead, the judgements reflect the statistical distribution, information transformation and processing of neural spikes.

# 6  Conclusion

The limited precision of neuronal information processing implemented into Lick-lider's model for periodicity detection is responsible for a better fit of the model to the psychophysics of the human ear. It is able now to react on consonant musical intervals that are out of tune in a way we are accustomed to in music hearing. As of yet, the measured hearing tolerances have not been able to be derived from the neural processes quantitatively. But they can be explained by the statistical behavior of the neurons. The tolerances in the face of detuning that occur in the models of Tramo et al. (2001) and Ebeling (2007) are a consequence of the wide impulses used in their models. This is not sufficient for the experimentally determined tolerances. Moreover, it could be shown that beats and roughness are of no relevance concerning the judgement of intervals.

**Acknowledgements** I would like to thank Oliver Fricke for his statistical work, Christoph Reuter for his help preparing Figs. 1 and 2, and Michael Oehler for the English translation.

# References

Abeles M (1994) Firing rates and well-timed events in the cerebral cortex. In: Domany E, Hemmen J, Schulten K (eds) Models of neural networks II, Physics of neural networks. Springer, New York, pp 121–140

Békésy Gv (1947) The variation of phase along the basilar membrane with sinusoidal vibrations. J Acoust Soc Am 19:452–460

Dudel J (1990) Erregungsübertragung von Zelle zu Zelle. In: Schmidt RF, Thews G (eds) Physiologie des Menschen, Springer, Berlin, pp 45–63

Ebeling M (2007) Verschmelzung und neuronale autokorrelation als grundlage einer konso-nanztheorie. Lang, Peter, Frankfurt/Main.

Eldredge DH (1974) Inner ear-cochlear mechanics and cochlear potentials. In: Handbook of sensory physiology, vol V/1. Springer, Berlin, pp 549–584

Fricke JP (1988) Klangbreite und Tonempfindung. Bedingungen kategorialer Wahrnehmung aufgrund experimenteller Untersuchung der Intonation. In: Musikpsychologie. Jahrbuch der Deutschen Gesellschaft für Musikpsychologie, Wilhelmshaven, pp 67–87

Fricke JP (2005) Classification of perceived musical intervals. In: Weihs C, Gaul W (eds) Classification - The ubiquitous challenge. Springer, Berlin, pp 585–592

Kandel E (2006) Auf der Suche nach dem Gedächtnis. Siedler, München

Keidel WD (1989) Biokybernetik des Menschen. Wissenschaftliche Buchgesellschaft, Darmstadt

Koch C (1999) Biophysics of computation. Information processing in single neurons. Oxford University Press (Computational Neuroscience), New York

Langner G, Schreiner CE (1988) Periodicity coding in the inferior colliculus of the cat II. Topographical organization. J Neurophysiol 60(6):1823–1840

Licklider JCR (1951) A duplex theory of pitch perception. Experientia 7:128–134

Singer W (2002) Der Beobachter im Gehirn. Suhrkamp, Frankfurt/Main

Tramo MJ, Cariani PA, Delgutte B, Braida LD (2001) Neurobiological foundations for the theory of harmony in western tonal music. In: The Biological Foundations of Music, Annals of the New York Academy of Sciences, 930, pp 92–116

# Applying Multiple Kernel Learning to Automatic Genre Classification

Hanna Lukashevich

**Abstract** In this paper we demonstrate the advantages of multiple-kernel learning in the application to music genre classification. Multiple-kernel learning provides the possibility to adaptively tune the kernel settings to each group of features independently. Our experiments show the improvement of classification performance in comparison to the conventional support vector machine classifier.

## 1 Introduction

During recent years the scientific and commercial interest in automatic methods for music genre classification has tremendously increased. Stimulated by the ever-growing availability and the size of digital music collections, music genre classification has been identified as an increasingly important means to aid convenient exploration of large music catalogs. Automatic audio genre classification is commonly based on extracted acoustical feature vectors. Each feature is designed to correlate with one of the aspects of perceptual similarity, e.g. timbre, tempo, loudness, harmony etc. The distinct acoustical features are joined together into so called acoustical feature vectors. While temporal changes in one feature correspond to temporal changes in the other feature (for instance, timbre might be changing along with loudness), the individual dimensions of feature vectors can often be strongly correlated or/and redundant. Thus feature selection methods are applied to choose more informative features and to decrease feature space dimensionality.

Due to the fair performance and the global and unique solution the Support Vector Machine (SVM) (Vapnik 1998) became the state-of-the-art technique for

H. Lukashevich (✉)
Fraunhofer Institute for Digital Media Technology, Ehrenbergstr. 31, 98693 Ilmenau, Germany
e-mail: lkh@idmt.fraunhofer.de

W. Gaul et al. (eds.), *Challenges at the Interface of Data Analysis, Computer Science, and Optimization*, Studies in Classification, Data Analysis, and Knowledge Organization, DOI 10.1007/978-3-642-24466-7_40, © Springer-Verlag Berlin Heidelberg 2012

genre classification. The optimization of the SVM parameters is performed in a cross-validation scenario, where these parameters are set simultaneously for all feature dimensions. However, acoustic feature vectors often have various origin, nature and background, thus the optimal kernel parameters for different feature sets might vary significantly. In this paper we apply a recently proposed Multiple Kernel Learning (MKL) technique that has been successfully used in the fields of computational biology (Sonnenburg et al. 2005), image information retrieval (Nakajima et al. 2009), and audio tag annotation (Barrington et al. 2008). In contrast to classic SVMs, MKL provides a possibility of weighting over different kernels depending on a feature set. We compare the results obtained in a MKL scenario to the classification results achieved in a state-of-the-art genre classification system.

## 2   Mathematical Background

In a music genre recognition task we consider the dataset $(\mathbf{x}_i, y_i)$ of data samples $\mathbf{x}_i$ and corresponding labels $y_i$, where $i = 1, \ldots, l$ and $l$ is a number of data samples.

### 2.1   Support Vector Machine

SVM is a binary discriminative classifier, attempting to generate an optimal decision plane between feature vectors of the training classes (Vapnik 1998). Given the training vectors $\mathbf{x}_i \in R^d$, $i \in [l]$ in $d$-dimensional feature space, the model is estimated in the following way:

$$\min_{\mathbf{w}, b, \xi} \quad \frac{1}{2} \|\mathbf{w}\|_2^2 + C \sum_{i=1}^{l} \xi_i \tag{1}$$

$$\text{subject to} \quad y_i \left( \mathbf{w}^T \phi(\mathbf{x}_i) + b \right) \geq 1 - \xi_i, \ \xi_i \geq 0,$$

where $\|\cdot\|_2$ denotes the $l^2$ norm, $b$ specifies the distance from the decision hyperplane $\mathbf{w}$ to the origin, $\xi_i$ are slack variables, and $C > 0$ is a cost parameter. When system (1) is solved, classification is conducted according to the following decision function: $\text{sign}(\sum_{i=1}^{l} y_i \alpha_i K(\mathbf{x}_i, \mathbf{x}) + b)$, where $K(\mathbf{x}_i, \mathbf{x}_j) \equiv \phi(\mathbf{x}_i)^T \phi(\mathbf{x}_j)$ is a kernel function, equivalent to a dot product of the feature vectors mapped into the higher dimensional space; and $\alpha$ is a vector with Lagrange multipliers, needed to solve (1). Here we use the most common type of kernel, namely Radial Basis Function (RBF): $K(\mathbf{x}_i, \mathbf{x}_j) = \exp\left(-\gamma \|\mathbf{x}_i - \mathbf{x}_j\|^2\right)$.

## 2.2 Multiple Kernel Learning

The ideas of MKL have been firstly proposed by Lanckriet et al. (2004) and extended to a large scale solution[1] by Sonnenburg et al. (2006). Let $K_1(\mathbf{x}_i, \mathbf{x}_j), \ldots, K_M(\mathbf{x}_i, \mathbf{x}_j)$ be $M$ kernel matrices obtained from different features or with different kernel settings, MKL extends the regular SVM by learning a linear mixture of kernels: $K(\mathbf{x}_i, \mathbf{x}_j) = \sum_{m=1}^{M} \beta_m K_m(\mathbf{x}_i, \mathbf{x}_j)$ with $\beta_m \geq 0$ and $\sum_{m=1}^{M} \beta_m = 1$. The model is modified as follows:

$$\min_{\mathbf{w}_m, b, \xi} \quad \frac{1}{2} \left( \sum_{m=1}^{M} \|\mathbf{w}_m\|_2 \right)^2 + C \sum_{i=1}^{l} \xi_i \tag{2}$$

$$\text{subject to} \quad y_i \left( \sum_{m=1}^{M} \mathbf{w}_m^T \phi_m(\mathbf{x}_i) + b \right) \geq 1 - \xi_i, \ \xi_i \geq 0,$$

with a decision function defined as: $\text{sign}(\sum_{i=1}^{l} y_i \alpha_i \sum_{m=1}^{M} \beta_m K_m(\mathbf{x}_i, \mathbf{x}) + b)$.

## 3 Evaluation Setup

In this section we provide some details on the evaluation setup. First of all, we describe the musical dataset involved in the experiments. Afterwards, we briefly introduce audio features used for compact and informative representation of audio tracks.

In our experiments we use the ISMIR2004 Audio Description Contest Dataset[2]. This data-set includes 1,422 music tracks that are manually subdivided into 6 genre categories: Classical, Electronic, Jazz&Blues, Metal&Punk, Rock&Pop, and World. The dataset is split in two about equal parts to form the training and test sets.

## 3.1 Audio Features

We utilize a broad palette of low-level acoustic features and several mid-level representations (Bello and Pickens 2005). These mid-level features are computed on 5.12 s excerpts and observe the evolution of the low-level features. With the

---

[1] Available online: www.shogun-toolbox.org

[2] Available online: http://ismir2004.ismir.net/genre_contest/index.htm

help of mid-level representations, timbre texture (Tzanetakis and Cook 2002) can be captured by descriptive statistics as well as by including additional musical knowledge. To facilitate an overview the audio features are subdivided to three categories by covering the timbral, rhythmic and tonal aspects of sound.

Although the concept of *timbre* is still not clearly defined with respect to music signals, it proved to be very useful for automatic music signal classification. To capture timbral information, we use Mel-Frequency Cepstral Coefficients (MFCC), Spectral Crest Factor (SCF), Audio Spectrum Centroid (ASC), Spectral Flatness Measurement (SFM), Octave-based Spectral Contrast (OSC), Zero-Crossing Rate (ZCR), Normalized Loudness and Log-Loudness. In addition, modulation spectral features (Atlas and Shamma 2003) are extracted from the aforementioned features to capture their short term dynamics. We applied a cepstral low-pass filtering to the modulation coefficients to reduce their dimensionality and to decorrelate them as described in (Dittmar et al. 2007).

All *rhythmic* features used in the current setup are derived from the energy slope in excerpts of the different frequency-bands of the Audio Spectrum Envelope feature. These comprise the Percussiveness (Uhle et al. 2003) and the Envelope Cross Correlation (ECC). Further mid-level features (Dittmar et al. 2007) are derived from the Auto-Correlation Function (ACF). In the ACF, rhythmic periodicities are emphasized and phase differences annulled. Thus, we compute also the ACF Cross-Correlation (ACFCC). The difference to ECC again captures useful information about the phase differences between the different rhythmic pulses. In addition, the log-lag ACF and its descriptive statistics are extracted according to (Gruhne et al. 2009).

*Tonality* descriptors are computed from a Chromagram based on Enhanced Pitch Class Profiles (EPCP) (Lee 2006).The EPCP undergoes a statistical tuning estimation and correction to account for tunings deviating from the equal tempered scale. Pitch-space representations as described in (Gatzsche et al. 2007) are derived from the Chromagram as mid-level features. Their usefulness for audio description has been shown in (Gruhne and Dittmar 2009).

Classifying music tracks that are described with a set of audio features having different time resolution still remains a challenging task. To tackle this issue, we model each feature dimension of one song following so called "bag-of-frames" approach (Aucouturier et al. 2007). Here feature values for each dimension are modeled by a single Gaussian, so that each feature dimension within a song is represented by the sample mean and standard deviation of the feature values. In addition, for each dimension of low-level and mid-level features we calculate the differences between the neighbor frames. This forms so called *delta* features that have already proved their efficiency for MFCCs. In addition, each feature dimension is normalized by inter quantile range. All in all, each music track is represented with a feature vector having 2296 feature dimensions.

To provide an illustrative example, Fig. 1 presents the box plots for the mean over the first dimension of the Log-Loudness feature. This feature can be roughly interpreted as a signal energy in the low frequency band. This frequencies are hardly involved for classical music, but become inevitable for metal and punk.

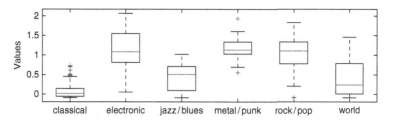

**Fig. 1** Box plots for the first dimension of Log-Loudness feature for all 6 classes

### 3.1.1 Feature Selection

Not all of the acoustical feature vectors are equally informative for the automatic genre classification task. In this work we utilize a univariate filter feature selection method called Inertia Ratio Maximization using Feature Space Projection (IRMFSP) firstly proposed by Peeters and Rodet (2003). This feature selection algorithm is motivated by the ideas similar to Fisher's discriminant analysis. On each iteration of the algorithm we look for the feature maximizing the ratio of between-class inertia to the total-class inertia. To avoid that on the next iteration the next chosen feature could bring the same information, all features are orthogonalized to already selected ones. The algorithm can be stopped, when the desired number of features is chosen, or when the relative change of observed inertia ratio fulfills predefined conditions. In this evaluation we use the IRMFSP algorithm with the modifications proposed in (Essid 2005). Each of the experiments is run over the following quantities of features: 16, 32, 64, 128, 256, 512, 1,024, and 2,048. This log-line scale is used considering that improvement is more significant when a small number of features is used.

### 3.1.2 Groups of features

As described in Sect. 2.2 Multiple-Kernel Learning provides a possibility to tune the kernel settings independently for several feature groups. In contrast to (Barrington et al. 2008) we do not consider every feature as a single source of information and thus do not construct a separate kernel for each feature in the MKL scenario. We form four distinct groups of features which are later used in Experiment 3 of Sect. 3.2. Groups of features are set up based on some heuristics on the musical information impact of the features and on the statistical properties of feature dimensions. First two groups cover timbral features: the first one includes SFM, MFCC and ASC; the second one comprises SCF, OSC, NormLoud and LogLoud. The other two groups include rhythmical and tonality features correspondingly. The groups are not well-balanced according to the number of feature dimensions. The feature selection described above is conducted disregarding the group information.

## 3.2   Experiments

In order to compare the results obtained in a MKL scenario to the classification results achieved in a state-of-the-art music genre classification system, we design the following three experiments:

1. *SVM, one kernel:* We apply a classical SVM with a RBF kernel. Here both kernel parameter $\gamma$ and cost parameter $C$ are optimized by a 5-fold cross-validation (CV) over the training set. One kernel with the optimized parameters is used for all features.
2. *MKL, all features:* Instead of optimizing the kernel parameter $\gamma$ in a CV scenario, here we consider the whole family of RBF kernels with various $\gamma$. The optimal parameters are chosen by finding optimal weighting factors $\beta_k$. This weighted sum of RBF family kernels is applied to all features simultaneously.
3. *MKL, grouped features:* Here we add one more degree of freedom to the classification algorithm by considering the RBF kernel family with various $\gamma$ for each group of features independently. Note, that in this case we obtain four weighted sums of RBF kernels – for each of four feature groups – that are applied only to the features in the corresponding groups.

For both conventional SVM and MKL, one-vs-one decomposition with subsequent voting is utilized to enable multi-class classification.

## 4   Results

The results of all three experiments are presented in Fig. 2. The conventional state-of-the-art one kernel SVM approach achieves the accuracy of 74% for 64 chosen feature dimensions. The lower performance of the algorithm for the higher number of feature dimensions can be explained by the noisy non-dicriminative nature of the additional feature vectors and by the restricted number of samples available for training.

In the second experiment we apply the MKL to choose an optimal kernel as a weighted combination of RBF kernels. Due to the norm on the weighting coefficients $\beta_m$ the MKL tends to assign a high value weight only to one of the sub-kernels. In fact the MKL in this case chooses the RBF kernel with the optimal $\gamma$ parameter. The performance of the algorithm is comparable to the performance of the conventional SVM.

The third experimental scenario brings the significant improvement. Here the optimal kernel settings are chosen for each feature group individually. In this case the maximal accuracy of 79% is reached for 64 chosen feature dimensions. Note, that this scenario avoids the drastic decrease of the classification accuracy for the higher number of selected features.

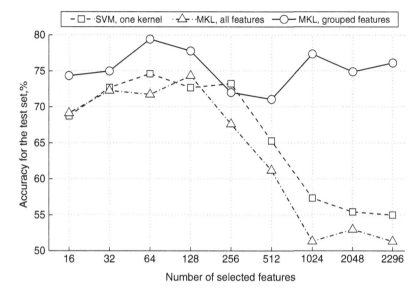

**Fig. 2** Classification accuracy (in percent) for the test part of the dataset for the three experimental scenarios as described in Sect. 3.2

## 5   Conclusions and Outlook

In this paper we applied a recently proposed Multiple Kernel Learning technique to an automatic music genre classification.

Multiple-kernel learning becomes apparent as a powerful method to improve automatic music genre classification. Individual kernels for various feature groups bring additional freedom to the classifier while making the results better interpretable. Our future work will be directed towards the choice of optimal sets of features based both on the statistical properties of involved features and their musical meaning. The feature groups in the current study are not well balanced, so additional weighting has to be incorporated in the future. Among the RBF kernel family, other custom kernels for feature similarity will be integrated.

**Acknowledgements** This work has been partly supported by the German research project *GlobalMusic2One*[3] funded by the Federal Ministry of Education and Research (BMBF-FKZ: 01/S08039B). Additionally, the Thuringian Ministry of Economy, Employment and Technology supported this research by granting funds of the European Fund for Regional Development to the project *Songs2See*[4], enabling transnational cooperation between Thuringian companies and their partners from other European regions.

---

[3] see http://www.globalmusic2one.net

[4] see http://www.songs2see.net

# References

Atlas L, Shamma S (2003) Joint acoustic and modulation frequency. EURASIP J Appl Signal Proces 7:668–675

Aucouturier JJ, Defreville B, Pachet F (2007) The bag-of-frames approach to audio pattern recognition: A sufficient model for urban soundscapes but not for polyphonic music. J Acoust Soc Am 122(2):881–891

Barrington L, Yazdani M, Turnbull D, Lanckriet G (2008) Combining feature kernels for semantic music retrieval. In: Proc. of the 9th Intl. Conf. on Music Information Retrieval (ISMIR), pp 614–619

Bello JP, Pickens J (2005) A robust mid-level representation for harmonic content in music signals. In: Proc. of the 6th Int. Conf. on Music Information Retrieval (ISMIR), London, UK, pp 304–311

Dittmar C, Bastuck C, Gruhne M (2007) Novel mid-level audio features for music similarity. In: Proc. of the Int. Conf. on Music Communication Science (ICOMCS), Sydney, Australia

Essid S (2005) Classification automatique des signaux audio-fréquences: Reconnaissance des instruments de musique. PhD thesis, l'Université Pierre et Marie Curie, Paris, France

Gatzsche G, Mehnert M, Gatzsche D, Brandenburg K (2007) A symmetry based approach for musical tonality analysis. In: Proc. of the 8th Int. Conf. on Music Information Retrieval (ISMIR), Vienna, Austria, pp 207–210

Gruhne M, Dittmar C (2009) Comparison of harmonic mid-level representations for genre recognition. In: Proc. of the 3rd Workshop on Learning the Semantics of Audio Signals (LSAS), Graz, Austria, pp 91–102

Gruhne M, Dittmar C, Gaertner D (2009) Improving rhythmic similarity computation by beat histogram transformations. In: Proc. of the 10th Int. Society for Music Information Retrieval Conf. (ISMIR), Kobe, Japan

Lanckriet G, Cristianini N, Bartlett P, Ghaoui LE, Jordan M (2004) Learning the kernel matrix with semidefinite programming. J Mach Learn Res 5:27–72

Lee K (2006) Automatic chord recognition from audio using enhanced pitch class profile. In: Proc. of the Int. Computer Music Conf. (ICMC), New Orleans, USA, pp 306–313

Nakajima S, Binder A, Müller C, Wojcikiewicz W, Kloft M, Brefeld U, Müller KR, Kawanabe M (2009) Multiple kernel learning for object classification. Tech. rep., Information-Based Induction Sciences, Fukuoka, Japan

Peeters G, Rodet X (2003) Hierarchical gaussian tree with inertia ratio maximization for the classification of large musical instruments databases. In: Proc. of the 6th Intl. Conf. on Digital Audio Effects (DAFx)., London, UK

Sonnenburg S, Rätsch G, Schäfer C (2005) Learning interpretable SVMs for biological sequence classification. In: Miyano S, Mesirov J, Kasif S, Istrail S, Pevzner P, Waterman M (eds) Research in Computational Molecular Biology, Lecture Notes in Computer Science, vol 3500, Springer, Berlin/Heidelberg, pp 389–407

Sonnenburg S, Rätsch G, Schäfer C, Schölkopf B (2006) Large scale multiple kernel learning. J Mach Learn Res 7:1531–1565

Tzanetakis G, Cook P (2002) Musical genre classification of audio signals. IEEE Trans Speech Audio Process 10(5):293–302

Uhle C, Dittmar C, Sporer T (2003) Extraction of drum tracks from polyphonic music using independent subspace analysis. In: Proc. of the 4th Int. Symposium on Independent Component Analysis (ICA), Nara, Japan, pp 843–848

Vapnik V (1998) Statistical learning theory. Wiley, New York

# Multi-Objective Evaluation of Music Classification

Igor Vatolkin

**Abstract** Music classification targets the management of personal music collections or recommendation of new songs. Several steps are required here: feature extraction and processing, selection of the most relevant of them, and training of classification models. The complete classification chain is evaluated by a selected performance measure. Often standard confusion matrix based metrics like accuracy are calculated. However it can be valuable to compare the methods using further metrics depending on the current application scenario. For this work we created a large empirical study for different music categories using several feature sets, processing methods and classification algorithms. The correlation between different metrics is discussed, and the ideas for better algorithm evaluation are outlined.

## 1 Introduction

Audio music classification belongs to the research area music information retrieval (MIR) (Downie 2008; Tzanetakis and Cook 2002). In the recent years many advanced and complex methods have been developed (Ras and Wieczorkowska 2010). The different classification approaches (see e.g. Duda et al. (2001) for a method overview) are trained on features like low-level audio signal characteristics, metadata information, or playlist analysis. Typically, standard metrics like accuracy or error rate are used for comparison between classification algorithms. However, more metrics are existing which can play important roles for different application scenarios and user preferences.

For music a balance between safety and surprise can be adjustable. Higher specificity ensures that only a few number of negative songs will be classified as

I. Vatolkin (✉)
TU Dortmund, Dortmund, Germany
e-mail: igor.vatolkin@udo.edu

W. Gaul et al. (eds.), *Challenges at the Interface of Data Analysis, Computer Science, and Optimization*, Studies in Classification, Data Analysis, and Knowledge Organization, DOI 10.1007/978-3-642-24466-7_41, © Springer-Verlag Berlin Heidelberg 2012

belonging to the corresponding category. On the other side it may remain high for a very small rate of true positives. If only tracks of the same artist will be recommended to a user, the accuracy might be indeed very high, but the playlist becomes boring and highly predictable. The exact balance of preferences are certainly dependent on a concrete person. It can be even thinkable to limit the accuracy performance to a certain level with the aim to present a listener different music from time to time. If the music data set is strongly unbalanced due to the current categorization task, some combinations of recall and specificity may provide a fair evaluation of the algorithms.

Another tradeoff is classification quality against amount of indexing disc space. Further, a large number of music features used for training may imply slower classification (e.g. on a mobile device) or the number of tracks can be simply too high. The task is to prune the data (smaller feature number, sparse analysis of song parts) while saving the acceptable classification performance.

## 2  Background

### 2.1  Method Chain

The complete method chain used in music classification is outlined in Fig. 1. In our work we use only the features extracted from the audio data, since they can be automatically extracted for any music piece. Then the features are preprocessed, i.e. they are aggregated for a given part of a song (called partition). In this study we select the features only from the time frames between the previously extracted onset times to reduce the data dimensionality. After the processed features are combined with the given label, the classification models can be trained. The application of these models on the separate test data set provides an opportunity to measure the algorithm performance with different metrics.

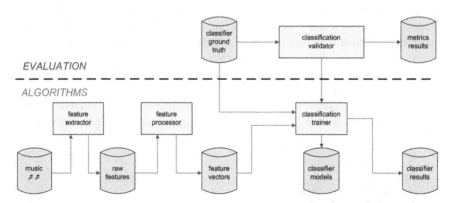

**Fig. 1** Algorithms and Evaluation

## 2.2 Evaluation Metrics

For the comparison of algorithms we use a set of different metrics. Well-known measures which are based on the confusion matrix data for binary classification are accuracy, precision, recall, specificity and $F_1$-measure ($TP$, true positives, is the number of the positive instances which are correctly predicted; $TN$, true negatives, is the number of the negative instances which are correctly predicted; $FP$, false positives, is the number of the negative instances which are wrongly predicted as positives, and $FN$, false negatives, is the number of the positive instances which are wrongly predicted as negatives (Witten and Frank 2005)):

$$accuracy = \frac{TP + TN}{TP + FP + TN + FN} \tag{1}$$

$$preicision = \frac{TP}{TP + FP} \tag{2}$$

$$recall = \frac{TP}{TP + FN} \tag{3}$$

$$specificity = \frac{TN}{FP + TN} \tag{4}$$

$$F_1 = \frac{2 \cdot precision \cdot recall}{precision + recall} \tag{5}$$

For the comparison of these metrics in the MIR domain see e.g. Lidy and Rauber (2005).

Absolute relative error and mean squared error (MSE) are also commonly applied for data mining tasks (Witten and Frank 2005):

$$E_r = \frac{\sum_{i=1}^{N} |p_i - l_i|}{N} \tag{6}$$

$$E^2 = \sqrt{\frac{\sum_{i=1}^{N} (p_i - l_i)^2}{N}}, \tag{7}$$

where $\mathbf{p}$ and $\mathbf{l}$ are the binary vectors with predicted and labeled relationships and $N$ is the overall number of the input instances.

The metrics which are useful for unbalanced sets can be calculated from both the recall and specificity as motivated in Sokolova et al. (2006). Youden's index measures the algorithm's ability to correctly label both positive and negative data samples:

$$\gamma = recall + specificity - 1 \tag{8}$$

Positive and negative likelihoods measure mean better performance on positive and negative classes:

$$l_+ = \frac{recall}{1 - specificity}; \quad l_- = \frac{1 - recall}{specificity} \tag{9}$$

The geometric means is high when both recall and specificity are high and the difference between them is low:

$$g = \sqrt{recall \cdot specificity} \tag{10}$$

The last two measures used in our study refer to correlation between the ordered labeled instances (ground truth) and the predicted instances. The first one is the standard correlation coefficient $\rho$ and the second one is Spearman's Rho rank coefficient (a special case of the Pearson product-moment correlation coefficient) (Hogg and Craig 1995):

$$c_\rho = \frac{\sum_i^N (R(p_i) \cdot R(l_i)) - N\left(\frac{N+1}{2}\right)^2}{\sqrt{\left(\sum_i^N (R^2(p_i)) - N\left(\frac{N+1}{2}\right)^2\right) \cdot \left(\sum_i^N (R^2(l_i)) - N\left(\frac{N+1}{2}\right)^2\right)}} \tag{11}$$

Another metric calculation issue, especially for the song categorization, is the metric label. Since the features aggregated for song partitions and not the complete songs are used as training instances, the metric calculated on the *partition level* outputs a precise information about the algorithm performance. On the other side, for a real-world situation the complete song must be assigned to a category – if e.g. a short intro of a rock song is identified as classic, but the remaining partitions are classified correctly, the performance on the partition level should be averaged for the song:

$$s(k, i) = \frac{1}{P} \sum_{p=1}^{P} c(k, i, p) \tag{12}$$

with $c(k, i, p) = 1$ if partition $p$ of track $k$ belongs to category $i$, or $c(k, i, p) = 0$ otherwise.

## 3 Empirical Study: Metric Correlation Analysis

### 3.1 Setup

For the comparison of the metric coherence we designed a large empirical study predicting the 6 genres and 8 styles based on AllMusicGuide labels (AllMusicGuide 2010) for our music database (Music Test Database of the CI Research Group 2010). For each category the training set was limited to 10 and 20 prototype songs, half of them provided the positive song partitions and another half the negative partitions.

Five feature sets were used: a 12-dimensional CMRARE feature vector (Martin and Nagathil 2009), MFCCs and Delta MFCCs (52 dims.), chroma features (38 dims.), a large set of features from Theimer et al. (2008) (230 dims.) and the previous set extended with many MIR Toolbox functions (Lartillot et al. 2008) (646 dims.).

We applied three different feature processing methods based on onset and structure information and trained the classification models with four algorithms: decision tree C4.5, random forest, $k$-Nearest Neighbors and Naive Bayes. Omitting the five cases where not of all metrics could be calculated (e.g. if all instanced were classified as one class), the complete number of initial experiments was 1,675.

For the visualization of correlation between each pair of metrics, we created the correlation matrix $C$:

$$\forall i, j \in \{1, \ldots, M\} : C_{ij} = \left| \rho(m_i, m_j) \right|, \tag{13}$$

where $M$ is the metric number, and $\rho(m_i, m_j)$ a correlation coefficient between metrics $m_i$ and $m_j$.

For the triangular matrix, the mean correlation coefficient can be calculated:

$$\mu = \sum_{i=1,\ldots,M; j=i+1,\ldots,M} C_{ij} \cdot \frac{2}{M^2 - M} \tag{14}$$

$\mu$ discards the high safe-correlation and describes the averaged pair-calculated correlation between all metrics: if it is closer to 1, it means that all metrics are similar and it is not required to design multi-objective evaluation scenarios; if it is closer to 0, it means, that the metrics are highly uncorrelated and the calculation of several metrics can provide better evaluation of algorithm performance.

## 3.2 Discussion of Results

Figure 2 presents the calculated matrix $C$ for metrics on song level (left) and partition level (right). The interpretation of this figure leads to some significant suggestions. Some of the metrics are highly correlated (e.g. accuracy and relative error, both correlation coefficients, Youden's index and geometric means), which can be also derived in some cases from the metric definitions. However some of them are very independent (accuracy and recall, precision and recall, negative likelihood and MSE). It means that the best designed algorithm chain due to one metric does not mean that it will outperform other algorithm chains responding to other metrics. And as it was discussed before in the introduction, a set of different metrics makes sense in music classification. Roughly speaking, if the algorithm tuning is done to optimize an accuracy value (a typical situation in the MIR literature), it will not guarantee the best performance on metrics like recall, likelihoods, precision and so on.

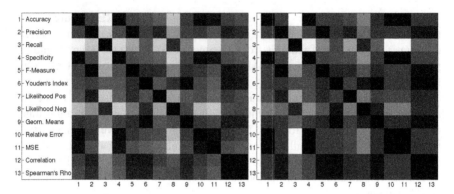

**Fig. 2** Correlation Matrices for Song and Partition Level Metrics. Dark cells: high (anti)correlation; white cells: low (anti)correlation

**Table 1** Metrics which are less correlated with other metrics

| Level | 1st | 2nd | 3rd |
|-------|-----|-----|-----|
| Song | *recall* | $l_-$ | *precision* |
| | $(\mu_r = 0.333)$ | $(\mu_r = 0.484)$ | $(\mu_r = 0.601)$ |
| Partition | *recall* | *precision* | $l_-$ |
| | $(\mu_r = 0.360)$ | $(\mu_r = 0.551)$ | $(\mu_r = 0.617)$ |

Another interesting aspect is the comparison between two levels. The both correlation coefficients are very similar on the partition level, but if the classification results are averaged for the complete songs, they are not dependent so high any more. Also such very closely related metrics by definition as absolute error and MSE are less correlated on the song level. And the song level evaluation is more important for real-word classification tasks where small errors on outlier partitions can be accepted and the main performance on complete songs plays the role for a listener. If we calculate the mean correlation coefficient $\mu$ for the lower triangular matrix as provided in (14), it is equal to 0.637 for song-level based metrics and 0.674 for partition-level based metrics.

The mean relative correlation coefficient of a metric can be estimated for the search of the metrics which are high or less correlated with all other in average:

$$\mu_r(m_i) = \frac{\sum_{j \in \{1,...,M\} \setminus i} |\rho(m_i, m_j)|}{M - 1} \quad (15)$$

The three less and most correlated metrics are given in Tables 1, 2. For the choice of the metrics for the algorithm evaluation it can be recommended to calculate only one from the most correlated metrics, and several further from the less correlated ones.

Figure 3 compares the correlation matrices for the different classification methods. Here only the song-level metrics are calculated, and we removed two well-balanced categories. The difference between the figures is not very high, however

**Table 2** Metrics which are mostly correlated with other metrics

| Level | 1st | 2nd | 3rd |
|---|---|---|---|
| Song | $\rho$ ($\mu_r = 0.756$) | $g$ ($\mu_r = 0.733$) | $\gamma$ ($\mu_r = 0.731$) |
| Partition | $\rho$ ($\mu_r = 0.804$) | $c_\rho$ ($\mu_r = 0.804$) | $g$ ($\mu_r = 0.780$) |

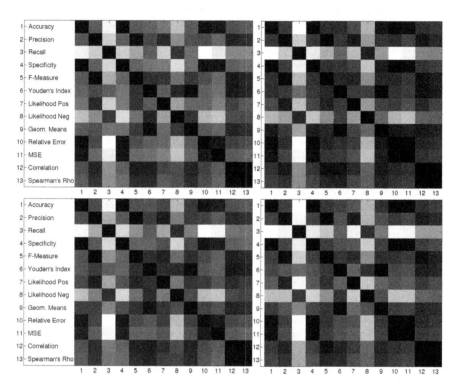

**Fig. 3** Correlation Matrices for experiments with C4.5 (top left), random forest (top right), $k$NN (bottom left) and Naive Bayes (bottom right). Dark cells: high (anti)correlation; white cells: low (anti)correlation

the matrices are not similar: for decision tree C4.5 the metrics are at least correlated ($\mu = 0.66$) and for random forest at most correlated ($\mu = 0.691$). In other words, if a concrete classifier algorithm $a_1$ outperforms another algorithm $a_2$ responding to some chosen criterion, it does not mean that it will outperform it due to other criteria, and for the same algorithm again some metrics have a very low dependency from each other.

The similar observation holds for three examined feature processing methods, however the difference is smaller: from $\mu = 0.664$ for onset reduction methods (only time frames between onset events are analyzed) to $\mu = 0.685$ for interval

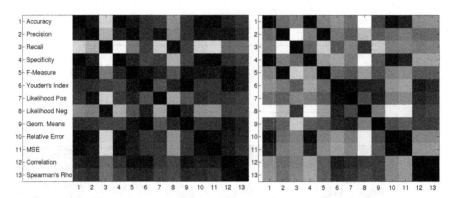

**Fig. 4** Correlation matrices for experiments for three "easy" categories (left) and three "hard" categories (right). Dark cells: high (anti)correlation; white cells: low (anti)correlation

selection method (only features from 30 s of each music track are aggregated) and $\mu = 0.671$ for onset method extended by structural reduction (here only several partitions from each song segment are proceeded to the classifier, where the segments are extracted by the algorithm from Paulus and Klapuri (2008)).

Another very interesting analysis result is given in Fig. 4. Here we calculated the category complexities based on a large number of trials with randomly selected features. The mean accuracy was used to distinguish between the classes: the easiest three classification tasks were to classify classic against other genres (trained on 10 and 20 songs) and jazz against other genres (trained on 20 songs). The three hardest categories were R&B, ClubDance and ProgRock (trained on 10 songs). For the easiest classification tasks $\mu = 0.742$ and the metrics correlate stronger. For the hardest tasks $\mu = 0.545$, and the dependency between metric pairs is rather weak. So it can be suggested, that especially for hard classification tasks the impact of more metrics must be analyzed. Also the concrete $C_{ij}$ values are sometimes very different: accuracy and recall are rather independent (as can be also seen in Figs. 2, 3), but have some weak correlation for hard categories. So the main conclusion from this figure is, that the dependency of evaluation metrics is not similar for different classification tasks. One cannot rely on the suggestion that the evaluation due to some metric is not relevant for a certain case.

## 4 Conclusion and Outlook

Summarizing the results from the empirical study, it can be affirmed, that the evaluation of the music classification chain should be very carefully planned and depends on different factors from the categorization task itself to the method parameters. Certainly this statement holds also for other classification problems, however some specific design aspects are especially important for music recommendation tasks.

The tradeoff between high performance on positive or negative examples, desired surprise effect or safety, impact of balance in the given music collection can be chosen depending on the music listener preferences or application scenario.

In the further work we are going to examine the different metrics using the optimization of the feature selection. One possibility is to continue our previous work on single-objective optimization by evolutionary strategies (Vatolkin et al. 2009; Bischl et al. 2010) and to analyze the development of different metrics and its correlation to the main optimization metric. Another possibility is to apply multi-objective heuristics, which search for the Pareto set of optimal tradeoff solutions regarding to several criteria (Deb 2001).

**Acknowledgements** We thank the Klaus Tschira Foundation for financial support.

# References

AllMusicGuide (2010) Webpage. URL http://www.allmusic.com, cited Feb 2011

Bischl B, Vatolkin I, Preuss M (2010) Selecting small audio feature sets in music classification by means of asymmetric mutation. In: Proceedings of the 11th International Conference on Parallel Problem Solving From Nature (PPSN), Krakow, pp 314–323

Deb K (2001) Multiobjective optimization using evolutionary algorithms. Wiley-Interscience, Chichester, UK

Downie JS (2008) The music information retrieval evaluation exchange (2005-2007): A window into music information retrieval research. Acoust Sci Technol 29(4):247–255

Duda RO, Hart PE, Stork DG (2001) Pattern classification. Wiley, New York

Hogg RV, Craig AT (1995) Introduction to mathematical statistics. Macmillan, New York

Lartillot O, Toiviainen P, Eerola T (2008) A Matlab toolbox for music information retrieval. In: Preisach C, Burkhardt H, Schmidt-Thieme L, Decker R (eds) Data analysis, machine learning and applications, studies in classification, data analysis, and knowledge organization. Springer, Berlin, pp 261–268

Lidy T, Rauber A (2005) Evaluation of feature extractors and psycho-acoustic transformations for music genre classification. In: Proceedings of the 6th International Conference on Music Information Retrieval (ISMIR), London, UK, pp 34–41

Martin R, Nagathil A (2009) Cepstral modulation ratio regression (CMRARE) parameters for audio signal analysis and classification. In: Proceedings of the International Conference on Acoustics, Speech and Signal Processing (ICASSP), Taipei, Taiwan, pp 321–324

Music test database of the CI research group (2010) http://ls11-www.cs.tu-dortmund.de/rudolph/ mi#music_test_database, cited Feb 2011

Paulus J, Klapuri A (2008) Music structure analysis using a probabilistic fitness measure and an integrated musicological model. In: Bello JP, Chew E, Turnbull D (eds) Proc. of the 9th International Conference on Music Information Retrieval, pp 369–374

Ras ZW, Wieczorkowska A (2010) Advances in music information retrieval. Springer, Berlin

Sokolova M, Japkowicz N, Szpakowicz S (2006) Beyond accuracy, F-score and ROC: A family of discriminant measures for performance evaluation. In: Sattar A, Kang BH (eds) Proceedings of the AAAI'06 workshop on Evaluation Methods for Machine Learning, Springer, Berlin, pp 1015–1021

Theimer W, Vatolkin I, Eronen A (2008) Definitions of Audio Features for Music Content Description, Algorithm Engineering Report TR08-2-001. Technische Universität Dortmund, Germany

Tzanetakis G, Cook P (2002) Musical genre classification of audio signals. IEEE Trans Speech Audio Process 10:293–302
Vatolkin I, Theimer W, Rudolph G (2009) Design and comparison of different evolution strategies for feature selection and consolidation in music classification. In: Proceedings of the 2009 IEEE Congress on Evolutionary Computation (CEC 2009), IEEE Press, Piscataway, NJ, pp 174–181
Witten IH, Frank E (2005) Data mining. Elsevier, Amsterdam

# Partition Based Feature Processing for Improved Music Classification

Igor Vatolkin, Wolfgang Theimer, and Martin Botteck

**Abstract** Identifying desired music amongst the vast amount of tracks in today's music collections has become a task of increasing attention for consumers. Music classification based on perceptual features promises to help sorting a collection according to personal music categories determined by the user's personal taste and listening habits. Regarding limits of processing power and storage space available in real (e.g. mobile) devices necessitates to reduce the amount of feature data used by such classification. This paper compares several methods for feature pruning – experiments on realistic track collections show that an approach attempting to identify relevant song partitions not only allows to reduce the amount of processed feature data by 90% but also helps to improve classification accuracy. They indicate that a combination of structural information and temporal continuity processing of partition based classification helps to substantially improve overall performance.

## 1 Introduction

Listening to music has not lost any bit of its attraction to consumers during recent times - people have collected large amounts of music tracks and carry around huge music collections on their mobile devices. Attempts to sort music by predefined genres in general suffer from the absence of a genre taxonomy (Pachet and Cazaly

I. Vatolkin (✉)
TU Dortmund, Dortmund, Germany
e-mail: igor.vatolkin@udo.edu

W. Theimer
Research In Motion, Bochum, Germany
e-mail: wolfgang.theimer@ieee.org

M. Botteck
e-mail: martin.botteck@ieee.org

W. Gaul et al. (eds.), *Challenges at the Interface of Data Analysis, Computer Science,*     411
*and Optimization*, Studies in Classification, Data Analysis, and Knowledge Organization,
DOI 10.1007/978-3-642-24466-7_42, © Springer-Verlag Berlin Heidelberg 2012

2000). Consequently, music listeners often tend to relate subgenres to main genres in a rather individual fashion and prefer to decide between subgenres according to their personal taste. Hence, a more sophisticated approach of music selection seeks to express commonalities between tracks by their allocation to personal categories rather than "genres". Such classification based on perceptual features shall serve to take into account musical taste and, eventually, listening habits. Previous investigations have shown that respective approaches allow attractive and novel UI concepts (Mörchen et al. 2005; Theimer et al. 2008a).

This paper presents several approaches to this selection and is organized as follows: Sect. 2 outlines the general concept of perceptual features as well as new concepts for identifying relevant song partitions. It also introduces methods to smooth classification across song partitions. Section 3 describes a set of exhaustive experiments comparing a large variety of feature data reduction and classification methods. Section 4 concludes with a summary of results.

## 2  Music Processing Concepts

### 2.1  Audio Features for Music Classification

Music is a unique type of media content. It is a highly structured audio signal which can be described by formal musical terms: Timbre, harmony, melody, rhythm, tempo and structure. Whereas the spectral properties of sound (timbre) can be described on a time scale of approx. 20 ms, harmonies require the simultaneous presence of tones for several 100 ms. Melody and rhythm can only be classified in longer intervals (some seconds). The structure of a music track requires the longest intervals (on the order of tens of seconds) to provide meaningful classification results.

During analysis a set of musical features is extracted from each music track. The $N$ raw features $f_i, i \in [1, N]$ describe different characteristics of music. We used a set of up to 115 features defined in Theimer et al. (2008b), focusing on timbre, harmonic and rhythm properties. They are extracted in a sequence of time windows for one track of music. The set of computed features for a one time window forms a feature vector

$$\mathbf{f} = (f_1, f_2, ..., f_N)^T \qquad (1)$$

Concatenation of the feature vectors for all $M$ time windows creates an $N \times M$ feature matrix $\mathbf{F}$.

### 2.2  Feature Data Reduction

The dimensionality of the raw feature matrix $\mathbf{F}$ is prohibitive for classification purposes if the number of training examples is significantly smaller: A music track of 3 minutes duration and 100 features for each 23 ms non-overlapping time window e.g. is represented by approximately one million feature values. The set of features must be pruned so that the representation is more compact than the original music signal. This decreases the computational load for music analysis and avoids the danger of overfitting classifier models. Many statistical and heuristic methods exist which aim to reduce the number $N$ of features (Vatolkin et al. 2009). In this work we concentrate on the algorithms which operate on the columns of $\mathbf{F}$.

Music is composed of notes which are the events at which the sound changes. A short 1/16 note at 100 beats/minute has a duration of 150 ms. This is significantly longer than the basic time window interval of 23 ms. Therefore it is justified to compute the music features only at a rate determined by the duration of the shortest note (tatum period). The absolute tatum time depends on the (time-varying) tempo of the music track (Klapuri et al. 2006). Figure 1 shows the development of time signal RMS for 512 sample frames. Corresponding beat, tatum and onset events are marked by strokes. Onsets correspond to the starting notes and higher energy amplitudes. Beat events are also placed at or near energy peaks most of the time, but are not related to every peak. Tatum times describe the smallest perceived entity of periodic signal impulses and do not match exactly the played notes.

The knowledge of these events may be used for time window selection. In our experiment study we investigated the following methods with the aim to reduce the input data for classification: The first algorithm group processes only time frames to which onset/beat/tatum events belong to $(O, B, T)$. Another group $(O', B', T')$ uses time frames exactly between these events (it can be assumed that the sound is more stable between notes). Intervals forward 30 s of the time signal to the classifier, either from the beginning of the music track $(I_1)$, or exactly from the middle $(I_2)$, or after the first minute $(I_3)$ - for a wide range of popular radio songs one can be sure that the intro is over after the first minute). More sophisticated

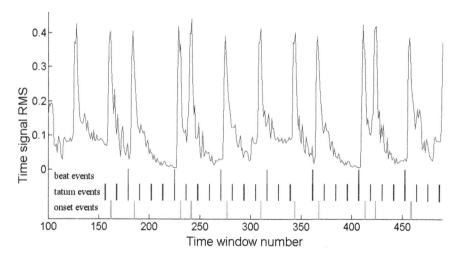

**Fig. 1** Time events and music signal (Excerpt from "Everything Counts"-Depeche Mode)

methods incorporate the information about song segments. They correspond to the intervals with very different distributions of audio features. The proposed structural pruning methods $S_1, S_2, S_3$ analyze only one, two or four partitions from the middle of each recognized song segment. Further structural pruners $SO_1, SO_2, SO_3$ use additionally the previously described onset pruning method.

Several comparison methods are also applied. A baseline processing $N$ forwards all features from all time frames. Data sampling algorithms select each $k$-th frame. $k$ can be set depending on the number of onset times $o$, e.g. $k = 1.5 \cdot o$, however this algorithm is not independent of time information any more.

## 3  Partition Based Classification

### 3.1  Definition of Partition Based Features

Let us now assume that the features are only computed at times $t_1, \ldots, t_M$ which are the outcome of the time window reduction outlined in Sect. 2.2. A human listener can determine the musical character of a song within a couple of seconds. So from the perception point of view it makes sense to accumulate the musical statistics over this time interval. At the same time this reduces the amount of features further and makes them more robust since they are based on several observations. In the following this interval is called partition. The feature values $f_n(t_i)$ in each partition are described by a statistical model, e.g. a Gaussian mixture model.

### 3.2  Processing of Partition Based Classification Results

In our work we consider binary classification results: A partition is either classified to belong to the category for which the classifier is trained or not. Since we have several partitions per music track, we can compute a similarity $s(k, i) \in [0, 1]$, how close the track $k$ is to the category $i$, by averaging the results from all $P$ partitions:

$$s(k, i) = \frac{1}{P} \sum_{p=1}^{P} c(k, i, p) \tag{2}$$

with $c(k, i, p) = 1$ if partition $p$ of track $k$ belongs to category $i$, or $c(k, i, p) = 0$ otherwise.

In many cases classifiers are trained for individual categories and give binary results. A classifier that provides a 1 out of $I$ categories result can only be trained easily when the number $I$ of categories does not change dynamically. Still we can post-process the partition classification results for mutually exclusive categories (e.g. music genres) by discouraging partitions with conflicting classifier outputs (i.e. more than one positive classification):

$$s(k,i) = \frac{2}{P} \sum_{p=1}^{P} \frac{c(k,i,p)}{1 + \sum_{i'=1}^{I} c(k,i',p)} \tag{3}$$

The confidence in the classification results for an individual category is higher when the classification results are stable, i.e. show the same binary value for each partition. Intervals of neighboring partitions with toggling results should be deemphasized since they do not provide reliable classification results:

$$s(k,i) = \frac{\sum_{p=1}^{P} c(k,i,p)[1 - z(k,i,p)]}{\sum_{p=1}^{P} [1 - z(k,i,p)]} \tag{4}$$

where the function $z(k,i,p)$ models the local toggling frequency of the classifier results in an interval $[p - Z, p + Z]$ around partition $p$:

$$z(k,i,p) = \frac{1}{2Z} \sum_{p'=p-Z}^{p+Z-1} |c(k,i,p'+1) - c(k,i,p')| \tag{5}$$

if $p - Z \geq 0 \wedge p + Z - 1 \leq P$ , and $z(k,i,p) = 0$ otherwise.

## 4  Experiments and Results

### 4.1  Design of Experiments

For the classification experiments we created the database of 120 CDs listed in the Music Test Database of the CI Research Group (2010). For practical reasons we created categories based on genre and style data from AllMusicGuide (2010) thus relying on the personal taste of several music critics: 6 genres and 8 styles shall resemble the same amount of "personal categories" in the experiments. For the validation of classification performance we used a test set of 120 songs, each song randomly dropped from each album of our collection (omitting the songs used for training).

We used the decision tree algorithm C4.5 for classification. It is a fast and robust method which achieves acceptable results without a complex tuning process. Secondly, it has been shown in different studies (e.g. Blume et al. (2008)), that feature extraction design is the basis for successful classification. Consequently, the experiments aim to show the optimization potential by suitable feature processing – choosing optimal classifiers is not in the focus of the presented research work.

Revising the methods introduced in Sect. 2.2, we analyzed 6 methods based on onset/beat/tatum events (Klapuri et al. 2006; Lartillot et al. 2008), 6 methods based on structural information (Paulus and Klapuri 2008), 3 interval selection methods, a baseline without data reduction and a large number of comparison data sampling and onset data sampling methods. Classification inputs were built by mean and standard deviation of each feature for 5 s song partitions with 50% overlap.

For the evaluation of results we calculated classification accuracy and pruning ratio. For each of the 28 categories we sorted the algorithms due to accuracy metric and estimated the algorithm ranks (1 for the best algorithm and 28 for the worst). For the comparison of algorithms across the complete set of categories we use the mean accuracy rank.

## 4.2 Discussion of Results

The classification performance for the robust onset reduction method can be seen in Fig. 3. Figure 2 gives an overview about the tradeoff between pruning ratio and accuracy rank. It can be treated as a multi-objective problem where the 1st criterion is to reduce the data used for learning (saving indexing space, algorithm processing time and limiting the danger of overfitting) and the second one is to keep or improve the classification performance.

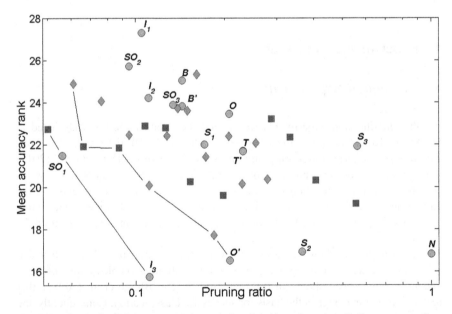

**Fig. 2** Pruning ratio and mean accuracy rank for algorithms investigated. Squares: sampling methods without time structure information; diamonds: sampling with factor depending on onset event number; circles: methods based on time structure information

The main achievement is that it is possible to save and even improve accuracy whilst strongly reducing the data amount. Two non-dominated fronts with the best balanced solutions are marked by lines. Concerning accuracy, the best method is the interval selector $I_3$, where 30 s after the first minute are processed. The pruner $O'$ which selects frames between onset events is the second best. Both algorithms provide reasonable data reduction rates, especially compared to the baseline method $N$. For the strongest pruning ratio, the structural pruner $SO_1$ is a good decision. Here one partition from each segment is selected after onset pruning. However the accuracy rank suffers from such strong data reduction effect.

Some further conclusions are: Selection of 30 s from the beginning of the song (method $I_1$) is worse than selection of 30 s from the middle $I_2$. Both methods perform significantly weaker than $I_3$. Both beat pruners $B$, $B'$ reduce data better than the tatum pruners $T$, $T'$, but have lower accuracy. Comparison methods with no or less time structure knowledge (marked by squares and diamonds) have pruning rates similar to some of the more complex methods, but only one of them belongs to the five best with regard to accuracy rank.

However, the results must be treated carefully. The music classification chain is very complex and depends on many parameters which may interact – music categories, classification method parameters, used features, frame and partition sizes, choice of evaluation metric etc. Although a definite recommendation is not possible, some guidelines can be indeed provided. It is obvious, that data reduction does not necessarily mean the loss of classification performance. Some methods, like interval selection or processing of time frames between onset events, can be a good start choice, also for experiment studies with other focus areas.

Figure 3 shows that the concept of consolidating classification results across all partitions of a music track and selecting the category with the highest classification rate is quite powerful, even for the small amount of training samples used here. The classification accuracy on the partition level is 69.7%. Using majority voting on the

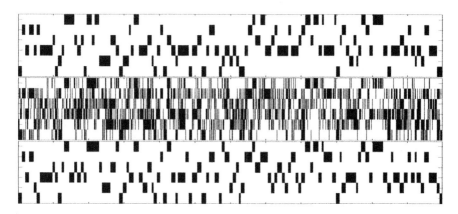

**Fig. 3** Binary relationships of all partitions (horizontal axis) to six genre categories Classical, Electronic, Jazz, Pop/Rock, R&B, Rap (vertical axis). Top subfigure: ground truth; middle: raw classifications; bottom: classification with majority voting

track level according to (3) the individual classifier performance increases to 88.5% and outliers are eliminated.

## 5 Conclusion and Outlook

The experiments carried out indicate that classification performance indeed improves with a suitably reduced amount of feature data. It can be shown that methods reducing the temporal density of features deliver superior performance in several cases. Furthermore, we have been able to further advance the overall classification performance by introducing smoothing functions on classification results for track partitions. Although some of the presented methods clearly rate better than others, these methods may underperform in selected configurations or classifications. Hence, generalization of the results remains somewhat difficult.

Ongoing and future research tackles this aspect by addressing extended data and parameter configurations (more personal categories to classify, new features). Specifically, selection of a suitable classification promises to further improve overall performance. Results shall be ranked by a multi-objective approach allowing a tradeoff between ranges of evaluation criteria. Furthermore, the integration of high-level descriptors like instrument, harmony or melody features promises to be more understandable and may lead to faster learning.

**Acknowledgements**  We thank the Klaus Tschira Foundation for financial support.

## References

AllMusicGuide (2010) Webpage. URL http://www.allmusic.com, cited Feb 2011
Blume H, Haller M, Botteck M, Theimer W (2008) Perceptual feature based music classification - a DSP perspective for a new type of application. In: Proceedings of the SAMOS VIII Conference, (IC-SAMOS), Samos, Greece, pp 92–99
Klapuri AP, Eronen AJ, Astola JT (2006) Analysis of the meter of acoustic musical signals. IEEE Trans Audio Speech Audio Process 14(1):342–355
Lartillot O, Toiviainen P, Eerola T (2008) A Matlab toolbox for music information retrieval. In: Preisach C, Burkhardt H, Schmidt-Thieme L, Decker R (eds) Data analysis, machine learning and knowledge organization. Springer, Berlin
Mörchen F, Ultsch A, Nöcker M, Stamm C (2005) Databionic visualization of music collections according to perceptual distance. In: Proceedings of the 6th International Conference on Music Information Retrieval, London, UK, pp 396–403
Music Test Database of the CI Research Group (2010) http://ls11-www.cs.tu-dortmund.de/rudolph/mi#music_test_database. Cited 30 Jul 2010
Pachet F, Cazaly D (2000) A taxonomy of musical genres. In: Proc. Content-Based Multimedia Information Access, (RIAO), Paris, France
Paulus J, Klapuri A (2008) Music structure analysis using a probabilistic fitness measure and an integrated musicological model. In: Proc. of the 9th Int'l Conf. on Music Information Retrieval

Theimer W, Vatolkin I, Botteck M, Buchmann M (2008a) Content-based similarity search and visualization for personal music categories. In: Proc. of the Sixth International Workshop on Content-Based Multimedia Indexing, London

Theimer W, Vatolkin I, Eronen A (2008b) Definitions of Audio Features for Music Content Description. Technical Report TR08-2-001, Chair of Algorithm Engineering, TU Dortmund

Vatolkin I, Theimer W, Rudolph G (2009) Design and comparison of different evolution strategies for feature selection and consolidation in music classification. In: Proc. of the 2009 IEEE Congress on Evolutionary Computation (CEC), IEEE Press, Piscataway, NJ, pp 174–181

# Software in Music Information Retrieval

**Claus Weihs, Klaus Friedrichs, Markus Eichhoff, and Igor Vatolkin**

**Abstract** Music Information Retrieval (MIR) software is often applied for the identification of rules classifying audio music pieces into certain categories, like e.g. genres. In this paper we compare the abilities of six MIR software packages in ten categories, namely operating systems, user interface, music data input, feature generation, feature formats, transformations and features, data analysis methods, visualization methods, evaluation methods, and further development. The overall rankings are derived from the estimated scores for the analyzed criteria.

## 1 Introduction

Music Information Retrieval (MIR) (Ras and Wieczorkowska 2010) is the research area which provides an extended and automatic analysis of music data, e.g. for automatic transcription, harmony and melody recognition, instrument detection etc. One of the most prominent applications is the classification of audio music pieces by means of automatically identified so-called low level properties of the music time series. For MIR a large number of different software packages are currently available (see e.g. Typke et al. (2005)). Considering the advantages and drawbacks of the existing tools, we aim to improve the development of our newly developed package AMUSE. For the comparison we concentrate on the, in our view, most important MIR-task, the classification of audio signals, here into genres. For this benchmark we have selected six actual and open-source packages. Note,

C. Weihs (✉) · K. Friedrichs · M. Eichhoff
Department of Statistics, TU Dortmund University, Germany
e-mail: weihs@statistik.tu-dortmund.de; friedrichs@statistik.tu-dortmund.de;
eichhoff@statistik.tu-dortmund.de

I. Vatolkin
Department of Computer Science, TU Dortmund University, Germany
e-mail: igor.vatolkin@cs.tu-dortmund.de

W. Gaul et al. (eds.), *Challenges at the Interface of Data Analysis, Computer Science,*     421
*and Optimization*, Studies in Classification, Data Analysis, and Knowledge Organization,
DOI 10.1007/978-3-642-24466-7_43, © Springer-Verlag Berlin Heidelberg 2012

however, that not all compared software packages support classification. In this case an additional backend is needed. Note, moreover, that we will compare software products for special tasks (e.g. MusicMiner) and all-round tools (e.g. RapidMiner). For the comparison a data set with 354 songs are used, 120 of them classical and 234 non-classical according to AllMusicGuide (http://www.allmusic.com/), all in MP3 format. The main task is the separation of classical from non-classical music.

In Sect. 2 the software products are introduced and the software test is performed for the criteria operating systems, user interface, music data input, feature generation, feature formats, transformations and features, data analysis methods, visualization methods, evaluation methods, and further development. In Sect. 3 the overall ranking is discussed and in Sect. 4 a conclusion is given.

## 2 Software Products

The software packages compared are the MIR Toolbox developed by the University of Jyväskylä (Lartillot and Toiviainen 2007), CLAM (C++ Library for Audio and Music) developed by the University Pompeu Fabra of Barcelona (Amatriain 2007), MusicMiner developed by the University of Marburg (Mörchen et al. 2005), RapidMiner developed by the TU Dortmund and the software company Rapid I (Mierswa et al. 2006), jMIR (Java MIR) developed by the McGill University, Montreal (McKay 2010), as well as AMUSE (Vatolkin et al. 2010) newly developed by the TU Dortmund. In the comparison, RapidMiner 5.0 is assessed, the additional abilities of the older version 4.6 concerning music time series analysis are not considered since further support of these abilities is not planned. On the other hand, the abilities of MATLAB are taken into account for the MIR Toolbox. Please refer also to the web addresses of the software packages in the software references. Note that this comparison is not claiming to be complete. There are certainly other MIR-software packages. Examples are Marsyas (Music Analysis, Retrieval and Synthesis for Audio Signals: http://marsyas.info/) and M2K (Music to Knowledge: http://www.music-ir.org/evaluation/m2k/).

*Software test*: The software products are scored in 10 categories with a maximum of 100 points in each category. In Sect. 3 different weighting schemes of categories will be used for different kinds of users. Weighted averages of points will be used as overall scores and the ranking will be given according to these scores. We carried out the test in July 2010.

Note that the first author fixed the criteria and the second and third authors carried out the evaluations. Only the 4th author is involved in the software AMUSE. So, no conflict of interests is to be expected. Moreover, on the one hand there was no attempt to objectify the valuation by means of a larger sample of raters. On the other hand, criteria as well as scores are all explicit. And finally, the weights assigned to the subcategories of the different categories in Sect. 2 as well as to the categories in Sect. 3 and the corresponding scores can easily be adapted to one's own taste.

*Operating systems*: The first criterion is the availability of the software packages for the three most important operating systems, Windows (60 points), Linux / Unix (30 points) and Mac OS (10 points). Here, all packages got the maximum score, except AMUSE which was only available on Linux/Unix.

*User interface*: Let us first illustrate the graphical part of the User Interface (GUI). In CLAM there are graphical connections between processing units (analysis, transformations, synthesis, visualizations). This allows a clear handling even without programming skills (see Fig. 1). In version 5.0 of RapidMiner additionally there are interactive help and suggestions in an extra section (see Fig. 2). In jAudio, the feature extraction tool of jMIR, the handling for feature generation is intuitively list based. For ACE (Autonomous Classification Engine), the music classification tool of jMIR, the GUI is still under construction. Currently, it has mainly just one

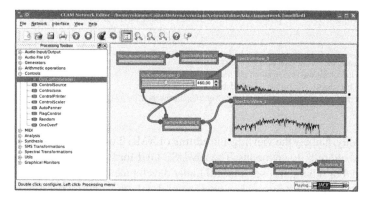

**Fig. 1** CLAM GUI

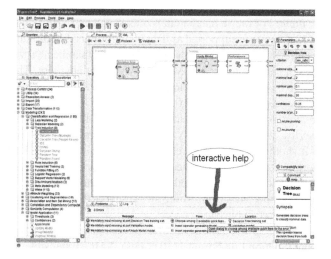

**Fig. 2** GUI RapidMiner 5.0

**Table 1** Scores: User interface

| Software | GUI | Batch mode | Documentation | Intuitive handling | Score |
|---|---|---|---|---|---|
| MIR Toolbox | 0 | 20 | 20 | 10 | 50 |
| CLAM | 20 | 20 | 15 | 15 | 70 |
| MusicMiner | 15 | 20 | 10 | 10 | 55 |
| RapidMiner | 30 | 20 | 20 | 20 | **90** |
| jMIR | 10 | 20 | 20 | 15 | 65 |
| AMUSE | 10 | 20 | 5 | 15 | 50 |

**Table 2** Scores: Music data input

| Software | MP3 | WAV | AU | Score |
|---|---|---|---|---|
| MIR Toolbox | x | x | AU, AIFF | 90 |
| CLAM | x | x | AU, VOC, AIFF, Ogg-Vobis | **95** |
| MusicMiner | x | x | – | 80 |
| RapidMiner | – | – | MP3, WAV via version 4.6 | 20 |
| jMIR | x | x | – | 80 |
| AMUSE | x | x | – | 80 |

AU: Simple audio file format (Sun Microsystems)
VOC: Audio file format used by Creative Labs hardware (SoundBlaster)
AIFF: Audio Interchange File Format (standard for storing sound data)
Ogg-Vobis: Free, open standard container format (Xiph. Org Foundation)

functionality, namely the viewing and editing of XML files. The GUI of MusicMiner uses good old pull-down menus. The AMUSE GUI for feature extraction is similar to jAudio. For other purposes it is still under development.

However, in the category User Interface not only the GUI is assessed (max. 30 points), but also the batch mode (20 points), the documentation (max. 20 points), and the intuitive handling (max. 30 points). This leads to score Table 1. RapidMiner scored distinctly best.

*Music data input*: For music data input, scores are given if MP3-files (40 points) or WAV-files (40 points) are supported as data input formats. For the support of files in other relevant formats max. 20 points are given. Note that the new version 5.0 of RapidMiner is not capable to read in MP3 or WAV files anymore. Thus the older version 4.6. has to be used. This leads to the score Table 2. Obviously, all packages scored similarly, except RapidMiner 5.0.

*Feature generation*: Concerning feature generation, four criteria are assessed. Music time series are typically analyzed in small windows not less than 512 observations. It should be possible to choose the length of the windows and their overlap proportion, to choose specified or random parts of the time series for analysis, to import and export the treated time series, and to extend the feature set by easily programming new ones. Concerning these criteria, the MIR Toolbox appeared to be distinctly best (see Table 3).

*Feature formats*: The file formats considered as standard for saving data are ARFF (Attribute Relation File Format) (30 points), XML (Extensible Markup Language) (30 points), and the Excel format (20 points). ARFF is much more memory efficient (see Figs. 3 and 4) and XML is well readable and standardized. The

**Table 3** Scores: Feature generation

| Maximum score | 30 | 30 | 20 | 20 | 100 |
|---|---|---|---|---|---|
| Software | Window length and Overlap | Parts/Random | Import/Export | Extendable | Score |
| MIR Toolbox | Yes | Yes | 15 | Yes | **95** |
| CLAM | No | No | 10 | Yes | 30 |
| MusicMiner | No | No | 0 | No | 0 |
| RapidMiner | No | No | 20 | Yes | 40 |
| jMIR | Yes | No | 15 | Yes | 65 |
| AMUSE | No | Yes/No | 10 | Yes | 50 |

**Fig. 3** ACE XML format

```
<?xml version="1.0"?><!DOCTYPE feature_vector_file [ <!ELEMENT feature_vector_file
(comments, data_set+)> <!ELEMENT comments (#PCDATA)> <!ELEMENT
data_set (data_set_id, section*, feature*)> <!ELEMENT data_set_id (#PCDATA)>
<!ELEMENT section (feature+)> <!ATTLIST section start CDATA ""        stop
CDATA ""> <!ELEMENT feature (name, v+)> <!ELEMENT name (#PCDATA)>
<!ELEMENT v (#PCDATA)>]><feature_vector_file>  <comments></comments>
                        <data_set>
                        <data_set_id>C:\user\friedrichs\dsam_matlab\MAP1_6\wavFileStor
e\drums1.wav</data_set_id>
                                <feature>
                                                <name>Zero Crossings Overall Standard
Deviation</name>
                                                <v>6.49E1</v>
                                </feature>
                                <feature>
                                                <name>Method of Moments Overall Standard
Deviation</name>
                                                <v>5.913E-1</v>
                                                <v>3.128E1</v>
                                                <v>1.335E3</v>
                                                <v>1.357E5</v>
                                                <v>9.988E6</v>
```

**Fig. 4** ARFF format

```
@relation jAudio
@ATTRIBUTE "Zero Crossings Overall Standard Deviation0" NUMERIC
@ATTRIBUTE "Method of Moments Overall Standard Deviation0" NUMERIC
@ATTRIBUTE "Method of Moments Overall Standard Deviation1" NUMERIC
@ATTRIBUTE "Method of Moments Overall Standard Deviation2" NUMERIC
@ATTRIBUTE "Method of Moments Overall Standard Deviation3" NUMERIC
@ATTRIBUTE "Method of Moments Overall Standard Deviation4" NUMERIC
@ATTRIBUTE "Zero Crossings Overall Average0" NUMERIC
@ATTRIBUTE "Method of Moments Overall Average0" NUMERIC
@ATTRIBUTE "Method of Moments Overall Average1" NUMERIC
@ATTRIBUTE "Method of Moments Overall Average2" NUMERIC
@ATTRIBUTE "Method of Moments Overall Average3" NUMERIC
@ATTRIBUTE "Method of Moments Overall Average4" NUMERIC
@DATA
6.49E1,5.913E-1,3.128E1,1.335E3,1.357E5,9.988E6,5.593E1,8.746E-
1,5.385E1,3.837E3,2.738E5,6.102E7
```

special ACE-XML in jMIR is designed for MIR, supports multiple-classification and logical grouping of features. Other formats than ARFF, XML or Excel can score a maximum of 20 points. Other important formats are the internal MATLAB format and the LRN format in MusicMiner with a unique key and the features in a single tab separated line. It appears that concerning feature formats RapidMiner is best (see Table 4), supporting not only the standard formats but also many others.

*Transformations and features*: When more relevant features are automatically considered by the MIR software, the chance for achieving a good classification improves. Note that all the software packages assume that there is no a-priori knowledge about the relevant features for a classification problem. We have

**Table 4** Scores: Feature formats

| Software | Format | Score |
|---|---|---|
| MIR Toolbox | MAT, ARFF, XML, Excel | 90 |
| CLAM | XML, SDIF | 40 |
| MusicMiner | LRN | 10 |
| RapidMiner | ARFF, XML, Excel, … | **100** |
| jMIR | ARFF, XML | 60 |
| AMUSE | ARFF | 30 |

**Table 5** Scores: Transformations and features

| Maximum Score | 10 | 10 | 10 | 20 | 10 | 25 | 15 | 100 |
|---|---|---|---|---|---|---|---|---|
| Software | Dynamics | Rhythm/ Tempo | Timbre | Pitch | Tonality | Spectral (e.g. DFT) | MFCC | Score |
| MIR Toolbox | 3 | 8 | 10 | 15 | 10 | 25 | 15 | 86 |
| CLAM | 5 | 5 | 8 | 15 | 10 | 20 | 15 | 78 |
| MusicMiner | 3 | – | 8 | 15 | – | 20 | 15 | 61 |
| RapidMiner | – | – | – | – | – | – | – | 0 |
| jMIR | 5 | 3 | – | 10 | – | 25 | 15 | 58 |
| AMUSE | 7 | 10 | 10 | 18 | 10 | 25 | 15 | **95** |

checked whether a package generates features from a music time series representing dynamics, tempo/rhythm, timbre, pitch, and tonality in the analyzed time series window. Moreover, spectral transformations (e.g. DFT, Digital Fourier Transformation) should be available, and the MFCCs (Mel Frequency Cepstral Coefficients). Concerning these criteria, AMUSE definitely scored best (see Table 5). Note that RapidMiner 5.0 cannot generate any music features automatically, the corresponding abilities of RapidMiner 4.6 are not supported anymore.

*Classification*: For successful classification not only different (supervised) classification methods (as e.g. k-NN (k Nearest Neighbors), LDA (Linear Discriminant Analysis) or SVM (Support Vector Machine)) should be available in the software package, but also powerful pre-processing, e.g. for feature extraction and reduction (like PCA (Principal Component Analysis) or ICA (Independent Component Analysis)), resampling methods for the assessment of the classification result (like subsampling methods, the bootstrap, and cross-validation), and optimization tools for feature selection and optimization of free parameters, e.g. by evolutionary algorithms (EAs). Note that RapidMiner is completely integrated as library in AMUSE as well as the R-package mlr (machine learning in R), a very powerful evaluation tool (cp. Bischl et al. (2010)), that the MIR Toolbox is connected with MATLAB, and that RapidMiner is a general data analysis tool including many generic methods. Also, we checked whether the packages include cluster analysis methods for unsupervised learning. Concerning all these criteria, RapidMiner scored best (see Table 6). Note that CLAM does not include any classification tool, yet, and MusicMiner does mainly include clustering.

As an example for optimization, let us consider feature selection. AMUSE (with mlr) offers the $(\mu + \lambda)$ EA and $(\mu , \lambda)$ EA with asymmetric mutation, a rank search (Greedy Forward Search, GFS), a random Monte Carlo search, and a combination

**Table 6** Scores: Classification

| Max. Score | 30 | 30 | 15 | 15 | 10 | 100 |
|---|---|---|---|---|---|---|
| Software | Pre-processing | Classification | Resampling | Optimization | Clustering | Score |
| MIR Toolbox | MATLAB | k-NN, GMM, MATLAB | MATLAB | MATLAB | k-means, MATLAB | 70 |
| CLAM | – | – | – | – | – | 0 |
| Music-Miner | PCA | – | – | – | ESOM | 20 |
| Rapid-Miner | PCA, ICA, smoothing | Bayes, SVM, boosting, bagging, decision tree, ANN, LDA, k-NN | subsampling, bootstrap, cross valid. | EAs, grid search, holdout | EM, k-means | **100** |
| jMIR | PCA | decision tree, Bayes, k-NN, SVM, ANN, Adaboost, bagging | – | EAs, experimental meta learning | – | 50 |
| AMUSE | quartiles, PCA, outlier removal | decision trees, Bayes, k-NN, LDA, SVM, mlr (R) | subsampling, mlr (R) | EAs | hierarchical clustering | 85 |

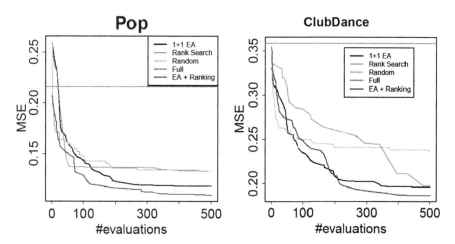

**Fig. 5** Feature selection: AMUSE with mlr; Pop / ClubDance vs. the rest

of EA and ranking for this purpose. Figures 5 show that the combination of EA and rank search results in the lowest mean error rates for the classification of Pop songs and ClubDance songs against all other songs, respectively, for up to 500 EA-evaluations (Bischl et al. 2010).

**Table 7** Scores: Visualization

| Maximum Score | 20 | 20 | 15 | 20 | 25 | 100 |
| Software | Temporal | Spectral | Harmonics | Classification | Evaluation | Score |
|---|---|---|---|---|---|---|
| MIR Toolbox | x | x | Keyspace | Class. Tree | – | 50 |
| CLAM | x | x | Keyspace, Tonnetz | – | – | 55 |
| MusicMiner | – | – | – | Music Map (ESOM) | – | 20 |
| RapidMiner | x | – | – | SOM-Plot, Class. Tree | Lift Chart, ROC Curve | **60** |
| jMIR | – | – | – | – | – | 0 |
| AMUSE | – | – | – | – | – | 0 |

**Fig. 6** Vizualisaion
MusicMiner

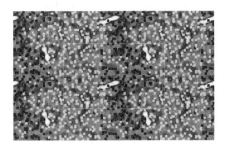

*Visualization methods*: Visualization of results gets more and more important in a more and more visually oriented society. Unfortunately, visualization is often neglected in the MIR software packages. We checked whether there are visualizations for the classification and evaluation results as well as for the temporal, spectral and harmonic development of a music piece. Here, RapidMiner, CLAM, and the MIR Toolbox convinced most (see Table 7), whereas jMIR and AMUSE have no visualization at all. MusicMiner is only strong in clustering using Emergent Self-Organizing Music Maps (ESOMs), an (interactive) 2D-visualization of the high dimensional data space (see Fig. 6). Note that neighboring music pieces in the music map might be similar in the feature domain (e.g. for timbre, dynamic and pitch range etc.), but not necessarily regarding to human judgement. CLAM only displays the spectrogram, a so-called Tonnetz, i.e. the played tones, the chroma peaks, and the song's segmentation into best fitting chords (see Fig. 7).

*Evaluation methods*: For evaluation of classifications at least an estimator of the misclassification rate has to be available, e.g. the mean over many subsamples. A better impression of the classification quality can be achieved by an interval representing an uncertainty region of the, e.g., mean. From a confusion matrix even more quality measures could be derived which might be included by proper customizing. Here, again RapidMiner scored best (100%) (see Table 8).

Let us exemplify the results of classification and evaluation for the genre classification problem mentioned in the introduction. 120 classical vs. 234 non-classical songs should be classified by means of 46 features for each song,

**Fig. 7** Visualization: CLAM
Chordata

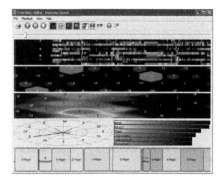

**Table 8** Scores: Evaluation methods

| Maximum Score | 40 | 30 | 20 | 10 | 100 |
|---|---|---|---|---|---|
| Software | Estimator | Uncertainty Interval | Confusion Matrix | Customize | Score |
| MIR Toolbox | x | – | x | 0 | 60 |
| CLAM | – | – | – | 0 | 0 |
| MusicMiner | – | – | – | 0 | 0 |
| RapidMiner | x | x | x | 10 | **100** |
| jMIR | x | – | x | 0 | 60 |
| AMUSE | x | x | x | 5 | 95 |

extracted by RapidMiner 4.6. We compared decision trees automatically produced by AMUSE (with mlr), RapidMiner, and the MIR Toolbox, and assessed the results by means of subsampling with 30 iterations. Figure 8 shows the corresponding decision tree of AMUSE. The means and standard deviations of the classification errors show that the the quality of all methods is similar with AMUSE (with mlr) performing best with $mean = 6.53\%$ and $std.dev. = 2,08\%$, followed by MIR toolbox ($mean = 6.84\%, std.dev. = 2,24\%$), and RapidMiner ($mean = 8.35\%, std.dev. = 2,65\%$).

*Further development*: Last but not least we checked whether the software packages are planned to be further developed in the future. We scored 100 points for new developments based on the experiences of the older packages and for regular further development. For casual further development 50 points were given. The MIR Toolbox, jMIR and AMUSE scored 100 points, CLAM and RapidMiner only 50 points. RapidMiner was downgraded, since there is no further development for audio file processing. Only MusicMiner is not developed anymore.

## 3 Overall Ranking

For the overall ranking, weights can be assigned to the previously discussed criteria. Two example weightings representing possible user and developer preferences are given in Table 9. Based on these weights two different mean scores are calculated leading to overall rankings (see Table 10). Note that these rankings are similar

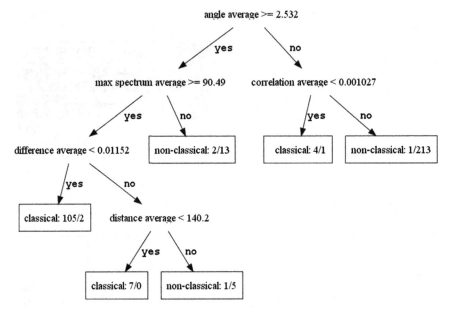

**Fig. 8** Decision tree: AMUSE with mlr

**Table 9** Different weighting schemes

| Evaluation Category | User | Developer |
|---|---|---|
| Supported Operating Systems | 2 | 2 |
| User Interface | **5** | **1** |
| Music Data Input | **4** | 2 |
| Feature Generation | 3 | 3 |
| Features Formats | **1** | **4** |
| Transformations and Features | 3 | 3 |
| Data Analysis Methods | 4 | 3 |
| Visualization Methods | **4** | **1** |
| Evaluation Methods | 4 | 3 |
| Further Development | 2 | **5** |

for the two weightings. Only the middle places 2 – 4 differ. Overall, the MIR Toolbox clearly won, followed by a group built by AMUSE, jMIR, and RapidMiner. RapidMiner 5.0 would have been much better ranked if music feature processing would have been supported as in version 4.6. CLAM and MusicMiner suffer from the fact that they are not supporting (supervised) classification. Note that from the individual rankings the best package cannot be easily identified (see Fig. 9) because the performance is not uniform in the different categories. Also note that the MIR Toolbox but also all the other packages are downgraded corresponding to visualization, meaning that in this respect there is much room for improvement.

**Table 10** Mean weighted scores and overall rankings

| Software | 1 | 2 | 3 | 4 | 5 | 6 | 7 | 8 | 9 | 10 | User | Developer | Rank |
|---|---|---|---|---|---|---|---|---|---|---|---|---|---|
| MIR Toolbox | 100 | 50 | 90 | 95 | 90 | 86 | 70 | 50 | 60 | 100 | **73,8** | **84,2** | 1 |
| CLAM | 100 | 70 | 95 | 30 | 40 | 78 | 0 | 55 | 0 | 50 | 50,4 | 46,3 | 5 |
| MusicMiner | 100 | 55 | 80 | 0 | 10 | 61 | 20 | 20 | 0 | 0 | 35,9 | 26,6 | 6 |
| RapidMiner | 100 | 90 | 20 | 40 | 100 | 0 | 100 | 60 | 100 | 50 | 65,3 | 65,2 | 2–4 |
| jMIR | 100 | 65 | 80 | 65 | 60 | 58 | 50 | 0 | 60 | 100 | 59,8 | 69,0 | 2–4 |
| AMUSE | 30 | 50 | 80 | 50 | 30 | 95 | 85 | 0 | 95 | 100 | 63,0 | 69,1 | 2–4 |

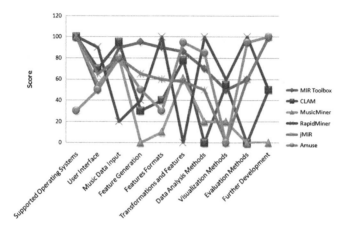

**Fig. 9** Rankings

**Table 11** Strengths and weaknesses

| Software | Positive | Negative |
|---|---|---|
| MIR Toolbox | Easy to Combine with other Matlab Toolboxes | Requires Matlab and programming skills (no GUI) |
| CLAM | Visualizations | No classification |
| MusicMiner | Music Map | Limited capabilities |
| RapidMiner | GUI, classification | No further development for audio files |
| jMIR | data format: ACE XML | No visualizations |
| AMUSE | Generation of many features | No visualizations |

# 4 Conclusion

Concluding our investigations for tool comparison, we can state that the MIR Toolbox clearly won overall, but does not have a GUI. On the other hand, a powerful GUI does not mean richness of the integrated algorithms and many well-chosen classification algorithms must not imply a sufficient output visualization. The choice must be carefully done depending on the preferences of the current user. The newly developed package AMUSE as well as jMIR especially lack visualization. For the special abilities and disadvantages of all the tools see Table 11.

# References

Amatriain X (2007) CLAM: A framework for audio and music application development. IEEE Softw 24(1):82–85

Bischl B, Vatolkin I, Preuss M (2010) Selecting small audio feature sets in music classification by means of asymmetric mutation. In: PPSN XI: Proceedings of the 11th International Conference on Parallel Problem Solving from Nature, Springer, Lecture Notes in Computer Science, vol 6238

Lartillot O, Toiviainen P (2007) MIR in Matlab (II): A toolbox for musical feature extraction from audio. In: Proc. 8th International Conference on Music Information Retrieval (ISMIR 2007), Vienna, pp 127–130

McKay C (2010) Automatic music classification with jMIR. PhD thesis, McGill University

Mierswa I, Wurst M, Klinkenberg R, Scholz M, Euler T (2006) YALE: Rapid prototyping for complex data mining tasks. In: KDD '06: Proceedings of the 12th ACM SIGKDD international conference on Knowledge discovery and data mining, ACM, New York, NY, pp 935–940

Mörchen F, Ultsch A, Nöcker M, Stamm C (2005) Databionic visualization of music collections according to perceptual distance. In: Proceedings of the 6th International Conference on Music Information Retrieval (ISMIR 2005), London, UK, pp 396–403

Ras ZW, Wieczorkowska A (eds) (2010) Advances in music information retrieval. Springer, Berlin

Typke R, Wiering F, Veltkamp RC (2005) A survey of music information retrieval systems. In: Proceedings of the 6th International Conference on Music Information Retrieval (ISMIR 2005), London, UK, pp 153–160

Vatolkin I, Theimer W, Botteck M (2010) AMUSE (Advanced MUSic Explorer) – A multitool framework for music data analysis. In: Proceedings of the 11th International Society for Music Information Retrieval Conference (ISMIR 2010), Utrecht, Netherlands, pp 33–38

*Software references*:

- MIR Toolbox:(free) https://www.jyu.fi/hum/laitokset/musiikki/en/research/coe/materials/mirtoolbox
- RapidMiner: (free and enterprise versions) http://rapid-i.com/
  - Java MP3 Plugin in version 4.6:
    http://java.sun.com/javase/technologies/desktop/media/jmf/mp3/download.html/

    · Copy the plugin into "JAVA_HOME/lib/ext"
    · In Windows, the JAVA_HOME of RapidMiner is the "jre" - directory!
- MusicMiner: (free) http://musicminer.sourceforge.net/
- CLAM: (free) http://clam-project.org/
- jMIR: (free) http://jmir.sourceforge.net/
- AMUSE: (free) http://sourceforge.net/projects/amuse-framework/

# Part IX
# Data Analysis in Banking and Finance

# Conditional Factor Models for European Banks

Wolfgang Bessler and Philipp Kurmann

**Abstract** The objective of this study is to analyze the risk factors and their time-variability that may be well suited to explain the behavior of European bank stock returns. In order to test for the relative importance of risk factors over time, we employ a novel democratic orthogonalization procedure proposed by Klein and Chow (Orthogonalized equity risk premia and systematic risk decomposition. Working Paper, West Virginia University, 2010). The time-variability in estimated coefficients is further modeled by conditional regression specifications that incorporate macroeconomic as well as stock market based information variables. In a final step, these conditional multifactor models are evaluated on their ability to capture return information related to traditional cross-sectional variables. Overall, we provide empirical evidence on time-varying relative factor contributions for explaining European bank stock returns. Moreover, we conclude that conditional multifactor models explain a significant portion of the size, value and momentum effects in cross-sectional regressions.

## 1 Introduction

The banking industry is one of the most important sectors in European capital markets. Banks play a vital role in transmitting monetary policy and in financing growth opportunities for industrial firms worldwide. However, during the recent financial crisis the overall liquidity and solvency of banks became a major public concern causing an immense stress on banks, industrial firms and the real economy.

W. Bessler (✉) · P. Kurmann
Center for Finance and Banking, University of Giessen, Licher Strasse 74, 35394 Giessen
e-mail: wolfgang.bessler@wirtschaft.uni-giessen.de; philipp.kurmann@wirtschaft.uni-giessen.de

W. Gaul et al. (eds.), *Challenges at the Interface of Data Analysis, Computer Science, and Optimization*, Studies in Classification, Data Analysis, and Knowledge Organization, DOI 10.1007/978-3-642-24466-7_44, © Springer-Verlag Berlin Heidelberg 2012

With the emergence of considerable balance sheet problems caused by some new asset classes, banks' share prices dropped significantly, thereby generating substantial losses for equity investors worldwide and highlighting the potential risks of the banking industry.

Because banks are considered to be different from non-financial firms in terms of their business activities, accounting policies and regulatory oversight, their stock returns should be determined by specific factors that are characteristic to the banking industry. Traditional studies focusing on the return determinants of banks emphasized the importance of macroeconomic variables such as interest rates and exchange rates. In terms of non-financial firms, however, including certain firm characteristics for explaining the cross-sectional variation of stock returns has become a standard approach. Fama and French (1993) introduced the mimicking factor portfolios *SMB* and *HML* to account for the historically documented outperformance of small and value stocks, respectively. This three-factor model was extended by incorporating an additional factor (*UMD*) designed to capture the momentum effect (Carhart 1997). Hence, the four-factor Carhart model (4FM) controls for the most important stock return patterns in empirical finance.

Banks and financial firms are usually excluded from studies of asset-pricing models due to their relatively high leverage and specific business model. This questions the validity of particular firm characteristics such as the book-to-market ratio as cross-sectional return determinants for banks. However, Foerster and Sapp (2005) argue that the financial sector is no longer alone in its relatively high level of debt and constitutes a major part of most countries' capital markets, thus, underlining the importance to understand the risk factors driving bank stock returns. This view is supported by Barber and Lyon (1997) who argue that there exists no a priori reason to expect the variables firm size and book-to-market ratio to have different meanings for financial and non-financial firms. Hence, Viale et al. (2009) investigate the pricing relevance of *SMB*, *HML* and *UMD* for size- and book-to-market sorted US bank portfolios. Besides an additional factor capturing innovations to the term spread, however, only the momentum factor appears to provide weak incremental explanatory power in cross-sectional regressions.

Based on this relatively scarce empirical evidence, we analyze the pricing performance of the 4FM for a sample of European bank portfolios. We contribute to the literature on potential risk factors affecting bank stock returns in the context of European capital market integration. Beginning with a democratic orthogonalization procedure allowing for simultaneous factor transformations and, therefore, eliminating the discretionary choice on starting vector and orthogonalization sequence, we decompose banks' return variances over time. In a next step, we analyze the performance of a conditional representation of the 4FM, thereby, explicitly controlling for time-varying coefficients. Further cross-sectional analyses on banks' risk-adjusted returns are conducted in order to detect potential mispricings that remain even after applying the conditional 4FM.

## 2 Data and Methodology

The present study covers a sample of commercial banks, investment banks, mortgage banks as well as security brokers and dealers from Germany, France, and the UK between December 1989 and February 2010. Thus, we analyze the major players in European banking over a sufficiently long time period incorporating the introduction of the Euro and the recent financial crisis. The national banking sectors are represented by value-weighted total return portfolios that alleviate concerns of investability and illiquidity often attributed to equally-weighted portfolios. In terms of the 4FM, *MKT* denotes the national value-weighted market return less the respective one-month interbank rate. The mimicking factor portfolios *SMB*, *HML* and *UMD* are constructed following Fama and French (1993) and Carhart (1997), respectively. To incorporate conditioning information, we rely on variables that formerly have been shown to predict future stock market returns such as dividend yield (*DivYield*), term spread (*Term*), US default spread (*Def*) and lagged excess market return ($MKT_{t-1}$). By incorporating *Def* we assume a sufficient degree of capital market integration which seems to be a reasonable assumption for our country sample. Throughout the paper we take the perspective of a European investor so that all returns are denoted in Euro as provided by Thomson Reuters Datastream.

The unconditional empirical regression model is:

$$R_{P,t} = \beta_{P,0} + \beta_{P,1}\text{MKT}_t + \beta_{P,2}\text{SMB}_t + \beta_{P,3}\text{HML}_t + \beta_{P,4}\text{UMD}_t + \varepsilon_{P,t}. \quad (1)$$

Based on this regression, we decompose bank portfolio return variances by implementing the democratic orthogonalization procedure of Klein and Chow (2010) using rolling window estimations. In comparison to the traditional sequential approach, the democratic procedure is not sensitive to the order in which factors are selected for orthogonalization. Thus, we are not subject to the selection of a particular starting vector and resulting sequence biases. Following Klein and Chow (2010), the matrix of mutually uncorrelated and variance-preserving factors is denoted by:

$$F^{\perp}_{T \times K} = \overleftrightarrow{F}_{T \times K} + 1_{T \times 1} \bar{F}_{1 \times K} S_{K \times K}, \quad (2)$$

where $\overleftrightarrow{F}_{T \times K}$, $1_{T \times 1}$, $\bar{F}_{1 \times K}$, and $S_{K \times K}$ are the demeaned factor matrix, a vector of ones, the mean of the original factor matrix $F_{T \times K}$ and a $K \times K$ matrix that can be interpreted as the inverse of the correlation matrix between the original and orthogonalized factors, respectively.

In a next step, conditional regressions are conducted to explicitly control for time-varying model coefficients. We closely follow Bessler et al. (2009) by assuming a linear form for $\beta$:

$$\beta_{P,k,t} = \beta_{P,k}(Z_{t-1}) = \beta_{0,P,k} + B'_{P,k}z_{t-1}, \quad (3)$$

where $z_{t-1} = Z_{t-1} - E(Z)$ is the normalized vector of deviations of the information variables from their unconditional means, and $B_{P,k}$ is a coefficient vector with the same dimension as $Z_{t-1}$. Based on this conditional version of (1), we analyze time-series regressions and compute risk-adjusted returns for cross-sectional analyses that mitigate the well-known errors-in-variables (*EIV*) problem in the traditional Fama and Macbeth (1973) two-step procedure.

# 3  Empirical Results

## 3.1  Descriptive Statistics

In Table 1, we report descriptive statistics for the European bank portfolios. While each portfolio exhibits positive albeit insignificant average returns, the Jarque-Bera test clearly rejects the null hypothesis of normally distributed returns at any conventional significance level. This evidence might be attributable to our sample period including the recent financial crisis in which bank stock returns are characterized by severe tail risks. However, as the average portfolio size varies between 21 banks for the UK and 30 banks for Germany, we infer that idiosyncratic risk should be negligible in our test assets.

## 3.2  Unconditional Time-Series Regressions

In a first step, we analyze the performance of the four-factor models augmented by a separate factor capturing returns on a diversified European bank portfolio (*BANKS*) consisting of roughly 50 European banking stocks. We expect the latter to exhibit a time-varying but increasing importance for explaining national bank stock returns in the context of European capital market integration. In order to gain insights into the relative importance of these factors over time, we apply the democratic orthogonalization procedure considering a rolling window estimation of 60 months.

Figure 1 documents the time-varying nature of the factor-specific contributions for explaining European bank stock returns. For Germany and France *MKT* seems

**Table 1**  Descriptive Statistics for European bank portfolios

| Variable | DE | FR | UK |
|---|---|---|---|
| Mean in % | 0.48 | 0.70 | 0.37 |
| SD in % | 7.34 | 7.73 | 7.25 |
| Skewness | 0.20 | −0.26 | 0.18 |
| Kurtosis | 6.52 | 5.31 | 6.15 |
| Jarque-Bera | 120.09[a] | 53.77[a] | 96.50[a] |
| Average No. | 30 | 27 | 21 |

[a] denotes statistical significance at the 0.01 level

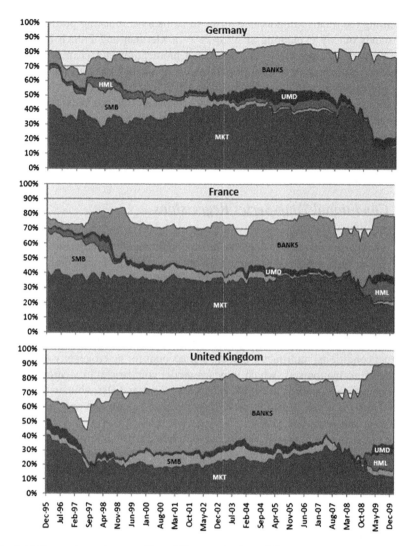

**Fig. 1** Rolling variance decomposition of European bank portfolios

to be the most important factor over the full sample period explaining on average 36% and 35%, respectively. Not surprisingly, *BANKS*' explanatory power remains relatively small before the introduction of the Euro while it exhibits a considerable increase afterwards. UK banks even show an increased co-movement with the European bank index beginning at the end of 1997. This evidence might be justified by the rising economic and monetary integration during our sample period. Moreover, *BANKS*' share on bank return variances features a significant increase during the recent financial crisis. This pattern is evident across countries and might be related to the emergence of sector-specific effects during this time period.

Concerning the relative importance of the three factor-mimicking portfolios, only *SMB* stands out by explaining at most 28% of German and French bank stock return variances, respectively. The value and momentum factors merely provide incremental explanatory power during the recent financial crisis for French and UK banks while their contribution is more diverse for German banks.

Overall, we infer that *MKT* and *BANKS* represent the major determinants of European bank stock returns. The mimicking factor portfolios SMB, HML, and UMD provide additional information on the return-generating process, however, mostly limited to particular time periods.

## 3.3 Conditional Time-Series Regressions

To incorporate the previously documented unconditional model dynamics we have to rely on certain variables that capture significant information regarding the future state of the economy as proxied by *MKT*. Hence, we regress its realization in period *t* on lagged realizations of the information variables.

Table 2 reports the multi-predictor results incorporating all four variables including a constant. It becomes evident that the set of statistically significant predictive variables is not homogeneous across countries. Hence, we have to construct conditional multifactor models that account for heterogeneity in national investors' information sets, which impedes model comparison to a certain extent. For Germany, we condition the 4FM on *DivYield*, *Term* and *Def* while in the case of France *DivYield* is substituted by $MKT_{t-1}$ which is, in turn, the only information variable for analyzing UK banks.

The conditional regression results are outlined in Table 3. Across countries, the average coefficient estimates exhibit similar patterns in terms of their sign and statistical significance. Not surprisingly, *MKT* is the most significant factor with estimated exposures between 0.94 for Germany and 1.32 for the UK. The coefficients on *SMB* are always negative, indicating a tilt towards large cap stocks. However, the coefficients are insignificant for Germany. *HML* is consistently significantly positive, thereby, suggesting that our test assets behave like value stocks over the sample period. Concerning the momentum factor, we document significantly negative coefficient estimates for Germany and the UK. This might

**Table 2** Predictive regressions of instrumental variables

| Variable | DE | FR | UK |
|---|---|---|---|
| Alpha | 0.003 | 0.004 | 0.002 |
| DivYield | $0.029^c$ | 0.019 | 0.017 |
| Term | $0.010^b$ | $0.008^c$ | 0.005 |
| Def | $-0.023^b$ | $-0.015^c$ | $-0.011$ |
| $MKT_{t-1}$ | 0.088 | $0.159^b$ | $0.221^a$ |

[a,b,c] denote statistical significance at the 0.01, 0.05, 0.10 level, respectively

**Table 3** Conditional regression results

|  |  | $\beta_0$ | MKT | SMB | HML | UMD |
|---|---|---|---|---|---|---|
| DE | Constant | 0.001 | $0.94^a$ | $-0.14$ | $0.26^a$ | $-0.14^b$ |
|  | DivYield | $-0.018$ | 0.61 | $1.09^b$ | 0.06 | 0.44 |
|  | Term | $-0.004$ | $-0.24^b$ | $-0.34^b$ | $0.51^a$ | $-0.23^b$ |
|  | Def | 0.002 | $-0.63^b$ | $-1.00^a$ | $-0.59$ | $-0.79^a$ |
| FR | Constant | $-0.002$ | $1.07^a$ | $-0.24^b$ | $0.47^a$ | 0.11 |
|  | Term | $-0.002$ | $-0.17$ | 0.07 | 0.18 | $-0.05$ |
|  | Def | $-0.007$ | $-0.35$ | $-0.05$ | $0.61^a$ | $-0.31^c$ |
|  | $MKT_{t-1}$ | $-0.020$ | $-2.03$ | $-0.72$ | $-0.28$ | $-0.01$ |
| UK | Constant | 0.002 | $1.32^a$ | $-0.14^b$ | $0.33^b$ | $-0.15^b$ |
|  | $MKT_{t-1}$ | 0.001 | $-0.92$ | 1.20 | $-0.68$ | $-1.41$ |

$^{a,b,c}$ denote statistical significance at the 0.01, 0.05, 0.10 level, respectively

be at least partly attributable to the strong underperformance of banks during the recent financial crisis.

The importance of conditioning information, i.e. the interaction terms in our multifactor regressions designed to control for time-varying effects in $\beta$'s, is merely observable for the latter estimates. Regarding the German bank portfolio, *Term* stands out as it captures significant time-variation across all factors while *Def* comprises time-varying effects on *MKT*, *SMB* and *UMD*. For the French bank portfolio only *Def* provides further insights into the structure of time-varying betas for *HML* and *UMD*. For UK banks, we are not able to document time-varying effects which might be attributable to the relatively sparse formulation of conditioning information based solely on $MKT_{t-1}$.

Overall, the interest-rate related information variables are most significant and provide valuable insights into the time-varying factor exposures of European banks. Therefore, our results are in line with former studies that emphasize the importance of interest rates for explaining bank stock returns. However, we provide additional evidence on their ability to capture time-varying effects in the context of the 4FM.

## 3.4 Cross-Sectional Regressions

In a final step, we analyze the ability of the conditional 4FM to control for the effects of well known cross-sectional return determinants by employing the two-step procedure of Fama and Macbeth (1973). Since the 4FM is motivated on the basis of particular stock return anomalies the corresponding firm characteristics, i.e. size, book-to-market and past returns over the lagged second to third, fourth to sixth, and seventh to twelfth month, respectively, should not exhibit any statistical significance after returns are risk-adjusted.

The results in Table 4 reveal that the majority of firm characteristics is significantly related to the cross-section of banks' excess returns (*Ex*). For risk-adjusted returns (*Adj*) the number of significant parameter estimates is remarkably reduced.

**Table 4** Cross-sectional regressions

|    |     | Const | Size | B/M | Ret23 | Ret46 | Ret712 |
|----|-----|-------|------|-----|-------|-------|--------|
| DE | Ex  | $0.194^a$ | $-0.025^a$ | $0.238^a$ | $-0.114$ | $-0.045$ | $-0.126^a$ |
|    | Adj | $0.009^a$ | $-0.003^b$ | $0.004$ | $0.020$ | $0.003$ | $-0.002$ |
| FR | Ex  | $0.005^a$ | $-0.002^a$ | $-0.002^a$ | $-0.010^a$ | $-0.011^a$ | $-0.010^a$ |
|    | Adj | $0.010^a$ | $-0.003^a$ | $0.009^a$ | $0.053^b$ | $0.027$ | $0.021^c$ |
| UK | Ex  | $0.000$ | $-0.001^a$ | $0.004^a$ | $-0.004^b$ | $-0.007^a$ | $-0.003^a$ |
|    | Adj | $0.011^a$ | $-0.002$ | $0.020^a$ | $0.005$ | $0.021$ | $-0.004$ |

[a,b,c] denote statistical significance at the 0.01, 0.05, 0.10 level, respectively

Hence, we conclude that the 4FM is able to capture a substantial portion of the cross-sectional effects of size, book-to-market, and past returns. However, the risk-adjustment is imperfect in that coefficients on certain firm characteristics remain significant.

# 4 Conclusions

In this study, we analyze the risk factors and the time-variability of risk factors that explain the behavior of European bank stock returns for the last two decades. By employing a novel, democratic orthogonalization procedure we document a time-varying nature of relative factor contributions for explaining bank stock return variances. Moreover, macroeconomic information variables seem to capture a significant portion of this time-variation in conditional multifactor regressions. Further analyses on risk-adjusted returns indicate that the conditional 4FM controls for a significant proportion of the size, value and momentum effects in cross-sectional regressions.

# References

Barber BM, Lyon JD (1997) Firm size, book-to-market ratio, and security returns: A holdout sample of financial firms. J Finance 52:875–883

Bessler W, Drobetz W, Zimmermann H (2009) Conditional performance evaluation for german equity mutual funds. Eur J Finance 15:287–316

Carhart MM (1997) On persistence in mutual fund performance. J Finance 52:57–82

Fama EF, French KR (1993) Common risk factors in the returns on stocks and bonds. J Financ Econ 33:3–65

Fama EF, Macbeth JD (1973) Risk, return, and equilibrium. J Polit Econ 81:607–636

Foerster SR, Sapp SG (2005) Valuation of financial and non-financial firms: A global perspective. J Int Financ Market Inst Money 15:1–20

Klein RF, Chow KV (2010) Orthogonalized equity risk premia and systematic risk decomposition. Working Paper, West Virginia University

Viale AM, Kolari JW, Fraser DR (2009) Common risk factors in bank stocks. J Bank Finance 33:464–472

# Classification of Large Imbalanced Credit Client Data with Cluster Based SVM

**Ralf Stecking and Klaus B. Schebesch**

**Abstract** Credit client scoring on medium sized data sets can be accomplished by means of Support Vector Machines (SVM), a powerful and robust machine learning method. However, real life credit client data sets are usually huge, containing up to hundred thousands of records, with good credit clients vastly outnumbering the defaulting ones. Such data pose severe computational barriers for SVM and other kernel methods, especially if all pairwise data point similarities are requested. Hence, methods which avoid extensive training on the complete data are in high demand. A possible solution is clustering as preprocessing and classification on the more informative resulting data like cluster centers. Clustering variants which avoid the computation of all pairwise similarities robustly filter useful information from the large imbalanced credit client data set, especially when used in conjunction with a symbolic cluster representation. Subsequently, we construct credit client clusters representing both client classes, which are then used for training a non standard SVM adaptable to our imbalanced class set sizes. We also show that SVM trained on symbolic cluster centers result in classification models, which outperform traditional statistical models as well as SVM trained on all our original data.

R. Stecking (✉)
Department of Economics, Carl von Ossietzky University Oldenburg, D-26111 Oldenburg, Germany
e-mail: ralf.w.stecking@uni-oldenburg.de

K.B. Schebesch
Faculty of Economics, Vasile Goldiş Western University Arad, Romania
e-mail: kbschebesch@uvvg.ro

W. Gaul et al. (eds.), *Challenges at the Interface of Data Analysis, Computer Science, and Optimization*, Studies in Classification, Data Analysis, and Knowledge Organization, DOI 10.1007/978-3-642-24466-7_45, © Springer-Verlag Berlin Heidelberg 2012

# 1   Motivation and Background

Credit client scoring on medium sized credit client data sets can be accomplished very successfully by means of support vector machines (SVM), which are a powerful and robust machine learning method for classification. However, most real life credit client data sets contain several ten or hundred thousand credit client records. Such large data sets impose severe computational barriers for nonlinear SVM, and they may even lead to computational intractability in practice. Such intractability is a general and highly relevant problem in other related classification domains which use large data sets like recommender systems in e-commerce. Hence, new effective solutions which avoid direct extensive training on the complete data are in high demand. In order to design such a non-extensive training method for our credit client data we propose to divide the large data set into a much smaller, but still sufficient number of homogeneous clusters.

These clusters are symbolic objects that will be represented in many alternative ways e.g. by intervals, means, frequency tables or histograms. Classification of such more involved data representations requires an adequate data coding to be combined with appropriate multivariate analysis tools. We show how to construct good and bad credit client behavior clusters that can be easily classified with any SVM variant. Hence, very large credit client data sets can be made accessible to SVM with non linear kernels but also to SVM which handle non standard situations of classification with highly asymmetric class sizes and misclassification costs, both very important in practice. Finally, we forecast future credit client behavior with our model by coding the newly arriving credit client records accordingly.

Generically, we are given a set of $N > 0$ training examples $\{x_i, y_i\}, i = 1, \ldots, N$, with $x_i \in \mathbb{R}^m$ the $m$ credit client features, and their associated labels $y_i \in \{-1, 1\}$, standing for "non-defaulting" and "defaulting" behavior. Support vectors are those training examples which are near the class boundaries of a SVM solution and which permit training of a classifier with the same out-of-sample performance as a classifier trained on all the other (redundant, etc.) training examples as well. In order to follow our exposition in the sequel, it suffices to know that the SVM finally produces a decision rule (a separating function) of the type $y^{pred} = \mathbf{sign}(s(x)) = \mathbf{sign}\left( \sum_{i=1}^{N} y_i \alpha_i^* k(x_i, x) + b^* \right)$, with $0 \leq \alpha_i^* \leq C$ and $b^*$ the result of the SVM optimization. Important is to note that $\alpha_i^* > 0$ (a support vector $i$ referred to as "SV" if $\alpha_i^* < C$ and as "BSV" if $\alpha_i^* = C$) actively contribute to the decision function by invoking the $i$th training example via a user defined kernel $k(,)$, which in most cases is selected to be a semi-positive, symmetric and distance dependent function.

Different aspects of combining clustering with SVM are treated in the more recent literature: Li et al. (2007) introduce a Support Cluster Machine (SCM) as a general extension of the SVM with the RBF kernel, where cluster size and cluster covariance information is incorporated into the kernel function. Evgeniou and Pontil (2002) propose a special clustering algorithm for binary class data, that tends to produce large clusters of training examples which are far away from the boundary between the two classes and small clusters near the boundary.

Yuan et al. (2006) concentrate on large *imbalanced* data sets. They partition the examples from the bigger negative class into disjoint clusters and train an initial SVM with RBF kernel using positive examples and cluster representatives of the negative examples. Yu et al. (2003) construct two micro-cluster trees from positive and negative training examples, respectively, and train a linear SVM on the centroids of the root nodes. Subsequently, training examples near the class boundary are identified and split up into finer levels using the tree structure. Wang et al. (2008) generalize this approach for solving nonlinear classification problems. None of the works mentioned employ classification modeling for credit client data. In a working paper (Stecking and Schebesch 2009) we presented first experiments of clustering credit data for classification with SVM. In this previous work we did not consider imbalanced data sets.

But in general, and also in our case, credit client data is highly imbalanced, i.e. there are much more non defaulting than defaulting credit clients. A survey of the literature on classification methods with unequal class sizes is given by Weiss (2004). An example from the credit industry for this issue is identifying fraudulent credit card transactions (Chan and Stolfo 2001). However, neither Chan and Stolfo (2001) nor Weiss (2004) use SVM as a classification method.

# 2 Extracting Information from Large Data Sets

In applications like credit scoring, the number of cases or clients can be expected to further grow in the future, whereas the number of easily accessible features per client can be expected to remain approximately the same (barring the use of some new and controversial information on health scores, on social networking, etc). Dealing with large enough $N$ can pose problems to almost any classification procedure. In Sect. 3 we describe credit client classification on a large empirical data set which already poses serious problems to nonlinear SVM models which we successfully used in previous work for much smaller credit client data sets.

Apart from seeking to alleviate numerical problems with large data sets, one may wonder whether the information from such large credit client data sets can be "compressed", leading to new models on such compressed data, which are comparable to the performance of models on the much smaller empirical data sets just mentioned.

A possible choice for the compression step is clustering and using cluster prototypes as the new training examples for SVM models. Clustering itself comes in many variants (see e.g. Jain et al. (1999)) employing a multitude of procedures for selecting distances, similarities and other relations between pairs and other subsets of points. By selecting adequately, one can exploit knowledge about geometries and other characteristics of data sets and improve clustering results substantially (e.g. Basu et al. (2009)).

Without having reliable sources of such knowledge for credit scoring data, one would use in a first approximation a simple and fast clustering procedure. Hence, in

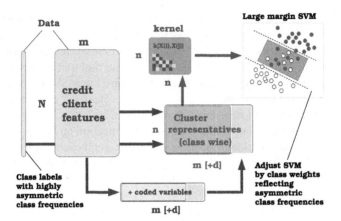

**Fig. 1** Reduce and augment your data set. Build a new training set by replacing the $N$ original cases (clients) by $0 < n \ll N$ prototypes of client clusters

the sequel, the method of choice will be k-means clustering, which, by avoiding the exhaustive computation of pair wise distances, is computationally efficient for very large data sets.

As a general strategy one can attempt to increase or to stabilize the performance of classifiers on large data sets by reducing (or compressing) the number of training examples and, possibly, by adding more derived input features to the $m$ primary inputs, and/or by modifying the representation of inputs, which is summarized in Fig. 1.

In our case of supervised learning the clustering from Fig. 1 is is done for both client classes (i.e. "defaulting" and "non-defaulting") separately. The new training examples are now the cluster representatives. Input feature information representation of these new training examples may be enhanced by adding some dimensions $d$, which contain some additional aggregate or other coded information about the respective cluster. A comparatively small $n \times n$ kernel matrix results for the final SVM classifier. The ubiquitous problem of highly asymmetric relative class frequencies is addressed by using a non-standard SVM with class weights.


## 3   Credit Client Data Set Description

The data set used consists of 139,951 clients for a building and loan credit. There are twelve variables per client of which eight are categorical (with two up to five categories) and four are quantitative. Input variables include personal attributes like *client's profession*, loan related attributes like *repayment rate* and object related attributes like *house type*. The default rate within one year is 2.6% (3,692 out of 139,951). The data set is large and in general not (or hardly) tractable for standard SVM. Furthermore, extremely unequal class sizes will make classification difficult. Former experiments suggest using a small (with equal sized classes) sample for SVM model building. This however leads to just average out-of-sample classification results for the large data set (Stecking and Schebesch 2006). In the present work, an alternative way of dealing with such large data sets will be shown: First, a cluster analysis is performed and then, based on special cluster representations, the SVM is estimated.

The advantage of a preprocessing type cluster analysis is massive down-sizing of the large data set of $N$ cases into a much smaller set of $n \ll N$ clusters. These cluster solutions may include a weighting scheme, e.g. using a balanced number of "good" and "bad" credit client clusters. If there is noise in the data it will be averaged out through clustering. Clusters can be represented as *symbolic objects*. It will be shown, that symbolic object coding is a comprehensive way to preprocess credit client data that easily can incorporate mixed (i.e. categorical *and* quantitative) variable types.

## 4    Generating Data Clusters

In a first step, the data set is divided into "good" and "bad" credit clients. Subsequently, unsupervised k-means clustering (MacQueen 1967) is used, partitioning the large data set into equal numbers of clusters from "good" and "bad" classes respectively, while preserving the class labels. For each of the labeled clusters the relative frequency of all categories per qualitative variable is recorded. Quantitative variables are divided into four equally sized intervals (with quartiles of the full variable range as interval borders). Relative frequencies for these intervals are also recorded. A data cluster finally is represented by 43 inputs between zero and one. Experiments include runs of k-means clustering with 100 clusters.

## 5    Symbolic Coding of Data Clusters

In contrast to "classical" data analysis where elements are individuals with single variable outcomes (*"first degree objects"*), in symbolic data analysis elements, respectively *objects* in general, are *classes of individuals* or *categories of variables* that usually will be described by more than one outcome per variable (*"second degree objects"*) (Bock and Diday 2000). Clusters of credit clients can be seen as such symbolic objects with e.g. an interval representation of the amount of credit that was given to the cluster members. However, a special data description is needed to represent the variable outcomes of a symbolic object. A complete overview of symbolic variable types can be found in Billard and Diday (2006). In the present work the clusters (the symbolic objects) are described by *modal variables* where categories or intervals (of categorical or quantitative) variables appear with a given probability. Each symbolic variable is represented by a vector of probabilities, e.g. $X_u = \{\eta_{u1}, p_{u1}; \ldots; \eta_{us}, p_{us}\}$ where outcome $\eta_{uk}$ (of object $u$ relating to variable $X$ with $k = 1, \ldots, s$ outcomes) is given with a probability $p_{uk}$ and $\sum_{k=1}^{s} p_{uk} = 1$. Outcomes $\eta_{uk}$ may either be categories or (non-overlapping) intervals. Therefore, categorical as well as quantitative variables are represented by a vector of *standardized* values that are strictly between 0 and 1. Furthermore, we are able to refer to past work on the algebra of symbolic variables (in the framework of symbolic data analysis, e.g. Billard and Diday (2006), including descriptive statistics, dependency measures and distance functions as well as diverse multivariate data analysis approaches.

# 6 Classification Results

The initial data set of 139,951 cases is randomly divided into a training and a validation set in a 2 : 1 relation. K-means clustering is used to partition the training set into 100 clusters with equal numbers of "good" and "bad" clusters each. This procedure is repeated ten times resulting in ten different clustered training sets with a related hold-out validation set. Each cluster is represented by modal variables (as outlined before) and can be given as input to SVM with linear and to SVM with RBF kernel. Linear SVM needs hyperparameter tuning for the bound $C$ of the dual variables $\alpha$. Here, setting $C = 100$ leads to a SVM that sufficiently separates "good" from "bad" clusters. In case of SVM with RBF kernel at first the width parameter $\sigma$ of the kernel function $k(u, v) = \exp(-g\|u - v\|^2)$, where $g = \frac{1}{\sigma^2}$, is initialized to the mean Euclidian distance of pairwise comparisons of all input examples. Grid search optimization of width parameter $\sigma$ and model capacity control parameter $C$, which is the bound of the dual variables $\alpha$, then leads to values of $\sigma = 2.58$ and $C = 25$. Furthermore, the cluster representations are of *equal class sizes*, i.e. $\pi_T^- = \pi_T^+ = 0.5$ with $\pi_T^-$ ($\pi_T^+$) as the ratio of good (bad) credit clients in the training set. The validation set on the other hand is *highly imbalanced*, i.e. there are much more good than bad clients with $\pi_V^- = 0.974$ and $\pi_V^+ = 0.026$. SVM dealing with such data situations require just one additional parameter $L$ describing the proportions of class membership in the training and in the validation set (Lin et al. 2002). We define an additional control parameter

$$L = \frac{C^{(+)}}{C^{(-)}} = \frac{\pi_T^- \cdot \pi_V^+}{\pi_T^+ \cdot \pi_V^-} = \frac{0.5 \cdot 0.026}{0.5 \cdot 0.974} = 0.0267,$$

that allows for class wise different bounds $C^{(+)}$ and $C^{(-)}$ instead of $C$. This leads to a different classification accuracy for both classes.

How can cluster based SVM be used to predict the behavior of unknown credit clients? For this purpose, we introduce *relative frequency coding* for the ten hold-out validation sets. Each credit client of the validation set is assigned to a category (or an interval) with a probability of either zero or one. In this way data descriptions for individuals match with those of the data clusters. Subsequently, future credit

**Table 1**    Area under curve (AUC) statistics computed for ten randomly selected validation sets with $N = 46650$. Models are trained on 100 cluster representations

| | AUC (Validation sets with $N = 46650$) $n = 100$ Cluster | | | |
| | Mean | Std.Dev. | Minimum | Maximum |
|---|---|---|---|---|
| SVM Linear | 0.667 | 0.012 | 0.651 | 0.687 |
| SVM RBF | 0.685 | 0.007 | 0.676 | 0.697 |
| Lin. Disc. analysis | 0.623 | 0.025 | 0.564 | 0.647 |
| Logistic regression | 0.597 | 0.021 | 0.556 | 0.614 |

client behavior can be predicted by a classification function that was estimated on a clustered training set before. In the sequel four classification models are evaluated: linear SVM, SVM with RBF kernel, Linear Discriminant Analysis (LDA) and Logistic Regression. LDA is the most traditional (Durand 1941), Logistic Regression the most common approach in the credit scoring industry (Thomas et al. 2005). Owing to the extremely unequal class sizes of our large data set standard measures of accuracy are impractical for models to be estimated on such data and must be replaced by a more complex measure. A good candidate is the *ROC* (Receiver operating characteristics) curve (Hanley and McNeil 1982) which offsets true and false positive rates of a classification model using model predictions against reference output (i.e. the true labels of the data set). The area under curve *(AUC)* then denotes the area between a "pure chance" diagonal and the *ROC* curve as a measure for the cut-off independent classification power of a decision function.

We used training sets with 100 clusters from ten random splits of the large data set. In order to assess the performance of the four different models *ROC curves* for hold-out validation sets were generated and the *area under curve (AUC)* statistics were calculated. Classification results are shown in Table 1. Both SVM clearly dominate the traditional methods, the SVM with RBF kernel being the best one. The minimum *AUC* of the worst SVM (0.651) is still higher than the maximum *AUC* of the best traditional method (0.647). In terms of standard deviation the *AUC* results of both SVM are more stable than the ones of the traditional methods. LDA and Logistic Regression used the same training and validation cycle as the SVM in order to allow for a fair comparison. In our former experiments a small data sample (instead of data clusters) was used for model building. This lead to AUC of 0.576 (SVM RBF), 0.583 (Linear SVM), 0.582 (LDA) and 0.584 (Logistic Regression), see Stecking and Schebesch (2006).

# 7    Conclusions

Our exploratory experimental approach indicates that cluster based SVM models used on very large credit client data sets are successful in describing and predicting credit client defaulting behavior. Furthermore, symbolic coding of clusters is introduced, which enables representation of more complex cluster information. This leads to competitive classification results with regard to *ROC* properties of the classification models, which are superior to traditional models as well as to former experiments using sample based data sets. A model building process using cluster based SVM is a suitable way for treating highly asymmetric credit client data containing an overwhelming number of non-defaulting cases, most relevant for banking practice. For large data sets, computationally less demanding clustering algorithms like k-means must be employed. The a priori choice of distance measures on the data points and the preselection of the symbolic encoding may limit or bias the expression of complex class boundaries. As yet, the effect of using different class-wise clusters numbers or even clusterings is also unexplored. Further work will study validation procedures, which allow for adapting symbolic cluster encodings and similarity measures to our credit client data.

# References

Basu S, Davidson I, Wagstaff K (2009) Constrained clustering: Advances in algorithms, theory, and applications. Data mining and knowledge discovery series. Chapman Hall/CRC Press, Boca Raton, FL

Billard L, Diday E (2006) Symbolic data analysis. Wiley, New York

Bock HH, Diday E (2000) Analysis of symbolic data: Exploratory methods for extracting statistical information from complex data. Springer, Berlin

Chan PK, Stolfo SJ (2001) Toward scalable learning with non-uniform class and cost distributions: A case study in credit card fraud detection. In: Proceedings of the Fourth International Conference on Knowledge Discovery and Data Mining, pp 164–168

Durand D (1941) Risk elements in consumer installment financing. National Bureau of Economic Research, New York

Evgeniou T, Pontil M (2002) Support vector machines with clustering for training with very large datasets. Lect Notes Artif Intell 2308:346–354

Hanley A, McNeil B (1982) The meaning and use of the area under a receiver operating characteristics (ROC) curve. Diagn Radiol 143:29–36

Jain AK, Murty MN, Flynn PJ (1999) Data clustering: A review. ACM Comput Surv 31(3):264–323

Li B, Chi M, Fan J, Xue X (2007) Support cluster machine. In: Proceedings of the 24th International Conference on Machine Learning, New York, pp 505–512

Lin Y, Lee Y, Wahba G (2002) Support vector machines for classification in nonstandard situations. Mach Learn 46(1–3):191–202

MacQueen JB (1967) Some methods for classification and analysis of multivariate observations. In: Proceedings of the Fifth Symposium on Math, Statistics and Probability, University of California Press, Berkeley, CA, pp 281–297

Stecking R, Schebesch KB (2006) Variable subset selection for credit scoring with support vector machines. In: Haasis HD, Kopfer H, Schönberger J (eds) Operations research proceedings. Springer, Berlin, pp 251–256

Stecking R, Schebesch KB (2009) Clustering large credit client data sets for classification with SVM. In: Credit Scoring and Credit Control XI Conference, CRC Edinburgh, p 15 ff.

Thomas LC, Oliver RW, Hand DJ (2005) A survey of the issues in consumer credit modelling research. J Oper Res Soc 56(9):1006–1015

Wang Y, Zhang X, Wang S, Lai KK (2008) Nonlinear clustering–based support vector machine for large data sets. Optim Meth Software Math Programm Data Mining and Machine Learning 23(4):533–549

Weiss GM (2004) Mining with rarity: A unifying framework. SIGKDD Explorations 6(1):7–19

Yu H, Yang J, Han J (2003) Classifying large data sets using SVMs with hierarchical clusters. In: Proceedings of the ninth ACM SIGKDD international conference on Knowledge discovery and data mining, ACM, New York, KDD '03, pp 306–315

Yuan J, Li J, Zhang B (2006) Learning concepts from large scale imbalanced data sets using support cluster machines. In: Proceedings of the ACM International Conference on Multimedia. ACM, New York, pp 441–450

# Fault Mining Using Peer Group Analysis

David J. Weston, Niall M. Adams, Yoonseong Kim, and David J. Hand

**Abstract** There has been increasing interest in deploying data mining methods for fault detection. For the case where we have potentially large numbers of devices to monitor, we propose to use peer group analysis to identify faults. First, we identify the peer group of each device. This consists of other devices that have behaved similarly. We then monitor the behaviour of a device by measuring how well the peer group tracks the device. Should the device's behaviour deviate strongly from its peer group we flag the behaviour as an outlier. An outlier is used to indicate the potential occurrence of a fault. A device exhibiting outlier behaviour from its peer group need not be an outlier to the population of devices. Indeed a device exhibiting behaviour typical for the population of devices might deviate sufficiently far from its peer group to be flagged as an outlier. We demonstrate the usefulness of this property for detecting faults by monitoring the data output from a collection of privately run weather stations across the UK.

## 1  Introduction and Related Work

Databases consisting of multiple time series present new challenges in data mining. These challenges include classification, regression and clustering. One way to approach such problems is to view the collection of time series as realizations from

D.J. Weston (✉)
The Institute for Mathematical Sciences, Imperial College London, UK
e-mail: david.weston@imperial.ac.uk

N.M. Adams · D.J. Hand
Department of Mathematics, Imperial College London, UK
e-mail: n.adams@imperial.ac.uk, d.j.hand@imperial.ac.uk

Y. Kim
Industrial Systems Engineering, Yonsei University, Seoul, South Korea
e-mail: yoonseong@yonsei.ac.kr

W. Gaul et al. (eds.), *Challenges at the Interface of Data Analysis, Computer Science, and Optimization*, Studies in Classification, Data Analysis, and Knowledge Organization, DOI 10.1007/978-3-642-24466-7_46, © Springer-Verlag Berlin Heidelberg 2012

a population. This implies that all the time series are subject to the same dynamics. However, it is easy to conceive of databases that consist of collections of time series subject to different dynamics. Transaction histories in plastic card finance databases are an example, the behaviours of account holders are likely to be qualitatively dissimilar. The potential heterogeneity of a collection of time series is a serious problem. One approach is to treat each series separately. A more refined recent approach, called peer group analysis (Bolton and Hand 2001; Weston et al. 2008), attempts to find and utilise groups of time series that exhibit similar behaviour.

In this work we wish to detect faults in many devices simultaneously. Crucially we assume that for each device there are others that co-evolve to a certain degree – this is the peer group principle. We compare a device's measurements with those identified as co-evolving. This has two interesting properties. First, we do not need a model to describe the device's behaviour since the co-evolving time series automatically take this into account. Second, an outlier from these locally co-evolving time series need not be a global outlier to the population, as distinct from anomaly detection methods.

Venkatasubramanian et al. (2003a,b,c) provide an extensive review that splits the fault detection and diagnosis literature into model based methods and process history based methods. The model based approach uses either quantitative approaches (Venkatasubramanian et al. 2003c), such as modelling the process using a Kalman filter, or qualitative approaches (Venkatasubramanian et al. 2003a), such as using causal models. The process history methods rely on using knowledge derived from past data (Venkatasubramanian et al. 2003b). These methods too can be split into quantitative and qualitative approaches, where statistical methods and expert systems are respective examples.

In the following section we present peer group analysis (PGA) in more detail. In Sect. 3 we describe the data set. This data consists of weather measurements from real weather stations. The data is not labelled with occurrences of faults, so in Sect. 4 we introduce a definition of a fault. Section 4 continues with a description of the experiments performed and the results obtained. Finally, Sect. 5 discusses conclusions and describes future work.

## 2   Peer Group Analysis

PGA is an unsupervised method for monitoring behaviour over time (Bolton and Hand 2001). Its uses include fraudulent behaviour detection in plastic card transactions (Bolton and Hand 2001; Weston et al. 2008).

We assume we have $m$ time series, $x_1, \ldots, x_m$. Each time series is real valued multivariate with dimension $V$. Here, the time series are all time aligned and of equal length. $x_i(t, v)$ denotes the value of the $v$th variable at time $t$ for the $i$th time series. The interval $t \in \{1, \ldots, D\}$ is the historical data used to build peer groups.

PGA is split into two parts. First for each times series, or *target*, a peer group for each is found. This can be achieved in many ways, for example using apriori information, or by comparing the time series themselves. Second, after identifying the peer groups, we track the behaviour of each target with respect to its peer group.

## 2.1 Peer Group Membership

In the case of measuring daily meteorological conditions, Sharif and Burn (2007) use the Mahalanobis distance where the covariance matrix is measured over the set of candidate days. This suggests a simple way to compare time series consisting of $D$ days of daily meteorological data, which is to sum the square of the Mahalanobis distance for each day. We follow the method described in Weston et al. (2008) which is essentially this distance measure with the inclusion of a simple dimensionality reduction step. The method proceeds as follows. We first subdivide the historical data for each time series into $N$ non-overlapping windows. For each time series we calculate the mean of the data within each window. For the time series $x_i$, which has length $D$, we construct a new times series of length $N$,

$$w_i(1, \tfrac{D}{N}), \ldots, w_i((n-1)\tfrac{D}{N} + 1, n\tfrac{D}{N}) \ldots, w_i((N-1)\tfrac{D}{N} + 1, D)$$

where $w_i(s, e)$ is a column vector of length $V$ representing the average value of the $i$th time series from $t = s$ to $t = e$, i.e. $s, e$ are the start and end days of each window. To simplify notation we use $w_i(n)$ to denote the $n$th window ($n = 1, \ldots, N$).

The distance $d(x_i, x_j)$ between two time series, $x_i$ and $x_j$, is the square root of sum of the squared Mahalanobis distances for each window.

$$d(x_i, x_j) = \sqrt{\sum_{n=1}^{N}(w_i(n) - w_j(n))^T C_n^{-1}(w_i(n) - w_j(n))},$$

where $C_n$ is the covariance matrix of the $n$th window over the population of time series and $T$ is the transpose. We use the Mahalanobis distance to handle changes in population behaviour between windows, as well as to handle variable scaling.

Once we have a measure of similarity between devices, we can build the peer groups. For time series $x_i$ we sort the remaining time series in order of increasing distance (decreasing similarity), yielding $x_{\pi(1),n}, \ldots, x_{\pi(p-1),n}$. A peer group of size $k$ for the time series $x_i$ is simply the first $k$ time series in this ordering.

Weston et al. (2008) set the peer group size manually. Here we propose a method to automatically set the peer group size.

## 2.2 Peer Group Size

In order to set the peer group size for a target time series, we measure how closely that time series is tracked by the mean of its peer group over the training data.

We choose the size of the peer group that minimizes the overall tracking distance

$$(x_{i,t} - \mu(t,k))^T C_t^{-1} (y_{i,t} - \mu(t,k)), \tag{1}$$

where $\mu(t,k)$ is the mean of $k$th closest time series at time $t$ and $C_t$ is the covariance of the population at time $t$.

The intuition behind this idea is, if we assume there exists an optimal size of the peer group for a time series, then peer groups that are larger than this optimal size will contain behaviours that do not track the time series so well. Consequently the distance between the time series and the mean of the peer group will tend to diverge. On the other hand, if the peer group is smaller than the optimal size it is likely to be more sensitive to noise. This is analogous to the problem of bandwidth selection in kernel density estimation. We set the peer group size to be

$$\operatorname*{argmin}_{k} \sum_{t=1}^{D} (x_{i,t} - \mu(t,k))^T C_t^{-1} (y_{i,t} - \mu(t,k)). \tag{2}$$

Having obtained peer groups for each time series, we are in a position to monitor subsequent behaviour. Here, we are interested, in particular, with outlier behaviour.

## 2.3  Peer Group Outlier Detection

We monitor how closely each time series is tracked by its peer group. There are a number of possibilities to achieve this goal. Weston et al. (2008) use the Mahalanobis distance of the target from the mean of its peer group. The covariance matrix used for the Mahalanobis distance is the covariance matrix of the peer group members. We cannot follow this approach, since in our case the automatically sized peer groups can be too small to robustly measure their covariance matrices. We use the tracking distance Equation 1 to monitor a target from its peer group. Each time a target is further from its peer group than some externally set threshold we flag it as an outlier.

## 3  Weather Data

Weather Underground Inc.[1] delivers weather information from a variety of sources including one project that collects data from privately owned weather stations around the world. Each weather station has its own dedicated web page where all

---

[1]http://www.wunderground.com/

data uploaded is available for public view. The uploaded data are sensor readings for a number of weather variables. Information about the location and type of weather station is also available.

Not all weather stations measure the same variables. We selected 200 weather stations from the United Kingdom that had active sensors for humidity, temperature, wind speed and dew point. We obtained data for a period of 71 days starting from January 1st 2007. The data consisted of two sets. The first set is the *raw data* that consisted of the values uploaded to the website for the sensors described above. The second set is the *daily average data* generated by the website from the raw data.

## 4 Experiments and Results

The experiments are split into three sections. As noted above we do not have fault labelled data. In the first section we describe how we analysed the raw data to generate fault labels for the daily average data. The next section describes a method to remove global outliers. Finally we describe the experimental details and results of the PGA algorithms.

The first 50 days of the data were used to build peer groups and the remaining 21 days were used to evaluate the performance of the method.

### 4.1 Fault Definition

To evaluate the fault detection performance of our method we need to know where and when faults occurred. Unfortunately, we do not have fault labelled data. Instead we define a fault based on two types of error that occur in the daily average data.

An offline weather station is represented as missing values in the raw data. The algorithm used to generate the daily average data on the website assumed the missing values to be zero. Consequently the daily average will typically be lower than the true value. The longer the station is offline the larger the error in the daily average.

Another type of error occurs in the daily average data when a weather station gets locked into a cycle of repeatedly submitting the same value. This results in the daily average being either higher or lower than the true value, depending on the weather conditions at that time. Of course, there is the possibility that the weather conditions have not changed and the repetitions are therefore legitimate.

We say a fault has occurred on a weather station on day $d$, if there has been greater than or equal to $t$ hours of missing data (not necessarily contiguous) for the variables temperature, humidity and dew point. From the data, we cannot distinguish between a calm day of zero wind speed and missing data, so we do not consider an extended period of missing wind speed as a fault.

A fault also is said to have occurred if there are greater than or equal to $t$ contiguous hours of repeated data on any of the variables temperature, dew point and non-zero wind speed. We do not consider extended periods of constant humidity as a fault since it can legitimately remain the same value (to the precision of the recorded measurements) for long intervals. Note, there are other forms of fault such as systematic errors in which a device has been incorrectly calibrated which we do not address here.

The value of $t$ was set by investigating the number of days classified as containing a fault for different values of $t$ over the test data, as shown in Fig. 1. We assume that faults are relatively rare, therefore we have selected the elbow point as a cut-off for the value of $t$. For $t < 4$ it is likely we will observe a preponderance of fault days. For $t = 4$, 141 days were labelled as faults out of a total 4,200 (this is equivalent to 200 weather stations and 21 test days).

As noted above, there is the possibility that a repetition is legitimate. However, the longer the duration of measurement repetition the more likely it is to be an erroneous reading (especially in the UK). Note also that the longer the period of repetition the more likely the daily average value datum will deviate strongly from its true value and therefore is more likely to be detected as an outlier. We therefore decided to investigate how well our method can detect faults defined for periods, $t \geq 4, 8, 16, 20$.

Obviously our fault definition provides us with a method for fault detection. Monitoring the raw output from the weather stations would be a straightforward way to detect these specific faults. However, monitoring only the daily average data provides a more challenging problem for fault detection and solutions to which, we hope, have more general applicability.

## 4.2   Global Outlier Removal

As we have noted previously, one advantage of PGA is that outliers to the peer group are not necessarily outliers to the population. This is useful in the context of fault

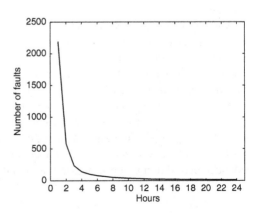

**Fig. 1** Number of days classified as having a fault for different values of $t$

**Fig. 2** Number of days
classified as a global outlier
for different values of $c$

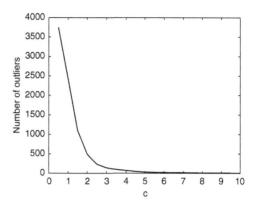

**Table 1** Number of outliers
that are faults for 4 h and 20 h
fault definitions

| 4 h | Faults | Non-faults | Total |
|---|---|---|---|
| Outliers | 21 | 121 | 142 |
| Non-outliers | 120 | 3,938 | 4,058 |
| Total | 141 | 4,059 | 4,200 |
| 20 h | Faults | Non-faults | Total |
| Outliers | 1 | 141 | 142 |
| Non-outliers | 14 | 4,044 | 4,058 |
| Total | 15 | 4,185 | 4,200 |

detection only if the faulty sensor readings are not always outliers to the population of sensor readings. We investigate how effective PGA is at finding "inlying outliers" and how often a fault is a global outlier.

A simple outlier detector is used to find global outliers from the test data, described as follows. Each sensor variable is standardised independently such that on each day the set of measurements from all of the weather stations has mean zero and variance one. This makes all the sensor variables on each day commensurate. A weather station is a global outlier on a particular day if any of its sensor readings is more than $c$ standard deviations from the mean. Figure 2 shows the number of days that are declared as outliers for different values of $c$. Since global outliers should be rare, we set $c = 3$. Examples for the number of faults before and after global outlier removal are shown in Table 1. Note that most of the fault days, for each definition of fault, are not global outliers.

## 4.3  Peer Group Analysis Experiments

The data used to build the peer groups is not fault free. The effects of those faults on measuring the similarity between time series is attenuated by using the dimensionality reduction step described in Sect. 2.1. The 50 days used to build peer groups were subdivided into 5 windows of length 10 days. The peer group sizes for each weather station was selected automatically as described in Sect. 2.2.

A weather station is flagged as having a fault should the distance between the weather station and its peer group exceed a user set threshold. To define this threshold requires the specification of decision costs. To avoid this we use ROC curves to measure the performance of our method. ROC analysis allows us to measure the performance of our method over all threshold values.

The fault detection performance of PGA is shown in Fig. 3. We see that performance decreases as the fault definition allows shorter periods to be defined as faults. This is as we would expect since the effect of the repeated data will have a smaller effect on the daily average.

If we remove those days from the performance evaluation for each weather station that are global outliers, we obtain Fig. 4. We see the performance is

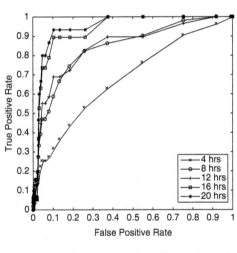

**Fig. 3** Fault detection performance including outliers days

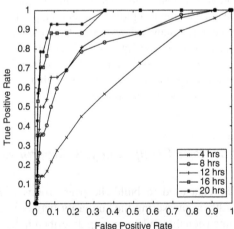

**Fig. 4** Fault detection performance with outliers days ignored

consistently slightly worse than with the global outliers. This demonstrates that outliers to peer groups are not necessarily outliers to the population and we can use this to detect faults that are themselves not outliers to the population of weather stations.

## 5 Conclusions and Future Work

In this paper we have demonstrated the use of PGA for mining faults in multiple time series. We have shown from a synthetic definition of a fault that faulty sensor readings may not necessarily be global outliers. As such, the utility of our method, which can detect more local phenomena, has been demonstrated.

One further noteworthy point regarding PGA is that we only need to specify a measure of distance rather than building a model of usual behaviour in order to detect anomalies. From a purely data mining perspective, this is very useful since the peer group method can be deployed to discover unusual behaviour using very little domain knowledge.

**Acknowledgements** We would like to express appreciation to Weather Underground, Inc. for use of their data. The work of David Weston was supported by grant number EP/C532589/1 from the UK Engineering and Physical Sciences Research Council. The work of Yoonseong Kim was supported by the Korea Research Foundation Grant funded by the Korean Government (MOEHRD) (KRF-2006-612-D00100). The work of David Hand was partially supported by a Royal Society Wolfson Research Merit Award.

## References

Bolton RJ, Hand DJ (2001) Unsupervised profiling methods for fraud detection. In: Conference on Credit Scoring and Credit Control 7, Edinburgh, UK, 5-7 Sept
Sharif M, Burn DH (2007) Improved k-nearest neighbor weather generating model. J Hydrolog Eng 12:42
Venkatasubramanian V, Rengaswamy R, Kavuri SN (2003a) A review of process fault detection and diagnosis Part II: Qualitative models and search strategies. Comput Chem Eng 27(3):313–326
Venkatasubramanian V, Rengaswamy R, Kavuri SN, Yin K (2003b) A review of process fault detection and diagnosis Part III: Process history based methods. Comput Chem Eng 27(3):327–346
Venkatasubramanian V, Rengaswamy R, Yin K, Kavuri SN (2003c) A review of process fault detection and diagnosis Part I: Quantitative model-based methods. Comput Chem Eng 27(3):293–311
Weston DJ, Hand DJ, Adams NM, Whitrow C, Juszczak P (2008) Plastic card fraud detection using peer group analysis. Adv Data Anal Classif 2(1):45–62

# Part X
# Data Analysis in Health and Environment

# Discovering Possible Patterns Associations Among Drug Prescriptions

Joana Fernandes and Orlando Belo

**Abstract** The constant growth of data storage associated to medication prescriptions allows people to get powerful and useful information by applying data mining techniques. The information retrieved by the patterns found in medication prescriptions data can lead to a wide range of new management solutions and possible services optimization. In this work we present a study about medication prescriptions in northern region of Portugal. The main goal is to find possible relations among medication prescriptions themselves, and between the medication prescribed by a doctor and the lab associated with those medications. Since this kind of studies is not available in Portugal, our results provide valuable information for those working in the area that need to make decisions in order to optimize resources within health institutions.

## 1 Introduction

The constant growth of all kinds of data and the advance in technology occurred in the last years have allowed companies to get huge amounts of useful information, giving them important market opportunities based on task-oriented knowledge (Sumathi and Sivanandam 2006). Such knowledge can be obtained by the extraction of trends, patterns or other possible anomalies found in data, which are in the most of the cases not visible by simple data analysis processes. Therefore, that kind of need, made companies focusing on tasks related to *Knowledge Discovery in Databases* (KDD) and Data Mining fields (Fayyad et al. 1996a,b). That was not different for public institutions involved in the clinical practice sector, that sooner or later recognized the utility of those methods and strategies for discovering associations. Therefore, the use of electronic records in clinical practice, containing all the steps

J. Fernandes · O. Belo (✉)
Department of Informatics, School of Engineering, University of Minho, Gualtar, Braga, Portugal
e-mail: joanapfernandes@gmail.com; obelo@di.uminho.com

W. Gaul et al. (eds.), *Challenges at the Interface of Data Analysis, Computer Science, and Optimization*, Studies in Classification, Data Analysis, and Knowledge Organization, DOI 10.1007/978-3-642-24466-7_47, © Springer-Verlag Berlin Heidelberg 2012

of a medical appointment, can be used to help decision making processes and give an entire view of all the health services provided from all points of view (Tomás et al. 2008), supplying enriched elements for better management and decision-making processes. Based on that, the main goal of this work is to find possible relations among medication prescriptions. This study was limited to the northern region of Portugal and, more precisely, just to some of the health centers in each district. The final dataset used had a total of approximately 2 millions records and each record is defined by only one prescribed medication. Three different studies were carried out. In the first, the main goal was to find associations between the prescriptions of medication. In the second study, the main goal was to find possible relations with the doctor's prescriptions and the labs of the prescribed medication. In this case, the results were not very conclusive due to a problem in the original data. The third study was an overall study of the prescriptions pattern or more precisely of the dataset characteristics. With this work we concluded that the scope of progression is huge in this area and there are still a lot of possible studies that can be done. During the next sections we will present and discuss all the processes and tasks related to all the studies carried out.

## 2   Medication Prescription Patterns

### 2.1   A Real World Scenario

The potential and useful information that we can get with data mining applications like this one, and the lack of this kind of studies in Portugal were the main reasons for us to do this work. Despite the small number of initiatives similar to this one that have been made in Portugal until now, all of them provided valuable and potential information for decision-making (Carvalho 2008). All the data used in this work were provided by *Administração Regional de Saúde do Norte* (ARSN), the public company responsible for all the health centers and services in the north of Portugal. The data represents all the records about medication prescriptions since 1 of January of 2008 to 31 of December of 2008, for all health centers in North Portugal. In total, the data set had approximately 21 million of records. Although data were collected from all the health centers, we used only a part of it, containing a specific selection of records. This choice was based on the population characteristics as representative of all the existing population, like people abilities or its social-economic level. After the selection, the final data set was approximately 10% of the initial data volume, consisting of a total of approximately 2 million records.

### 2.2   From Raw Data to Patterns

The large amounts of data that are daily stored in operational systems are now easier to analyze and to collect due to the information systems technologies advances

(Fayyad et al. 1996b). The process of analysis is normally defined as the discovering of trends, patterns or possible anomalies in the data, which cannot be easily found by conventional methods (Lloyd-Williams et al. 1995). The need of finding those trends and patterns is made by using data mining mechanisms that allow the extraction of useful information by applying different algorithms to a considerable amount of data. It is essential that the information source is presented as single, separate, clean, integrated and self-consistent (Connolly and Begg 1998). This means that before the application of any data mining technique, a serious quality review and evaluation of the available data should be done.

# 3   Discovering Relevant Patterns

## 3.1   The Knowledge Discovering Process

For a greater quality and impact in the application of data mining techniques to this work, all the steps of the KDD process were carefully and meticulously followed in this work: from the selection, pre-processing and transformation of the data, to the data mining application and interpretation and evaluation of results. In a first phase an exploratory analysis was made in order to find what was the real quality of the data for each axes of analysis of medical prescription correlations. In a second phase, based on the results of the first phase and on our final goal, we applied some cleaning and selection options, where some of the records were removed from the data set due to some incongruities or because the medication were completely irrelevant to this study. After this, we decided to implement a specific data structure that could easily respond to the data mining application needs and to be appropriated to receive the several axes of analysis presented by the managers of the organization we were working with. For that reason, a data warehouse was implemented with the ability to support such structure and all the future analysis and decision making processes. The data warehouse has only a single star scheme (Fig. 1) integrating a total of six dimensions. Beyond these six dimensions, the fact table has also 9 degenerated dimensions, and one single measure representing the total number of medication for each stored record. The grain of the data mart corresponds to one medication prescribed by a doctor, to a user, in a certain local, and in a specific period of time.

## 3.2   Mining Models and Techniques

In order to get some relevant knowledge from the data about prescriptions, we used three data mining techniques, each one for a different knowledge extraction goal. Defining an association model and discovering process to the selected data set were the most relevant part of our work. In this case, our main goal was to find

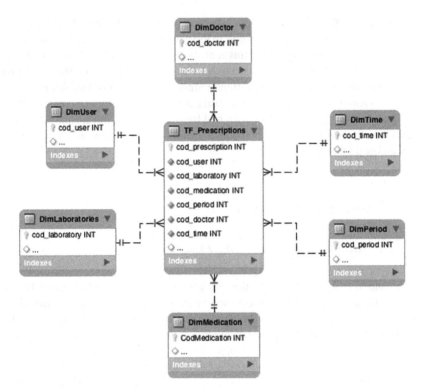

**Fig. 1** Data Warehouse star scheme

associations between the prescriptions of medication and to find possible relations
with the doctor's prescriptions and the correspondent labs of the prescribed medication. However, in order to get a general description about how prescriptions are
applied in Portugal, we used clustering algorithms to get clusters of similar patterns
and then a classification algorithm on those clusters to get a better description and
characterization of them. As we know, association rules are normally used to find
the nature of the causalities between values of different attributes, more precisely
to find relationships between different attributes in large customers databases
(Aggarwal and Yu 1999), although our data set was not exactly like a database of
customers company's. We define each prescription made as a transaction and the
medication prescribed as the products bought. Seeing the problem in this manner,
we can use our dataset in a way where the application of association rules is quite
ordinary.

The algorithm used for finding frequent *itemsets* was the FP-Growth (*Frequent
Pattern Growth*) (Han et al. 2000). The fact that this algorithm presents some
advantages related to the implementation and functioning when comparing to the
Apriori algorithm (Han et al. 2000) was the main reason to use it. The efficiency of
the FP-Growth algorithm is achieved by the application of three different methods:

(1) a large database is compressed in a much smaller and condensed data structure, which avoids repeated scans to the database that usually are very costly; (2) the FP-Tree structure adopts a pattern fragmented growth that avoids additional costs in the creation of large number of candidates sets; and (3) the partitioning method based in the divide-and-conquer algorithm reduces dramatically the search space. One of the major advantages of the FP-Growth algorithm relies on the fact that it needs only two scans over the transaction database in order to construct the FP-Tree. In the first scan, all the frequent *itemsets* and their supports are calculated and sorted in a descending order of support count for each transaction. In the second scan, the items in each transaction are kept together into a prefix tree and common nodes that appear in different transactions are counted. Common nodes are linked together by a pointer called node-link. Due to the fact that the items are sorted descending by their support value, the nodes closer to the root of the prefix tree are the most common in the transactions (Han et al. 2000). In conclusion, the choice of FP-Growth is justified essentially by the fact that, comparatively to Apriori, FP-Growth presents lower computational costs and is a very reliable algorithm in finding frequent patterns with all sizes (Wu et al. 2008; Hipp et al. 2000).

As we said previously, we also applied clustering and classification algorithms to the final dataset. The main goal of clustering problems is to group similar records together in a large database of multidimensional records. In this way we created sets of medication prescription clusters that had similar patterns. For creating the clusters we used the *k*-means algorithm. In a basic way, the *k*-means algorithm aims to partition the dataset in *k* clusters in which each observation belongs to the cluster with the nearest mean. This is done in an iterative method between two steps till convergence (when the assignments of new *centroids* no longer change or hit the maximum number of steps) (Han et al. 2000): (a) each data point is assigned to its closest *centroid* (initial *k* clusters); and (b) for each cluster a new *centroid* is relocated to the center (mean) of all data points assigned to it. The fact that is a simple, easily understandable and scalable algorithm was the main reason for choosing it. Its ability to deal with large datasets in a reduced time was also another important reason.

After, we applied a classification algorithm to the set of clusters returned, in which the class label was the clusters descriptions. A classification problem is referenced as a supervised learning, when the training data is used in order to model the relationships between the attributes and the class label (Aggarwal and Yu 1999). As our main goal for using classification problems was to simplify the final description of the clusters obtained, we decided to use decision tree algorithms. The algorithm used was the C4.5. In a brief way, this algorithm is based in the divide-and-conquer algorithm (Han et al. 2000). More precisely, partitioning recursively the dataset until each partition contains mostly of the examples of a particular class. Each non-leaf node of the tree contains a split point that is a condition to decide how the data should be partitioned. The fact that decision trees are extremely easy to read (and to understand) and produce normally very accurate results were the main reasons for using that kind of algorithms.

# 4 Working Results and Evaluation

As referred, three main studies were carried on, where the third study was specially oriented to provide a general description about eventual patterns of medication prescriptions in Portugal. In the studies for association rules some pre-processing steps had to be done. More precisely, we had to group the medication prescribed for each prescription and then transform the nominal data into binomial data (using the *split* operator from *RapidMiner*). In that way we could apply the FP-Growth algorithm straightforward, as it was a transactional database. As we saw previously, the first study was carried out to find associations between medication prescriptions. Now, for this case, we defined a very low minimum support as 0.001, due to the high number of records and different medication descriptions. The measure used was the lift, and only rules that had a lift greater than 50 were presented. This means that in a rule $A \Rightarrow B$, the appearance of $A$ is associated with the appearance of $B$. We must note that lift is a symmetric measure, unlike the confidence measure. While the confidence is an estimate of $P(C|A)$, the probability of observing $C$ given $A$, the lift measures how far from independence are $A$ and $C$. Lift values close to 1 imply that $A$ and $C$ are independent (Hahsler and Hornik 2007).

In Table 1 it is possible to see the associations found between medication prescriptions for each selected district. Based on this, we can conclude that are some medications that are normally prescribed together. In the second study, the main goal was to find possible relations between doctor's prescriptions and laboratories of the prescribed medication. In this case the support defined was also 0.001 for the same reasons mention above. However, in this case, the measure used was confidence, because now we wanted to know if there is any relation between the prescription of the doctor and the laboratories of the medication prescribed. More precisely,

**Table 1** Associations between the medication prescribed

| Districts | Rules | Support | Lift |
|---|---|---|---|
| Aveiro | Adalgur N $\Rightarrow$ Arthtotec | 0.001 | 67.92 |
| | Vigantol $\Rightarrow$ Cebiolon | 0.002 | 444.1 |
| | Ben-U-Ron $\Rightarrow$ Brufen Suspensão | 0.002 | 55.15 |
| Braga | Atrovent Unidose $\Rightarrow$ Ventilan | 0.001 | 94.22 |
| | Vigantol $\Rightarrow$ Cebiolon | 0.001 | 248.5 |
| Bragança | Vigantol $\Rightarrow$ Cebiolon | 0.001 | 436.5 |
| | Ben-U-Ron $\Rightarrow$ Brufen Suspensão | 0.002 | 54.26 |
| | Atrovent Unidose $\Rightarrow$ Ventilan | 0.002 | 73.99 |
| Porto | Relmus $\Rightarrow$ Voltaren | 0.002 | 113.0 |
| | Vigantol $\Rightarrow$ Cebiolon | 0.001 | 248.5 |
| | Profenid $\Rightarrow$ Relmus | 0.001 | 69.72 |
| | Atrovent Unidose $\Rightarrow$ Ventilan | 0.001 | 104.2 |
| Viana do Castelo | Relmus $\Rightarrow$ Voltaren | 0.002 | 172.7 |
| | Profenid $\Rightarrow$ Relmus | 0.001 | 188.5 |
| Viana Real | Atrovent Unidose $\Rightarrow$ Ventilan | 0.002 | 105.8 |
| | Vigantol $\Rightarrow$ Cebiolon | 0.002 | 302.7 |

**Table 2** Association between doctors and laboratories of the medication prescribed

| Rules | Support | Confidence |
|---|---|---|
| Med_819674 ⟹ Sanofi Aventis | 0.001 | 0.200 |
| Med_219975 ⟹ Sanofi Aventis | 0.001 | 0.200 |
| Med_458534 ⟹ Sanofi Aventis | 0.001 | 0.203 |
| Med_13734 ⟹ Sanofi Aventis | 0.001 | 0.203 |
| Med_989878 ⟹ Sanofi Aventis | 0.001 | 0.207 |
| Med_366277 ⟹ Sanofi Aventis | 0.001 | 0.207 |
| Med_6370 ⟹ Sanofi Aventis | 0.001 | 0.208 |
| Med_162131 ⟹ Sanofi Aventis | 0.001 | 0.208 |
| Med_77941 ⟹ Sanofi Aventis | 0.001 | 0.209 |
| Med_812629 ⟹ Sanofi Aventis | 0.001 | 0.211 |
| Med_576377 ⟹ Sanofi Aventis | 0.002 | 0.212 |
| (...) | (...) | (...) |

in a $A \Rightarrow B$ rule, means that the doctor $A$ prescribes the laboratory $B$ with $c\%$ of confidence, independently of the rest of the prescriptions of the laboratory $B$ by other doctors. Some results can be seen in Table 2 – in order to preserve their identity the doctors' codes were encrypted.

Finally, in the third study we applied the $k$-means algorithm to the dataset in order to get a set of clusters with similar relations. First of all, we transform the nominal data to numerical data, and then in order to reduce the range of values, we normalize the data with a *Z-transformation* function. After some evaluation of the best $k$ (number of clusters) using the Davies-Bouldin index (Davies and Bouldin 1979), we defined the value $k$ as 13. We also defined the max number of optimization steps as 500, and the max number of runs as 50. After all the 13 clusters were created we applied the *C4.5* algorithm in order to get a better description of the clusters created. In this step, the label class was the description of the clusters; this means a total of 13 different classes. The results obtained showed us some patterns that can be designated as valuable information for resource optimization within health institutions: (1) the districts of Braga, Porto and Bragança have similar prescriptions characteristics, as well as Aveiro, Vila Real and Viana do Castelo; (2) usually, women go more often to health centers on Monday and Wednesday; (3) in Winter the health centers have normally more older people with health problems, when compared with Summer that has normally more younger people going to the health centers; and (4) the number of prescriptions for people under 15 years are normally related to the districts of Braga, Porto, Vila Real.

## 5  Conclusions and Future Work

In this work we presented a set of studies about how to find prescription patterns using data mining techniques. From all the data that was supplied from ARSN, involving a selected group of health centers of the northern region of Portugal, we just used a total of 10% of the all data once we selected a restricted set of target districts. All the results were discussed with the ARSN, and we realized that some of the found associations were expected; others were new and were considered as useful, namely the ones related to the medication prescription association study. Results were truly satisfactory and demonstrated some of the reality that exists in medication prescriptions in the target regions. Therefore, it would be very interesting to expand this first study to the rest of the districts to find out the main differences between each district. In the second study carried out, the results were not very conclusive due to some problems detected in the data provided to us – for example, one lab can be associated with different health areas, like oncology and cardiology, and although they both belong to the same lab, they should be differentiated. In this case, the analysis of the results turned out to be more difficult, but still, we were able to get good results and obtain new information. In the third study, we could get a general description of the overall patterns of the medication prescriptions in the northern region of Portugal. This final study was very interesting to see some curious characteristics of some of the medication patterns for some districts. Finally, we concluded that the scope of progression for this work is huge and there are still a lot of possible studies that can be done approaching the field of medication prescriptions.

**Acknowledgements** We would like to express our gratefulness to ARSN for their support in all the phases of this work, and specially for all the data supplied which made this work possible.

# References

Aggarwal CC, Yu PS (1999) Data mining techniques for associations, clustering and classification. In: PAKDD 1999: Proceedings of the Third Pacific-Asia Conference on Methodologies for Knowledge Discovery and Data Mining, pp 13–23

Carvalho MC (2008) Prescrição de antibióticos nos centros de saúde da região de saúde do norte: Padrão e variabilidade geográfica. Universidade do Porto, Porto

Connolly TM, Begg CE (1998) Database systems: A practical approach to Design, implementation and management, 2nd edn. Addison-Wesley, Longman

Davies DL, Bouldin DW (1979) A cluster separation measure. IEEE Trans Pattern Anal Mach Intell 2:224–227

Fayyad U, Piatetsky-Shapiro G, Smyth P (1996a) From data mining to knowledge discovery in databases. AI Mag 13:37–54

Fayyad U, Piatetsky-Shapiro G, Smyth P (1996b) The KDD process for extracting useful knowledge from volumes of data. Comm ACM 39:27–34

Hahsler M, Hornik K (2007) New probabilistic interest measures for association rules. Intell Data Anal 11:437–455

Han J, Jian P, Yin Y (2000) Mining frequent patterns without candidate generation. In: SIGMOD 2000: Proceedings of the 2000 ACM SIGMOD International Conference on Management of Data, pp 1–12

Hipp J, Guntzer U, Nakhaeizadeh G (2000) Algorithms for association rule mining – a general survey and comparison. ACM SIGKDD Explorations Newsletter 2:58–64

Lloyd-Williams M, Jenkins J, Howden-Leach H, Mathur R, Cook I, Morris C (1995) Knowledge discovery in an infertility database using artificial neural networks. In: IEE Colloquium on knowledge discovery in databases, vol 21, pp 1–3

Sumathi S, Sivanandam SN (2006) Introduction to data mining and its applications. Springer, New York

Tomás A, Broeiro P, Faria-Vaz A (2008) Os sistemas de prescrição electrónica. Associação Portuguesa dos Médicos de Clínica Geral 24:632–640

Wu X, Kumar V, Quinlan JR, Ghosh J, Yang Q, Motoda H, McLachlan G, Ng A, Liu B, Yu P, Zhou ZH, Steinbach M, Hand D, Steinberg D (2008) Top 10 algorithms in data mining. Knowl Inform Syst 14:1–37

# Cluster Analytic Strategy for Identification of Metagenes Relevant for Prognosis of Node Negative Breast Cancer

Evgenia Freis, Silvia Selinski, Jan G. Hengstler and Katja Ickstadt

**Abstract** Worldwide, breast cancer is the second leading cause of cancer deaths in women. To gain insight into the processes related to the course of the disease, human genetic data can be used to identify associations between gene expression and prognosis. Moreover, the expression data of groups of genes may be aggregated to metagenes that may be used for investigating complex diseases like breast cancer. Here we introduce a cluster analytic approach for identification of potentially relevant metagenes. In a first step of our approach we used gene expression patterns over time of erbB2 breast cancer MCF7 cell lines to obtain promising sets of genes for a metagene calculation. For this purpose, two cluster analytic approaches for short time-series of gene expression data – DIB-C and STEM – were applied to identify gene clusters with similar expression patterns. Among these we next focussed on groups of genes with transcription factor (TF) binding site enrichment or associated with a GO group. These gene clusters were then used to calculate metagenes of the gene expression data of 766 breast cancer patients from three breast cancer studies. In the last step of our approach Cox models were applied to determine the effect of the metagenes on the prognosis. Using this strategy we identified new metagenes that were associated with metastasis-free survival patients.

E. Freis (✉)
Department of Statistics, Dortmund University of Technology, Dortmund, Germany

Leibniz Research Centre for Working Environment and Human Factors (IfADo),
Dortmund University of Technology, Dortmund, Germany
e-mail: freis@statistik.tu-dortmund.de

S. Selinski · J.G. Hengstler
Leibniz Research Centre for Working Environment and Human Factors (IfADo),
Dortmund University of Technology, Dortmund, Germany

K. Ickstadt
Department of Statistics, Dortmund University of Technology, Dortmund, Germany

W. Gaul et al. (eds.), *Challenges at the Interface of Data Analysis, Computer Science, and Optimization*, Studies in Classification, Data Analysis, and Knowledge Organization, DOI 10.1007/978-3-642-24466-7_48, © Springer-Verlag Berlin Heidelberg 2012

# 1 Introduction

To date complex diseases are supposed to be influenced by groups of co-regulated genes related to biological processes rather than by single genes. Usually, the expressions of a group of genes with similar expression patterns in diseased tissue are aggregated to a so-called metagene that reflects the "mean" expression of these genes. The metagenes in turn may then be used as prognostic factors of cancer patients (Rody et al. 2009; Schmidt et al. 2008).

There exist a number of possible approaches to obtain promising sets of genes for metagene calculation (Freis et al. 2009; Schmidt et al. 2008; Winter et al. 2007). In this paper we introduce a cluster analytical strategy to identify metagenes relevant for the prognosis of node negative breast cancer. Our approach is based on cluster and survival analysis and differs from common approaches, that define metagenes by, e.g., clustering the RNA expression of a given set of genes (Winter et al. 2007). In a first step we used gene expression patterns over time of erbB2 breast cancer cell lines. ErbB2 is a proto-oncogene that can be induced to identify groups of genes that might be related to breast cancer prognosis since their overexpression in breast cancer is associated with worse prognosis (Slamon et al. 1987). Co-regulated genes are genes with similar expression patterns over time. In the present study they were identified by a cluster analysis approach for short time-series of gene expression data. We next focussed on gene groups with an overrepresentation of GO categories or common transcription factor binding sites in their promoter regions. These clusters were used in turn to calculate metagenes of the gene expression data of 766 breast cancer patients. As a last step of our approach, Cox proportional hazards models were applied to determine the effect of the metagenes on the prognosis in terms of metastasis-free survival. Several potentially relevant metagenes were detected based on this strategy.

# 2 Data

The following two types of data were used in our cluster analytic approach.

In the breast cancer MCF7 cell line, the oncogenic variant of erbB2 NeuT was induced by doxycycline (for details, see Hermes 2007; Trost et al. 2005). Three independent batches of MCF7/NeuT cells were exposed to doxycycline for periods of 0, 6, 12 and 24 h as well as for 3 and 14 days in independent experiments. Gene expressions at these six time points were assessed using the Affymetrix Gene Chip HG-U133A-plus_2 and analyzed at the microarray core facility of the Interdisziplinäres Zentrum für klinische Forschung Leipzig (Faculty of Medicine, University of Leipzig). For the proposed cluster analysis, we selected 2,632 probe sets differentially expressed at least at one time point using moderate t-tests and the False Discovery Rate (for details, see Krahn 2008).

For the breast cancer data we studied the metastasis-free survival interval (MFI) of 3 individual cohorts (Schmidt et al. 2008) consisting of 766 node negative breast cancer patients, who did not receive any systemic therapy. The MFI was computed from the date of diagnosis to the date of diagnosis of distant metastasis. The first study cohort consists of 200 consecutive lymph node negative breast cancer patients treated at the Department of Obstetrics and Gynecology of the Johannes Gutenberg-University Mainz (for patients' characteristics, see Schmidt et al. 2008). Gene expression profiling of the patients' RNA was performed using the Affymetrix HG-U133A array and the GeneChip System. Two breast cancer Affymetrix HG-U133A microarray data sets were additionally downloaded from the National Center for Biotechnology Information Gene Expression Omnibus (GEO) data repository. The Rotterdam cohort represents 286 lymph node negative patients and the Transbig cohort consists of 280 samples from breast cancer patients. Appendant GEO sample numbers are listed in the Supplementary Tables previously published by Schmidt et al. (2008).

## 3   Methods

The cluster analytic strategy proposed here (see Fig. 1) differs from common approaches in that it combines information about time-dependent gene expressions in BRC cell lines and biological a priori knowledge to obtain meaningful gene sets for the subsequent metagene calculation and survival analysis in a unique way. Other approaches, e.g., Schmidt et al. (2008) perform hierarchical cluster analysis but

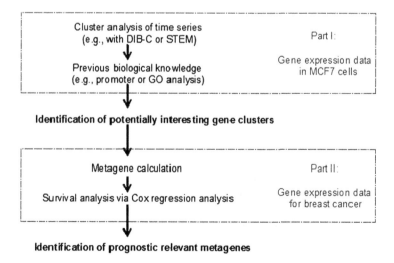

**Fig. 1** Cluster analytic strategy consisting of two parts leading to identification of prognostic relevant metagenes. Underlying part I are gene expressions in cell line data, underlying part II are gene expressions in breast cancer data

without considering the temporal dynamics and biological backround information. Winter et al. (2007) cluster the RNA expression of biological related genes without considering gene regulation over time. Thus, in contrast to (Schmidt et al. 2008) and (Winter et al. 2007) the present approach identifies gene clusters with similar gene expression profiles and common biological features that are likely to be co-regulated. In the following we will describe the different steps of the strategy in more detail.

## 3.1 Cluster Analysis

One of the goals of temporal gene expression analysis is to identify genes with similar expression patterns over time. There exist a number of possible approaches to analyze short-time series with simplification-based strategies, which usually characterize the temporal profiles with a set of symbols indicating different states or trends of the gene expression time series (Wang et al. 2008). Here, the following two approaches were used in the first step of our strategy.

The Short Time-Series Expression Miner (STEM) (Ernst et al. 2005) is based on predefined model profiles that describe the potentially different gene expression patterns. In a first step the gene expression of each probe set at time point 0, $y_{i0}$, is subtracted from the gene expression at the following time points. For each probe set $i$ and each time point $t$ the repetitions are summarized to a median gene expression time series $\tilde{y}_{it}$. The set of model profiles is defined by two parameters $c$ and $m$, where $c$ describes the amount of change of the gene expression between two successive time points and $m$ defines the maximum number of representative clusters. The subset of model profiles is chosen so that the minimum distance between any two model profiles in the subset is maximized. The median gene expression profiles $\tilde{y}_{it}$ are assigned to the cluster with the minimum distance $d(\tilde{y}_{it}, m_{kt})$ with $m_{kt}$ being the time-dependent $k$-th model profile. The distance $d$ is defined as $d = 1 - \rho$ where $\rho$ is the correlation coefficient of Bravais-Pearson between the gene expression profile and the model profile

$$\rho((\tilde{y}_{it})_{t=1,\dots,T}, (m_{kt})_{t=1,\dots,T}) = \frac{\sum_{t=1,\dots,T}(\tilde{y}_{it} - \bar{\tilde{y}}_i)(m_{kt} - \bar{m}_k)}{\sqrt{\sum_{t=1,\dots,T}(\tilde{y}_{it} - \bar{\tilde{y}}_i)^2 \sum_{t=1,\dots,T}(m_{kt} - \bar{m}_k)^2}}$$

with bars indicating arithmetic means. Significant model profiles are identified by comparing the number of assigned gene expression profiles to their random distribution over the set of model profiles.

The Difference-Based Clustering Algorithm (DIB-C) (Kim and Kim 2007) is based on moderate t-tests and assigns for all adjacent time points a sequence of symbols indicating the direction and the rate of change of the gene expression time series. The algorithm characterizes each gene expression pattern of $T$ time points by $(T - 1) + (T - 2)$ symbols. The first order difference is defined as the gradient between two time points (denoted by Increase (**I**), Decrease (**D**) or No change (**N**)) and the second order difference is specified by the shape of the expression time curve between three adjacent time points (denoted by conVex (**V**), concAve (**A**) and **N**o change (**N**)). To reduce the number of small clusters groups that are smaller than a pre-defined threshold are dissolved and the corresponding probe sets are then assigned to the nearest cluster with the most correlated genes. For the distance metric $d$ the Bravais-Pearson empiric correlation coefficient (cf. STEM) of medians over all objects is used.

## 3.2   Identification of Potentially Relevant Cluster

The result of the cluster analysis is a number of sets of co-expressed genes. These co-expressions might be the result of common regulatory mechanisms. Thus, in the next step of our strategy we identified sets of genes with common biological features for a subsequent metagene calculation. For this purpose, we applied promoter or GO analysis.

Promoter analysis searches for transcription factors (TF) whose binding sites are significantly enriched in a given set of genes compared to a predefined background set of genes (here all human genes) (Elkon et al. 2003). We applied the **PR**omoter Integration in **M**icroarray **A**nalysis (**PRIMA**) program (Elkon et al. 2003) that runs under the **EXP**ression **AN**alyzer and **D**isplay**ER** 2.0 software package (**EXPANDER**) (Shamir et al. 2005) to all clusters of genes with similar expression profiles.

An alternative approach for cluster selection is the use of **G**ene **O**ntology (**GO**) criteria instead of promoter analysis as applied, for instance, by Glahn et al. (2008). Gene ontology analysis aims to identify GO groups that are significantly enriched in a given set of genes compared to a predefined background set of genes (here also all human genes). It is performed using **TANGO** (**T**ool for **AN**alysis of **GO** enrichments) (Tanay 2005) that runs under EXPANDER, too.

## 3.3   Metagene Calculation

In the following step, the identified gene clusters with an enrichment of particular TF binding sites or with overrepresented GO categories were used to calculate metagenes of the breast cancer data.

There exist a number of possible approaches to calculate metagenes of gene expression data (Schmidt et al. 2008; Winter et al. 2007, e.g.,). In (Freis et al. 2009) we compared the approach of (Schmidt et al. 2008) with three modifications thereof. Here we applied a robust version of the commonly used standardization with median gene expression value $\tilde{x}_i$ in the respective probe set $i, i = 1, \ldots, n$, and the median absolute deviation $MAD(x_i)$

$$x_{ij\_norm} = \frac{x_{ij} - \tilde{x}_i}{MAD(x_i)}.$$

Using this normalization, a metagene $z$ for patient $j$, $j = 1, \ldots, m$, and gene cluster $C$ was calculated as a median value of all normalized gene expressions for this person and corresponding probe sets

$$z_{j,C} = \tilde{x}_{ij\_ij\_norm}, i \in C.$$

## 3.4 Survival Analysis

The Cox proportional hazards model is a well-known semi-parametric method for modelling survival data. It relates the time of an event to a number of explanatory variables known as covariates. In the last step of our strategy we applied Cox models to determine the effect of the metagenes $z = (z_1, \ldots, z_j)$ as covariates on the prognosis $h(t, z)$ in terms of metastasis-free survival time $t$, $h(t, z) = h_0(t)exp(\beta z)$, where $h_0(t)$ is the baseline hazard rate and $\beta = (\beta_1, \ldots, \beta_j)$ are the parameters. Survival time was compared with the two-sided Log-rank test at a significance level of $\alpha = 0.05$ with correction for multiple testing.

## 4 Results

The aim of our paper was to introduce a cluster analytical strategy including additional biological knowledge to identify metagenes relevant for the prognosis of node negative breast cancer. For this purpose, in a first part we used MCF7 breast cancer cell lines with the induced oncogene erbB2 and in a second part the gene expression data and survival of 766 breast cancer patients.

Due to arbitrary and restricted selection of basis model profiles used STEM, we mainly focussed on the DIB-C algorithm. In the first step of our approach this algorithm detected 692 gene clusters in MCF7 cell lines with similar expression patterns over time. After Bonferroni-adjustment 36 of them were significant in the combined cohort, 5 in the Mainz and the Rotterdam cohorts, and 1 in the Transbig collective.

The subsequent promoter analysis using PRIMA or the GO analysis using TANGO reduced the non-adjusted number of significant clusters to 21 and 60 potentially relevant gene groups, respectively. The number of significant clusters after corrections for multiple testing fell to zero in three individual cohorts. But considering clusters significant in the combined cohort after Bonferroni-adjustment and in at least two individual cohorts without adjustment yields 23 gene groups.

In the next step of our cluster analytic approach we used the resulting clusters from part I to calculate metagenes of the gene expression data of breast cancer patients. Here we focus on one gene cluster characterized by NNNDN, NNNV in the DIB-C code, where gene expressions decreased 3 days after induction of erbB2 by doxycycline. This gene group was significant in the combined cohort as well as in the Mainz cohort after adjustment and in the two other collectives without adjustment. The expressions of associated genes of breast cancer in 766 patients were in turn used to calculate metagene of this gene group.

In the last step of our approach we analyzed with Cox models, how the metagene obtained for this cluster predicts prognosis in breast cancer. Here the high metagene expression is associated with worse prognosis for metastasis-free survival (combined cohort: $p_{adj} = 0.004$, HR $= 1.710$). This observation was validated analyzing the three cohorts individually. A more detailed report on the results is in preparation.

All analyses were carried out using R 2.11.1. (R Development Core Team 2010).

# 5   Discussion

In this paper we present a strategy for identifying prognostic relevant metagenes in breast cancer data. Note that in the different steps of the strategy the particular methods used may be supplemented by other similar approaches. So, e.g., the modified STEM algorithm as well as the model based or spherical cluster analysis as applied, for instance, in Dortet-Bernadet and Wicker (2008) will be used as alternative cluster algorithm for the subsequent metagene calculation and survival analysis.

Besides promoter and GO analysis further prior knowledge about biological processes may be used for cluster selection.

Using the resulting clusters from part I we calculate metagenes of the gene expression data of breast cancer patients. In this paper, we focussed on one new gene set that was associated with MFI by analyzing 766 breast cancer patients. The receptor tyrosine kinase ErbB2 is generally considered as an adverse prognostic factor (Petry et al. 2010). Six of the fourteen genes (RAD51, GINS2, FAM64A, CDCA8, LOC146909, ASF1B) in this cluster were associated with worse prognosis. GINS2 and CDCA8 play a role in proliferation (DNA replication and mitotic spindle stability, respectively), RAD51 and ASF1B are involved in DNA repair, whereas the precise functions of LOC146909 and FAM64A are not yet known. Thus not surprisingly, the calculated metagene was also associated with worse prognosis when using the Cox model in the last step of the strategy. For $\alpha = 0.05$ and correction for multiple testing no TF bindings sites and GO criteria were overrepresented here. The corresponding and more detailed report on this cluster and their biological validation and interactions is in preparation.

Nevertheless, further experiments are needed to understand the exact roles of the identified gene sets in tumour development.

**Acknowledgements**  We would like to thank Ulrike Krahn, Marcus Schmidt, Mathias Gehrmann, Matthias Hermes, Lindsey Maccoux, Jonathan West, and Holger Schwender for collaboration and numerous helpful discussions.

# References

Dortet-Bernadet JL, Wicker N (2008) Model-based clustering on the unit sphere with an illustration using gene expression profiles. Biostatistics 9(1):66–80

Elkon R, Linhart C, Sharan R, Shamir R, Shiloh Y (2003) Genome-wide in silico identification of transcriptional regulators controlling the cell cycle in human cells. Genome Res 13:773–780

Ernst J, Nau GJ, Bar-Joseph Z (2005) Clustering short time series gene expression data. Bioinformatics 21(1):159–168

Freis E, Selinski S, Weibert B, Krahn U, Schmidt M, Gehrmann M, Hermes M, Maccoux L, West J, Schwender H, Rahnenfhrer J, Hengstler J, Ickstadt K (2009) Effects of metagene calculation on survival: An integrative approach using cluster and promoter analysis. In: Sixth International Workshop on Computational Systems Biology, Tampere, Finland, TICSP series 48, pp 47–50

Glahn F, Schmidt-Heck W, Zellmer S, Guthke R, Wiese J, Golka K, Hergenroder R, Degen GH, Lehmann T, Hermes M, Schormann W, Brulport M, Bauer A, Bedawy E, Gebhardt R, Hengstler JG, Foth H (2008) Cadmium, cobalt and lead cause stress response, cell cycle deregulation and increased steroid as well as xenobiotic metabolism in primary normal human bronchial epithelial cells which is coordinated by at least nine transcription factors. Arch Toxicol 82:513–24

Hermes M (2007) Konditionale Expression von Her2/NeuT: Einfluss auf die Zell- und Tumorenentwicklung. PhD thesis, University of Leipzig

Kim J, Kim JH (2007) Difference-based clustering of short time-course microarray data with replicates. Bioinformatics 8:253

Krahn U (2008) Identifikation von Clustern in Gene-Expressions-Zeitreihen zur Analyse der Zellentwicklung. Diploma thesis, TU Dortmund

Petry IB, Fieber E, Schmidt M, Gehrmann M, Gebhard S, Hermes M, Schormann W, Selinski S, Freis E, Schwender H, Brulport M, Ickstadt K, Rahnenfuhrer J, Maccoux L, West J, Kolbl H, Schuler M, Hengstler JG (2010) ERBB2 induces an antiapoptotic expression pattern of Bcl-2 family members in node-negative breast cancer. Clin Cancer Res 16(2):451–460

R Development Core Team (2010) R: A language and environment for statistical computing. Vienna, Austria, URL http://www.R-project.org

Rody A, Holtrich U, Pusztai L, Liedtke C, Gaetje R, Ruckhaeberle E, Solbach C, Hanker L, Ahr A, Metzler D, Engels K, Karn T, Kaufmann M (2009) T-cell metagene predicts a favourable prognosis in estrogen receptor negative and HER2 positive breast cancers. Breast Cancer Res 11:R15

Schmidt M, Bohm D, von Torne C, Steiner E, Puhl A, Pilch H, Lehr HA, Hengstler JG, Kolbl H, Gehrmann M (2008) The humoral immune system has a key prognostic impact in node-negative breast cancer. Cancer Res 68:5405–5413

Shamir R, Maron-Katz A, Tanay A, Linhart C, Steinfeld I, Sharan R, Shiloh Y, Elkon R (2005) EXPANDER - an integrative program suite for microarray data analysis. BMC Bioinformatics 6:232

Slamon DJ, Clark GM, Wong SG, Levin WJ, Ullrich A, McGuire WL (1987) Human breast cancer: Correlation of relapse and survival with amplification of the HER-2/neu oncogene. Science 235:177–182

Tanay A (2005) Computational analysis of transcriptional programs: Function and evolution. PhD thesis, Tel Aviv University

Trost TM, Lausch EU, Fees SA, Schmitt S, Enklaar T, Reutzel D, Brixel LR, Schmidtke P, Maringer M, Schiffer IB, Heimerdinger CK, Hengstler JG, Fritz G, Bockamp EO, Prawitt D, Zabel BU, Spangenberg C (2005) Premature senescence is a primary fail-safe mechanism of ERBB2-driven tumorigenesis in breast carcinoma cells. Cancer Res 65:840–849

Wang X, Wu M, Li Z, Chan C (2008) Short time-series microarray analysis: Methods and challenges. BMC Syst Biol 2:58

Winter SC, Buffa FM, Silva P, Miller C, Valentine HR, Turley H, Shah KA, Cox GJ, Corbridge RJ, Homer JJ, Musgrove B, Slevin N, Sloan P, Price P, West CM, Harris AL (2007) Relation of a hypoxia metagene derived from head and neck cancer to prognosis of multiple cancers. Cancer Res 67(7):3441–3449

# Part XI
# Analysis of Marketing, Conjoint, and Multigroup Data

# Image Clustering for Marketing Purposes

Daniel Baier and Ines Daniel

**Abstract** Clustering algorithms are standard tools for marketing purposes. For example, in market segmentation, they are applied to derive homogeneous customer groups. However, recently, the available resources for this purpose have extended. So, e.g., in social networks potential customers provide images – and other information as e.g. profiles, contact lists, music or videos – which reflect their activities, interests, and opinions. Also, consumers are getting more and more accustomed to select or upload personal images during an online dialogue. In this paper we discuss, how the application of clustering algorithms to such uploaded image collections can be used for deriving market segments. Software prototypes are discussed and applied.

## 1 Introduction

The popularity of classification methods in marketing research can be easily seen from their detailed description in nearly all available marketing and marketing research textbooks as well as their heavy usage in statistical software packages like SAS or SPSS, both by practitioners and academic researchers. So, e.g., in market segmentation, hierarchical or partitioning algorithms are applied to sociodemographic, psychographic, preference, or usage descriptions of potential customers in order to derive homogeneous groups of them.

However, recently, through the success of social network services (e.g. facebook, flickr) available resources for this purpose have extended. So, in social networks, potential customers meet and share their interests, activities, and opinions with other

D. Baier (✉) · I. Daniel
Institute of Business Administration and Economics, Brandenburg University of Technology
Cottbus, Postbox 101344, 03013 Cottbus, Germany
e-mail: daniel.baier@tu-cottbus.de

W. Gaul et al. (eds.), *Challenges at the Interface of Data Analysis, Computer Science, and Optimization*, Studies in Classification, Data Analysis, and Knowledge Organization, DOI 10.1007/978-3-642-24466-7_49, © Springer-Verlag Berlin Heidelberg 2012

network members by providing, e.g., self-describing profiles, contact lists, images, music, or videos. The actual ARD/ZDF-online survey (ARD/ZDF 2010) assesses that in 2010 already 69.4% of the Germans aged 14 or more are at least occasional internet users and that 45% of them use social networks daily, additional 42% of them at least weekly. 69% of these 49 Mio. German internet users have already uploaded an image, 15% a video (ARD/ZDF 2010). Mostly, these images and videos reflect their individual activities, opinions, and interests, e.g., w.r.t. their living conditions as well as their spare time and holiday experiences or their favorite stars, songs, films, or pictures. Consequently these different media types could be used as a basis for market segmentation where personal activities, interests, and opinions play a major role. In this paper, a first step in this direction is investigated where the grouping of customers is done using uploaded images. The paper is organized as follows: In Sect. 2 the main idea behind the usage of image clustering for marketing purposes is introduced and two own prototypes for analyzing image similarities are discussed. In Sect. 3 available feature extraction and grouping algorithms for image clustering are reflected. Section 4 discusses the viability of the new approach.

## 2   Usage of Image Clustering for Marketing Purposes

Market segmentation is still the primary use of clustering methods in marketing research (see, e.g., Wedel and Kamakura 2000). Since Smith's well-known intro-ductory article (Smith 1956), this application task is associated with the division of a heterogeneous market into a number of smaller homogeneous markets, in order to provide a possibility to satisfy the customers' desires more precisely. The following four steps are usual for this purpose (see, e.g., Baier 2003, Baier and Brusch 2008 for more details): In a first step ("probing" in Wedel and Kamakura's 2000 termi-nology), the most important determinants of buying behavior are identified. In a second step ("partitioning"), homogeneous groups of potential buyers are identified by applying an adequate segmentation method to an adequate segmentation base usually consisting of collected data from a representative sample of respondents that reflect their activities, interests, and opinions. In a third and fourth step, the segmentation results are used for selecting target segments ("prioritizing") and for developing an attractive strategy for these selected market segments ("positioning"). In principle, using images or image collections from respondents in the second step could also yield valuable segmentation results, if the used segmentation method is able to group consumers accordingly. However, the usage of images as a segmentation base raises additional questions: (1) Does the uploaded visual content really reflect the consumers' activities, interests, and opinions? (2) Which clustering algorithms are able to group the respondents according to their uploaded images?

The view that uploaded images in social networks really reflect consumers' activities, interests, and opinions (question 1) is supported by findings in visual sociology, where usually four different social reasons for sharing images (especially personal photographies) have been identified (see, e.g., van House 2007): Images are uploaded to remember and to construct narratives of one's live and a sense of self and identity (memory, narrative, and identity), to maintain relationships with others by sharing places, events, or activities (relationships), to ensure that others see one as one wishes to be seen by showing self-portraits, images of one's friends, family, possessions, activities, and so forth or by demonstrating one' aesthetic sense, humor, or skill (self-representation), and to reflect one's unique point of view, creativity, or aesthetic sense (self-expression).

In social networks (especially in flickr) uploading images is in this sense often seen as a compact substitute for more direct forms of interaction using emails or blogs. Here, the visual information of the uploaded images indeed reflects consumers' activities, interests, and opinions. Instead of asking the respondents whether they like to go to the sea or they like to go to the mountains during their vacancies, one could ask them to upload a picture of their favorite holiday destination. Instead of asking them whether their apartment or house is traditionally furnished or in a more modern way, one could ask them to upload a picture of their living room. Instead of asking them whether they like to go to parties or they like to stay at home in the evening, one could ask them to upload a picture of their usual evening activities. These pictures are assumed to contain valuable visual information. So, e.g., Sinus Sociovision, a German specialist for socio-cultural research and the development of consumer typologies, uses pictures of typical living rooms to describe differences between their found social milieus (e.g., Sinus Sociovision 2009).

The problem of clustering such uploaded images (question 2) is closely related to the problem of grouping web image search results, a currently intensively studied problem in content-based multimedia information retrieval or image retrieval (see, e.g., Chen et al. 2003, Schmitt 2005, Lew et al. 2006). Here, often the problem occurs how to organize image search results in order to facilitate user's image retrieval. The automatic grouping of images according to similar low-level features (e.g. based on extracted color distributions, textures, or forms) seems to be promising in specific domains of applications (e.g. vacation, outdoor/indoor, landscape images). So, e.g., Rodden et al. (2001) showed through experiments, that the automatic arrangement based on visual similarity helped the participants to divide the search results into expected simple genres with small inter-individual differences. For the automatic arrangement a visual similarity index was used that took colors, textures, and the broad spatial layout of image regions into account. An MDS-like procedure was used for presenting the image search results to the participants. However, the domain-independent semantically precise grouping or arrangement without user involvement (in content-based image retrieval terms: relevance feedback) is still an open problem (see, e.g., Datta et al. 2008).

Moreover, in most statistical software packages (especially SAS or SPSS) algorithms for image feature extraction and clustering are not available. Even in more flexible systems like MATLAB or R image clustering is only possible with additional packages. So, the EBImage and the biOps packages provide procedures for reading, writing, processing, and analyzing images in R. However, their usage is restricted to specific fields, different from our image clustering

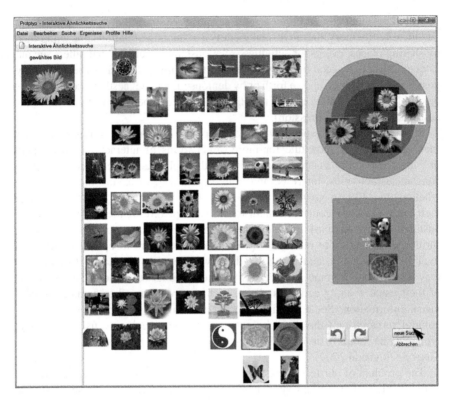

**Fig. 1** Print screen of a software prototype for searching similar images in an image database

purposes. Consequently, a new project has been initiated at Brandenburg University
of Technology Cottbus (BTU) to provide statistical software for this purpose and to
investigate whether these image clustering algorithms could be used for marketing
purposes. The project is funded by the German Federal Ministry of Education and
Research (BMBF). Actually two software prototypes using the same segmentation
methods for different usage scenarios are developed at the chair of marketing and
innovation management and the chair of database and information systems: One
for searching similar images in an image database (see Fig. 1 for a screenshot),
another one for analyzing image databases statistically, including algorithms for
feature extraction, clustering and classification (see Fig. 2 for a screenshot). The
implemented feature extraction, clustering, and classification algorithms base on
the available R- and C++-packages as well as own research in these fields.

Figure 1 shows a typical scenario when an image database has to be searched
for similar pictures to a pre-specified one, here: the sun flower image top left of the
screen. Based on a weighted similarity index w.r.t. extracted low-level features and
an MDS-like procedure pictures are found and spatially arranged according to their
similarity. If the selection and arrangement does not reflect the user's perceived
similarity, he/she is asked for a relevance feedback. He/she can indicate wrongly

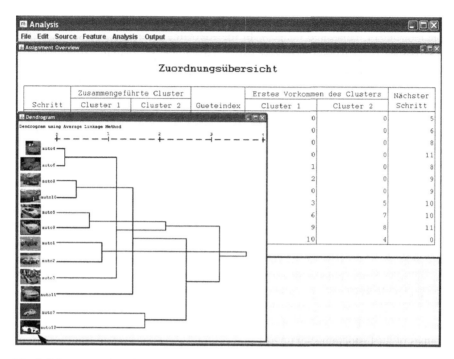

**Fig. 2** Print screen of a software prototype for analyzing image databases statistically

placed pictures by moving them to the right of the screen (drag and drop): Dissimilar pictures that are too closely placed to the pre-selected one are moved to the red box, similar pictures that are too distantly placed are moved to the green dartboard. With this additional information, the weights can be modified and an improved selection and arrangement can be displayed until the user is satisfied with the search results.

Additionally, Fig. 2 shows – as another image grouping scenario – the results of a hierarchical clustering w.r.t. 12 pre-selected pictures. Here, again, extracted low-level features and a weighted similarity index are used. In the following sections some of the available algorithms are discussed and applied in order to show their usefulness for marketing purposes.

# 3 Image Clustering Algorithms

As already mentioned, for image clustering two approaches are popular: an approach where the user is asked to evaluate (and modify) derived similarities and groupings (see Fig. 1) and an (completely automated) approach based on image feature extraction, calculation of similarities, and clustering (see Fig. 2). Whereas the first approach has proven to be useful in more general application fields, the second approach seems to be useful in specific domains of application (e.g. vacation, outdoor/indoor, landscape images). In the following we restrict ourselves to the discussion of the second approach. We discuss low-level feature extraction first, then similarity indices and clustering algorithms.

Extracted features should characterize their image by quantifiable properties (see, e.g., Choras 2007, Maheshwari et al. 2009). Typical are distributions of color intensities, textures, or forms. According to the abstraction-level, these features can be further divided into pixel-level features (w.r.t. to pre-defined pixels of the image), local features (w.r.t. to pre-defined parts of the image) and global features (w.r.t. to the entire image). Besides such general features also domain-specific features could be extracted such as human faces, fingerprints, and conceptual features (bottles, trees ...) which are often derived from other features. Features that are derived from other features are often named high-level features whereas features derived directly from the image are named low-level features.

Most popular in image retrieval are low-level features derived from color distributions. The reasons of this popularity are their similarity to the human recognition system but also their simplicity and low storage requirements. So, e.g., if the picture is RGB (red-green-blue) 8-bit coded, the color histogram is only a 768-dimensional vector containing the frequencies of pixels with specific red, green, and blue intensities, each ranging from 0 to 255. After normalization (that the 768 values of the histogram sum up to 1) the color histogram is even independent from the size of the image, besides its independence from the form and from rotations. Alternatively or additionally the color moments (mean, variance, and skewness of the color distributions) could be used as even more compact color-based low-level features.

Besides color based features, texture and form features can help to specify image similarities. So, e.g., for grey-valued (or transformed) images, the co-occurrence matrix can be used: The co-occurrence of pixels with pre-specified gray values at a given distance can be condensed to one-dimensional indicators such as energy, inertia, correlation, difference moment and entropy in order to distinguish textured from non-textured images. In a similar way, the existence of predefined forms in the image (circles, boxes, ...) can be detected using shape descriptors or spatial moments (e.g. Zernike moments, see, e.g., Choras 2007).

For applying clustering algorithms to these feature descriptions of images, a similarity index has to be defined. Here, weighted Minkowski distances w.r.t. color histograms – either bin-by-bin or using quadratic forms – as well as the so-called earth-mover-distance are most popular (see, e.g., Schmitt 2005, Choras 2007). When using (many) different low-level features, the problem of defining adequate weights is getting severe. Law et al. (2004) have proposed simultaneous feature selection and clustering algorithms for this purpose. They use the so-called feature salience concept for this purpose by iteratively searching for best distinguishing feature sets. An application of the discussed algorithms is discussed in the following.

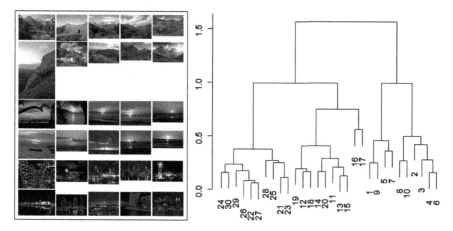

**Fig. 3** Holiday pictures (1-10: mountains, 11-20: sun and sea, 21-30: city lights, numbering starts from the upper left, row-by-row) and results using RGB color histograms for feature extraction, Euclidean distances, an average linkage for clustering

# 4   Sample Application

For testing the image clustering algorithms various image data bases have been used: At BTU a database of about 1.000.000 flickr images (see salt.informatik.tu-cottbus.de:8080/retrieval_flickr for an image search demonstration) with holiday, apartment, spare time, friends, and partner pictures is available which could be used for testing different feature extraction and distance indices. Also, images from former marketing experiments could be used, e.g., the cognac and cigarette print ad data from Gaul and Baier (1994) where various print ad (image) similarity, attitude, and preference data were collected from respondents. The results were promising since collected similarities from respondents and similarities w.r.t. low-level image features correlated.

In the following we restrict our discussion to a sample of 30 holiday pictures where three different vacancy destinations are displayed: mountains (picture 1-10), sun and sea (picture 11-20), city lights (picture 21-30). The 30 pictures were found using google image search w.r.t. these key words. They are display in Fig. 3 together with the results using RGB color histograms for feature extraction, Euclidean distances and average linkage for clustering. Again, the clustering results seem promising: The different holiday destinations could be grouped as expected.

# 5   Conclusions and Outlook

This paper discussed a new approach for deriving market segments from consumer data using the visual information of uploaded images as a segmentation base and new software prototypes for performing such analyses. Even though the research is still at the beginning, first results – as well as comparable approaches in related research fields – look promising: Images reflect consumers' activities, interests, and opinions, clustering algorithms are able to reproduce expected groupings using only the visual information in the images. Of course further research is needed. The ongoing BMBF research project will be used for this purpose.

# References

ARD/ZDF (2010) ARD/ZDF-Onlinestudie 2010. www.ard-zdf-onlinestudie.de

Baier D (2003) Classification and marketing research. Taksonomia 10:21–39

Baier D, Brusch M (2008) Marktsegmentierung. In: Herrmann A, Homburg C, Klarmann M (eds) Handbuch Marktforschung, Methoden - Anwendungen - Praxisbeispiele. Gabler, Wiesbaden, pp 769–790

Chen Y, Wang JZ, Krovetz R (2003) Content-based image retrieval by clustering. In: Proceedings of the 5th ACM SIGMM International Workshop on Multimedia Information Retrieval (MIR 2003), Berkeley, CA, USA, pp 193–200

Choras RS (2007) Image feature extraction techniques and their applications for CBIR and biometrics systems. Int J Biol Biomed Eng 1(1):6–16

Datta R, Joshi D, Li J, Wang JZ (2008) Image retrieval: Ideas, influences, and trends of the new age. ACM Comput Surv 40(2):5:1–5:60

Gaul W, Baier D (1994) Marktforschung und Marketing Management: Computerbasierte Entscheidungsunterstützung. Oldenbourg, München

van House N (2007) Flickr and public image-sharing: Distant closeness and photo exhibition. In: Conference on Human Factors in Computing Systems, San Jose, CA, USA, pp 2717–2722

Law M, Figueiredo M, Jain AK (2004) Simultaneous feature selection and clustering using mixture model. IEEE Trans Pattern Anal Mach Intell 26(9):1154–1166

Lew MS, Sebe N, Djeraba C, Jainl R (2006) Content-based multimedia information retrieval: State of the art and challenges. ACM Trans Multimedia Comput Comm Appl 2(1):1–19

Maheshwari M, Silakari S, Motwani M (2009) Image clustering using color and texture. In: First International Conference on Computational Intelligence, Communication Systems and Networks, July 23–25, 2009, Indore, India, pp 403–408

Rodden K, Basalej W, Sinclair D, Wood KR (2001) Does organisation by similarity assist image browsing? In: Proceedings of Human Factors in Computing Systems, March-31-April 5, 2001, Seattle, Washington, USA, pp 190–197

Schmitt I (2005) Ähnlichkeitssuche in Multimedia-Datenbanken - Retrieval, Suchalgorithmen und Anfragebehandlung. Oldenbourg, München

Sinus Sociovision (2009) Informationen zu den Sinus-Milieus 2009. Sinus Sociovision GmbH, Heidelberg.

Smith W (1956) Product differentiation and market segmentation as alternative marketing strategies. J Market 21:3–8

Wedel M, Kamakura WA (2000) Market segmentation: conceptual and methodological foundations. Kluwer, Dordrecht

# PLS-MGA: A Non-Parametric Approach to Partial Least Squares-based Multi-Group Analysis

Jörg Henseler

**Abstract** This paper adds to an often applied extension of Partial Least Squares (PLS) path modeling, namely the comparison of PLS estimates across subpopulations, also known as multi-group analysis. Existing PLS-based approaches to multi-group analysis have the shortcoming that they rely on distributional assumptions. This paper develops a non-parametric PLS-based approach to multi-group analysis: PLS-MGA. Both the existing approaches and the new approach are applied to a marketing example of customer switching behavior in a liberalized electricity market. This example provides first evidence of favorable operation characteristics of PLS-MGA.

## 1 Introduction

For decades, researchers have applied partial least squares path modeling (PLS, see Tenenhaus et al. 2005; Wold 1982) to analyze complex relationships between latent variables. In particular, PLS is appreciated in situations of high complexity and when theoretical explanation is scarce (Chin 1998) – a situation common for many disciplines of business research, such as marketing, strategy, and information systems (Henseler 2010). In many instances, researchers face a heterogeneity of observations, i. e. for different sub-populations, different population parameters hold. For example, institutions releasing national customer satisfaction indices may want to know whether model parameters differ significantly between different industries. Another example would be cross-cultural research, in which the culture or country plays the role of a grouping variable, thereby defining the sub-

J. Henseler (✉)
Institute for Management Research, Radboud University Nijmegen,
Thomas van Aquinostraat 3, 6525 GD Nijmegen, The Netherlands
e-mail: j.henseler@fm.ru.nl

W. Gaul et al. (eds.), *Challenges at the Interface of Data Analysis, Computer Science, and Optimization*, Studies in Classification, Data Analysis, and Knowledge Organization, DOI 10.1007/978-3-642-24466-7_50, © Springer-Verlag Berlin Heidelberg 2012

populations. As these examples show, there is a need for PLS-based approaches to multi-group analysis.

The predominant approach to multi-group analysis was brought foreward by Keil et al. (2000) and Chin (2000). These authors suggest to apply an unpaired samples $t$-test to the group-specific model parameters using the standard deviations of the estimates resulting from bootstrapping. As Chin (2000) notes, the parametric assumptions of this approach constitute a major shortcoming. As PLS itself is distribution-free, it would be favorable to have a non-parametric PLS-based approach to multi-group analysis.

The main contribution of this paper is to develop a non-parametric PLS-based approach to multi-group analysis in order to overcome the shortcoming of the current approach. The paper is structured as follows. Next to this introductory section, the second section presents the existing approach and elaborates upon its strengths and weaknesses. The third section develops the new approach and describes its characteristics. The fourth section presents an application of both the existing and the new PLS-based approach to multi-group analysis to an example from marketing about the consumer switching behavior in a liberalized electricity market. Finally, the fifth section discusses the findings of this paper and highlights avenues for further research.

## 2 The Chin/Keil Approach to Multi-Group Analysis

In multi-group analysis, a population parameter $\theta$ is hypothesized to differ for two or more subpopulations. At first, we limit our focus on the case of two groups, and will generalize in the discussion.

Typically, multi-group analysis consists of two steps. In a first step, a sample of each subpopulation is analyzed, resulting in groupwise parameter estimates $\tilde{\theta}^g$. In a second step, the significance of the differences between groups is evaluated.

Chin (2000) as well as Keil et al. (2000) propose to use an unpaired samples $t$-test in order to test whether there is a significant difference between two group-specific parameters. They suggest comparing the parameter estimate of the first group, $\tilde{\theta}^{(1)}$, with the parameter estimate of the second group, $\tilde{\theta}^{(2)}$. The test statistic is as follows (see Chin 2000):

$$t = \frac{\tilde{\theta}^{(1)} - \tilde{\theta}^{(2)}}{\sqrt{\frac{\left(n^{(1)}-1\right)^2}{n^{(1)}+n^{(2)}-2} \cdot se_{\theta^{(1)}} + \frac{\left(n^{(2)}-1\right)^2}{n^{(1)}+n^{(2)}-2} \cdot se_{\theta^{(2)}}} \cdot \sqrt{\frac{1}{n^{(1)}} + \frac{1}{n^{(2)}}}} \tag{1}$$

This statistic follows a $t$-distribution with $n^{(1)} + n^{(2)} - 2$ degrees of freedom. The subsample-specific parameter estimates are denoted as $\tilde{\theta}^{(g)}$ (with $g$ as a group index), the sizes of the subsamples as $n^{(g)}$, and the standard errors of the parameters as resulting from bootstrapping as $se_{\tilde{\theta}^{(g)}}$. Instead of bootstrapping, sometimes jackknifing is applied (e.g. Keil et al. 2000).

The $t$-statistic as provided by Equation 1 is known to perform reasonably well if the two empirical bootstrap distributions are not too far away from normal and/or the two variances $n^{(1)} \cdot se^2_{\tilde{\theta}^{(1)}}$ and $n^{(2)} \cdot se^2_{\tilde{\theta}^{(2)}}$ are not too different from one another. If the variances of the empirical bootstrap distributions are assumed different, Chin (2000) proposes to apply a Smith-Satterthwaite test. The modified test statistic becomes (see Nitzl 2010):

$$t = \frac{\tilde{\theta}^{(1)} - \tilde{\theta}^{(2)}}{\sqrt{\frac{n^{(1)}-1}{n^{(1)}} se^2_{\tilde{\theta}^{(1)}} + \frac{n^{(2)}-1}{n^{(2)}} se^2_{\tilde{\theta}^{(2)}}}} \tag{2}$$

Also this statistic follows a $t$-distribution. The number of the degrees of freedom $v$ for the $t$-statistic is determined by means of the Welch-Satterthwaite equation (Satterthwaite 1946; Welch 1947)[1]:

$$v(t) = \frac{\left(\frac{n^{(1)}-1}{n^{(1)}} se^2_{\tilde{\theta}^{(1)}} + \frac{n^{(2)}-1}{n^{(2)}} se^2_{\tilde{\theta}^{(2)}}\right)^2}{\frac{n^{(1)}-1}{n^{(1)2}} se^4_{\tilde{\theta}^{(1)}} + \frac{n^{(2)}-1}{n^{(2)2}} se^4_{\tilde{\theta}^{(2)}}} - 2 \tag{3}$$

## 3  A New PLS-Based Approach to Multi-Group Analysis

It is obvious that the aforementioned approaches to group comparisons with their inherent distributional assumptions do not fit PLS path modeling, which is generally regarded as being distribution-free. Taking into account this criticism against the available approaches, this paper presents an alternative approach to PLS-based group comparisons that does not rely on distributional assumptions. The working principle of the novel PLS multi-group analysis (PLS-MGA) approach is as follows: Just like within the parametric approaches, the data is divided into subsamples according to the level of the grouping variable, and the PLS path model is estimated for each subsample. Moreover, each subsample becomes subject to a separate bootstrap analysis. The novelty of the new approach to PLS-based multi-group analysis lies in the way in which the bootstrap estimates are used to assess the robustness of the subsample estimates. More specifically, instead of relying on distributional assumptions, the new approach evaluates the observed distribution of the bootstrap outcomes. It is the aim of this section to determine the probability of a difference in group-specific population parameters given the group specific estimates and the empirical cumulative distribution functions (CDFs). Let $\tilde{\theta}^{(g)}$ $(g \in \{1,2\})$ be the group-specific estimates. Without loss of generality, let us assume that $\tilde{\theta}^{(1)} > \tilde{\theta}^{(2)}$. In order to assess the significance of a group effect, we are looking for $P\left(\theta^{(1)} \leq \theta^{(2)} \mid \tilde{\theta}^{(1)}, \tilde{\theta}^{(2)}, \text{CDF}(\theta^{(1)}), \text{CDF}(\theta^{(2)})\right)$.

---

[1]This notation of the Welch-Satterthwaite equation was derived by Nitzl (2010). Note that the formula proposed by Chin (2000) is incorrect.

Let $J$ be the number of bootstrap samples, and $\tilde{\theta}_j^{(g)*}$ ($j \in \{1, \ldots, J\}$) the bootstrap estimates. In general, the mean of the bootstrap estimates differs from the group-specific estimate, i. e. the empirical distribution of $\theta^{(g)}$ does not have $\tilde{\theta}^{(g)}$ as its central value. In order to overcome this, we can determine the centered bootstrap estimates $\tilde{\theta}_j^{(g)\bar{*}}$ as:

$$\forall g, j : \qquad \tilde{\theta}_j^{(g)\bar{*}} = \tilde{\theta}_j^{(g)*} - \frac{1}{J} \sum_{i=1}^{J} \tilde{\theta}_i^{(g)*} + \tilde{\theta}^{(g)}. \qquad (4)$$

Making use of the bootstrap estimates as discrete manifestations of the CDFs we can calculate

$$P\left(\theta^{(1)} \leq \theta^{(2)} \mid \tilde{\theta}^{(1)}, \tilde{\theta}^{(2)}, \text{CDF}(\theta^{(1)}), \text{CDF}(\theta^{(2)})\right) = P\left(\theta_i^{(1)\bar{*}} \leq \theta_j^{(2)\bar{*}}\right) \qquad (5)$$

Using the Heaviside step function $H(x)$ as defined by

$$H(x) = \frac{1 + \text{sgn}(x)}{2}, \qquad (6)$$

Equation (5) transforms to

$$P\left(\theta^{(1)} \leq \theta^{(2)} \mid \tilde{\theta}^{(1)}, \tilde{\theta}^{(2)}, \text{CDF}(\theta^{(1)}), \text{CDF}(\theta^{(2)})\right) = \frac{1}{J^2} \sum_{i=1}^{J} \sum_{j=1}^{J} H\left(\tilde{\theta}_j^{(2)\bar{*}} - \tilde{\theta}_j^{(1)\bar{*}}\right). \qquad (7)$$

Equation (7) is the core of the new PLS-based approach to multi-group analysis. The idea behind it is simple: Each centered bootstrap estimate of the second group is compared with each centered bootstrap estimate of the first group. The number of positive differences divided by the total number of comparisons (i.e., $J^2$) indicates how probable it is in the population that the parameter of the second group is greater than the parameter of the first group.

## 4   A Marketing Example

We illustrate the use of both the existing and the new PLS-based approach to multi-group analysis on the basis of a marketing example, namely customer switching behavior in a liberalized energy market. Prior studies and marketing theory (c. f. Jones et al. 2000; de Ruyter et al. 1998) suggest that customers are less likely to switch their current energy provider if they are satisfied or if they perceive high switching costs. From the Elaboration Likelihood Model it can be derived that consumer behavior is contingent on the level of involvement (Bloemer and Kasper 1995; Petty and Cacioppo 1981).

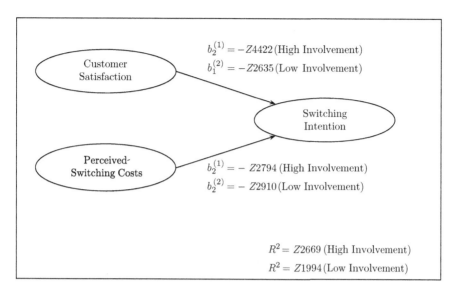

**Fig. 1** Structural model with groupwise parameter estimates (standardized PLS path coefficients)

A cross-sectional study among consumers was conducted in order to test the proposed hypotheses. The data at hand stems from computer-assisted telephone interviews with 659 consumers. 334 consumers indicated to be highly involved in buying electricity, while 325 consumers said to have a low involvement. Customer satisfaction, switching costs, and customer switching intention were measured by multiple items using mainly five-point Likert scales.

We create a PLS path model as depicted in Fig. 1. This model captures the two direct effects of customer satisfaction and perceived switching costs on customer loyalty. In order to account for the moderating effect of involvement, we estimate the model separately once for the group of highly involved consumers and once for the group of consumers having low involvement. Figure 1 also reports the standardized path coefficients per group as estimated by means of the PLS software SmartPLS (Ringle et al. 2007).

Moreover, we conduct bootstrap resampling analyses with 500 bootstrap samples per group. Based on the estimates, the bootstrap estimates and their standard deviations, we calculated the $p$-values for group differences in the effects of customer satisfaction and perceived switching costs on switching intention. Table 1 contrasts the results of the different PLS-based approaches to multi-group analysis, i. e. the parametric test with equal variances assumed (homoskedastic), the parametric test with equal variances not assumed (heteroskedastic), and the non-parametric PLS-MGA.

The different PLS-based approaches to multi-group analysis provide similar results. All approaches find a significant difference in strength of the effect of customer satisfaction on switching intention ($\alpha = .05$). This means, for highly

**Table 1** Comparison of statistical tests on group differences

| Hypothesis | Statistical test | $p$-value (one-sided) |
|---|---|---|
| Customer satisfaction | Parametric, homoskedastic | .0123 |
| ↓ | Parametric, heteroskedastic | .0056 |
| Switching intention | PLS-MGA (non-parametric) | .0056 |
| Perceived switching costs | Parametric, homoskedastic | .4308 |
| ↓ | Parametric, heteroskedastic | .4428 |
| Switching intention | PLS-MGA (non-parametric) | .5528 |

involved consumers, the level of customer satisfaction is a stronger predictor of switching behavior than for consumers with low involvement. Moreover, all approaches reject a group effect in the impact of perceived switching costs on switching intention. Despite the general convergence of findings, there seem to be notable differences in statistical power between the approaches. For instance, both the parametric test with equal variances not assumed and PLS-MGA are able to detect the group effect on a .01 significance level, whereas the parametric test with equal variances assumed is not.

## 5 Discussion

It was the aim of this contribution to introduce a non-parametric approach to PLS-based multi-group analysis. The new approach, PLS-MGA, does not require any distributional assumptions. Moreover, it is simple to apply in that it uses the bootstrap outputs that are generated by the prevailing PLS implementations such as SmartPLS (Ringle et al. 2007), PLS-Graph (Soft Modeling, Inc. 1992–2002), PLS-GUI (Li 2005), or SPAD (Test and Go 2006).

Technically, the new approach to PLS-based multi-group analysis, PLS-MGA, is purely derived from bootstrapping in combination with a rank sum test, which makes it conceptually sound. Still, its use has only been illustrated by means of one numerical example. Future research should conduct Monte Carlo simulations on PLS-MGA in order to obtain a better understanding of its characteristics, such as for instance its statistical power under various levels of sample size, effect size, construct reliability, and error distributions.

Further research is also needed to extend PLS-MGA to analyze more than two groups at a time. As a quick solution, multiple tests with a Bonferroni correction could be performed. Alternatively, an adaptation of the Kruskal-Wallis test (Kruskal and Wallis 1952) to PLS-based multi-group analysis might be promising.

Finally, PLS-based multi-group analysis has been limited to the evaluation of the structural model so far, including this article. However, PLS path modeling does not put any constraints on the measurement model so that measurement variance could be an alternative explanation for group differences. Up to now, no PLS-based approaches for examining measurement invariance across groups have been proposed yet. Given its ease and robustness, PLS-MGA may also be the point of departure for the examination of group differences in measurement models.

# References

Bloemer JMM, Kasper HDP (1995) The complex relationship between customer satisfaction and brand loyalty. J Econ Psychol 16:311–329

Chin WW (1998) The partial least squares approach to structural equation modeling. In: Marcoulides GA (ed) Modern methods for business research. Lawrence Erlbaum Associates, Mahwah, NJ, pp 295–336

Chin WW (2000) Frequently Asked Questions – Partial Least Squares & PLS-Graph. URL http://disc-nt.cba.uh.edu/chin/plsfac/plsfac.htm

Henseler J (2010) On the convergence of the partial least squares path modeling algorithm. Comput Stat 25(1):107–120

Jones MA, Mothersbaugh DL, Beatty SE (2000) Switching barriers and repurchase intentions in services. J Retail 76(2):259–274

Keil M, Tan BCY, Wei KK, Saarinen T, Tuunainen V, Wassenaar A (2000) A cross-cultural study on escalation of commitment behavior in software projects. Manag Inform Syst Q 24(2):299–325

Kruskal W, Wallis W (1952) Use of ranks in one-criterion variance analysis. J Am Stat Assoc 47(260):583–621

Li Y (2005) PLS-GUI - Graphic User Interface for Partial Least Squares (PLS-PC 1.8) Version 2.0.1 beta. University of South Carolina, Columbia, SC

Nitzl C (2010) Eine anwenderorientierte Einführung in die Partial Least Square (PLS)-Methode. Universität Hamburg, Institut für Industrielles Management, Hamburg

Petty R, Cacioppo J (1981) Attitudes and persuasion: classic and contemporary approaches. W.C. Brown, Dubuque, Iowa

Ringle CM, Wende S, Will A (2007) SmartPLS 2.0 M3. University of Hamburg, Hamburg, Germany, URL http://www.smartpls.de

de Ruyter K, Wetzels M, Bloemer JMM (1998) On the relationship between perceived service quality, service loyalty and switching costs. Int J Serv Ind Manag 9(5):436–453

Satterthwaite F (1946) An approximate distribution of estimates of variance components. Biometrics Bull 2(6):110–114

Soft Modeling, Inc (1992–2002) PLS-Graph version 3.0. URL http://www.plsgraph.com

Tenenhaus M, Vinzi VE, Chatelin YM, Lauro C (2005) PLS path modeling. Comput Stat Data Anal 48(1):159–205

Test & Go (2006) SPAD version 6.0.0. Paris, France

Welch B (1947) The generalization of 'Student's' problem when several different population variances are involved. Biometrika 34(1/2):28–35

Wold HOA (1982) Soft modelling: the basic design and some extensions. In: Jöreskog KG, Wold HOA (eds) Systems under indirect observation. Causality, structure, prediction, vol II. North-Holland, Amsterdam, New York, Oxford, pp 1–54

# Love and Loyalty in Car Brands: Segmentation Using Finite Mixture Partial Least Squares

Sandra Loureiro

**Abstract** This study seeks to understand the relationship among brand love, inner self, social self, and loyalty perceived by users of three car brands. The model estimation includes structural equation analysis, using the PLS approach and applying the finite mixture partial least squares (FIMIX-PLS) to segment the sample. The research findings showed that area of residence and age are the main difference that characterizes the two uncovered customer segments. Car users of the large segment live mainly in the big city Oporto and are younger than car users of the small segment. For this small group, social self doesn't contribute to enrich the brand love, they don't give very much importance to what others think of them, and so, the social aspects and the social image are not a key factor to create a passion and an attraction to the car brand. Indirectly, the social identification is not important to reinforce the intention to recommend and to buy a car with the same brand in the future. On the other hand, the cosmopolitan car users of the large segment consider that the car brand image should fit their inner self and the social group of belonging in order to improve the love to the brand and the intention to recommend and to buy a car with the same car brand in the future.

## 1 Introduction

In their seminal work, Oliver, Oliver et al. (1997) highlight the importance of delighting the customer as an extension of providing basic satisfaction and propose a model of delight and satisfaction. In their work, delight is conceptualized as a function of surprising consumption, arousal, and positive affect. Fournier (1998) also points out the importance of love in consumers' long-term relationships. Later,

S. Loureiro (✉)
University of Aveiro, Campus de Santiago, 3810193 Aveiro, Portugal
e-mail: sandra.loureiro@ua.pt

W. Gaul et al. (eds.), *Challenges at the Interface of Data Analysis, Computer Science, and Optimization*, Studies in Classification, Data Analysis, and Knowledge Organization, DOI 10.1007/978-3-642-24466-7_51, © Springer-Verlag Berlin Heidelberg 2012

Ahuvia (2005) found that many consumers do have intense emotional attachments
to some objects as they were people.

Following on the foregoing, Carroll and Ahuvia (2006, p.81) proposed the
concept of brand love, defined as "the degree of passionate emotional attachment
a satisfied consumer has for a particular trade name". Based on such findings,
we propose a model that links self expressive brand, brand love, and loyalty. The
concept of self expression brand has two dimensions: the brand reflects the inner
self and the brand enhances the social self.

The major objective of this paper is to test a model of brand love and loyalty,
using data from 329 car users of the three car brands (Toyota, Renault, and
Ford), applying the PLS (Partial Least Squares) technique and the finite mixture
partial least squares (FIMIX-PLS), proposed by Hahn et al. (2002), to segment the
sample. This approach combines a finite mixture procedure with an expectation-
maximization (EM)-algorithm specifically coping with the ordinary least squares
(OLS)-based predictions of PLS and permits reliable identification of distinctive
customer segments, with their characteristic estimates for relationships of latent
variables in the structural model.

## 2    Literature Review and Hypothesis

Nowadays, due to increased competition, researchers and practitioners found out
that simply to satisfy the consumers may not be sufficient for continuing success
in today's competitive marketplace. Thus, when customers received a service
or good that not only satisfies, but provides unexpected value or unanticipated
satisfaction, such an experience can lead to the reaction of customer delight
(Chandler 1989). This pleasant emotional experience can lead to a more powerful
emotional experience like the love for a product or brand. Indeed, past research has
underlined the importance of affect intensity in both interpersonal love (e.g., Regan,
Kocan, and Whitlock 1998) and consumer behavior (e.g., Oliver, Rust, and Varki
1997).

The emotional attachments to a brand can be the result of a consumer's long-
term relationship with the brand. Thus, Carroll and Ahuvia (2006) state that brand
love includes a willingness to declare love (as if the brand were a person) and
involves integration of the brand into the consumer's identity. The love feelings
of a consumer to a brand are greater for brands that play a significant role in shaping
their identity. So, self-expressive brand is the consumer's perception of the degree
to which the specific brand enhances one's social self and/or reflects one's inner
self (Carroll and Ahuvia 2006, p.82) and has two dimensions: inner-self and social
self. Consumers satisfied with the brand and who also love the brand are expected
to be more willing to repurchase and to recommend the product with such a brand
to others.

The above considerations lead us to expect that inner self and social self have
a positive direct effect on car brand love and that the car brand love exercises a
positive direct effect on loyalty:

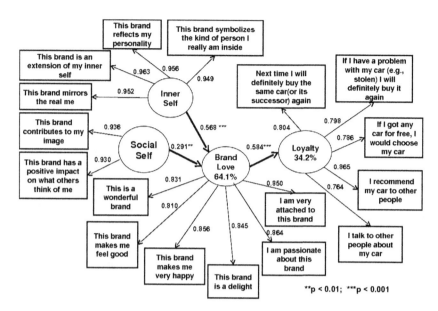

**Fig. 1** Structural results of the Brand Love and Loyalty model

H1: Inner self has a positive effect on car brand love
H2: Social self has a positive effect on car brand love
H3: Car brand love has a positive effect on loyalty

# 3 Method

Drawing from the literature review in Sect. 2, the research model links inner self and social self to brand love and this to loyalty as shown in Fig. 1. The questionnaire that operationalized the latent variables as well as the socio-demographic variables was pre-tested by six car users personally interviewed at three car dealers. Then 329 car users of three brands (Toyota, Renault, and Ford), satisfied with their cars and who had bought a new car 2 years ago answered the questionnaire during May to September 2009. The two year delay between the purchase and the survey was necessary to make sure that the customers had gained enough experience with the car as well as the services offered. The reasons for selecting the brands above are due to the fact they were the car brands most sold during 2007 and 2008 (ACP 2009) and have their origin (the name) in three different countries.

## 3.1 Measurement

The four latent variables in this study were measured by means of multi-item scales. Inner self was measured using four items, social self was operationalized using two items, and brand love with six items (Carroll and Ahuvia 2006). Finally, loyalty is assessed by five items adapted from scales previously developed and used to measure loyalty (Zeithaml et al. 1996; Loureiro and Miranda 2008). Loyalty was measured with two types of components: behavior and attitude.

# 4  Results

First the model was estimated. The Blindfolding technique was used to calculate the $Q^2$ (Fornell and Bookstein 1982) and the nonparametric bootstrap approach for estimating the precision of the PLS estimates. So, 500 samples sets were created in order to obtain 500 estimates for each parameter in the PLS model. Each sample was obtained by sampling with replacement in the original data set (Fornell and Larcker 1981; Chin 1998). Then, the FIMIX-PLS was applied to segment. Finally, a parametric analysis, through a t-test, was used to determine if the segments are statistically different. For each segment the model was estimated again and the precision of the PLS estimates are analyzed, too.

The research model is shown in Fig. 1 and the results were estimated using the statistical software SmartPLS 2.0. To evaluate the PLS model, this study follows the suggestions proposed by Chin (1998). Thus, first the adequacy of the measures is assessed by evaluating the reliability of the individual measures and the discriminant validity of the constructs. Then, the structural model is evaluated.

Assessing the empirical results, all factor loadings (Fig. 1) of reflective constructs approach or exceed 0.707, which indicates that more than 50% of the variance in the manifest variable is explained by the construct (Carmines and Zeller 1979). Composite reliability was used to analyze the reliability of the constructs since this has been considered a more exact measurement than Cronbach's alpha (Fornell and Larcker 1981). The constructs are reliable since the composite reliability values exceed the threshold of 0.7 and even the strictest one of 0.8 (Nunnally 1978). Furthermore, all Cronbach's alpha exceed the 0.8 (see Table 1). The measures also demonstrates convergent validity as the average variance of manifest variables extracted by constructs (AVE) is at least 0.5 (Fornell and Larcker 1981). To assess discriminant validity the square root of AVE should be greater than the correlation between the construct and other constructs in the model (Fornell and Larcker 1981) (see Table 1). The results show that in all cases, the square root of AVE is greater than the variance shared between each construct and the opposing constructs. Consequently, we can also consider a high degree of discriminant validity for all constructs in this study.

Central criteria for the evaluation of the inner model comprise $R^2$ and the Goodness-of-Fit (GoF) index by Tenenhaus et al. (2005). The model demonstrated a high level of predictive power for brand love, as the modeled constructs explained 64.1% of the variance in brand love and a moderate value for loyalty (34.2%). The average communality of all reflective measures is relatively high, leading to a satisfactory GoF outcome (0.621). All values of the cross-validated redundancy

**Table 1** Reliability and discriminant validity

| Construct | C. Alpha | C.R. | AVE | Correlation BL | of IS | Construct L | SS |
|---|---|---|---|---|---|---|---|
| Brand Love (BL) | 0.968 | 0.976 | 0.912 | 0.955* | | | |
| Inner Self (IS) | 0.851 | 0.931 | 0.870 | 0.711 | 0.933* | | |
| Loyalty (L) | 0.919 | 0.936 | 0.710 | 0.774 | 0.694 | 0.843* | |
| Social self (SS) | 0.863 | 0.901 | 0.646 | 0.475 | 0.506 | 0.585 | 0.804* |

* Diagonal elements in the 'correlation of constructs' matrix are the square root of average variance extracted (AVE). For adequate discriminant validity, diagonal elements should be greater than corresponding off-diagonal elements.

**Table 2** Model selection

| No. of segments | $K = 2$ | $K = 3$ | $K = 4$ |
|---|---|---|---|
| AIC (Akaike's Information Criterion) | 1589.622 | 1777.886 | 2009.567 |
| BIC (Bayesian Information Criterion) | 1631.379 | 1842.419 | 2096.876 |
| CAIC (Consistent AIC) | 1631.412 | 1842.471 | 2096.946 |
| EN (Normed Entropy Statistic) | 0.496 | 0.413 | 0.292 |

measure $Q^2$ (Stone-Geisser-Test) are positive, indicating high predictive power of the exogenous constructs (Chin 1998).

In the next analytical step, the FIMIX-PLS module of SmartPLS 2.0 was applied to segment based on the estimated scores for latent variables. FIMIX-PLS results were computed for two, three, and four classes. The results reveal that the choice of two segments is appropriate for customer segmentation purposes. The EN evaluation criteria considerably decrease in the ensuing numbers of segments (see Table 2) and each additional segment has only a small size, which explains a marginal portion of heterogeneity in the overall set of data. This criterion indicates the degree of separation for all observations and their estimated membership probabilities on a case-by-case basis, and it subsequently reveals the most appropriate number of latent classes for segmentation (Ramaswamy et al. 1993). Next, observations are assigned to each segment according to the segment membership's maximum a posteriori probability. The first segment represents 82% of the sample and the second segment 18%.

Table 3 presents the global model and FIMIX-PLS results for two latent segments. Before evaluating goodness-of-fit measures and inner model relationships, all outcomes for segment-specific path model estimations were tested with regard to reliability and discriminant validity. The analysis showed that all measures satisfy the relevant criteria for model evaluation (Chin 1998).

All path coefficients are significant at a level of 0.001 or 0.01, except the relationship between social self and brand love for the second segment. In this study the nonparametric Bootstrap approach was used for estimating the precision of the PLS estimates. The differences between the samples of the two segments were compared using a parametric analysis through a t-test of m+n+2 degrees of freedom (Chin 2011). All differences between two segments are significant in each of the three structural paths.

**Table 3** Global model and disaggregate results for two latent segments

| Structural Paths | Global | FIMIX-PLS | | t[mgp] |
|---|---|---|---|---|
| | | 1st. Segment | 2nd. Segment | |
| Brand Love - >Loyalty | 0.584*** | 0.770*** | 0.334** | 23.421*** |
| Inner Self - >Brand Love | 0.568*** | 0.574*** | 0.408** | 6.837*** |
| Social Self - >Brand Love | 0.291*** | 0.403*** | −0.075ns | 16.863*** |
| Segment size | 1.000 | 0.818 | 0.182 | |
| R² Brand Love | 0.641 | 0.831 | 0.134 | |
| R² Loyalty | 0.342 | 0.593 | 0.127 | |
| GoF | 0.621 | 0.756 | 0.304 | |

**p <0.01; ***p <0.001; ns = not significant. t[mgp] = t-value for multi-group comparison test
(see expression 1)

**Fig. 2** Segmentation-tree
results for Exhaustive
CHAID

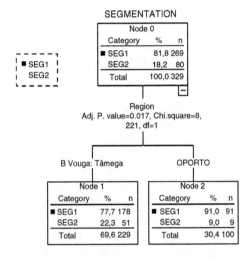

The final step involves the identification of explanatory variables that best
characterize the two uncovered customer segments. We consequently applied the
Exhaustive CHAID (Biggs et al. 1991) algorithm using SPSS. As a dependent
variable we used the results of the partition from the application of the FIMIX-
PLS technique. The independent variables were the socio-demographic variables.
The result is shown in Fig. 2.

Segment one comprises car users that live mainly in Oporto and segment two
consists of customers of the sub-regions near by Oporto. Region could be the main
variable to explain the segmentation, but the analysis of the other demographic
variables reveal that car users of the first segment are younger (most with less than
40 years) than the car users of the second segment.

For the car users of the second segment social self doesn't contribute significantly
to explain brand love (see Table 3) and inner self explains only 13.4% of the variance
in brand love (the passion and the attraction to the brand). Thus, the model is more
accurate to explain the behavior of the car users of the first segment.

# 5 Conclusions

This study provides an application of the finite mixture partial least squares (FIMIX-PLS) method for capturing heterogeneity in PLS path modeling of brand love and loyalty. For this study, users of three car brands were approached in car concessionaires. This approach enables us to identify two segments with distinct inner model path estimates that differ substantially from the aggregate-level analysis. Car users of the larger segment are young adults, most younger than 40 years, and they live mainly in a big city, like Oporto. The smaller segment gathers adults that live mainly in Baixo Vouga (a less cosmopolitan area than Oporto). For this group, social self doesn't contribute to enrich the brand love. These results seem to show that these car users don't give very much importance to what others think of them, and so, the social aspects and the social image are not a key factor to create a passion and an attraction for the car brand. Indirectly, the social identification is not important to reinforce the intention to recommend and to buy a car with the same car brand in the future. The inner self, in other words, the fit between the car image and the self image, reveals to be more effective in order to involve users in the attraction and passion for the car brand than the social self. Nevertheless, for customers in the big city, more cosmopolitan and younger, social self shouldn't be ignored. The car users of the large segment need to feel that his/her car and the brand car are accepted by the social group they belong to or by the social group that they wish to belong to.

The results show that the place where customers live and the age can make a difference in the way they identify with the brand, even when the places of residence are not far from each other. The brands and their sellers should pay attention to this insight.

In the future, more latent variables should be introduced in order to better understand the antecedents of brand love. The live style of the car users, in addition to the socio-demographic variables, may provide a better understanding of the car user and owner behavior.

# References

ACP Automovel Clube de Portugal (2009) Qashqai carro do ano 2008. URL http://www.acp.pt/index.php?template_id=6165

Ahuvia AC (2005) Beyond the extended self: Loved objects and consumers identity narratives. J Consum Res 32:171–184

Biggs D, de Ville B, Suen E (1991) A method of choosing multiway partitions for classification and decision trees. J Appl Stat 18(1):49–62

Carmines EG, Zeller RA (1979) Reliability and validity assessment. Ed. Sage Publications, London

Carroll B, Ahuvia A (2006) Some antecedents and outcomes of brand love. Market Lett 17:79–89

Chandler CH (1989) Quality: Beyond customer satisfaction. Qual Progr 22:30–32

Chin WW (1998) The partial least squares approach to structural equation modeling. In: Marcoulides GA (ed) Modern methods for business research. Lawrence Erlbaum Associates Publisher, Mahwah, NJ, pp 295–358

Chin WW (2011) Frequently asked questions partial least squares and PLS-graph, 2000. URL http://disc-nt.cba.uh.edu/chin/plsfaq/plsfaq.htm

Fornell C, Bookstein FL (1982) Two structural equation models: LISREL and PLS applied to consumer exit-voice. J Market Res 19:440–452

Fornell C, Larcker DF (1981) Evaluating structural models with unobservables variables and measurement error. J Market Res 28:39–50

Fournier S (1998) Consumers and their brands: Developing relationship theory in consumer research. J Consum Res 24:343–373

Hahn C, Johnson MD, Herrmann A, Huber F (2002) Capturing customer heterogeneity using a finite mixture PLS approach. Schmalenbach Bus Rev 54(3):243–269

Loureiro SM, Miranda FJ (2008) The importance of quality, satisfaction, trust, and image in relation to rural tourist loyalty. J Travel Tour Market 25(2):117–136

Nunnally J (1978) Psychometric theory, 2nd edn. McGraw-Hill, New York

Oliver RL, Rust RT, Varki S (1997) Customer delight: Foundations, findings and managerial insight. J Retail 73(3):311–336

Ramaswamy V, DeSarbo WS, Reibstein DJ, Robinson WT (1993) An empirical pooling approach for estimating marketing mix elasticities with PIMS data. Market Sci 12(1):103–124

Regan PC, Kocan ER, Whitlock T (1998) Aint love grand! A prototype analysis of the concept of romantic love. J Soc Personal Relation 15:411–420

Tenenhaus M, Vinzi VE, Chatelin YM, Lauro C (2005) PLS path modeling. Comput Stat Data Anal 48:159–205

Zeithaml V, Berry L, Parasuraman A (1996) The behavioural consequences of service quality. J Market 60(2):31–46

# Endogeneity and Exogeneity in Sales Response Functions

Wolfgang Polasek

**Abstract** Endogeneity and exogeneity are topics that are mainly discussed in macroeconomics. We show that sales response functions (SRF) are exposed to the same problem if we assume that the control variables in a SRF reflect behavioral reactions of the supply side. The supply side actions are covering a flexible marketing component which could interact with the sales responses if sales managers decide to react fast according to new market situations. A recent article of Kao et al. (Evaluating the effectiveness of marketing expenditures, Working Paper, Ohio State University, Fisher College of Business, 2005) suggested to use a class of production functions under constraints to estimate the sales responses that are subject to marketing strategies. In this paper we demonstrate this approach with a simple SRF(1) model that contains one endogenous variable. Such models can be extended by further exogenous variables leading to SRF-X models. The new modeling approach leads to a multivariate equation system and will be demonstrated using data from a pharma-marketing survey in German regions.

## 1 Introduction

Kao et al. (2005) have proposed a simultaneous estimation of sales in dependence of marketing inputs. The new idea behind this approach is that the (optimal) expenditures, which are inputs in the SRF, might depend on the current sales and should be estimated endogenously.

Polasek (2010) has introduced a family of SRF(k) for a cross-sectional sample where the parameter k denotes the number of endogenous input variables (like marketing expenditures or sales related covariates). In this paper we show how

W. Polasek (✉)

IHS, Stumpergasse 56, A-1060 Vienna, Austria

e-mail: polasek@ihs.ac.at

W. Gaul et al. (eds.), *Challenges at the Interface of Data Analysis, Computer Science,* 511
*and Optimization*, Studies in Classification, Data Analysis, and Knowledge Organization,
DOI 10.1007/978-3-642-24466-7_52, © Springer-Verlag Berlin Heidelberg 2012

to estimate a SRF(1) model that explains sales output by a Cobb-Douglas type of function of marketing variables.

The main point of the current approach is the extension of the traditional estimation of a SRF to a system of observations, because some input variables are jointly determined by the output. This problem was solved by system estimation in macroeconomics over the last decades. New is the assumption that the endogeneity of the inputs stems from an implied (stochastic) optimality consideration, which allows the first partial derivative of the SRF to follow a normal distribution. We call this crucial behavioral assumption the stochastic partial derivative (SPD) assumption. More extensions of these modeling considerations lead to a model choice problem. The details of this model choice problem need to be more worked out in future, but we will concentrate in the next section on the model estimation part and we will use the marginal likelihood criterion for the Bayesian model choice.

If the first partial derivative of a response model is jointly determined with the dependent variable, then these 2 equations imply a simultaneous equation system, since the stochastic restrictions imply an endogeneity of the input variables. There are 3 main implications that constitute a SRF-SPD model:

1. Sales (demand) model: a (production) function of the input variables $X$ explain ⇒ output sales variables $y$ plus noise.
2. Behavioral (supply) model + SPD assumptions: results of marketing efforts are proxied by first derivatives and ⇒ follow a normal distribution.
3. Conditional on SPD and known SRF coefficients: The SPD assumption leads to ⇒ stochastic regressors.

In the next section we describe the SRF(1) model together with a Bayesian (MCMC) estimation procedure. Section 1 introduces the SRF(1)X(q) model (a notation similar to notations in multiple time series) to indicate that a SRF model does not need to have only endogenous variables, but can also include q exogenous covariates. Section 2 explains the estimation procedure and Sect. 3 discusses a regional sales response model example that involves a German pharma-marketing data set for the years 2008 and 2007. In a final section we conclude.

## 2   The SRF(1) Model with Endogeneity

We consider the simplest possible model to demonstrate the endogeneity effects in the SRF(1) sales response function with one input variable $x$

$$y = \beta_0 x^{\beta_1} e^{\epsilon}, \tag{1}$$

where $\epsilon$ is a $N[0, \sigma_y^2]$ distributed error term. By taking logs for the n cross-sectional observations we get the following linear regression model

$$ln\ y \sim N[\mu_y = X\beta, \sigma_y^2 I_n] \tag{2}$$

with the regression coefficients $\beta = (ln\ \beta_0, \beta_1)$ and the regressor matrix $X = (1_n : ln\ x)$ where $1_n$ is a vector of ones and $x$ is the cross-sectional decision variable that will influence the sales $y$ (a $n \times 1$ vector) in the n regions. Thus the model is of the type of a log linear production function as it is used in macro-economics. The new assumption is the stochastic partial derivative (SPD) assumption which implies an additional equation for the supply side behavior:

$$\partial y / \partial x = \beta_0 \beta_1 x^{\beta_1 - 1} e^\epsilon$$

Next, we make the assumption that the log derivatives across all units

$$ln\ y_x = log(\beta_0 \beta_1) + (\beta_1 - 1) ln\ x \tag{3}$$

are normally distributed with parameters that can be estimated:

$$ln\ y_x \sim N[ln\ \lambda, \sigma_\lambda^2] \quad \text{or } ln\ y_x \sim N[ln\ \lambda 1_n, \sigma_\lambda^2 I_n]. \tag{4}$$

This means that the sales responses $y$ and the decision variable $x$ might follow a restriction that allocates resources according to the first derivative of the SRF but the empirical observations across the n regions reveal some noise. These stochastic fluctuations across regions are captured by the mean response $\lambda$, and the variance $\sigma_\lambda^2$ of the constraint measures the looseness or strength of this optimality behavior in the model and can be interpreted as a tightness parameter for the SPD restriction.

Adding the stochastic partial derivative (SPD) restrictions for the $x$ regressor into the SRF model imposes the behavioral assumption that the marketing efforts are allocated in such a way that the marketing results (via the SRF derivative) should be equal across the regional units: Additionally, the above SPD assumption leads to a normal distribution of the regressor $x$:

$$ln\ x \sim N[\mu_x, \sigma_x^2 I_n] \tag{5}$$

where the parameters $\theta_x = (\mu_x, \sigma_x^2)$ are determined by the SRF and SPD parameters. This approach actually implies the endogeneity of $x$ in the SRF model and shows how the SPD assumption (i.e. the feedback from the sales to the marketing control variables) influences the inference process to yield a simultaneous estimation model.

To see how the SPD assumption translates to an assumption about the $x$ variable, we re-write the exponent of the SPD density (4) and use the log derivative (3)

$$N[ln\ \lambda, \sigma_\lambda^2] \propto (ln\ y_x - ln\lambda)^2 / \sigma_\lambda^2$$
$$= (log(\beta_0 \beta_1) + (\beta_1 - 1) ln\ x - ln\lambda)^2 / \sigma_\lambda^2 =$$

$$= \left(\frac{log(\beta_0\beta_1/\lambda)}{1-\beta_1} - ln\ x\right)^2 (\beta_1 - 1)^2/\sigma_\lambda^2$$

$$\propto N[ln\ x\ |\ \mu_x, \sigma_x^2 = \sigma_\lambda^2/(\beta_1 - 1)^2]$$

with the mean $\mu_x = \frac{log(\beta_0\beta_1/\lambda)}{1-\beta_1}$ and variance $\sigma_x^2 = \frac{\sigma_\lambda^2}{(\beta_1-1)^2}$ of $log(x)$. (Note that the Jacobian for $ln\ x$ from $\lambda$ is just $1/|\beta_1 - 1|$.)

Finally, we define the SRF(1)-SPD model by the following sequence of (conditional) normal densities:

**Definition 1.** The SRF(1)-SPD model $y = \beta_0 x^{\beta_1} e^u$ is defined as the following set of 3 log-normal densities:

$$ln\ y\ |\ SRF, \beta \sim N[ln\ \beta_0 + \beta_1 ln\ x, \sigma_y^2]$$

$$\Rightarrow ln\ x\ |\ SPD, \beta \sim N[(ln\ \beta_0 + ln\ \beta_1 - ln\ \lambda)/(\beta_1 - 1), \sigma_x^2]$$

$$ln\ y_x\ |\ SPD, \theta_x \sim N[ln\ \lambda, \sigma_\lambda^2].\qquad\qquad(6)$$

where "SRF" stands for the functional form of the model and "SPD" stands for $y_x$, the first derivative of the SRF(1) model and with the parameters of the model given by $\theta = (\theta_x, \theta_\lambda) = (\beta, \sigma_y^2, \lambda, \sigma_\lambda^2)$. "$\Rightarrow$" denotes the derived distribution for $ln\ x$.

An alternative way of writing the SRF(1) model is:

$$Demand:\quad N[ln\ y\ |\ \mu_y, \sigma_y^2],\quad \mu_y = ln\ \beta_0 + \beta_1 ln\ x;$$

$$\Rightarrow Supply:\quad N[ln\ x\ |\ \mu_x, \sigma_x^2],\quad \mu_x = \frac{log(\beta_0\beta_1/\lambda)}{1-\beta_1};$$

$$Behavior:\quad N[ln\ y_x\ |\ ln\ \lambda, \sigma_\lambda^2].$$

It is important to note that while we assume independence between the 3 densities in Definition 1, the parameters of the $x$ distribution are not independent and are functions of the parameters of the demand and behavioral equations.

The supply side of the model reacts to the market by "targeting" the first derivative of the SRF model to get approximately equal results across all regions. Making the assumption that the first derivative of the SRF model with respect to the control variable $x$ is the marketing target of the supply side has 2 implications for the modeling process: (1) the control variable $x$ is for the SRF estimation and, therefore, a simultaneous system for x and y has to estimated. (2) The parameters of the x-density depend on the functional form and the parameters of the behavioral equation, $\theta_\lambda$. Because the results of marketing efforts on y can not be observed directly, this variable is a latent variable and has to be proxied by the first derivative that depends on the functional form. Thus, the behavioral equation has to be included into the model and is part of the simultaneous estimation process.

For statistical inference we can estimate the parameter vector $\theta$ of the system by maximum likelihood or by MCMC, assuming a prior density given by $p(\theta)$. In the next section we outline the MCMC procedure.

## 2.1 MCMC Estimation in the SRF(1) Model

The MCMC estimation of the SRF(1) model requires the likelihood function for $\mathscr{D} = (ln\ y, ln\ x)$ given by

$$l(\mathscr{D} \mid \theta) = N[ln\ y \mid \mu_y, \sigma_\epsilon^2 I_n] N[ln\ x \mid \mu_x, \sigma_x^2 I_n] * J \tag{7}$$

and J being the Jacobian from the transformation of $\theta_\lambda$ to $\theta_x$.

$$\mu_y = ln\ \beta_0 + \beta_1 ln\ x \text{ and } \mu_x = (ln\ \beta_0 + ln\ \beta_1 - \mu_\lambda)/(\beta_1 - 1)$$

The prior density is assumed to be

$$p(\theta) = N[\beta \mid \beta_*, H_*] N[\lambda \mid \lambda_*, \sigma_{\lambda*}^2] \prod_{j \in \{y, \lambda\}} \Gamma[\sigma_j^{-2} \mid \sigma_{j*}^2 n_{j*}/2, n_{j*}/2]. \tag{8}$$

Then the posterior distribution for $\theta$ is given by $p(\theta \mid \mathscr{D}) \propto l(\mathscr{D} \mid \theta) p(\theta)$.

**Theorem 1 (MCMC in the SRF(1)-SPD model).**
*The MCMC iteration in the SRF(1)-SPD model with the likelihood function (7) and the prior density (8) takes the following draws of the full conditional distributions (fcd):*

1. *Starting values: set $\beta = \beta_{OLS}$ and $\lambda = 0$*
2. *Draw $\lambda$ from $p(\lambda \mid \lambda_{**}, \tau_{**}^2)$*
3. *Draw $\beta$ by a Metropolis step (see Chib and Greenberg (1995)) from $p[\beta \mid b_*, H_*] l(\theta \mid y)$*
4. *Draw $\sigma_y^{-2}$ from $\Gamma[\sigma_y^{-2} \mid s_{y**}^2, n_{y**}]$*
5. *Draw $\sigma_\lambda^{-2}$ from $\Gamma[\sigma_\lambda^{-2} \mid s_{\lambda**}^2, n_{\lambda**}]$*
6. *Repeat until convergence.*

The proof can be found in Polasek (2010). The marginal likelihood of model $\mathscr{M}$ is computed by the Newton and Raftery (1994) formula

$$\hat{m}(\mathbf{y} \mid \mathscr{M})^{-1} = \frac{1}{n_{rep}} \sum_{j=1}^{n_{rep}} \left( \sum_{i=1}^{n} ln\ l(\mathscr{D}_i \mid \mathscr{M}, \theta_{(j)}) \right)^{-1}$$

$$\text{or} \quad \hat{m}_\alpha = \bar{m}_\alpha (l(\mathscr{D} \mid \mathscr{M}, \theta_{(j)})^{-1})$$

**Fig. 1** The beta coefficients
of visits and sales U2008 per
capita (pc)

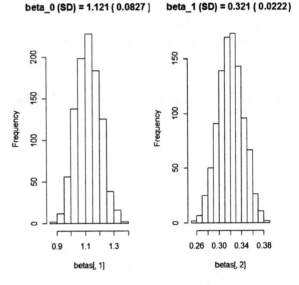

where $\mathscr{D}_i$ is the i-th data observation and with the likelihood given in (7) and $\hat{m}_\alpha$ is the $\alpha$-trimmed mean to avoid outliers.


## 3   Application: Sales Response in Pharma Marketing

We apply the SRF(1) model with and without exogenous variables to regional pharma sales in Germany for the year 2008. The model explains the regional whole sales of the product M11 by the total sales $U2008$ of the company across all regions. The posterior means (and SD) of the SRF(1) model for $M11_{pc} = ln(U2008M11/Pop)$ by $U2008_{pc}$ (pc: per capita) is:

$$M11_{pc} = 1.121 + 0.321\ U2008_{pc}\ ,\quad R^2 = 0.1502$$

$$\underset{(0.0827)\ \ (0.0222)}{}\qquad\qquad \underset{\sigma_y=1.2581}{}$$

with $\lambda = 27.8(1.678)$ and $\sigma_\lambda = 1.26(0.021)$; acceptance rate $= 50.8\%$. The OLS fit is quite similar to the posterior mean and the F-statistic $(df = 1, 1898) = 335.4101$ is highly significant (p-value $= 0$).

```
    OLS       Estimate Std.Err t-value Pr(>|t|); R^2=0.1502
    Intercept   1.1237  0.0673 16.6984          0
    X           0.3222  0.0176 18.3142          0
```

## 3.1 Exogenous Variables: A SRF(1) Model for Regional Pharma Sales

We extend the SRF(1) model for the sales of the M11 product $M11_{pc} = ln(U2008M11/Pop)$ by including 2 more regressors: the number of visits for the related product $V13_{pc} = ln(visitsM13/Pop)$ and the purchasing power potential $ln(PPP)$. The posterior mean (and SD) of the Bayesian regression estimation are:

$$M11_{pc} = 1.152 + 0.22\,U2008_{pc} + 0.19\,V13_{pc} + 0.19ln(PPP)$$
$$\qquad\quad (0.2507) \quad (0.0251) \qquad\quad (0.0239) \qquad\quad (0.0359)$$

with $\lambda = 22.478(2.268)$ and $\sigma_\lambda = 1.225(0.020)$ (acceptance rate = 51%).

The OLS fit is (with $R^2 = 0.1967$, $F(df = 3, 1896) = 154.76$, p-value=0) and $\sigma_y = 1.2238$:

$$M11_{pc} = 1.1510 + 0.2197\,U2008_{pc} + 0.1893\,V13_{pc} + 0.0962ln(PPP)$$
$$\qquad\quad (0.1987) \quad (0.0197) \qquad\qquad (0.0191) \qquad\qquad (0.0283)$$

The posterior distribution of the parameters is displayed in Figs. 2 and 3.

We see that the exogenous variables have improved the fit but the $R^2$ is still low, despite the significant coefficients. This is because the underlying dispersion across

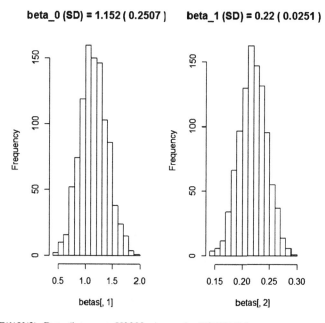

**Fig. 2** SRF(1)X(2): Betas(intercept, U2008pc) on sales U2008M11pc

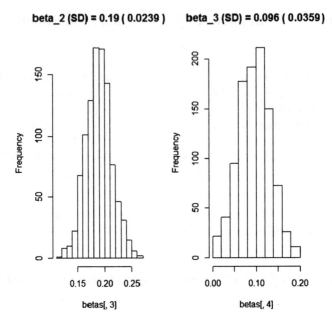

**Fig. 3** SRF(1)X(2): Betas(visitsM13pc,PPP) on sales U2008M11pc

regions is quite high, but positive slopes are present. Not all exogenous variables fit nicely with an SRF(1)-X model. The reasons for the high volatility of the SRF in dependence of covariates have to be further explored. If covariates can be found that reduce this volatility, more accurate forecasts on the effectiveness of varying the marketing control variable $x$ can be made. More on how regional heterogeneity can be modeled, can be found in Polasek (2010).

## 4 Summary

In this paper we have proposed the SRF(1) model family with endogenous variable stochastic derivative constraints, as it was suggested by Kao et al. (2005). The MCMC estimation of the model is quite straightforward, despite the fact that the constraints lead to a simultaneous sales model. The class of SRF models with stochastic partial derivative (SPD) constraints is quite flexible and allows the inclusion of exogenous variables denoted by SRF(1)-X models. Future research will concentrate on the model selection procedure and on the problem, if model choice can help to decide statistically what variables in a SRF model are endogenous or exogenous. We have demonstrated this approach by a regional pharmaceutical sale model for Germany. In Polasek (2010) it was shown that a whole SRF(p)-X(q) model family can be estimated by MCMC and the SRF approach can be extended to a spatial sales response model that takes the neighborhood structure of the observations or other regional variations into account (see Baier and Polasek 2010).

# References

Baier D, Polasek W (2010) Marketing and regional sales: Evaluation of expenditure strategies by spatial sales response functions. In: Bock HH, Gaul W, Schader M, Bodendorf F, Bryant PG, Critchley F, Diday E, Ihm P, Meulmann J, Nishisato S, Ohsumi N, Opitz O, Radermacher FJ, Wille R, Locarek-Junge H, Weihs C (eds) Classification as a tool for research, studies in classification, data analysis, and knowledge organization. Springer Berlin Heidelberg, pp 673–681
Chib S, Greenberg E (1995) Understanding the Metropolis-Hastings algorithm. Am Stat 49:327–335
Kao LJ, Chiu CC, Gilbride T, Otter T, Allenby GM (2005) Evaluating the effectiveness of marketing expenditures. Working Paper, Ohio State University, Fisher College of Business
Newton MA, Raftery AE (1994) Approximate Bayesian inference with the weighted likelihood bootstrap (with discussion). J Royal Stat Soc B 56:3–48
Polasek W (2010) Sales response functions (SRF) with stochastic derivative constraints. Working Paper, Institute of Advanced Studies, Wien

# Lead User Identification in Conjoint Analysis Based Product Design

**Alexander Sänn and Daniel Baier**

**Abstract** Nowadays, the lead user method [von Hippel, Manag Sci 32(7):791–805, 1986; Lüthje et al. (Res Pol 34(6):951–965, 2005)] and conjoint analysis [Green and Rao (J Market Res 8(3):355–363, 1971), Baier and Brusch (Conjointanalyse: Methoden - Anwendungen - Praxisbeispiele, Springer, Heidelberg, 2009)] are widely used methods for (new) product design. Both methods collect and analyze customers' preferences and use them for (optimal) product design. However, whereas the lead user method primarily creates breakthrough innovations [see von Hippel et al. (Harv Bus Rev 77(5):47–57, 1999)], conjoint analysis is more capable for incremental innovations [Helm et al. (Int J Manag Decis Making 9(3):242–26, 2008), Baier and Brusch (Conjointanalyse: Methoden - Anwendungen - Praxisbeispiele, Springer, Heidelberg, 2009)]. In this paper we extend conjoint analysis by lead user identification for the design of breakthrough innovations. The new procedure is compared to standard conjoint analysis in an empirical setting.

## 1 Introduction

The main goal of this paper is to measure the performance of traditional conjoint analysis against a lead user extended one. The preferences of lead users and non lead users are compared and the effect of the difference to the results of standard conjoint analysis is described. As a consequence of these observations conjoint analysis and the lead user method are combined in a new approach to increase the performances of both methods. The field of mountain biking is used as an application example. Lead users are expected to decline innovations not solving

A. Sänn (✉) · D. Baier
Institute of Business Administration and Economics, Brandenburg University of Technology
Cottbus, Postbox 101344, 03013 Cottbus, Germany
e-mail: alexander.saenn@tu-cottbus.de; daniel.baier@tu-cottbus.de

W. Gaul et al. (eds.), *Challenges at the Interface of Data Analysis, Computer Science, and Optimization*, Studies in Classification, Data Analysis, and Knowledge Organization, DOI 10.1007/978-3-642-24466-7_53, © Springer-Verlag Berlin Heidelberg 2012

their individual needs unlike non lead users preferring (incremental innovations) in general. Section 2 of the paper provides the theoretical background. Section 3 describes the empirical study, including survey preparation and accomplishment, and summarizes the results to prove the expected differences. In Sect. 4 the results of the study are discussed and the characteristics of lead user and non lead user groups are compared based on the empirical findings. Further an outlook to future research is given.

## 2 Known and New Approaches for Product Design

### 2.1 History

The historical background of currently used methods, schemes and techniques for (new) product design and development is characterized by a huge amount of innovation flops and high financial risks for the innovative company. Cooper and Kleinschmidt described new products as high-risk endeavors binding most of the company's resources to product development and commercialization. Approximately 46% of these resources go to unsuccessful projects and further 35% to products failing the commercialization (Cooper and Kleinschmidt 1987). The traditional scheme of the manufacturer active paradigm (MAP) classified the customer as a passive stakeholder, responding to new developments by rejection or acceptance. Von Hippel noted that the need information itself is a very complex construct and therefore gathering information about customers' needs is a hard process. Thus conventional market research techniques fail and a "... whole new approach is needed to produce products and services that accurately respond to users' needs" (von Hippel 2001). The customer active paradigm (CAP) dealt with the customer as an innovation supplier, developing own solutions based on common products or developing even new products. Today, there are several known methods to measure customers' preferences, e.g. conjoint analysis (Green and Rao 1971), as well as the customer integration in the innovation process itself, e.g. the lead user method introduced in 1986 (von Hippel 1986). On the one hand conjoint analysis is a widely used method and a state of the art technique for incremental innovations according to the MAP (Baier and Brusch 2009), but on the other hand the lead user method is the state of the art technique to generate breakthrough innovations according to the CAP (von Hippel et al. 1999; von Hippel 2005).

### 2.2 Conjoint Analysis

Conjoint analysis is a state of the art technique to measure customers' preferences (see Green and Rao 1971). Since it was developed in 1971 by Green and Rao the method has been improved and extended to several variants (e.g.

Wittink and Cattin 1989). Furthermore, conjoint analysis is still an actual topic in practice and science (Teichert and Shehu 2010). Traditional conjoint analysis itself follows an approach for measuring preferences on complete stimuli (a set of attributes) by decomposition. The stimuli are generated by predefined attributes and attribute-levels (Baier and Brusch 2009). Further there are limitations for using traditional conjoint analysis in the new product development process, e.g. because of the limitation of attribute numbers and attribute levels as well as the excessive demand from the customer itself. The lack of knowledge and experience of ordinary customers influences the reliability of conjoint analysis, because of the fact that consumers cannot fully express their needs, have latent needs and change their mind frequently (Slater and Narver 1998; Jeppesen 2005). Proceeding problems of conjoint analysis are the separation of high- and low-involved respondents as shown and the issue of a long-range forecast throughout a survey (Jeppesen 2005). The traditional conjoint analysis was chosen as a dominant methodical foundation since former empirical studies used it too (Helm et al. 2008).

## 2.3 Lead User Method

The lead user method itself was developed as a market-oriented technique to integrate customers in the development process according to the CAP (von Hippel 1986). A company focuses on special customers – so-called lead user – and collaborates with the lead user group within several development steps to generate breakthrough innovations (Lilien et al. 2002). Breakthrough innovations are described as discontinuous innovations leading to advanced technological capabilities or enhanced product capabilities by combining knowledge from different fields. Further, radical innovations are described by combining both attributes (e.g. Veryzer 1998). The lead user is describes as a customers that is ahead of the market trends and have needs that cannot be satisfied by current market-available products. The lead user is basically described by: "They are at the leading edge of an important market trend" and "...are currently experiencing needs that will be experienced by many users in that market" (von Hippel 1986; von Hippel 2005). In addition, lead user are said to posses competencies for solution generation, do research on problem related information, have a wide-ranged market overview and face extreme situations. Further, a lead user might be able to develop a personal need solution and may innovate on its own. As opposed to a random set of customers and typical respondents, dealing with real-world experiences and trying to integrate the innovation into a new usage context that does not even exist yet with common experiences, lead users are able to overcome this issue and think out of their experiences (or out of the box) (Lüthje and Herstatt 2004). Along with these main characteristics the lead user itself reveals his idea to the company and receives a needed solution as his individual benefit from co-operation. Nowadays the lead user method is the state of the art technique to generate breakthrough innovations with minor risks of commercialization (see Herstatt and von Hippel (1992),

Schreier and Prügl (2008)). A lead user project follows four steps (von Hippel 1986; Lilien et al. 2002; von Hippel 2005):

1. Goal generation and team foundation
2. Market trends and needs research
3. Lead user identification
4. Concept workshop

## 2.4   Implementing Lead User Identification in Conjoint Analysis

The new approach describes the extension of traditional conjoint analysis by a lead user identification part. Instead of using the results of all respondents for designing products, the results of the identified lead users are analyzed separately. The extended approach is applied within the mountain biking field, where user innovations are quite frequent (Franke and Shah 2003). Preliminary research was done by Helm et al. (2008) using traditional conjoint analysis and Lüthje et al. (2005) applying the lead user method in the field of mountain biking (see Helm et al. (2008), Lüthje et al. 2005). The usability of conjoint analysis for (breakthrough) innovations is restricted, so it is the main objective of the extension to identify leading customers by their use experience, their technical skills and their own idea development. The concatenation of conjoint analysis and lead user identification in the field of mountain biking may increases the performance of the measurement for generating breakthrough innovations throughout a survey.

## 3   An Application to Mountain Bike Product Design

### 3.1   Application Outline

The study started with trend analysis by gaining information from (scene) magazines (e.g. "bike", "mountain bike", "mtb" and "Fahrrad news"), commercial catalogs, discussion boards, dealers and professional mountain bike sportsmen as a proper foundation for the upcoming lead user identification process. Further, historical literature was screened for an overview of nineteenth and twentieth century patents in the bicycle market (see Herzog 1991). The captured innovations were aggregated in three runs into five innovation fields (e.g. transmission), shown in Table 1. With the help of experts, bicycle dealers and (semi-)professionals the trends were categorized in innovation levels and aggregated to five attributes for stimuli card rendering.

These cards consist of three triple leveled attributes (suspension, transmission, wheels & tires) and two double leveled attributes (eBike concept and the pedal

**Table 1** Selected attributes and attribute-levels for conjoint measurement

| Attribute | Standard level | Incremental level | Breakthrough level |
|---|---|---|---|
| Suspension | No suspension | Hardtail suspension | Full suspension |
| Transmission | Standard drive train | Carbon chain guide | Gear belt |
| Wheels & tires | Standard wheel | Runflat tires | Fiberglass wheel |
| eBike concept | No, w/o power assistance | Yes, with individual assistance | |
| Safety | No, w/o pedal lock | Yes, with pedal lock | |

lock as the breakthrough innovations), given in Table 1. The orthogonal design was applied to reduce the combinations to a set of sixteen stimuli cards and to lower the excessive demand from the customer. Based on previous research and experience in mountain biking a two-parted survey was further rendered. The first part covered customers' use experience along with the technical skills and the evaluation of predefined bicycle innovations (e.g. runflat tires) for check of reliability. The second part asked about the self-made innovation, the innovation depth (idea, concept, prototype or market status), the basic problems the user addresses and the used sources of solution information. In preparation of the survey accomplishment information of mountain biking hot spots, cycling clubs, bicycle dealers and local professional bikers were gathered as the target market. Since the lead user method concentrates on analog markets and foreign markets too, different cyclists (e.g. racing, bmx etc.) were interviewed as well. We expect that the separation of lead user respondents, by extending the conjoint analysis with a lead user identification part, will result in breakthrough innovations even within traditional conjoint measurement. Further we assume that the overall non lead users' preference differs from lead users' preference and influences the overall conjoint measurement resulting in more incremental innovations. According to previous research reports, user-innovators in the mountain biking field (about 38%) pretend to be individual riders with no major financial resources (Lüthje et al. 2005; Helm et al. 2008). Since lead users are pragmatists (Slater and Narver 1998), we predict that they decline the given innovations, which are not solving their individual needs leading to breakthrough innovations.

## 3.2 Data Collection

The survey accomplishment itself was divided into three general steps. The face-to-face interviewed respondents had to sort and rank the stimuli cards according to their individual preferences in the first step, fill out the questionnaire in the second step and rate the holdout cards at last. From time to time the order of the first and second step was changed to avoid order effects. The study itself was located in different cities to avoid a local search bias. Users of the target market of mountain biking, of analogue markets (e.g. bicycle racing) and of foreign market segments were identified according to the lead user method. The interviewers concentrated

**Table 2** Selected respondent groups in the target market, analog markets, and foreign market segments

| |
| --- |
| Extreme sports (e.g. downhill, stunt cyclists) |
| Professional sports (e.g. competing cyclists) |
| Semi-professional sports (e.g. club cyclists) |
| Business background (e.g. bike messengers) |
| Ordinary use context (e.g. freetime cyclists) |

**Table 3** Preferred attribute-levels across all respondents and for each user group

| Attribute | Overall | Lead user | Non lead user |
| --- | --- | --- | --- |
| Suspension | Hardtail | Hardtail | Hardtail |
| Transmission | Carbon | Standard | Carbon |
| Wheels & tires | Runflat | Standard | Runflat |
| eBike concept | No | No | No |
| Safety | Yes | No | Yes |

on cyclists in extreme, in professional and in business surroundings as well as on ordinary cyclists as assumed non lead users, given in Table 2.

## 3.3 Data Quality

The empirical research generated n = 123 complete surveys out of 140 interviews (87%). Among all completed surveys 96 were considered to work with for further research in reason of external quality requirements, leading to a response rate of 68% (Lüthje et al. (2005) with 42% response rate and Helm et al. (2008) with 94% response rate). The average age of all respondents was 26 years, categorized by 78% male and 22% female interview partners. The average time for the conjoint measurement was about 25 minutes per respondent. Overall 30 innovative ideas were gathered. Among all ideas 60% were shifted to further development status (4 ideas currently being on the market) leading to an idea ratio of 30% (Lüthje et al. (2005): 38%). The average Pearson $r = 0.986$ and Spearman $r = 0.816$ indicate a satisfactory internal and external validity (Helm et al. (2008): Spearman $r = 0.85$). A first comparison of lead user group and non lead user group shows differences in the validity measurement, as lead user seem to answer in a less consistent way (Pearson $r = 0.961$, Spearman $r = 0.736$ lead user group; Pearson $r = 0.989$, Spearman $r = 0.835$ non lead user group). Survey surroundings might caused this effect.

## 3.4 Results

Among lead user and non lead user groups the preferences of suspension and eBike concept differ in a decent significant way as transmission differs in a less significant way. Along with this issue the overall conjoint analysis is influenced by the higher amount of non lead users. The preferences are given in Table 3.

**Table 4** Part-worths of attribute-levels for all respondents and for each user group

| | Part-worths | | |
| --- | --- | --- | --- |
| Attribute-level | Overall (n = 96) Mean (Std. dev.) | Lead user (n = 18) Mean (Std. dev.) | Non lead user (n = 78) Mean (Std. dev.) |
| No suspension | −1.546 (3.416) | .116 (3.075)** | −1.930 (3.394) |
| Hardtail | 1.097 (1.834) | .914 (1.736) | 1.139 (1.864) |
| Full suspension | .449 (2.945) | −1.030 (2.762) | .790 (2.896)** |
| Standard | −.076 (1.198) | .569 (1.389)* | −.224 (1.106) |
| Carbon | .376 (1.313) | .139 (1.335) | .431 (1.311) |
| Gear belt | −.301 (1.218) | −.708 (1.284) | −.207 (1.191) |
| Standard | −.220 (1.220) | .144 (1.295) | −.303 (1.195) |
| Runflat | .171 (1.451) | −.190 (1.358) | .254 (1.467) |
| Fiberglass | .049 (1.375) | .046 (1.282) | .049 (1.404) |
| No eBike | .583 (1.790) | 1.483 (1.908)** | .375 (1.707) |
| Yes, eBike | −.583 (1.790) | −1.483 (1.908) | −.375 (1.707)** |
| No safety | −.193 (1.322) | .267 (1.294) | −.300 (1.313) |
| Yes, safety | .193 (1.322) | −.267 (1.294) | .300 (1.313) |

Significance: *p ≤ 0.1, **p < 0.05, ***p < 0.01

Further was expected that the separation of lead user and non lead user groups could reveal breakthrough innovations within conjoint analysis. As the study showed the extension led indeed to 3 of 5 different preferred attribute-levels (see Table 3). This provides valuable information for the product design process. As assumed the non lead user group prefers innovations in general in contrast to the pragmatical lead user group. The accepted attribute-levels need further investigation within a lead user workshop. Along with the expected results the survey provides 30 new ideas for additional research, among them 16 innovations by lead users.

Table 4 indicates the differences in preferences of the lead user segment and the non lead user segment in contrast to standard (overall) conjoint analysis.

# 4 Conclusions and Outlook

The aspect of different preferences from the lead user group to the non lead user group was shown in this paper. In contrast to the standard conjoint analysis the separation of lead user and non lead user group leads to different results and might reduces the risk of innovations. The lead user group is able to express their needs in a more concrete and reliable way (Jeppesen 2005). Further, the group seems to rely on basic developments and decline innovations (improvements as well as breakthrough innovations) not fitting own needs. Although they are said to have a wide-ranged market overview and to collect use experience with new products first, but they tend to be pragmatists (Slater and Narver 1998; Lüthje et al. 2005). Since this conjoint measurement used often market-available innovations, bad use experiences might caused the result. As could be seen, both breakthrough innovations (eBike and pedal lock) were rejected by the lead user group, the eBike concept was further rejected by the non lead user group. Overall the lead user identification increases the conjoint analysis' performance and the gathered ideas and further developed product concepts within this questionnaire provide valuable input for further measurements. An upcoming research has to prove the gathered lead user characteristics.

# References

Baier D, Brusch M (eds) (2009) Conjointanalyse: Methoden - Anwendungen - Praxisbeispiele. Springer, Heidelberg

Cooper RG, Kleinschmidt EJ (1987) New products: What separates winners from losers? J Prod Innovat Manag 4(3):169–184

Franke N, Shah S (2003) How communities support innovative activities: An exploration of assistance and sharing among end-users. Res Pol 32(1):157–178

Green PE, Rao VR (1971) Conjoint measurement for quantifying judgmental data. J Market Res 8(3):355–363

Helm R, Steiner M, Scholl A, Manthey L (2008) A comparative empirical study on common methods for measuring preferences. Int J Manag Decis Making 9(3):242–26

Herstatt C, von Hippel E (1992) : Developing new product concepts via the lead user method: A case study in a "low tech" field. J Prod Innovat Manag 9(3):213–221

Herzog U (1991) Fahrradpatente. Erfindungen aus zwei Jahrhunderten, 2nd edn. Moby Dick

von Hippel E (1986) Lead users: A source of novel product concepts. Manag Sci 32(7):791–805

von Hippel E (2001) Perspective: User toolkits for innovation. J Prod Innovat Manag 18:247–257

von Hippel E (2005) Democratizing innovation. MIT Press, Cambridge, MA

von Hippel E, Thomke S, Sonnack M (1999) Creating breakthroughs at 3M. Harv Bus Rev 77(5):47–57

Jeppesen LB (2005) User toolkits for innovation: Consumers support each other. J Prod Innovat Manag 22(4):347–362

Lilien GL, Morrison PD, Searls K, Sonnack M, von Hippel E (2002) Performance assessment of the lead user idea-generation process for new product development. Manag Sci 48(8):1042–1059

Lüthje C, Herstatt C (2004) The lead user method: An outline of empirical findings and issues for future research. R&D Manag 34(5):553–568

Lüthje C, Herstatt C, von Hippel E (2005) User-innovators and "local" information: The case of mountain biking. Res Pol 34(6):951–965

Schreier M, Prügl R (2008) Extending lead-user theory: Antecedents and consequences of consumers' lead userness. J Prod Innovat Manag 25(4):331–346

Slater SF, Narver JC (1998) Customer-led and market-oriented: Let's not confuse the two. Strat Manag J 19(10):1001–1006

Teichert T, Shehu E (2010) Investigating research streams of conjoint analysis: A bibliometric study. Bus Res 3(1):49–58

Veryzer RW (1998) Discontinuous innovation and the new product development process. J Prod Innovat Manag 15(4):304–321

Wittink DR, Cattin P (1989) Commercial use of conjoint analysis: An update. J Market 53(3):91–96

# Improving the Validity of Conjoint Analysis by Additional Data Collection and Analysis Steps

Sebastian Selka, Daniel Baier, and Michael Brusch

**Abstract** Depending on the concrete application field and the data collection situation, conjoint experiments can end up with a low internal validity of the estimated part-worth functions. One of the known reasons for this is the (missing) temporal stability and structural reliability of the respondents' part-worth functions, another reason is the (missing) attentiveness of the respondents in an uncontrolled data collection environment, e.g. during an online interview with many parallel web applications (e.g. electronic mail, newspapers or web site browsing). Here, additional data collection and analysis has been proposed as a solution. Examples of internal sources of data are response latencies, eye movements, or mouse movements, examples of external sources are sales and market data. The authors suggest alternative procedures for conjoint data collection that deal with these potential sources of internal validity. A comparison in an adaptive conjoint analysis setting shows, that the new procedures lead to a higher internal validity.

## 1 Introduction

Conjoint analysis (CA) is an old and established method for estimating the structure of consumer's preferences (see Green and Srinivasan (1978), Green et al. (2001)). It is widely used in different research areas such as modern product and service development (e.g. Brusch and Baier (2008)) as well as in classic and modern methodological variations (e.g. Louviere and Woodworth (1983), Johnson (1987)).

Collecting empirical data about consumers' preferences is typically connected with problems like the dynamic nature of consumers' preferences as well as

S. Selka (✉) · D. Baier · M. Brusch
Chair of Marketing and Innovation Management, Institute of Business Administration and Economics, Brandenburg University of Technology Cottbus, Postbox 101344, 03013 Cottbus, Germany
e-mail: sebastian.selka@tu-cottbus.de; daniel.baier@tu-cottbus.de; m.brusch@tu-cottbus.de

W. Gaul et al. (eds.), *Challenges at the Interface of Data Analysis, Computer Science, and Optimization*, Studies in Classification, Data Analysis, and Knowledge Organization, DOI 10.1007/978-3-642-24466-7_54, © Springer-Verlag Berlin Heidelberg 2012

measurement problems due to fatigue and boredom. Furthermore, there are learning effects during the data collection phase as well. A new approach of dealing with these problems will be suggested: The construction and analysis of additional or repeated questions. For this, an additional data collection phase at the end of the study will be appended. This can increase internal and maybe external validity.

This paper proposes this approach and is structured as follows. Section 2 gives an overview about typical problems of CA applications. Section 3 shows potential approaches of dealing with these problems. Estimation approaches of collecting empirical data will be shown and summarized. Section 4 comes up with a new approach of an additional data collection phase. Section 5 is used to present the findings of the new approach. Section 6 will show the issues and outlook for future research and applications of the newly developed data collection procedure based on Johnson's adaptive CA.

## 2 Problems with Conjoint Analysis Applications

CA is one of the most prominent tools in consumer preference measurement and widely used in marketing practice. Respondents are confronted in an interview with attribute-level-combinations and are asked to rate these stimuli w.r.t. preference on ordinal or metric response scales. Then, using ANOVA-like procedures, the importance of attributes and levels (part-worth functions) are derived from these ratings. However, an often stated problem of CA is dealing with a large number of attributes and the implications for possible attribute combinations. Also the dynamic of respondent's preference structures (see Pauwels (2004); Johnson and Orme (1996)), the (missing) temporal stability and structural reliability in respondent's part-worth functions (see McCullough and Best (1979)) as well as learning effects during an interview are potential sources for lower internal and external validity (see Netzer et al. (2008)).

Besides this known problems, typical CA studies take a long time to be answered. Some methods come up with a lot of choice decisions which could lead to some sort of information overload and respondent's fatigue or boredom (see e.g. Jacoby et al. (1974), Jacoby (1984), Chen et al. (2009), DeSarbo et al. (2005)). Furthermore, using parallel websites (e.g. digital newspapers, blogs) and web 2.0 applications (e.g. games, social networks, chats) during the self-administered data collection in an interview induces distractions and perturbations. In this context, e.g. Sethuraman et al. (2005) have shown that results in online and offline studies can be substantively different.

In this scope, distractions during questions and information overload are pointed out as a main problem of today's CA applications.

# 3 Known Approaches to Deal with These Problems

The field of estimation approaches for CA is wide and comprehensively, but insufficiently researched. There are methods which:

- Try to decrease the number of stimuli per interview, e.g.

  - Hierarchical CA (HCA, Louviere (1984)),
  - Hierarchical Bayes CA (HBCA, Allenby et al. (1995)),
  - Adaptive CA (ACA, Johnson (1987)),

- Try to activate respondent's attentiveness, e.g.

  - Incentive-aligned CA (ICA, Ding et al. (2005)),

- Use simpler questions, e.g.

  - Self-Explicated Methods (SEM, Srinivasan (1988)),
  - (Computerized) Customized Conjoint Analysis (CCA, Srinivasan and Park (1997); CCC, Hensel-Börner and Sattler (2006)), or
  - ACA,

- Ask for choice decisions instead of preference rankings, e.g.

  - Choice Based CA (CBC, Louviere and Woodworth (1983)),
  - Limit CA (LCA, Voeth and Hahn (1998)), or
  - Hierarchical Limit CA (HILCA, Voeth (2000)).

As we have seen in the prior section, there are many problems that arise from collecting data online. Existing estimation approaches deal with this problem in many cases, as shown in the listing above. Now, in this scope, a new approach will be proposed: An approach, which tries to deal with noisily answered questions by repeating them at the end of the self-administered online survey. Shown existing methods will be compared against the background of the ability for such an additional data collection.

Detecting and dealing with noisy answers requires just in time estimations of part-worth functions at the individual level. An optional data collection and analysis phase will be appended to an already widely applied method, to detect and - if necessary - repeat noisy or even unclear answered questions during an online survey.

Extendable methods must be applicable with common software tools in a practicable way. Furthermore, the base approach for appending an additional data collection phase must be widely applied in the marketing research area. Summarized results are given as an overview in Table 1.

All examined approaches try to increase the number of possible attributes or to get more valid results by each respondent. However, the traditional and modern methods neither try to detect noisy respondent feedback nor to deal with it. Even Johnson's ACA does not. Nevertheless, the evaluation tried to identify methods, which can be extended by an additional data collection phase. As shown in Table 1, ACA and CBC convince through their wide applications in practice and the dealing

**Table 1** Comparison of available estimation approaches

| Method | CBC | HCA | ACA | HBCA | CCA, CCC | LCA | HILCA | ICA |
|---|---|---|---|---|---|---|---|---|
| Just in time estimation | − | + | + | − | + | + | + | + |
| Data estimation at individual level | − | + | + | 0 | + | + | + | + |
| Awareness of the method | + | − | + | + | − | − | − | − |
| Availability of software | + | − | + | 0 | − | − | + | − |
| Dealing with large attribute numbers | + | + | + | 0 | + | + | + | − |
| Detecting and dealing with noisy answers | − | − | − | 0 | − | − | − | − |

(− means "does not apply", 0 means "depending", + means "applies")

with large number of attributes. However, just in time estimations are required for additional data collection and analysis steps. Based on this results, Johnson's approach could be identified as point of origin. The availability of software, the establishment of the approach and the possibility of dealing with a large number of attributes qualifies this method as starting point for the proposed research. In addition, Johnson's approach is web-based, which corresponds to the aim and scope of this research.

## 4 New Approach eACA to Improve the Validity of ACA

Johnson proposed five data collection phases for collecting data within an ACA study. The optional first phase (to eliminate unacceptable attributes) is not used in most cases (Johnson (1987)) and is not considered in this scope. So, it will be assumed, that ACA just has four phases. An optional fifth phase will be appended to this method to check and improve the internal validity of CA. Table 2 compares ACA and the extended ACA (eACA).

## 5 Comparison of the New Approach eACA to ACA

Two CA studies with exactly the same study design (same four attributes, same three levels each) were performed. First, it should be pointed out, that both studies were conducted to test whether the new approach eACA is superior to ACA with non-calculated question repetitions. In the first study (study A), 221 respondents were interviewed during a major German IT trade show (called CeBIT 2010) by personal interviewers. The second study (study B) was slightly different to study A, and just extended by two controlled violations (a non closable full screen video ad overlay on trade-off questions 1 and 4). In contrast to study A, the second study was done under controlled conditions (computer laboratory; without interviewers) by

**Table 2** Comparison of a traditional ACA and the eACA approach (based on Johnson (1987))

| Phases | ACA | eACA |
|---|---|---|
| Step 1 (Level Rating Phase): Collecting respondent's rankings of the levels for each attribute w.r.t. preference. The respondent chooses his most favorite level, next most favorite level, and so on. Attributes with a clear a priori order need not be asked. | yes | yes |
| Step 2 (Importance Rating Phase): Collecting respondent's importance values on a nine point likert scale between his most preferred and least preferred level for each attribute. Collected numeric values are used to indicate the importance of each attribute and its levels giving self-explicated estimates of the part-worth functions. This is the base for preliminary estimations of the respondent's utilities. | yes | yes |
| Step 3 (Trade-Off Phase): The CA part starts by asking a fixed number of (difficult) trade-offs. The method updates respondent's utilities after each question and chooses new trade-off questions based on these new information. Typical trade-offs should not contain more than three attributes. | yes | yes |
| Step 4 (Calibration Phase): Based on already given answers (Phase 1-3), concrete custom-designed concepts were created to being evaluated. Evaluation has to be given by a likelihood-of-buying rating for each concept. The collected data are used to get information about respondent's in the product category. | yes | yes |
| Intermediate step (Calculation): Estimate respondent's preference vector and calculate $R^2$ on the fly and just in time. Compare stated and estimated preference vetors. | no | yes |
| Step 5 (Repetition Phase): Repeat (potentially) noisy answered trade-off questions, if a specific internal validity ($R^2$) threshold is not fulfilled. New given answers replace former given answers inside the stated preference vector. | no | yes |

**Table 3** Comparing study A and B outlines

| | Study A ($n = 221$) | Study B ($n = 13$) |
|---|---|---|
| Max number of additional questions | 2 | 3 |
| Avg. number of additional questions | 1.5 | 1.62 |
| Choosing repetition question | Just last $1 - 2$ questions | Based on highest deviations |
| Extra data collection phase started | 79.2% | 61.5% |
| Improved data sets* | 29.4% | 46.2% |
| Deteriorated data sets* | 25.8% | 7.7% |
| Containing Holdouts | No, to keep it short | No, to keep it short |

(* based on Sawtooth-$R^2$)

just 13 respondents. Both studies were done by computer. Though, the additional data-collection phases were slightly different. The selection of noisy answered trade-off questions in study B was based on the highest deviation between estimated and stated preference. On the contrary, study A just repeated the last one or two trade-off questions (non deviation based). Study A's repetition procedure was based on the assumption that the end of the questionnaire comes up with a decreased respondent's attention and an increased boredom and fatigue. Table 3 outlines both studies and shows a first non-detailed overview.

**Table 4** Correlation coefficients of estimated and stated preferences for study A

|              | ACA Data | eACA Data | Difference | Improvement | Deterioration |
|--------------|----------|-----------|------------|-------------|---------------|
| Spearman     | .947     | .947      | .000       | 67          | 70            |
| Pearson      | .969     | .968      | −.001      | 72          | 76            |
| Kendall      | .867     | .866      | −.001      | 65          | 70            |
| $R^2$        | .953     | .952      | −.002      | 66          | 82            |
| PHP-$R^2$    | .950     | .945      | −.005      | 58          | 90            |
| Sawtooth-$R^2$ | .572   | .576      | .004       | 65          | 58            |

**Table 5** Correlation coefficients of estimated and stated preferences for study B

|              | ACA Data | eACA Data | Difference | Improvement | Deterioration |
|--------------|----------|-----------|------------|-------------|---------------|
| Spearman     | .965     | .971      | .006       | 5           | 2             |
| Pearson      | .975     | .980      | .005       | 7           | 1             |
| Kendall      | .892     | .907      | .015       | 6           | 1             |
| $R^2$        | .960     | .969      | .009       | 7           | 1             |
| PHP-$R^2$    | .960     | .972      | .012       | 5           | 3             |
| Sawtooth-$R^2$ | .854   | .872      | .018       | 6           | 1             |

**Table 6** Detailed overview of repetitions of the trade-off phase in study B

| Trade-Offs               | No. 1 | No. 2 | No. 3 | No. 4 | No. 5 |
|--------------------------|-------|-------|-------|-------|-------|
| Repeated in first step   | 5     | 0     | 0     | 3     | 0     |
| Repeated in second step  | 0     | 5     | 2     | 0     | 1     |
| Repeated in third step   | 0     | 3     | 0     | 2     | 0     |

As stated above, study A just repeated last questions based on the assumption, that an explicit hint could increase respondent's attention on the very last and potential noisy answered questions. But the given results did not support this assumption as shown in Table 4. All given correlations are calculated between the estimated and the stated preferences. The given correlation coefficients are just based on phase 1 (level rating) and phase 3 (trade-offs), except sawtooth software's given $R^2$, which is based on all phases. R as statistic tool, SSI Web (for Sawtooth-$R^2$) and an own implementation (PHP) based on Baier (1998) were used for the calculations.

Table 5 shows that study B has led to better results and improvements during the phase of data collection.

Moreover, study B was extended with specific violations (see above) on the first and the fourth trade-off question, based on the assumption, that specific violations will led to high deviations on these trade-offs. This assumption could be supported, based on the results shown in Table 6.

It could be stated, that the slight improvement in study B and ACA studies in general could led to a higher internal validity. Uncontrolled repetition of questions seems not to work as intended. Just-in-time calculation of internal validity and the explicit hint of existing problems in the respondent's answers to the respondent himself during the data collection process (e.g. "There are some inconsistencies in your answers, please consider the following question again.") could increase data quality as shown in study B.

# 6 Conclusion and Outlook

The given results have shown some minor improvements in data consistency and data correlation. Specific violations decrease respondent's attention and specific information of implausible answers to the respondent himself could increase consistency and internal validity. Further tests and applications should be done. Additional data collection should be based on the internal validity and the external validity as well as on the other phases of a common ACA study. In addition, all approaches should be tested in a field with a larger number of respondents.

Applications of additional data collection phases should be tested with other CA procedures such as Voeth's HILCA, which can be used with a large number of attributes and which generates more valid results, than ACA does (see Wildner et al. (2007)). Last but not least, it should be pointed out, that external validity has not been examined in this scope, although this is planned for the future.

# References

Allenby GM, Arora N, Ginter JL (1995) Incorporating prior knowledge into the analysis of conjoint analysis. J Market Res 32(2):152–162

Baier D (1998) Conjointanalytische Lösungsansätze zur Parametrisierung des House of Quality: Methodik und Anwendungsbeispiel. In: VDI-Gesellschaft Systementwicklung und Projektgestaltung (ed) QFD - Produkte und Dienstleistungen marktgerecht gestalten, pp 73–88

Brusch M, Baier D (2008) Conjoint analysis for complex services using clusterwise hierarchical Bayes procedures. Stud Classification 31:431–438

Chen YC, Shang RA, Kao CY (2009) The effects of information overload on consumers' subjective state towards buying decision in the internet shopping environment. Electron Commerce Res Appl 8(1):48–58

DeSarbo W, Fong D, Liechty J, Coupland J (2005) Evolutionary preference/utility functions: A dynamic perspective. Psychometrika 70(1):179–202

Ding M, Grewal R, Liechty J (2005) Incentive-aligned conjoint analysis. J Market Res 42(1):67–82

Green PE, Srinivasan V (1978) Conjoint analysis in consumer research: Issues and outlook. J Consum Res Interdiscipl Q 5(2):103–123

Green PE, Krieger AM, Wind Y (2001) Thirty years of conjoint analysis: reflections and prospects. Interfaces 31(3):56–73

Hensel-Börner S, Sattler H (2006) Ein empirischer Validitätsvergleich zwischen der Customized Computerized Conjoint Analysis (ccc), der Adaptive Conjoint Analysis (aca) und Self-Explicated-Verfahren. Zeitschrift für Betriebswirtschaft 70:705–727

Jacoby J (1984) Perspectives on information overload. J Consum Res 4:432–435

Jacoby J, Speller DE, Kohn CA (1974) Brand choice behavior as a function of information load. J Market Res 11(1):63–69

Johnson RM (1987) Adaptive conjoint analysis. In: Proceedings of the Sawtooth Software Conferece on Perceptual Mapping, Conjoint Analysis, and Computer Interviewing, Sawtooth Software, Sun Valley, pp 253–265

Johnson RM, Orme BK (1996) How many questions should you ask in choice-based conjoint studies? Tech. rep., Sawtooth Software, Sequim

Louviere JJ (1984) Hierarchical information integration: a new method for the design and analyis of complex multiattribute judgement problems. Adv Consum Res 11(1):148–155

Louviere JJ, Woodworth G (1983) Design and analysis of simulated consumer choice or allocation experiments: An approach based on aggregate data. J Market Res 20(4):350–367

McCullough J, Best R (1979) Conjoint measurement: Temporal stability and structural reliability. J Market Res 16(1):26–31

Netzer O, Toubia O, Bradlow E, Dahan E, Evgeniou T, Feinberg F, Feit E, Hui S, Johnson J, Liechty J, Orlin J, Rao V (2008) Beyond conjoint analysis: Advances in preference measurement. Market Lett 19(3):337–354

Pauwels K (2004) How dynamic consumer response, competitor response, company support, and company inertia shape long-term marketing effectiveness. Market Sci 23(4):596–610

Sethuraman R, Kerin RA, Cron WL (2005) A field study comparing online and offline data collection methods for identifying product attribute preferences using conjoint analysis. J Bus Res 58(5):602–610

Srinivasan V (1988) A conjunctive-compensatory approach to the self-explication of multiattributed preference. Decis Sci 19(2):295–395

Srinivasan V, Park CS (1997) Surprising robustness of the self-explicated approach to customer preference structure measurement. J Market Res 34(2):286–291

Voeth M (2000) Nutzenmessung in der Kaufverhaltensforschung: Die Hierarchische Individualisierte Limit Conjoint-Analyse (HILCA). Gabler, Wiesbaden, Germany

Voeth M, Hahn C (1998) Limit conjoint-analyse. Market ZFP 2(2):119–132

Wildner R, Dietrich H, Hölscher A (2007) HILCA: A new conjoint procedure for an improved portrayal of purchase decisions on complex products. Yearbook Market Consum Res 5:5–20

# The Impact of Missing Values on PLS Model Fitting

Moritz Parwoll and Ralf Wagner

**Abstract** The analysis of interactive marketing campaigns frequently requires the investigation of latent constructs. Consequently, structural equation modeling is well established in this domain. Noticeably, the Partial-Least-Squares (PLS) algorithm is gaining popularity in the analysis of interactive marketing applications which may be attributed to its accuracy and robustness when data are not normally distributed. Moreover, the PLS algorithm also appraises incomplete data. This study reports from a simulation experiment in which a set of complete observations is blended with different patterns of missing values. We consider the impacts on the overall model fit, the outer model fit, and the assessment of significance by bootstrapping. Our results cast serious doubts on PLS algorithms' ability to cope with missing values in a data set.

## 1 Introduction

Structural Equation Modeling (SEM) has changed marketing research substantially (Homburg and Baumgartner 1995). Each SEM is a directed acyclic graph with the parameters capturing the causal impacts assigned to the edges of the graph. Algorithms for fitting these models can be distinguished between covariance-based maximum likelihood maximization (e.g. Jöreskog 1969), heuristic optimization (e.g. Marcoulides and Drezner 2001), and variance-based deviation squares minimization (Siemsen and Bollen 2007). Besides the Least Absolute Deviation and the Unweighed Least Squares approaches, the Partial Least Squares (PLS) algorithm is considered as the most prominent representative of the variance-based techniques. According to Henseler et al. (2009), there has been an increase in the number

M. Parwoll (✉) · R. Wagner
SVI Endowed Chair for International Direct Marketing, DMCC - Dialog Marketing Competence Center, University of Kassel, Germany
e-mail: parwoll@wirtschaft.uni-kassel.de; rwagner@wirtschaft.uni-kassel.de

W. Gaul et al. (eds.), *Challenges at the Interface of Data Analysis, Computer Science, and Optimization*, Studies in Classification, Data Analysis, and Knowledge Organization, DOI 10.1007/978-3-642-24466-7_55, © Springer-Verlag Berlin Heidelberg 2012

of studies relying on PLS published in double blind reviewed journals in recent years. However, increased attention devoted to PLS model fitting is not restricted to international marketing or management research.

The ability to quantify networks of constructs meeting less stringent distribution assumptions according to the data, smooth convergence behavior, and the small sample size properties provide reasons for choosing PLS. Another advantage of Wold's (1985) PLS algorithm is the ability to exploit all information provided by the data, even in cases where some observations may be incomplete. Considering the PLS algorithm, Tenenhaus et al. (2005) argue "(...) there may be a small amount of data that are missing completely at random". However, no previous study provides researchers with a precise specification of an acceptable proportion of missing values. Worsening the situation, O'Loughlin and Coenders (2004) conjecture that casewise deletion could result in biased parameter estimates, even if the values are missing completely at random.

Although missing values (MV) in some observations are likely, rather than an exception, in social science (Decker and Wagner 2000) and the MV are usually not missing completely at random, the extent of the bias appears to be a blind spot in previous research. Studies of MV's impact on fitting SEMs usually refer to covariance-based likelihood maximization approaches (See Olinsky et al. 2003, for a review) or assume the data to be missing completely at random (See Asparouhov and Muthèn 2010, for an exception). Grasping this challenge, we:

- Provide experimental evidence of the PLS's estimation bias introduced by values (1) missing completely at random (MCAR), (2) missing completely at random in classes (MCARC).
- Assess the impact of these two patterns on the fit indices most commonly used to evaluate the model's fit after fitting with the PLS algorithm.

The remainder of this paper is structured as follows. In the next section, we describe the simulation study by introducing both PLS-related assessments of the model's fit and the application domain, the European Customer Satisfaction Index – ECSI (Juhl et al. 2002). We outline the impact of different proportions of missing values on fitting the model. In the third section, we derive our conclusion and provide a brief outlook on further research.

## 2   Simulation Study

### 2.1   Research Design

To assess MV's impact results on the construct measurement level, we consider both the Average Variance Extracted (AVE)

$$AVE = \frac{\sum \lambda_i^2}{\sum \lambda_i^2 + \sum Var(\varepsilon_i)}$$

and the Composite Reliability (CR)

$$CR = \frac{(\sum \lambda_i)^2}{(\sum \lambda_i)^2 + \sum Var(\varepsilon_i)},$$

where $\lambda_i$ denotes the weight assigned to the edge linking the $i^{th}$ manifest indicator to the construct and $Var(\varepsilon_i)$ is the error variance of the $i^{th}$ indicator.

At the structural level, we consider the redundancy $Q^2$

$$Q^2 = 1 - \frac{\sum_{\mathscr{D}} SSE_{\mathscr{D}}}{\sum_{\mathscr{D}} SSO_{\mathscr{D}}}$$

in addition to the proportion of variance explained $R^2$ and path coefficients with $t$-confidence intervals from bootstrapping. Here the $SSE$ denotes the sum of squares in the prediction errors, $SSO$ the sum of squares of the observations and $\mathscr{D}$ is the set of omissions in the blindfolding procedure (omission distance). The total number of indicators $I < |\mathscr{D}| < N$, with $N$ denoting the total number of observations.

The common procedures of MV handling in PLS model fitting are mean replacement and case wise replacement. In the study, we consider both. All calculations are done with Smart-PLS (Ringle et al. 2005).

In our simulation study, the ECSI is considered because it is a well established model frequently used for introductions to PLS model fitting. The ECSI is illustrated in Fig. 1.

The model consists of seven constructs. A set of $I = 24$ manifest variables captures the constructs on ten-point scales. For our experiment, $N = 250$ complete observations are available. Thus, the data matrix for our experiment embraces a total of 6,000 entries. The data are standardized with a mean of zero and a variance equal to one.

## 2.2   Missing Completely at Random

First, 100 (1.5%) MVs are introduced. In steps of 100, we increased the number of missing values up to 600 (10%). In addition, 750 and 1,250 (21%) MVs are considered. Figure 2 provides an overview of the impact on the model's assessment.

Obviously, mean replacement leads to a decrease in the $AVE$ and $R^2$. Not grasped in the figure, casewise deletion tends to increase these assessments. These results are not surprising, because mean replacement reduces variance in the raw data and consequently of the constructs artificially. The $CR$ appears to be less affected for most constructs. Only for "EXPECTATION" does the $CR$ decline to zero in the casewise deletion scenario. The path coefficients and their $t$-values do not follow a consistent pattern. The coefficients have substantial variations with casewise deletion. Actually, some paths change their signs and become negative.

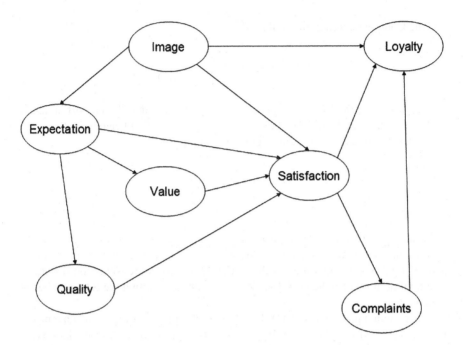

**Fig. 1** Structure of the ECSI

Considering the $t$-values in the case of mean replacement, non significant paths remain non significant and vice versa at a significance level of 95%. Casewise deletion results may lead to a change of the status of significance. Moreover, the prediction relevance ($Q^2$) is affected by MVs. In the case of mean replacement, the values decrease slightly, but remain above the critical level of 0. Casewise deletion results in greater variations even with a small number of MVs. Noticeably, some values decrease to or below zero, indicating the lack of prediction relevance of the construct.

## 2.3  Missing Completely at Random in Classes

The condition of the MCARC pattern is weaker than the MCAR condition. A distinction of classes in SEM is the assignment of manifest indicators to the constructs. Thus, the probability of MVs in the indicators of one construct might be related to the MV of another construct (or its indicators, respectively), but not to the observed indicators of the same construct (c.f. Decker and Wagner 2000 for a formal description). In the experiment, the relation between "SATISFACTION" and "LOYALTY" is considered. The three items which measure the SATISFACTION construct are randomly filled in with MVs. The mechanism is that in the indicator

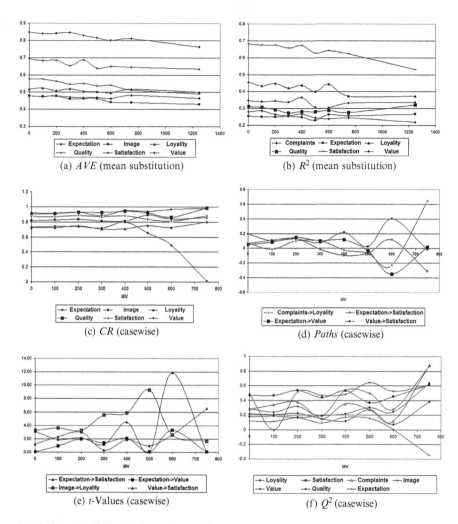

**Fig. 2** Impact of MCAR values on the model's assessments

set of the LOYALTY construct with a probability of 0.8 at least one MV occurs, if in this observation the indicator set of the satisfaction construct has at least one MV. In doing so, we filled in 300, 400, 500 and 600 MVs. Figure 3 provides an overview of the impact MCARC values have on the model's assessment.

As expected, only the constructs SATISFACTION and LOYALTY are affected by a decrease in their assessments ($R^2$, $AVE$, $CR$, and $Q^2$) in the case of mean replacement. In addition, the $R^2$ of "COMPLAINTS" decreases. This construct is a second order construct and determined by the two manipulated constructs within the ECSI framework (See Fig. 1).

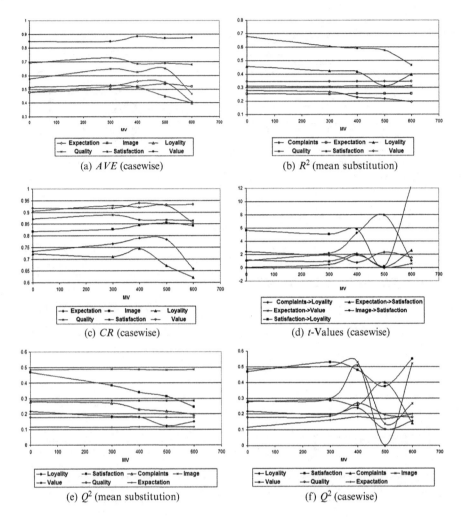

**Fig. 3** Impact of MCARC values on model's assessments

Casewise deletion leads to a change in all constructs. Noticeably, the $AVE$ in some cases undercuts the critical threshold of 0.5, which is considered as a limit referred to this quality criterion. The $CR$s oscillate above the critical level of 0.6. Worsening the situation, casewise deletion leads to a change in the sign for some of path coefficients. Both MV-algorithms can lead to a change in the status of significance, but the casewise deletion procedure results in stronger deviations. Considering the prediction relevance, only the affected constructs with mean replacement decrease little by little. One construct decreases to zero with 500 MV in the casewise deletion scenario.

## 3 Conclusion and Outlook

The results of this simulation experiment give cause for serious concerns with respect to fitting SEM using PLS-algorithms if the data have missing values. Obviously, the path coefficients and the $t$-values are affected substantially in both the MCAR and the MCARC cases. This has a high practical relevance because interpretations and managerial implications derived from the results are strongly based on the relations between the different constructs. Even with a small proportion of MVs a completely misleading assessment of a relationship is possible.

Furthermore, we argue that within the MCARC situation, non affected constructs show a consistent performance, but only in the mean replacement scenario. Generalizing the scenarios of our experiment, the mean replacement leads to more reliable results than casewise deletion. This is in contrast to recommendations derived from MV simulations in fitting SEM by maximizing the log-likelihood (e.g. Acock 2005). However, mean replacement introduces the problem of reducing the variance artificially.

Strong declines and variations were revealed in the casewise deletion scenarios. Particularly, for the path coefficients' $t$-values and constructs' $Q^2$ the results declined to assess at least the direction of the biases. Thus, we argue that these criteria are likely to be invalid if the data have missing values.

Further investigations should aim to broaden the external validity by considering a wider range of experimental conditions (particularly, alternating the sample size, number of constructs and indicators) systematically. These should include further assessments (e.g., the effect size criterion) as well as further MV imputations (e.g., multiple imputations, expectation maximization imputations). Furthermore, additional MV-patterns (e.g., censoring) should be considered within this context. Another challenging issue are the single-item constructs (Bergkvist and Rossiter 2007), which have become frequently used in recent publications. Providing complete Monte-Carlo simulation results seems to be most important. In doing so, alternative means and variances ought to be tested. Moreover, a theoretical or at least an intuitive explanation needs to be developed, particularly for the the case of MCAR data.

Finally, we are not aware of any study using the information provided by the MV patterns to improve the parameter estimation. Both mediating and moderating effects might be considered on the level of the structural model.

## References

Acock AC (2005) Working with missing values. J Marriage Family 67:1012–1028.
Asparouhov T, Muthèn B (2010) Multiple imputation with Mplus. MPlus Web Notes
Bergkvist L, Rossiter JR (2007) The predictive validity of multiple-item versus single-item measures of the same constructs. J Market Res 44:175–184

Decker R, Wagner R (2000) Fehlende Werte: Ursachen, Konsequenzen und Behandlung. In: C Homburg AH, Klarmann M (eds) Handbuch marktforschung. Gabler, Wiesbaden, pp 53–80

Henseler J, Ringle CM, Sinkovics R (2009) The use of partial-least-squares path modeling in international marketing. Adv Int Market 20:277–319

Homburg C, Baumgartner H (1995) Die Kausalanalyse als Instrument der Marketingforschung: Eine Bestandsaufnahme. Zeitschrift für Betriebswirtschaft 65:1091–1108

Jöreskog KG (1969) A general approach to confirmatory maximum likelihood factor analysis. Psychometrika 34:183–202

Juhl HJ, Kristensen K, Østergaard P (2002) Customer satisfaction and customer loyalty in European food retailing. J Retail Consum Serv 9:327–334

Lauro C, Vinzi VE (2002) Some Contributions To PLS Path Modeling and a System for European Customer Satisfaction. Università di Milano Bicocca, Milano, atti della XL1 riunione scientifica SIS

Marcoulides GA, Drezner Z (2001) Specification searches in structural equation models with a genetic algorithm. In: Marcoulides GA, Schumacker RE (eds) New developments and techniques in structural equation modeling. Lawrence Erlbaum, Mahwah, NJ, pp 154–164

Marcoulides GA, Saunders C (2006) PLS: A silver bullet? Manag Inform Syst Q 30:III–IX

Nelson PRC, Taylor PA, McGregor JF (1996) Missing data methods in PCA and PLS: Score evaluations with incomplete observations. Chemometr Intell Lab Syst 35:45–65

Olinsky A, Chen S, Harlow L (2003) The comparative efficacy of imputation methods for missing data in structural equation modeling. Eur J Oper Res 151:53–79

O'Loughlin C, Coenders G (2004) Estimation of the European customer satisfaction index: Maximum likelihood versus partial least squares: Application to postal services. Total Qual Manag 15:1231–1255

Ringle CM, Wende S, Will A (2005) SmartPLS. University Of Hamburg, Hamburg

Siemsen E, Bollen KA (2007) Least absolute deviation estimation in structural equation modeling. Sociol Methods Res 36:227–265

Tenenhaus M, Vinzi VE, Chatelin YM, Lauro C (2005) PLS path modeling. Comput Stat Data Anal 48:159–205

Wold H (1985) Partial least squares. In: Kotz S, Johnson NL (eds) Encyclopedia of statistical sciences. Wiley, New York, pp 581–591

# Part XII
# Data Analysis in Education and Psychology

# Teachers' Typology of Student Categories: A Cluster Analytic Study

Thomas Hörstermann and Sabine Krolak-Schwerdt

**Abstract** The present study demonstrates the application of cluster analysis to examine the typology of student categories of novice Luxembourgish teachers. Student categories are mental representations of groups of students in which teachers classify their students. The investigation of student categories is a relevant topic in education, because subsequent assessments of students may be biased by prior classification. Eighty two novice Luxemburgish teachers were asked to mention types of students they became acquainted with during teaching and described these types by characterizing attributes. Twenty types of students and 65 characterizing attributes were frequently mentioned by the teachers. These data formed the basis of a hierarchical-agglomerative cluster analysis, using average-linkage and complete-linkage clustering methods. The average-linkage-method resulted in 10 clusters, which were largely resembled by the resulting clusters of the complete-linkage-method. This indicates a clear structure in the student categories of Luxembourgish novice teachers. The clusters are compared to Hofer's (Informationsverarbeitung und Entscheidungsverhalten von Lehrern, Beiträge zu einer Handlungstheorie des Unterrichtens, Urban & Schwarzenberg, München, 1981) typology of student categories. The comparison leads to the assumption that the content of student categories may be partly influenced by educational and political discussion.

## 1 Stereotypes and Teacher–Student Interaction

Almost all forms of teaching in education include an intensive social interaction between teachers and their students. During this interaction, teachers perceive innumerable information about students. This information is in turn a necessary

T. Hörstermann (✉) · S. Krolak-Schwerdt
University of Luxembourg, Route de Diekirch, L-7220 Walferdange, Luxembourg
e-mail: thomas.hoerstermann@uni.lu; sabine.krolak@uni.lu

W. Gaul et al. (eds.), *Challenges at the Interface of Data Analysis, Computer Science, and Optimization*, Studies in Classification, Data Analysis, and Knowledge Organization, DOI 10.1007/978-3-642-24466-7_56, © Springer-Verlag Berlin Heidelberg 2012

component for subsequent teaching. Collecting information to assess students' capabilities, personalities, performance etc. is a prerequisite of a teacher's work. He can only act pedagogically and teach effectively when he is familiar with his students (Klauer 1978). For this information to be applicable in subsequent teaching the teacher has to encode and organize this information into a cognitive representation of students that is stored in memory. Assumptions about the structure of this cognitive representation can be derived from findings of cognitive psychology which state that repeated exposure to persons perceived as similar leads to the abstraction of a cognitive category in which common information about the persons is integrated. Person categories are also termed *stereotypes* in social cognition research (Aronson et al. 2004). Referring to the interaction between teachers and students, cognitive categories summarizing groups of students are termed *student stereotypes*. A student stereotype is the cognitive representation of students who are perceived as similar by the teacher in one or more attributes (Hofer 1981). In this paper, we present the findings of an empirical study on student stereotypes that teachers possess and the typological representation of such stereotypes in teachers' cognitions.

Using stereotypes in the information processing about students is of special interest for the investigation of the interaction between teachers and students, because it can exert influence on a teacher's subsequent grading of students' performance and can guide teachers' behavior towards the students. In general, this influence can be seen as highly adaptive. It allows the inference of missing information about a person from the corresponding stereotype and facilitates the formation of expectancies of a person's future behavior (Jussim 2005), thus leading to a more "predictable" world (Pendry 2007). Nevertheless, the use of stereotypes may lead to undesired effects, e.g. *confirmation biases* and *self-fulfilling prophecies* (Klayman and Ha 1987; Rosenthal and Jacobson 1968). Regarding the interaction between teachers and students, these effects might be especially problematic when the content of a student stereotype contains performance information (Rosenthal and Jacobson 1968). In this case, an inferred expectancy of low performance of a student may cause teachers to focus on further indicators of low performance (confirmation bias) and to alter their behavior accordingly, e.g. by giving less challenging tasks, that in turn can result in actual low performance of a student (self-fulfilling prophecy).

## *1.1 Empirical Findings on Student Stereotypes that Teachers Possess*

The existence of student stereotypes of experienced German teachers was demonstrated by a cluster analytical study of Hofer (1981). The study reports a typology of five student stereotypes which were labeled as *der Klassenprimus* (No. 1 in class), *der Arbeiter* (the working student), *der extrovertierte Schüler* (the extroverted student), *der introvertiert-sensible Schüler* (the introverted-sensitive student) and *der schlechte Schüler* (the bad student). The No. 1 in class characterizes intelligent and active students who receive good grades and are willing to invest much effort. The working student is a talented and industrious student working with high discipline. In social interaction, he tends to be calm and humble. The extroverted student shows outstanding social activity, whereas the introverted-sensitive student is shy and withdrawn. The bad student lacks ability and motivation. He shows bad working behavior and does not take part in classroom interactions.

## 1.2 Research Question

Hofer's study (Hofer 1981) illustrated the existence of student stereotypes of teachers and the person attributes characterizing these stereotypes. Nevertheless, the utility of Hofer's typology might be limited by now for further investigations of classroom interactions due to the fact that it reflects student stereotypes of teachers at the beginning of the 1980s. Actual discussions on the topic of classroom interactions frequently refer to aggressive or unattentive behavior of students as examples of possible new ways of categorizing students which can be found in the domain of pedagogy as well as in the media. This in turn may have changed student stereotypes that teachers use in judging students' performance and classroom behavior. Therefore, the aim of this study is the investigation of recent student stereotypes and to compare them to Hofer's typology.

## 2 Method

### 2.1 Participants and Procedure

Eighty two Luxembourgish teacher novices took part in the study. Seventy seven percent of the participants were female. The mean age of the participants was 22.6 years (SD = 4.7 years). All participants had acquired teaching experience in schools by obligatory practical training during teacher education. The participants were instructed to list types of students they got aware of during their practical teaching. For each of these types, the participants had to find a label that represents the type as accurate as possible. Additionally, the participants described each labeled type by characteristic person attributes. The participants were instructed that their description should be as detailed and complete as possible and that they can freely choose different aspects for their descriptions, e.g. personality, ability, interests, outer appearance or behavior patterns. The first page of the questionnaire contained a written form of the instruction and an example of a description. The example was intended to be unrelated to school settings and was a description of a depressive patient with some characterizing attributes, e.g. depressive mood, lack of interest and pleasure etc.

## 2.2 Analysis

The labels mentioned by the participants were analyzed by six raters for the occurrence of synonymous labels. If two labels were rated as synonymous, they were replaced by the most frequent of both labels. Then, frequently mentioned labels were identified by applying a cut-off criterion of being mentioned by at least six participants. This was accomplished as stereotypes are socially shared knowledge (Clark and Kashima 2007), and thus, we were only interested in those stereotypes which are used by teachers in a consensual manner, but not in idiosyncratic stereotypes. For these consensually labels, the characterizing attributes were listed and checked for synonyms. The cut-off frequency for identifying frequently mentioned attributes was set to at least four mentions. The cut-off values were set regarding previous research on stereotypes, where cut-off frequencies of at least 5% were used to exclude idiosyncratic mentions (Cantor and Mischel 1979; Eckes 1994). The 5% frequency was used for the attributes, whereas the cut-off frequency for the labels was raised to 7.5% to avoid that for labels just reaching the minimum frequency all attributes less mentioned than 100% for these labels are excluded from further analysis. Based on the label × attribute-matrix, a similarity matrix of the labels was computed. The $\varphi$-measure was chosen as similarity measure derived from the underlying frequency data with the advantage of a standardized value range (Seber 2004). For the classification of labels, a hierarchical agglomerative cluster analysis was performed using the average linkage clustering method. As the criteria for defining the number of clusters, the distance between adjacent fusion levels and the point-biserial correlation between the corresponding partition and the similarity data were applied. To ensure that the resulting cluster solution is not unique to the applied clustering method but represents a more stable solution, the analysis was repeated by use of the complete linkage method and results were compared to the results of the average linkage analysis (Milligan 1996).

## 3 Results

The 82 participants described 352 types of students in the study which corresponds to an average number of 4.29 student descriptions per participant. After the substitution of synonymous labels, the number of different labels was reduced to 106. Twenty of these labels passed the cut-off criterion of at least six mentions. The frequencies of the 20 labels are listed in Table 1.

These 20 labels were characterized by a total of 905 attributes, After correction for synonyms, 425 different attributes remained of which 65 were mentioned at least

**Table 1** Absolute frequencies of labels passing the cut-off criterion of at least six mentions

| Absolute frequency | Label | Absolute frequency | Label |
|---|---|---|---|
| 17 | Dreamer | 9 | Type with deviating behavior |
| 16 | Aggressive type | 8 | Know-it-all ("Besserwisser") |
| 16 | Clown in class | 8 | Industrious type |
| 14 | Shy type | 8 | Good student |
| 14 | Ambitious type ("Streber") | 8 | Type interested in nothing ("Null-Bock") |
| 13 | Outsider | 7 | Bright type |
| 13 | Dominant type | 7 | Motivated type |
| 13 | Withdrawn type | 6 | Lazy type ("Faulpelz") |
| 11 | Hyperactive type | 6 | Active type |
| 9 | Introverted type | 6 | Calm type |

If types were labelled by special German expressions, the original German label is provided in parenthesis

four times, thus passing the cut-off criterion. The frequencies of the 65 attributes were used to calculate the $\varphi$- coefficients of the labels. The pairwise similarities between the labels ranged from .49 (*ambitious type - know-it-all*) to 1.00 (for 41 pairs of labels). Based on this similarity matrix, agglomerative hierarchical cluster analyses using the average linkage and complete linkage method were performed. The resulting dendrograms from both methods are shown in Fig. 1.

Although the overall fusion process of both methods tends to be rather similar, they do not allow a clear determination of the number of clusters to be considered for interpretation. The number of resulting clusters indicated by the distances between the fusion levels of the average linkage method is ambiguous due to several rather equal distances between fusion levels. Also a combined consideration of both methods does not indicate a sufficiently clear number of clusters. To identify an appropriate number of clusters, the point-biserial correlation, which resembles the correlation between the pairwise similarities of the labels and their assignment to the same or different clusters, was taken into account as a stopping criterion. The point-biserial correlation coefficients for each fusion level of both methods are displayed in Fig. 2.

For the average linkage method, the point-biserial correlation criterion indicates three possible stopping points by rapid decreases between two fusion levels. The correlation coefficient decreases by .050 between the fusion levels indicating 15 and 13 clusters and by .033 between the fusion levels indicating 11 and 10 clusters and the fusion levels indicating 10 and 8 clusters. The point-biserial correlation for the complete linkage method shows a smooth, moderate decrease that is accelerated between the fusion levels corresponding to 11 and 9 clusters. This pattern indicates that the fusion levels with 10 and 11 clusters of the average linkage method should be considered as final solutions. For interpretation, the 10 cluster solution of the average linkage method is chosen with respect to the aim of data reduction. The selected cluster solution covers two clusters containing five student types, two

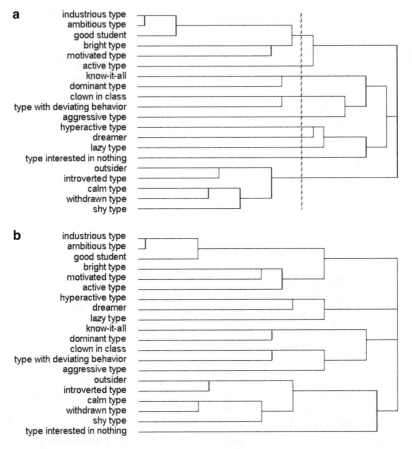

**Fig. 1** Dendrogram of the fusion process of (**a**) average linkage method and (**b**) complete linkage method. The dashed line indicates the interpreted solution

clusters containing two types and six singletons. Compared to the complete linkage solution with 11 clusters, the alignment of the student types to the clusters is identical except of two differences.

## 3.1   Interpretation of the Clusters

In addition to the labels of student types in each cluster, the characterizing attributes were considered as an additional source of information to identify the meaning of each cluster. To define the attributes which characterize each cluster, the mean frequency of each attribute for each cluster was computed and the resulting cluster × attribute matrix was standardized to its row and column sums. Then,

| number of clusters at fusion level | average linkage | complete linkage |
|---|---|---|
| 20 | .945 | .945 |
| 19 | .944 | .944 |
| 18 | .935 | – |
| 17 | .929 | .929 |
| 16 | .923 | .923 |
| 15 | .908 | – |
| 14 | – | .899 |
| 13 | .858 | – |
| 12 | – | .881 |
| 11 | .846 | .863 |
| 10 | .813 | – |
| 9 | – | .826 |
| 8 | .780 | – |
| 7 | .768 | – |
| 6 | .757 | .757 |
| 5 | – | .713 |
| 4 | .697 | .683 |
| 3 | .580 | – |

Missing coefficients for a specific number of clusters are due to the fusion of more than two clusters at the preceding fusion level

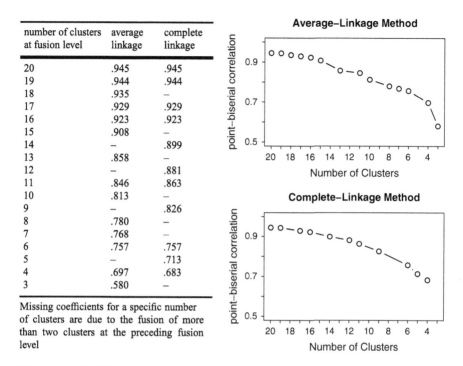

**Fig. 2** Point-biserial correlation coefficients for each fusion level of the average linkage and the complete linkage method

each attribute was used in the interpretation of the cluster for which it reached its maximum standardized frequency. The first large cluster contains the *industrious type, ambitious type, good student, bright student* and *motivated student*. All labels refer to high performance and motivation in class, thus describing a kind of *ideal student*. This is also reflected by the attributes, which describe these students as fast working, concentrated, intelligent, interested and receiving good grades. The *shy type, calm type, introverted type, withdrawn type* and the *outsider* belong to the second large cluster, which reflects mainly aspects of social reservation. The student types in this cluster were characterized as shy, avoiding other students, alone, not integrated and avoiding to attract attention. One of the two smaller clusters is formed by the *dominant type* and the *know-it-all*, types with a high self-esteem and the tendency to correct and command others. They are characterized as cheeky and dominant. The *clown in class* and the *type with deviating behavior* are covered by the second smaller cluster mainly focusing aspects of attention seeking and disturbing behavior. They are joking and disturbing class. The remaining six clusters were singletons, describing (a) an *aggressive type*, who is violent, short-tempered and beats others, (b) an *active type* talking much and loudly, (c) a nervous and screaming *hyperactive type*, who is easily distractable, (d) a *dreamer* with bad grades who acts slowly and is often lost in thought, (e) a *lazy type*, characterized as lazy and passive and receives bad grades although described as intelligent, and (f) a *type interested in nothing*, forgetting his school supplies and described as lethargic and completely unmotivated.

# 4 Discussion

The hierarchical agglomerative cluster analysis using the average linkage method resulted in ten clusters which were largely replicated by using the complete linkage method. The applied point-biserial correlation coefficient criterion for determining the number of clusters indicated this solution of ten clusters as an appropriate number of clusters to interpret. A comparison to Hofer's student types reveals a relatively high concordance with the present solution. Most of Hofer's types are represented in the clusters of the recent study. For example, Hofer's No. 1 in class matches well with the cluster describing the ideal student, whereas the introverted-sensitive student can be considered as comparable to the socially withdrawn types in the social reservation cluster and the bad student cluster can be compared to some degree to the type interested in nothing. Only Hofer's working student is not directly represented by a cluster of the present study, but its main aspects are partly reflected by the clusters covering social reservation and the ideal student. Although the main attribute domains characterizing student types are also in line with Hofer's results (performance, working behavior, interaction in class), some labels and attributes in the present study are not mentioned in Hofer's student types, e.g. hyperactivity, aggression and deviating behavior, indicating a time effect. In the 1980s, elements of current public discussions concerning educational problems like violence in schools or child-related psychological diseases, e.g. attention-deficit hyperactivity disorders, were largely unknown for most parts of the population. However, these aspects are clearly represented in the typology of the present study and may suggest that teachers direct attention to new patterns of students' behavior.

A methodological issue is raised by the fact that the characterizing attributes are not only used for the computation of the similarity measure, but also as an additional source of information for the interpretation of the clusters. This implicit alignment of attributes to the clusters may suggest the use of a two-mode rather than a one-mode clustering method, because two-mode methods are explicitly designed to allow a simultaneous clustering of both, in this case, labels and attributes. A reanalysis of the presented data is intended by the authors. Another possible limitation is due to the necessary preprocessing of the data before cluster analysis. Although the identification of synonyms was done by six raters, thus lowering the risk of individual subjectivity, the data underlying the cluster analysis still might be biased by the individual perspectives of these raters.

**Acknowledgements** The preparation of this paper was supported by grant KR 2162/4-2 & GR 1863/5-2 from the German Research Foundation (DFG) in the Priority Programme "Models of Competencies for Assessment of Individual Learning Outcomes and the Evaluation of Educational Processes" (SPP 1293)."

# References

Aronson A, Wilson TD, Akert RM (2004) Sozialpsychologie, 4th edn. Pearson, München

Cantor N, Mischel W (1979) Prototypes in person perception. In: Berkowitz L (ed) Advances in experimental social psychology, vol 12. Academic Press, New York

Clark AE, Kashima Y (2007) Stereotypes help people connect with others in the community: A situated functional analysis of the stereotype consistency bias in communication. J Pers Soc Psychol 93:1028–1039

Eckes T (1994) Explorations in gender cognition: Content and structure of female and male subtypes. Soc Cognition 12:37–60

Hofer M (1981) Schülergruppierungen in Urteil und Verhalten des Lehrers. [student groups in teachers' perception and behavior]. In: Hofer M (ed) Informationsverarbeitung und Entscheidungsverhalten von Lehrern. Beiträge zu einer Handlungstheorie des Unterrichtens, Urban & Schwarzenberg, München

Jussim L (2005) Accuracy in social perceptions. criticisms, controversies, criteria, components and cognitive processes. In: Zanna MP (ed) Advances in experimental social psychology, vol 37. Elsevier, Amsterdam

Klauer K (1978) Handbuch der Pädagogischen Diagnostik [Handbook of pedagogical diagnostics]. Schwann, Düsseldorf

Klayman J, Ha YW (1987) Confirmation, disconfirmation, and information in hypothesis testing. Psychol Rev 94:211–228

Milligan GW (1996) Clustering validation: Results and implications for applied analyses. In: Arabie P, Hubert LJ, De Soete G (eds) Clustering and classification. World Scientific, River Edge, NJ

Pendry L (2007) Soziale Kognition [social cognition]. In: Jonas K, Stroebe W, Hewstone M (eds) Sozialpsychologie. Springer, Heidelberg

Rosenthal R, Jacobson L (1968) Pygmalion in the classroom: Teacher expectation and pupils' intellectual development. Holt, Rhinehart & Winston, New York

Seber GAF (2004) Multivariate observations. Wiley, Hoboken, NJ

# ECTS Grades: Combination of Norm- and Criterion-Referenced Grading

Jonas Kunze and Andreas Geyer-Schulz

**Abstract** One element of the Bologna process are ECTS grades that should be reported together with national grades. While ECTS grades are norm-referenced, most national grading systems in the European Union are criterion-referenced. As "proper statistical methods" were one of the main requirements for establishing ECTS grades (ECTS User's Guide: European Credit Transfer and Accumulation System and the Diploma Supplement, Directorate-General for Education and Culture, European Commission, Brussels, 2005) and it was stated that ECTS grades are seldom reported (ECTS User's Guide: European Credit Transfer and Accumulation System and the Diploma Supplement, Directorate-General for Education and Culture, European Commission, Brussels, 2005), this article describes the process of ECTS grade assignment and addresses various issues that might be a reason for the missing process implementation of ECTS grades. Furthermore, two important steps of the assignment process are discussed in detail and quality measures are proposed that may be used in various forms to monitor ECTS grade assignment.

## 1 Introduction

Since the Bologna Declaration in 1998, the Bologna Process has influenced European Higher Education Systems in manifold ways. One of the measures to make degrees among European universities easily readable and comparable, are ECTS grades, "[...]to make national grading systems more transparent, to allow a smooth transfer of grades from one system to another, in order to serve mobile learners and graduates." (European Commission 2005) with the following definition:

J. Kunze (✉) · A. Geyer-Schulz
Lehrstuhl für Informationsdienste und Elektronische Märkte, Institut für Informationswirtschaft und -management, KIT, Germany, www.iism.kit.edu
e-mail: jonas.kunze@kit.edu; andreas.geyer-schulz@kit.edu

W. Gaul et al. (eds.), *Challenges at the Interface of Data Analysis, Computer Science, and Optimization*, Studies in Classification, Data Analysis, and Knowledge Organization, DOI 10.1007/978-3-642-24466-7_57, © Springer-Verlag Berlin Heidelberg 2012

Those [students] obtaining passing grades are divided into five subgroups: the best 10% are awarded an A-grade, the next 25% a B-grade, the following 30% a C-grade, the following 25% a D-grade and the final 10% an E-grade.

ECTS grades are therefore *norm-referenced*, i.e. the ECTS grade reflects the performance of a student relative to other students. In comparison, most European national grades are *criterion-referenced* that means that students evaluation is based on the percentage of correct answers in an exam.

In order to achieve the required transparency, further fundamental aspects of ECTS grades include that they were developed "to facilitate the understanding and comparison of grades" (European Commission 2005), and that they neither substitute local grades nor prescribe grade conversion.

This paper is structured as follows. In Sect. 2, some further characteristics and objections against ECTS grades are discussed. Afterwards in Sect. 3, assuming that a reference distribution is given, variants of the ECTS grade assignment function are presented that cope with the tying problem. In Sect. 4, robust methods for providing reference assessment data are explored, thereby identifying a pitfall in the ECTS Guide. Finally, the paper closes with a summary and an outlook.

## 2 ECTS Grades

Two characteristics of ECTS grades, that are of major importance for the intended transparency, are: (a) The description of the meaning of an ECTS grade is highly intuitive (e.g. "best 10%"). (b) An ECTS grade is easily depicted as it consists of only one letter. With these two characteristics, ECTS grades help the reader of any diploma to easily get an idea about the relative performance of the student.

The fact that the relative performance is measured, is subject to various concerns. Karran (2005, p. 9) objects that most European universities use criterion-referenced scales in the assessment, thereby "place[ing] these in some conflict with the norm-referenced rationale of the ECTS. On the other hand, it can be argued, that the combination of a norm-referenced and a criterion-referenced scale opens up new options for quality control. For example, if the second best grade in a local criterion-referenced scale gets assigned an ECTS grade "D", this might indicate that in this exam the assessment does not differentiate well between students. Having said this, it is clear that the question whether to publish ECTS grades along with local grades gets a political issue because shortcomings in assessment get publicly known. In fact, the implementation of ECTS grades are seldom completed (European Commission 2009). The European Commission (2009) argues that the determination of the five segments (i.e. best 10%, following 25%, and so on) out of collected assessment data was "far too ambitious and difficult to implement" and propose a simplified version. However, the simplified version loses the two aforementioned characteristics of ECTS grades. The objection, that the determination of the five segments is of high theoretical and technical complexity will be rejected in the next section.

Another peculiar observation that is made when reviewing relevant literature is that various authors assume that the ECTS grades require a normal distribution of local grades (Grosges and Barchiesi 2007; Webler 2010), but these assumptions are not based on citations. Formally, for the successful application of a norm-referenced scale, the only requirement is that the performances can be ordered. However, with real assessment data, this requirement is usually not fulfilled. This problem field is discussed in the following section.

From an organizational point of view, the decision to record and publish ECTS grades together with local grades has various requirements. First, grade recording is a process that is operated with very high frequency. Hence, the process of recording ECTS grades has to be highly robust and automated. Automated means that no manual interaction is needed and robust refers to the condition, that arbitrary input is handled in a meaningful way. Second, legal issues are of high importance. The possibility that a lawsuit could be filed against the way the ECTS grades are assigned, has to be as low as possible. Otherwise, the organizational costs in the legal department rise with the decision to record ECTS grades.

The used model for the calculation of ECTS grades consists of three steps: First, a reference distribution of local grades has to be identified (cf. Sect. 4). Second, from this reference distribution, an ECTS grade assignment function is determined (cf. Sect. 3). Third, the ECTS grade assignment function is used on assessment data, i.e. assigns to each local grade an ECTS grade.

## 3   ECTS Grades Assignment Function

The aim of this section is to discuss various approaches of how to get from a given reference distribution of local grades to an ECTS grade assignment function. Let $E = \{$ "A", "B", "C", "D", "E" $\}$ be the set of ECTS grades, and let $L$ be the set of local grades. An ECTS grade assignment function is a function $e$ that maps a local grade $l \in L$ to an ECTS grade $e(l) \in E$. An example of ECTS grade assignment for an exam where 20 was the best local grade and 10 was the worst local grade is depicted in Table 1.

Along with the definition of the ECTS grade assignment function, two assumptions were made. First, without loss of generality, we assume that the local grades are the highest differentiated assessment data available. If there is assessment data available that is higher differentiated than the local grades (e.g. such as achieved points in an exam), $L$ can be extended to describe this situation. Using highly differentiated data is generally advised as it helps to identify the required percentiles, but basing the calculation on local grades has various advantages, e.g. perceived non-discrimination and information availability at the central administration. Second, as $e$ assigns the same ECTS grade based on a local grade, the underlying assumption

**Table 1** Example of ECTS grades

| Local grade | Number of students | Percentile | ECTS grade |
|---|---|---|---|
| 20 | 2 | 2 | A |
| 19 | 4 | 6 | A |
| 18 | 4 | 10 | A |
| 17 | 9 | 19 | B |
| 16 | 16 | 35 | B |
| 15 | 19 | 54 | C |
| 14 | 11 | 65 | C |
| 13 | 7 | 72 | D |
| 12 | 8 | 80 | D |
| 11 | 10 | 90 | D |
| 10 | 10 | 100 | E |

is that all students with the same local grade receive the same ECTS grade. Under the aspect of non-discrimination, the authors are convinced that this approach is the only viable. The downside of this assumption is that the distribution of assigned ECTS grades may differ more from the theoretical distribution. An approach that relaxes this assumption is presented by Warfvinge (2008). His approach is to select randomly students with the same local grade and to assign them different ECTS grades.

As mentioned in the previous section, the general condition to apply a norm-based scale is that the performances can be ordered. However, if percentiles of the norm-based scale are given, as it is the case with ECTS grades, the condition can be reformulated: The specific condition of applying a norm-based scale with given percentiles, is that the given percentiles can be identified in the performance distribution. In the example shown in Table 1, the percentiles and the corresponding ECTS grades were unambiguously identified, but a complete order of performances is not possible as, for example, two students receive 20 points and it is unclear who of them was the best. When working with real data, an unambiguously identification of the required percentiles is an exception and the *percentile problem*, i.e. students with the same local grade "overlap" the given percentiles of ECTS grades, occurs. This problem is especially grave, if the number of local passing grades is lower than five, the number of ECTS grades. One example for this is the Swedish grading scale (Karran 2005).

The percentile problem is illustrated in Table 2. The performances of students with the local grade 18 lie between the 7th and 12th percentile. This means that the ECTS grade "A" as well as the ECTS grade "B" is reasonable. A similar situation is found at local grade 10. The most extreme form of the percentile problem is illustrated at local grade 13. Fifty percent of all students achieve this local grade. The covered percentiles refer to the ECTS grades "B", "C", and "D".

The three approaches to cope with this situation, are (a) upgrade the ECTS grade to the best one available, (b) downgrade the ECTS grade to the worst possible, or (c) take the average student and assign the corresponding ECTS grade. The

**Table 2** Percentile problem (example)

| Local grade | Number of students | Percentile | ECTS grades (possible) | ECTS grade (upgrade) | ECTS grade (downgrade) | ECTS grade (average) |
|---|---|---|---|---|---|---|
| 20 | 2 | 2 | A | A | A | A |
| 19 | 5 | 7 | A | A | A | A |
| 18 | 5 | 12 | A-B | A | B | A |
| 17 | 3 | 15 | B | B | B | B |
| 16 | 2 | 17 | B | B | B | B |
| 15 | 4 | 21 | B | B | B | B |
| 14 | 7 | 28 | B | B | B | B |
| 13 | 50 | 78 | B-D | B | D | C |
| 12 | 4 | 82 | D | D | D | D |
| 11 | 3 | 85 | D | D | D | D |
| 10 | 15 | 100 | D-E | D | E | E |

average student is found at the middle of the percentile interval corresponding to the local grade, e.g. as the performances of the students with the local grade 18 range from 7 to 12, the average student is at the percentile 9.5, hence falling into the best 10%. Therefore, the average approach would assign the ECTS grade "A" to all students with local grade 18. For the average approach, various variants are possible if looking at the special case that the average student is just at the border of two given percentiles (endpoint treatment).

Furthermore, there are two other approaches that may be applied, but that do not fit into the defined setting. First, as mentioned before, it is possible to assign randomly the required number of ECTS grades (Warfvinge 2008), e.g. assigning to 3 of the 5 students with local grade 18 an ECTS "A" and to 2 students with this local grade an ECTS "B". Second, one might think of extending the range of possible ECTS grades to ranges, i.e. assigning an "C-E" instead of "C", "D", or "E" if the local grade ranges over the corresponding ECTS percentiles. Another variant would be to take the upgrading approach and to add a sign that indicates overlapping percentiles, e.g. in the aforementioned example a "A-" would be assigned to students with local grade 18.

The presented approaches may be compared by various categories. First, the upgrade or downgrade approach, in comparison to the average approach, introduce a strong bias in the assignment of ECTS grades. Hence, when applying the upgrade approach, the percentage of assigned ECTS grades "A" will surely be higher than the given 10%. When talking about bias, it has to be noted, that there is an incentive for a higher education institution to assign better ECTS grades as these increase the externally perceived qualification of degree holders. However, the incentive is only present as long as this bias may not be revealed. Second, if the assumption that local grades are the highest differentiated assessment data available is relaxed, the upgrade approach is a good solution. Otherwise, higher differentiated assessment data may be used by the best performing students to file a lawsuit in order to acquire a better ECTS grade, resulting in high loads of administrative work for the organization as mentioned in the previous section.

## 4 Reference Distribution

Basically, there are two approaches to determine a reference distribution $\hat{f}$ in order to start the assignment process. The first way is to set $\hat{f}$ equal to the observed local grades of the exam for which a ECTS grade assignment is calculated. This is possible, if the number of observations is sufficiently high (e.g. European Commission (2005) recommends at least 30 individuals achieving passing grades). This method will be called *self referential*. The second way is using similar data, e.g. compile assessment data from several exams of similar level or over a various years in the past ("the distribution of marks over a five-year period is likely to produce balanced results." (European Commission 2005)). We call this method *foreign referential*.

Regarding the self referential determination of $\hat{f}$, it has to be noted, that this straight forward approach is not always viable. Problems include (a) small sample sizes, e.g. less than the recommended 30, (b) exam resits, (c) grade transfers that usually take place on an individual level, (d) and composed grades, such as a grade for a certain subject that consists of various module grades. Due to these problems, we will focus on foreign referential ECTS assignment.

The basic proposition in the texts of the European Commission is to compile the sample frequency distribution of various exams. If the sample frequency distribution should be determined by the assessment data of a five-year period, it can be described as follows: Let $f_i, i \in \{1, \ldots, t\}$ be the sample frequency distribution in period $i$ and $t$ be the actual period. Furthermore, let $n_i$ be the number of observations in period $i$. Hence, the first variant of $\hat{f}$, $\hat{f}^1$ is defined as:

$$\hat{f}_t^1(l) = \frac{n_{t-1} f_{t-1}(l) + \ldots + n_{t-5} f_{t-5}(l)}{n_{t-1} + \ldots + n_{t-5}} \tag{1}$$

But this method has severe disadvantages (cf. Kunze and Geyer-Schulz (2010a,b)):

- The number of observations has a linear influence in the formulation of $\hat{f}^1$. While this fact is meaningful for small observation, for larger observations this influence may lead to unintuitive changes in the development of $\hat{f}^1$ over time.
- The initialization of the process, i.e. definition of $\hat{f}_t^1, t \in \{1, \ldots, 5\}$ is difficult as the formula would refer to observations a $t \in \{-4, \ldots, 0\}$. Not only a sample frequency distribution has to be guessed but furthermore, the number of observations have to be defined.
- Although 5 periods are included, it is not assured that the total number of observations is always above 30.

As a consequence, the authors propose a modified exponentially smoothed development of $\hat{f}$: The function $\hat{f}_t^2$ has the parameters $w$, defining the maximal weight that a new observed frequency distribution has, and $m$, defining the minimal number of observations necessary to receive the maximal weight $w$. In comparison to exponentially weighted moving average, the weight is a function of the number of observations in the latest period. The weight grows linear with the number, but from a certain number $m$, the weight is fixed. The rationale is that with small numbers of observations, outliers are more probable. Therefore, a smaller weight is assigned than to samples with a sufficiently large number of observations.

$$\hat{f}_t^2(l) = w' f_{t-1}(l) + (1 - w') \hat{f}_{t-1}^2(l) \quad \text{with} \tag{2}$$

$$w' = \begin{cases} w & \text{if } n_{t-1} > m \\ w\frac{n_{t-1}}{m} & \text{else} \end{cases} \tag{3}$$

While $\hat{f}^1$ is based on absolute distributions, $\hat{f}^2$ uses only relative distributions and the number of observations only influences the update. Therefore, $\hat{f}^2$ has an implicit treatment of small sized observations. Furthermore, $\hat{f}^2$ smooths the observations and does not produce artifacts. Finally, $\hat{f}^2$ is simpler to initialize, as only one frequency distribution $\hat{f}_0^2$ has to be determined. For further details including examples and an extended discussion confer Kunze and Geyer-Schulz (2010a,b).

## 5 Quality Measures

The bias of the process of recording ECTS grades could be measured by recording the percentiles of all recorded ECTS grades. In the case of an unbiased process, these percentiles should be near to the percentiles of the definition of ECTS grades.

If one wants to compare different reference distributions $\hat{f}^1$, $\hat{f}^2$ to the observed distribution $f$, it is helpful to compare the corresponding ECTS grade assignment functions $e_{\hat{f}^1}, e_{\hat{f}^2}, e_f$, as only differences in the assignment function have an impact. Hence, we define the distance measure $s(e_1, e_2)$ to be the minimal number of incremental changes ("steps") that have to be made to $e_1$ so that it equals $e_2$. An incremental change refers to change the assignment of a certain local grade from the actual ECTS grade to the next ECTS grade (e.g. "A" to "B"). For example, the distance between the observed distributions of Table 1 $f^1$ and Table 2 $f^2$ is, using the average approach, $s(e_{f^1}, e_{f^2}) = 3$ as the ECTS grades for the local grades 15, 14, and 13 change from "C" to "B" or from "D" to "C". If the upgrade approach is used, the distance is 5 with local grade 13 changing by two steps from "D" to "B". Using the step distance measure, the speed of adoption to trends and the robustness of the reference distribution against outliers could be evaluated.

# 6  Summary and Outlook

This article discusses various aspects that are of major importance when implementing the process of ECTS grade reporting. The process is characterized by three steps: First, determining the reference distribution. Second, calculating an ECTS grades assignment function that depends on the reference distribution. Third, applying the ECTS grades assignment function on the assessment data for which ECTS grades should be reported. From a technical point of view, various methods for determining a reference distribution as well as for calculating the ECTS grade assignment function exist. However, in order to comply legal, organizational, and political requirements, most of them are not viable.

Referring to the determination of a reference distribution in the foreign-referential case, a robust method of modified exponential smoothing was presented. This method overcomes serious operative problems that are inherent by the proposed approach by the European Commission. However, open issues exist in this field. These include the combination of self- and foreign-referential aspects and the optimization of parameters.

Regarding the ECTS grade assignment based on a given distribution, various solutions were given regarding the percentile problem. However, due to legal restrictions of non-discrimination and incentives for strategic misrepresentation, there are strong incentives on an institutional level for choosing a biased method (upgrade approach). Using this approach on the European level would lead to a discrimination between countries as the bias is influenced by the number of local passing degrees available. The authors suggest a standardization of this process step between countries or an extension of the ECTS grades.

Finally, the authors think that the missing implementation of ECTS grades since the last 12 years can be attributed to the complex interaction of politics, legal aspects, organizational aspects, statistical issues, and IT requirements.

# References

European Commission (2005) ECTS User's Guide: European Credit Transfer and Accumulation System and the Diploma Supplement. Directorate-General for Education and Culture, European Commission, Brussels

European Commission (2009) ECTS User's Guide: European Credit Transfer and Accumulation System and the Diploma Supplement. Directorate-General for Education and Culture, European Commission, Brussels

Grosges T, Barchiesi D (2007) European credit transfer and accumulation system: An alternative way to calculate the ECTS grades. High Educ Eur 32(2,3):213–227

Karran T (2005) Pan-european grading scales: Lessons from national systems and the ECTS. High Educ Eur 30(1):5–22

Kunze J, Geyer-Schulz A (2010a) Die grundlegende Problematik der ECTS-Notenzuweisung: Ein Musterverfahren. Qualität in der Wissenschaft 4(1):24–28

Kunze J, Geyer-Schulz A (2010b) ECTS-Noten: Verfahren zur Kohortenberechnung. Tech. rep., emKIT. http://www.em.uni-karlsruhe.de/literatur/lit_show.php?id=15613. Cited 01 Aug 2010

Warfvinge P (2008) A generic method for distribution and transfer of ECTS and other norm-referenced grades within student cohorts. Eur J Eng Educ 33(4):453–462

Webler W (2010) Internationale Vergleichbarkeit von Noten im Hochschulbereich? Problematik der Notenvergabe, Referenzgrößen und der Verwendung der Gaußschen Normalverteilung. Qualität in der Wissenschaft 4(1):20–23

# Students Reading Motivation: A Multilevel Mixture Factor Analysis

**Daniele Riggi and Jeroen K. Vermunt**

**Abstract** Latent variable modeling is a commonly used data analysis tool in social sciences and other applied fields. The most popular latent variable models are factor analysis (FA) and latent class analysis (LCA). FA assumes that there is one or more continuous latent variables – called factors – determining the responses on a set of observed variables, while LCA assumes that there is an underlying categorical latent variable – latent classes. Mixture FA is a recently proposed combination of these two models which includes both continuous and categorical latent variables. It simultaneously determines the dimensionality (factors) and the heterogeneity (latent classes) of the observed data. Both in social sciences and in biomedical field, researchers often encounter multilevel data structure. These are usually analyzed using models with random effects. Here, we present a hierarchical extension of FA called multilevel mixture factor analysis (MMFA) (Varriale and Vermunt, Multilevel mixture factor models, Under review). As in multilevel LCA (Vermunt, Sociol Methodol 33:213–239, 2003), the between-group heterogeneity is modeled by assuming that higher-level units belong to one of $K$ latent classes. The key difference with the standard mixture FA is that the discrete mixing distribution is at the group level rather than at the individual level. We present an application of MMFA in educational research. More specifically, a FA structure is used to measure the various dimensions underlying pupils reading motivation. We assume that there are latent classes of teachers which differ in their ability of motivating children.

D. Riggi (✉)
Department of Statistics, University of Milan-Bicocca, Via Bicocca degli Arcimboldi 8, Ed. U7, Milan, Italy
e-mail: daniele.riggi@gmail.com

J.K. Vermunt
Department of Methodology and Statistics, Tilburg University, P.O. Box 90153, 5000 LE, Tilburg, The Netherlands
e-mail: j.k.vermunt@uvt.nl

W. Gaul et al. (eds.), *Challenges at the Interface of Data Analysis, Computer Science, and Optimization*, Studies in Classification, Data Analysis, and Knowledge Organization, DOI 10.1007/978-3-642-24466-7_58, © Springer-Verlag Berlin Heidelberg 2012

# 1 Introduction

In the past thirty years, in the field of applied and social science such as psychology, education, marketing, biology or medicine, it has been made possible to see a spread of many statistical tools for data analysis. These disciplines deal with unobserved concepts such as intelligence, skills, attitudes, medical conditions, personality traits, preferences or perceptions. In this context, researchers usually make assumptions about the existence of latent concepts from the observed items. Latent variables are hypothetical constructs that influence the phenomenon realizations and could be described as the *true* variables, while the observed variables are the indirect measures of that dimension (Skrondal and Rabe-Hesketh 2004).

Factor Analysis and Latent Class are two of the most used tools in the field of applied science. Factor Analysis describes the relationship (association) between manifest variables (indicators or items) and hidden continuous latent variables (factors) in the presence of population homogeneity. Factor models deal with continuous, dichotomous or ordinal indicators. The popularity of this approach increased in the 1950s and 1960s thanks to the development of statistical computing capacity. Nowadays, it represents one of the most popular tools for quantitative research in social science.

Latent class analysis is used in the presence of a population heterogeneity with categorical variables. These models were introduced by Lazarsfeld and Henry (1968) to explain differences in the way of responding to a survey with dichotomous items. Twenty years later, latent classes models were structured and carried on including nominal responses. In the same years, Day (1969) and Wolfe (1970) proposed the Finite Mixture model, based on the assumption that the observed data are generated by a finite mixture that identifies a subpopulation (McLahan and Peel 2000). In recent years, Latent Class and Finite Mixture are part of the most used toolboxes for applied research.

In many applied works, researchers often encounter multilevel structures, such as individuals with multiple response, repeated measures nested within groups (Vermunt 2003, 2008), multivariate repeated responses nested within individuals (Vermunt et al. 2008), or three level data sets (Vermunt 2007). Examples of multilevel data structure are patients nested with doctors or hospitals, students nested with teachers or schools, or repeated measures nested with subjects. With the hierarchical data structure, the assumption on the independence of observations in Factor Analysis, could not be considered valid for two reasons: the subject responses are influenced by external factors and the individual units share the same environment. To solve this problem, the multilevel techniques should be used to deal with both the relationship between observations, and between higher level factors.

Multilevel Mixture Factor Model (MMFM) integrates both Latent Class and Factor Analysis, because with this model it is both possible to cluster items in the presence of homogeneous population and to cluster individuals when the source of population heterogeneity is unknown. The added value is linked to the possibility to model between group differences, using $K$ latent classes or mixture components, instead of continuous group level factors or random effects.

## 2 Multilevel Mixture Factor Model

The multilevel mixture factor analysis could be described using the response models from the *Generalized Linear Modeling* family, referring to the notation used by Skrondal and Rabe-Hesketh (2004). Let $K$ be the total number of latent classes or mixture components and $k$ one of the latent classes ($k = 1, \ldots, K$). $y_{hij}$ denotes the observed response, where $h$ identifies items, $i$ subjects and $j$ groups. The superscripts (2) and (3) identify respectively the latent variables at the individual and the group level. The linear predictor for $y_{hij}$ is identified by $v_{hij}$ and the latent variables are denoted by $\eta_{ij}^{(2)}$ and $\eta_i^{(3)}$. The following equations describe the two level mixture factor models used in our application:

$$v_{ij} = \mu_j + \sum_{m=1}^{M} \lambda_m \eta_{ijm}^{(2)} \tag{1}$$

$$\eta_{ijm}^{(2)} = \sum_{k=1}^{K} \beta_{km} \eta_{kj}^{(3)} + \epsilon_{ijm} \tag{2}$$

The $\beta_{km}$ identify the group mean intercepts for the latent factors (2), while $\mu_j$ represents the item mean intercepts. The $\eta_{kj}^{(3)}$ is the mixture component at group level and it has a multinomial distribution and $\varepsilon_m$ has a multivariate normal distribution. A link function should be selected to link the response to the linear predictor. The *Cumulative Logit* function has been chosen because of the nature of items (Agresti 2002). Some restrictions on the factor score means (the sum over the $K$ mixture components is equal to zero) and on factor loadings (the first factor loading for each dimension is fixed to one) have been imposed to make the model estimable.

## 3 Application to Reading Motivation

In recent years, the educational context has increased in its importance. The reading ability became a fundamental aspect to maximize success in daily life and realize a person's own potential. A well literate population is essential to social growth and economic development. For such reasons, the International Association for the Evaluation of Educational Achievement (IEA) realized in *2006*, the Progress in International Reading Literacy Study (PIRLS) survey. This survey provides data for an international comparison of student reading achievement in primary school. The choice of the fourth grade students depends on the fact that this age is a transition point in reading child development. We focused our attention on reading motivation because it is an useful aspect to describe the children reading achievement. The data is the Italian subsample provided by INVALSI (National Institute for the evaluation of the educative system of instruction and formation).

**Table 1** Exploratory analysis – Rotated factor loadings

| Items | Enjoyment of reading | Value of reading | Self concept as reader |
|---|---|---|---|
| I Enjoy reading | **.81** | .18 | −.21 |
| Reading is boring | **−.76** | −.07 | .20 |
| Book as a present | **.70** | .18 | −.05 |
| Well for future | **.54** | .08 | .05 |
| Talk with friends | .06 | **.67** | −.06 |
| Talk with family | .07 | **.74** | −.06 |
| Reading for info | .16 | **.56** | −.06 |
| Read Aloud | .13 | **.59** | .06 |
| Reading is easy | .22 | .07 | **−.57** |
| Not as well as other | −.03 | −.06 | **.81** |
| Read slower than others | −.02 | .02 | **.80** |

In the past twenty years, many studies have been conducted to describe class situation and environmental influences (family or school) on motivation. In many articles reading motivation was often considered as a multidimensional construct. Guthrie and Wigfield (2000) studied reading development and described the factors that influence motivation: home environment, parental attitude, teacher involvement, teacher strategy of rewards and prices and evaluation. In the educational and psychological approaches, reading motivation is treated as a multidimensional construct and three dimensions have been selected to measure motivation (referring to the Motivation for Reading Scale (Saracho and Dayton 1989)): enjoyment of reading, value of reading and self concepts as a reader.

The students sample size is *3,581* (*1,742* girls and *1,839* boys) with *3,370* Italian students. The teacher sample size is *198*. The items for Enjoyment in Reading (ER) are related to statement towards reading. For the Value of Reading (VR), the items describe interest and value in daily life activities, while for the Self Concept as Reader (SCR), items measure reading ability. The exploratory factor analysis suggests a three factors structure (Table 1) and the confirmatory factor analysis supports this hypothesis (GFI = *.988* and RMSEA = *.035*).

The multilevel mixture factor analysis could be described as follows: the individual structure is composed by eleven items connected to three factors, while the group level is composed by a mixture component connected to the factor mean. In this application, where students share the same class environment, teachers could not be considered equal and for this reason a mixture component at group level is useful to model this heterogeneity.

For model selection, we refer to the work of Lukočienė, Varriale, and Vermunt (2010), where a simulation study shows that the best fit index is the *BIC* with sample size equal to the group level observations, because using the number of observations, it is possible to underestimate the number of mixture components, especially if the separation is weak or moderate (Lukočienė and Vermunt 2010).

**Table 2** Fit indexes for the Multilevel Mixture Factor Model

| Number of class | Log likelihood | BIC N obs | BIC N groups | AIC | Number of parameters |
|---|---|---|---|---|---|
| 3 | −40,415 | 81,281 | 81,121 | 80,940 | 55 |
| 4 | −40,381 | 81,245 | 81,074 | 80,880 | 59 |
| 5 | −40,352 | 81,220 | 81,037 | 80,830 | 63 |
| 6 | −40,336 | 81,220 | 81,026 | 80,805 | 67 |
| 7 | −40,324 | 81,229 | **81,024** | **80,790** | 71 |
| 8 | −40,320 | 81,254 | 81,037 | 80,791 | 75 |

**Table 3** Class mean factor scores for the Multilevel Mixture Factor Model

| Latent classes | ER | VR | SCR | Size |
|---|---|---|---|---|
| Very highly motivated | .56 | .18 | .69 | .05 |
| Highly motivated | .36 | .32 | −.01 | .16 |
| Motivated bad reader | .39 | .01 | −.16 | .21 |
| Medium motivated | .02 | −0.3 | −.12 | .24 |
| Unmotivated | −.39 | .12 | −.21 | .12 |
| Highly unmotivated | −.58 | −0.22 | −.09 | .13 |
| Very highly unmotivated | −0.36 | −.38 | −.34 | .09 |

Table 2 reports the fit indexes and the best solution is seven latent classes model ($BIC_{N_{groups}} = 81,024$).

In Table 3, the classes' size and factor means are reported. Positive values indicate higher level of motivation.

As it is possible to see from Table 3 and Fig. 1, there are seven different levels of motivation for the classrooms analyzed and this implies the presence of subpopulations in the sample.

# 4 Conclusion

Multilevel Mixture Factor Analysis provides a new approach to children's reading motivation and school context. The mixture component at group level, classifies teachers according to the latent structure measured at the individual level. Differences in motivation depend on teacher characteristics such as teaching abilities, practices, personal motivation, efforts and education, because students and family characteristics are not the only attributes that affect children motivation. The added value of the MMFA is the contemporaneous inclusion of factor analysis to classify items, and mixture components to classify subjects.

One of the possible extensions of MMFA is the introduction of covariates to explain factor score and class membership. The goal is to analyze both the effects of home and school environment on reading motivation. Taking inspiration from the educational and psychological theory (Cothern and Collins 1992), the following variables could be used to describe the home context: parental education, library

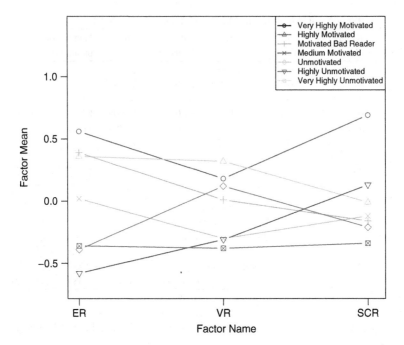

**Fig. 1** Factor means for the Multilevel Mixture Factor Model

use, sex, nationality, socio economic status, the importance of reading, and the pre-entrance reading ability. To describe the class context, the variables should be related to teachers, like the way of teaching and involving students in class activities.

Another extension of the MMFM, is the comparison between multilevel mixture factor analysis and mixture factor analysis (MFA) (Lubke and Muthén 2005). In MFA, the mixture component is at the individual level instead of being at the teacher level. In this way it is possible to classify students in groups ignoring the hierarchical structure and to analyze the class effect on children motivation and isolate the teacher effects on reading motivation.

# References

Agresti A (2002) Categorical data analysis. Wiley, New York, NY
Cothern NB, Collins MB (1992) An exploration: Attitude acquisition and reading instruction. Read Res Instruct 31:84–97
Day N (1969) Estimating the components of a mixture of normal distributions. Biometrika 56(3):463–474
Guthrie JT, Wigfield A (2000) Engagement and motivation in reading. In: Pearson PD, Barr R (eds) Handbook of reading research, vol 3. Erlbaum, New York
Lazarsfeld PF, Henry NW (1968) Latent structure analysis. Houghton, Mifflin, Boston

Lubke GH, Muthén BO (2005) Investigating population heterogeneity with factor mixture models. Psychol Meth 10:21–39

Lukočienė O, Vermunt JK (2010) Determining the number of components in mixture models for hierarchical data. In: Fink A, Lausen B, Seidel W, Ultsch A (eds) Advances in data analysis, data handling and business intelligence. Springer, Berlin, pp 241–249

Lukočienė O, Varriale R, Vermunt JK (2010) The simultaneous decision(s) about the number of lower- and higher-level classes in multilevel latent class analysis. Sociological Methodology 40(1):247–283

McLahan G, Peel D (2000) Finite mixture models. Wiley, New York

Saracho ON, Dayton CM (1989) A factor analytic study of reading attitudes in young children. Contemp Educ Psychol 14:12–21

Skrondal A, Rabe-Hesketh S (2004) Generalized latent variable modeling. Chapman and Hall, London

Vermunt JK (2003) Multilevel latent class models. Socio Meth 33:213–239

Vermunt JK (2007) A hierarchical mixture model for clustering three-way data sets. Comput Stat Data Anal 51:5368–5376

Vermunt JK (2008) Latent class and finite mixture models for multilevel dataset. Stat Meth Med Res 17:33–51

Vermunt JK, Tran B, Magidson JK (2008) Latent class models in longitudinal research. In: Menard S (ed) Handbook of longitudinal research: Design, measurement, and analysis. Academic Press, Burlington, MA, pp 373–385

Wolfe JH (1970) Pattern clustering by multivariate mixture analysis. Multivariate Behav Res 5:329–350

# Part XIII
# Analysis of Tourism Data

# Short Term Dynamics of Tourist Arrivals: What Do Italian Destinations Have in Common?

**Anna Maria Parroco and Raffaele Scuderi**

**Abstract** This work aims to detect the common short term dynamics to yearly time series of 413 Italian tourist areas. We adopt the clustering technique of Abraham et al. (Scand J Stat. 30:581–595, 2003) who propose a two-stage method which fits the data by B-splines and partitions the estimated model coefficients using a $k$-means algorithm. The description of each cluster, which identifies a specific kind of dynamics, is made through simple descriptive cross tabulations in order to study how the location of the areas across the regions or their prevailing typology of tourism characterize each group.

## 1 Introduction

Short term dynamics of tourist flows are of great importance for both private investors and policymakers. While the increase or decrease of the number of visitors has a relevant influence on the earnings of the former and consequently on the development of their economic activity, the analysis of such variations can better help the latter in order to address the policies of the public sector about tourism. This paper studies the short term yearly variation of tourist arrivals in Italian destinations, and specifically in tourist areas (*circoscrizioni turistiche*), that are sets of municipalities and the smallest territorial units for which Istat (the Italian Statistical Office) provides data on tourist flows – see Istat (2010).

The scope of the work is twofold. At first we aim to cluster the destinations according to the short term dynamics of tourist arrivals, and to identify the representative "average" time pattern characterizing each group of territories. We

A.M. Parroco · R. Scuderi (✉)

Department of Economics, Business and Finance, University of Palermo, Viale delle Scienze, ed. 13, 90128, Palermo, Italy

e-mail: parroco@unipa.it; scuderi@unipa.it

W. Gaul et al. (eds.), *Challenges at the Interface of Data Analysis, Computer Science, and Optimization*, Studies in Classification, Data Analysis, and Knowledge Organization, DOI 10.1007/978-3-642-24466-7_59, © Springer-Verlag Berlin Heidelberg 2012

adopt the approach of Abraham et al. (2003), who propose to group time series by first fitting B-splines and then applying a $k$-means algorithm to the resulting coefficients. Secondly, we study the groups' composition in terms of both the geographical location in a specific administrative region, and the prevailing typology of tourism according to the classification of the tourist areas of Istat (2010) (i.e., artistic places, seaside areas, hill areas, etc.).

The literature on tourism analysis mainly analyzes the long term patterns of the arrivals, where many studies test Butler's (1980) framework of the tourist areas lifecycle model, and the infra-annual dynamics of the seasonality (see Koenig (2005) for a review). Although time series analysis develops the issue of short time series and their forecasting (see Wang (2008) among the others), to our knowledge little attention has been paid by the studies on tourism about the short term temporal information, as well as to the clustering of territories according to the time pattern of an indicator.

The paper is organized as follows. In Sect. 2 we present the methodology, Sect. 3 describes the dataset and Sect. 4 discusses the results and draws a conclusion.

## 2 Methodology

Abraham et al. (2003) introduce an interesting algorithm for the clustering of time series, which is a combination of splines fitting and $k$-means clustering. Let $G_i$ be the $i$th time series of a set of $n$. The first step of the procedure consists in fitting piecewise polynomials, or *splines* (De Boor 1978; Schumaker 1981), to each observation $\{x_i^j, y_i^j\}_{j=1}^{m_i}$. Splines are very flexible and less sensitive to small changes in data than the interpolation of a unique polynomial to the series. Moreover the resulting coefficients better synthesize the information related to the succession of observations over time, than other classical time series clustering techniques such as the Euclidean distance and correlation's coefficients. Both of the latter are in fact invariant to transformations that alter the order of observations over time (i.e. the temporal indexing of data), and therefore they do not take into account the autocorrelation of the time series. Finally, piecewise polynomials do not necessarily require a large amount of observations in time, as instead the procedures based on stochastic processes do (see for instance the AR metrics in Piccolo (1990)).

Spline coefficients refer to specific time intervals, and this implies that they won't be comparable, if time points differ across the units or if there are missing data. Therefore, one should care about using the methodology with an appropriate dataset. This will be taken into consideration in Sect. 3, where a proper adjustment to the dataset is made.

Let $a = \xi_0 < \xi_1 < \cdots < \xi_K < \xi_{K+1} = b$ be the *knots*, that is a subdivision by $K$ points of the domain of $x \in [a, b]$. We define $s(x)$ as the spline function, that is a polynomial of degree $d$ on any interval $[\xi_{i-1}, \xi_i]$ with $d - 1$ continuous derivatives on $(a, b)$. For any fixed sequence of $K$ knots, the set of splines is a linear space of functions with $K + d + 1$ free parameters. If we consider a basis $(B_1, \ldots, B_{K+d+1})$

for such a linear space, $s(x)$ can be rewritten as

$$s(x, \beta) = \sum_{l=1}^{K+d+1} \beta_l B_l(x). \tag{1}$$

where $\beta = (\beta_1, \ldots, \beta_{K+d+1})'$ is the vector of spline coefficients and the basis can be usefully expressed by B-splines (or Basic splines: see Curry and Schoenberg (1966)).

The second step consists in clustering the $n$ time series in $k$ groups $\{C^1, \ldots, C^k\}$, from the set $\{\hat{\beta}^1, \ldots, \hat{\beta}^n\}$ of the estimated coefficients through the $k$-means procedure (Hartigan and Wong (1979); Ripley (1996)). Let $z = \{c^1, \ldots, c^k\}$ be a set of representatives of each cluster, where $c^i$ is called 'center' of $C^i$. The $k$-means aims at choosing $z$ so that it minimizes

$$\frac{1}{n} \sum_{i=1}^{n} \min_{c \in z} \|\hat{\beta}^i - c\|^2 \tag{2}$$

where $\| \cdot \|$ is the Euclidean norm. The optimal number $k$ of clusters is still an open question in the literature (see Chiang and Mirkin (2010)). In the following we adopt a heuristic criterion and consider different values of $k$ (2, 3 and 4). Although Abraham et al. (2003) prove that the algorithm is strongly consistent, no procedure guarantees that the global minimum of (2) is reached by the $k$-means algorithm, and thus any set of classes is not ensured to be the optimal partition.

The objective of this paper is related to the comparison of the arrivals variations over time, as caught by the shape of time series and synthesized through B-splines. In this sense the clustering of such series according to the mere number of tourists can be misleading since B-splines coefficients are affected by the unit of measurement and magnitude of the original series. For instance, think about the evolution of the flows in one city where tourism has traditionally been a relevant activity, and where different accommodation structures of any size are present, compared to the small mountain place where tourism is starting to develop. The comparability of the two series dynamics via B-spline is allowed if we transform the data properly. In the following we simply adopt the *relative change* of each series: if $x_{it}$ are the arrivals in the $i$th tourist area at time $t$, the relative change is given by $(x_{it} - \overline{x}_i)/\overline{x}_i$, where $\overline{x}_i$ is the mean of the arrivals in $i$ over time.

After we identify the groups and graphically represent the cluster centers, in order to associate a *representative* time series pattern to each set of tourist areas, we use simple cross-tabulations in order to analyze their composition. Specifically we will study whether the destinations located in a specific Italian region, or a determined typology of tourist areas, concentrate in a specific group and thus *characterize* the identified pattern.

## 3 Data

Before we present and discuss the empirical findings we introduce the dataset and point out some questions related to its use.

As already remarked above, our study identifies the tourist destination with the tourist area, that is the smallest territorial unit for which data on tourism are provided by Istat (2010). Data concerning the arrivals are published for both *hotels and similar establishments*, and *other collective establishments*, and range from 2000 to 2006. This paper considers the total number of arrivals, with no distinction between the typologies of accommodation structures. Some remarks on the data quality in terms of their comparability over time are necessary, especially for the implications on the comparison of the estimated spline coefficients (see Sect. 2):

• The number of the Italian tourist areas changed over time and increased from 542 in 2000–2005 to 554 in 2006.
• In some cases the composition of an area is not the same in all years. Fortunately, the yearly list of Italian municipalities, and the corresponding tourist area they belong to, is provided, and this allows to detect the changes that occurred.
• In some years only the estimates of the arrivals in some specific areas are reported, rather than the real values. This is mainly due to the missing data in one or more months of the considered year.
• It is not uncommon that data are provided jointly for two or more tourist areas, and thus statistical units are merged in temporary aggregations due to the European privacy regulations. Moreover every new unit originated by a merging is not composed by the same areas in all years. For instance, if at year $t$ areas A and B are merged into "A+B", and thus one finds tourists arrivals separately for "A+B" and, say, C, at $t + 1$ one can find data separately for A, B, C, while at $t + 2$ arrivals can be reported for A and "B+C".

For all these reasons we aggregated tourist areas and added the number of the arrivals accordingly, in order to obtain 413 new territorial units whose composition can be comparable over time.

For what concerns the cluster analysis instead, the technique we described in Sect. 2 is adopted from Abraham et al. (2003) to partition a set of 148 time series with a high number of 224 observations over time, where seven non-equidistant interior knots are opportunely chosen, through which the piecewise polynomials are interpolated. Due to the data availability our case study has the much lower number of 7 years. Our scopes are nevertheless different. As we already pointed out, since we aim to catch the time pattern given by the year-by-year variation of each series, we require that the piecewise polynomials pass through every observation.

The two variables we use to cross-tabulate the groups are the location of the destinations in the 20 Italian Regions, and the 10 typologies of tourist areas defined by Istat, and reported in Table 3. Although Istat assigns one category to each tourist area, the aggregation of the units we did led some *final* units to belong to more than one category, which were included in the residual category Other.

The R software was used to perform all the estimates of this paper.

## 4   Discussion and Conclusions

Figures 1, 2 and 3 report the clusters centers respectively at $k = 2$, $k = 3$, $k = 4$. In order to name each group we indicate the $\alpha$-th group when $k = \gamma$ as $g\alpha k\gamma$.

The clusters of $k = 2$ (Fig. 1) report two increasing trends with opposite substantially concave-convex pattern, where the units distribute almost equally. The time series forming the pattern $g1k2$ separate in $g1k3$ and $g2k3$ (Table 1), where the latter is also made of almost all the time series of the formerly $g2k2$ destinations, and presents a still convex trend but smaller yearly variations than $g1k2$. Just a very few units have such a peculiar pattern that forms the stationary trend of $g3k3$. The two well-defined concave and convex trends are also present when $k = 4$ ($g3k4$ and $g4k4$ – Fig. 3 and Table 2), where only a few members belong to the other two groups $g1k4$ and $g2k4$. The two predominant patterns substantially persisted in almost every iterations we ran, and even when we generated more than 4 groups

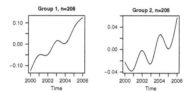

**Fig. 1** Clusters centers, $k = 2$

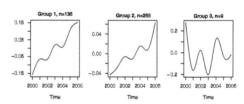

**Fig. 2** Clusters centers, $k = 3$

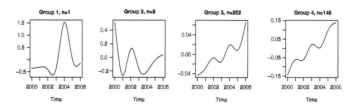

**Fig. 3** Clusters centers, $k = 4$

**Table 1** Transition matrix, from $k = 2$ to $k = 3$

|  | Group 1, $k = 3$ | Group 2, $k = 3$ | Group 3, $k = 3$ |
|---|---|---|---|
| Group 1, $k = 2$ | 0.654 | 0.346 | 0 |
| Group 2, $k = 2$ | 0 | 0.956 | 0.044 |

**Table 2** Transition matrix, from $k = 3$ to $k = 4$

|                    | Group 1, $k = 4$ | Group 2, $k = 4$ | Group 3, $k = 4$ | Group 4, $k = 4$ |
| ------------------ | ---------------- | ---------------- | ---------------- | ---------------- |
| Group 1, $k = 3$   | 0                | 0                | 0.007            | 0.993            |
| Group 2, $k = 3$   | 0                | 0                | 0.963            | 0.037            |
| Group 3, $k = 3$   | 0.111            | 0.556            | 0.333            | 0                |

(that is $k > 4$ – not reported) many units merged in two big clusters with such time features.

The robust increasing patterns highlight the well-known tendency of the arrivals in Italian tourist areas to grow. Two observations emerge for what concerns the differences between the two representative time series. The first one is related to the already mentioned global trends, whose different shapes and range seem to recall the uneven development of the areas at a different stage of the tourist destination lifecycle (see Butler (1980)). In this sense long term data would yet be necessary in order to compare the real pattern of arrivals with the theoretical evolution of the tourist area as a market product. The second observation strictly deals with the yearly variation, whose sign is opposite between the two series in most of the years: regardless of the *real* presence of time series having exactly this pattern, such variations probably suggest the presence of destinations which on the average are influenced by the short term dynamics *in two different ways*. The patterns of the clusters do not seem to be related to a particular macro-area of Italy (Table 3), although most of the destinations of regions in the Central and South Italy very frequently occur in what we defined as the convex and less variable cluster (Umbria, Latium, Sicily, Sardinia, where their percentage is greater than 70%) also as $k$ varies. Also the tourist areas located in Lombardy, Friuli-Venezia Giulia, Liguria and Marche occur frequently in such pattern, suggesting that the most relevant increase of tourism across the areas occurred over the very last years. Table 3 also does not suggest a prevalent concentration of the typologies of the destinations in a specific cluster when $k = 2$, except for the lake tourist areas which appear to be characterized by the convex pattern more frequently. As $k$ increases the percentage of the destinations in art, hill, religious, seaside and thermal (as well as the "other" and "other province seats" categories) in the convex pattern raise, while the concave one appears to be less *concentrated* across the considered categories.

Overall the evidence suggests that there can be found common dynamics between the short-term time series of the arrivals in the Italian tourist areas. If one wants to find a representative series for all destinations it would have an increasing trend. Such common driving element can be ideally decomposed into two opposite convex and concave patterns, which persist also when the number of the clusters increases. Despite the convex one would indicate that the tourism has increased in the last years, it ranges in a stricter interval than the latter, and it appears to be characterized by specific geographic counterparts and typologies of tourism. From such two points of view of the location in a region and the identification with a peculiar category of tourism, the concave pattern's larger range suggests that the faster development of

**Table 3** Groups composition by region and typology of tourist area (Istat classification)

| | $k = 2$ | | $k = 3$ | | | $k = 4$ | | | |
|---|---|---|---|---|---|---|---|---|---|
| | Gr.1 | Gr.2 | Gr.1 | Gr.2 | Gr.3 | Gr.1 | Gr.2 | Gr.3 | Gr.4 |
| Piedmont | 58.33 | 41.67 | 50.00 | 44.44 | 5.56 | 0.00 | 5.56 | 44.44 | 50.00 |
| Aosta Valley | 100.00 | 0.00 | 100.00 | 0.00 | 0.00 | 0.00 | 0.00 | 0.00 | 100.00 |
| Lombardy | 41.94 | 58.06 | 29.03 | 70.97 | 0.00 | 0.00 | 0.00 | 70.97 | 29.03 |
| Trentino Alto Adige | 70.00 | 30.00 | 50.00 | 50.00 | 0.00 | 0.00 | 0.00 | 48.00 | 52.00 |
| Veneto | 62.96 | 37.04 | 40.74 | 55.56 | 3.70 | 0.00 | 0.00 | 59.26 | 40.74 |
| Friuli Venezia Giulia | 45.45 | 54.55 | 27.27 | 72.73 | 0.00 | 0.00 | 0.00 | 72.73 | 27.27 |
| Liguria | 33.33 | 66.67 | 25.00 | 75.00 | 0.00 | 0.00 | 0.00 | 75.00 | 25.00 |
| Emilia-Romagna | 63.16 | 36.84 | 21.05 | 78.95 | 0.00 | 0.00 | 0.00 | 57.89 | 42.11 |
| Tuscany | 49.02 | 50.98 | 37.25 | 60.78 | 1.96 | 0.00 | 0.00 | 62.75 | 37.25 |
| Umbria | 25.00 | 75.00 | 0.00 | 100.00 | 0.00 | 0.00 | 0.00 | 100.00 | 0.00 |
| Marche | 48.00 | 52.00 | 20.00 | 80.00 | 0.00 | 0.00 | 0.00 | 76.00 | 24.00 |
| Latium | 27.78 | 72.22 | 11.11 | 72.22 | 16.67 | 0.00 | 16.67 | 66.67 | 16.67 |
| Abruzzo | 72.73 | 27.27 | 40.91 | 59.09 | 0.00 | 0.00 | 0.00 | 59.09 | 40.91 |
| Molise | 50.00 | 50.00 | 50.00 | 50.00 | 0.00 | 0.00 | 0.00 | 50.00 | 50.00 |
| Campania | 27.27 | 72.73 | 9.09 | 90.91 | 0.00 | 0.00 | 0.00 | 86.36 | 13.64 |
| Apulia | 50.00 | 50.00 | 44.44 | 55.56 | 0.00 | 0.00 | 0.00 | 61.11 | 38.89 |
| Basilicata | 75.00 | 25.00 | 75.00 | 25.00 | 0.00 | 0.00 | 0.00 | 25.00 | 75.00 |
| Calabria | 63.64 | 36.36 | 45.45 | 54.55 | 0.00 | 0.00 | 0.00 | 54.55 | 45.45 |
| Sicily | 25.93 | 74.07 | 11.11 | 81.48 | 7.41 | 3.70 | 0.00 | 77.78 | 18.52 |
| Sardinia | 27.27 | 72.73 | 18.18 | 81.82 | 0.00 | 0.00 | 0.00 | 81.82 | 18.18 |
| *Artistic* | 40.98 | 59.02 | 19.67 | 77.05 | 3.28 | 0.00 | 3.28 | 75.41 | 21.31 |
| *Hill* | 35.29 | 64.71 | 20.59 | 73.53 | 5.88 | 0.00 | 2.94 | 76.47 | 20.59 |
| *Lake* | 29.41 | 70.59 | 11.76 | 88.24 | 0.00 | 0.00 | 0.00 | 88.24 | 11.76 |
| *Mountain* | 74.68 | 25.32 | 64.56 | 32.91 | 2.53 | 0.00 | 1.27 | 32.91 | 65.82 |
| *Religious* | 0.00 | 100.00 | 0.00 | 100.00 | 0.00 | 0.00 | 0.00 | 100.00 | 0.00 |
| *Seaside* | 45.21 | 54.79 | 20.55 | 78.08 | 1.37 | 0.00 | 0.00 | 75.34 | 24.66 |
| *Thermal* | 45.83 | 54.17 | 33.33 | 66.67 | 0.00 | 0.00 | 0.00 | 62.50 | 37.50 |
| *Other, municipality* | 57.69 | 42.31 | 44.23 | 51.92 | 3.85 | 1.92 | 1.92 | 48.08 | 48.08 |
| *Other, province seat* | 50.00 | 50.00 | 25.00 | 75.00 | 0.00 | 0.00 | 0.00 | 75.00 | 25.00 |
| *Other* | 46.30 | 53.70 | 25.93 | 74.07 | 0.00 | 0.00 | 0.00 | 72.22 | 27.78 |

tourism in the last years appears to be more widespread across the territory and not specifically ascribable to a specific type of tourism.

The methodology we used, opportunely combined with other ones, can be applied also to the study of the effect of the short-term fluctuations on the tourist sector, such as the ones caused by crisis, structural breaks or shocks in the economic system. The spline interpolation of a piecewise polynomials which is derivable in all the points would request that data were *functional*, as also Abraham et al. (2003) stress. Nevertheless the use of the spline as a mere interpolation technique, with no aim to approximate a process that can be expressed by a function, makes it suitable to be used by who is in search of robust interpolation coefficients.

The results we obtained about tourism constitute an interesting starting point for the study of the short term dynamics of the arrivals across territories. In this sense, our work can be extended towards the understanding of common long term dynamics which characterize those destinations at a similar stage of their lifecycle through the use of a dataset with more observations over time. Moreover, the use of, and the comparison with, further methods such as mixture models which will be the object of future research would help to improve the knowledge of the common time patterns of the destinations.

**Acknowledgements** This research is part of the PRIN 2007 (Project of Relevant National Interest) *Mobility of regional incoming tourism. Socio-economic aspects of behaviors and motivations*, funded by the Italian Ministry for University and the University of Palermo. We thank the national coordinator of the Project, Prof. Franco Vaccina, for his precious suggestions and cooperation. Many thanks also to Daria Mendola, and the anonymous referees who gave us useful indications for the improvement of the paper.

# References

Abraham C, Cornillon PA, Matzner-Lober E, Molinari N (2003) Unsupervised curve clustering using b-splines. Scand J Stat 30:581–595

Butler RW (1980) The concept of a tourism area cycle of evolution: implications for the management of resources. Can Geogr 24:5–12

Chiang MMT, Mirkin B (2010) Intelligent choice of the number of clusters in k-means clustering: An experimental study with different cluster spreads. J Classification 27:3–40

Curry HB, Schoenberg IJ (1966) On Polya frequency functions. IV: The fundamental splines and their limits. J Anal Math 17:71–107

De Boor C (1978) A practical guide to splines. Springer, New York

Hartigan JA, Wong MA (1979) A k-means clustering algorithm. Appl Stat 28:100–108

Istat (Italian Statistical Office) (2010) Statistiche per politiche di sviluppo. Risorse turistiche. URL http://www.istat.it/ambiente/contesto/incipit/turistiche.html. Cited 14 Jun 2010

Koenig-Lewis N, Bischoff EE (2005) Seasonality research: The state of the art. Int J Tour Res 7:201–219

Piccolo D (1990) A distance measure for classifying ARIMA models. J Time Ser Anal 11:153–163

Ripley BD (1996) Pattern recognition and neural networks. Cambridge University Press, Cambridge

Schumaker LL (1981) Spline functions: Basic theory. Wiley, New York

Wang CH, Hsu LC (2008) Constructing and applying an improved fuzzy time series model: Taking the tourism industry for example. Expert Syst Appl 34:2732–2738

# A Measure of Polarization for Tourism: Evidence from Italian Destinations

**Raffaele Scuderi**

**Abstract** This paper proposes an index of polarization for tourism which links the axiomatic theory of Esteban and Ray with the classical hierarchical agglomerative clustering techniques. The index is aimed at analyzing the dynamics of the average length of stay across Italian destinations, and more specifically to detect whether the polarization within the set of clusters of places with similar values of the indicator has varied over time.

## 1 Introduction

Polarization has been a relatively recent issue of great interest for the study of the division of populations in groups according to their social and economic characteristics. The measurement of the *degree of clustering* of units with similar features has been formalized by the seminal work of Esteban and Ray (1994) – ER – who present the axiomatization of the concept and an index for its measurement in per capita income distributions. While the wealth per individual is the most investigated topic in the literature (see among the others Esteban and Ray (1994), Wolfson (1994), Esteban et al. (1999)), the measurement of the extent to which a population is divided in clusters can concern several fields, such as social conflicts, electoral behavior, marketing, etc. This work analyzes the issue of polarization in a peculiar aspect of tourism in Italy, which deals with the tendency of the mean average length of stay (ALOS)[1] in tourist destinations to reduce over time. Italy has

---

[1] The average length of stay is calculated by dividing the nights spent by the arrivals. According to EUROSTAT (2010), a night spent (or overnight stay) is each night that a guest actually

R. Scuderi (✉)
University of Palermo, Viale delle Scienze, ed. 14, 90128, Palermo, Italy
e-mail: scuderi@unipa.it

W. Gaul et al. (eds.), *Challenges at the Interface of Data Analysis, Computer Science, and Optimization*, Studies in Classification, Data Analysis, and Knowledge Organization, DOI 10.1007/978-3-642-24466-7_60, © Springer-Verlag Berlin Heidelberg 2012

traditionally been a relevant tourist place due to its historical, cultural and natural heritage, as much wide as spatially heterogeneous. While the number of tourists has increased in the last years, the average number of days spent in collective accommodations has nevertheless reduced. In this sense a measure of the overall polarization reflects whether such trend affected all tourist areas in the same way, and if the *structure* of territorial units with similar features evolved towards a common group or, on the contrary, diverged.

Starting from the methodology presented by Scuderi (2006) we propose an index based on the combination of ER's axiomatic theory and classical hierarchical agglomerative clustering techniques (HAC). At our knowledge the polarization framework has never been applied to the field of tourism, where the proposed measure can be useful in order to study the similarity of destinations. The proposed framework can be extended to all the fields where a synthesis of the dynamics of the clustering structure deriving from HAC is required. Moreover the study of this paper refers to the dynamics of grouping according to a single indicator, whereas our technique can be extended also to the multivariate (see Scuderi (2006)). The essentials of ER's framework are reported in Sect. 2, while the index is presented in Sect. 3. Sections 4 and 5 present the dataset, discuss the empirical evidence and conclude.

## 2 The Concept of Polarization and Its Measurement

In the analysis of territorial units the term *polarization* deals with two different notions. The first is wider and it is the one of ER concerning the shape features of distributions where the units concentrate around local *representative* values (i.e. the *poles*). The second meaning refers to those phenomena taking place in space (see Henderson (1988)), and specifically to the concentration of activities and people around portions of a physical place. As mentioned above the present study considers the first notion.

Polarization has been the object of a relatively recent and growing literature which especially analyzes the income distribution and more in general economic inequality (Esteban and Ray 1994; Wolfson 1994; Esteban et al. 1999). The earlier work by ER defines it as to the extent to which the population is clustered around distant poles, where:

- There can be found clusters exhibiting a high degree of within-group homogeneity and between-group heterogeneity.
- There is a small number of groups of significant size.
- The groups of insignificant size (e.g., isolated individuals) carry little weight.

---

spends (sleeps or stays) or is registered (his/her physical presence there being unnecessary) in a collective accommodation establishment or in private tourism accommodation; an arrival is defined as a person who arrives at a collective accommodation establishment or at a private tourism accommodation and who checks in.

Although ER claim that the study of such issue is relevant for many social and economic phenomena, such as the conflicts between social classes, labor market segmentation, distribution of firms size, etc., they explicitly analyze per capita income polarization as function of the *attitudes* of individuals, and specifically of:

1. The feelings of identification perceived by an individual towards her group (*identification function*).
2. How much an individual feels far from the other groups, which corresponds to the absolute distance between the representative income levels of each cluster (*alienation function*).

The three basic axioms of ER's framework are resumed as follows. Suppose to cluster a population into three groups, 0, $x$ and $y$, which correspond to income levels where $0 < x < y$:

- If $x$ and $y$ contain an equal number of members but smaller than 0, polarization raises if the two groups pool.
- If the groups have a different number of members and the smallest one is $x$ while the biggest is 0, polarization raises if $x$ gets closer to $y$.
- If the groups have a different number of members, and $x$ is the biggest group which is equally distanced from 0 and $y$, any equal shifting from the central to the lateral groups increases polarization.

ER show that an index which is consistent with their axioms is in the form

$$P(\pi, \mathbf{y}) = \kappa \sum_{i=1}^{n} \sum_{j=1}^{n} \pi_i^{1+\alpha} \pi_j |y_i - y_j|, \qquad (1)$$

where $\kappa$ is a scalar and has no bearing on the order of the measure, $i$ and $j$ are two groups from the set of $n$ clusters, $\pi_i$ is the number of individuals belonging to $i$, $y_i$ is the income of $i$, and $|y_i - y_j|$ assesses the alienation function. The parameter $\alpha \in (0, \alpha^*]$, where $\alpha^* \simeq 1.6$, measures the sensitivity to polarization and its use as exponent of $\pi_i$ reflects the form of the above mentioned identification function. Further axioms impart a minimal degree of polarization sensitivity by restricting the range of $\alpha^*$ to $[1, 1.6]$. Applied studies like the one of Esteban et al. (1999) employ three specific values of such range, and specifically $\alpha = 1, 1.3$ and $1.6$, as also we will do in the following.

The index (1) refers to the situation of *symmetric* polarization, where for instance the rich and the poor feel the same degree of identification with their group and of alienation from the other group. An alternative and more general way to mean polarization is the extension of (1) to a class of asymmetric indices where it is a function also of the observed characteristics of each cluster which empower the identification. The measure proposed by ER refers to situations like "the rich does not dream of becoming poor, but the poor wishes to become rich" and is in the form

$$P(\pi, \mathbf{y}) = \kappa \sum_{i=1}^{n} \sum_{j=1}^{n} \pi_i^{1+\alpha} \pi_j y_i^{\beta} |y_i - y_j|, \tag{2}$$

where the identification of the individuals in $i$ is empowered by their characteristics, exponentially weighted by $\beta$, $y_i^{\beta}$. Clearly when $\beta = 0$ (2) reduces to (1).

## 3   An Index of Asymmetric Polarization for Tourism

The index of polarization for the ALOS we propose is based on the extension of the ER index to the hierarchical agglomerative clustering techniques proposed by Scuderi (2006), who presents a measure of the economic convergence of countries. Consistent with the latter we mean to quantify the *degree of polarization* of the ALOS at each time $t$, and then observe the evolution of the index over the time in order to study polarization dynamics. The measure makes use of the notion of asymmetric polarization in (2) and it is in the form

$$P(\pi, \mathbf{l}) = k \sum_{i=1}^{n} \sum_{j(\neq i)=1}^{n} p_i^{1+\alpha} p_j \delta(l_i, l_j). \tag{3}$$

The set of the ALOS across destinations is represented by $\mathbf{l}$. Each group's population $p_i$ corresponds to the number of beds[2] of tourist areas belonging to the cluster $i$. The constant $k = (\sum_{i=1}^{n} p_i)^{-(2+\alpha)}$ simply normalizes the population by the total one of the whole set of clusters. Finally $\delta(l_i, l_j)$ is an *asymmetric* measure of the distance of $i$ from $j$. The questions that are yet to be explained are *how* groups of tourist destination with similar ALOS are determined, and mostly *why* an asymmetric measure, and *which* elements to consider for such a measure. The answers are related each other.

Among the wide variety of the methods to cluster units, for the scope of the present research we make use of the hierarchical agglomerative clustering techniques, and specifically of Ward's (1963) minimum variance method. The choice is consistent with the aim to detect how a set of groups evolves, and thus we require that no apriori number of groups is determined in order to allow units, for instance, to converge to a unique cluster as time passes, or on the contrary to a set of a higher number of groups. Consistent with this goal we fix the cutting line $\lambda_0$ of the dendrogram at the initial year and let it be the same over time, so that $\lambda_0$ constitutes a kind of "observation point" from which we detect how groups evolve.

---

[2]While in (1) and (2) the cluster population $\pi_i$ is the number of persons belonging to each group, that is the number of units over which the variable *per capita income* takes place, here each cluster's size $p_i$ is the total number of the beds in official accommodation collective structures of $i$, which analogously are the potential 'occasions' over which the tourists stay takes place.

**Fig. 1** An example of dendrogram and the distances considered by the index

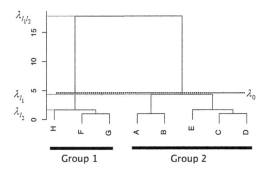

The choice of $\lambda_0$ is made by simply maximizing the between group distance (see Fig. 1).

Once the groups are determined $\delta(l_i, l_j)$ is obtained as follows – see Fig. 1. Let the heights $\lambda_{l_1}$ and $\lambda_{l_2}$ be respectively the linkage distances of groups 1 and 2. According to how groups are formed in hierarchical agglomerative clustering, $\lambda_{l_1}$ is the distance at which the members of $l_1$ reduce their distance to zero and as result they merge in one cluster. Thus $\lambda_{l_1,l_2}$ is the height at which the distance between $l_1$ and $l_2$ is zero since they merge in one bigger group. The measure expressing how far $l_1$ is from $l_2$ ($\delta(l_1, l_2)$) then equals the height at which $l_1$ and $l_2$ have no distance for they begin to be part of the same group ($\lambda_{l_1,l_2}$), less the one at which $l_1$ 'begins to exist' ($\lambda_{l_1}$). Extending such measures to $i$ and $j$ we obtain

$$\delta(l_i, l_j) = \lambda_{l_i,l_j} - \lambda_{l_i}. \tag{4}$$

Of course, since $\delta(l_j, l_i) = \lambda_{l_i,l_j} - \lambda_{l_j}$, and $\delta(l_i, l_j) = \delta(l_j, l_i)$ and $\lambda_{l_i} \neq \lambda_{l_j}$ by construction, we have $\delta(l_i, l_j) \neq \delta(l_j, l_i)$. Only when both $i$ and $j$ are made of one single unit, $\delta(l_i, l_j) = \delta(l_j, l_i)$, because $\lambda_{l_i} = \lambda_{l_j} = 0$. Thus $\delta(l_i, l_j)$ is not a distance, but rather an asymmetric measure of dissimilarity. In analogy with (2) we can ideally decompose each $\delta(l_i, l_j)$ in a first part related to the alienation and corresponding to the distance between $l_i$ and $l_j$, which in (2) is given by $|y_i - y_j|$, plus a second part related to the characteristics of $i$, empowering the identification of units in $i$, that is $y_i^{\beta}$ in (2). Indeed the terms *identification* and *alienation* are related to the behaviors of individuals: in the case of territorial units we can simply say that since one group formed at a lower height than the other, its members are *closer* to each other than the ones of the other group forming *later*. The different inner variability inside each group is just put into account by the different values of $\delta(l_i, l_j)$ and $\delta(l_j, l_i)$.

The plot of (3)'s time series allows to detect, how polarization changed over time. The index equals zero, when all units are part of the same group, and increases as polarization raises.

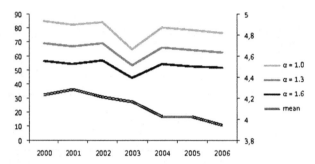

**Fig. 2** ALOS, whole set of destinations: polarization index (left axis) and average value across tourist areas (right axis)

## 4 Dataset

In the following Italian tourist destinations are identified with the tourist areas (*circoscrizioni turistiche*), that is the smallest territorial units for which data on tourism are provided by the Italian Statistical Office – Istat (2010). Tourist areas are partitions of the administrative provinces and are constituted by sets of municipalities. Data are provided separately for hotels and similar establishments, and other collective establishments, and for each of these two categories a further distinction is made between the number of Italian and foreign incoming tourists. Data cover only seven years, from 2000 to 2006: this limit does not allow to analyze the long term evolution of polarization dynamics, but it suffices to detect its short term evolution. The original number of areas changed from 542 in 2000–2005 to 554 in 2006; due to the different composition of the sets, as well as to other problems related to data quality, we reduced the number to 413 by merging the areas in order to obtain territorial units which are comparable over time. Results show the time series of the mean ALOS across destinations and the plot of $P(\pi, \mathbf{y})$ series for $\alpha = 1.0$, 1.3 and 1.6 – see Sect. 2. We separately report the evidence concerning the ALOS for the whole set (Fig. 2) and for four subsets: hotels and similar establishments (Fig. 3) and other collective accommodations (Fig. 4); Italian (Fig. 5) and foreign tourists (Fig. 6). All estimates were performed with the software R.

## 5 Discussion and Conclusions

The above mentioned decreasing trend of the mean ALOS emerges from all the sets, while we find interesting differences in terms of polarization variability over time. The overall decreasing trend of $P(\pi, \mathbf{1})$ in the whole set of destinations (Fig. 2) indicates that the polarization has reduced, but the tendency of the set of destinations to be similar has substantially been stationary within the two complementary subsets of *hotels and similar establishments* and *other collective accommodations*

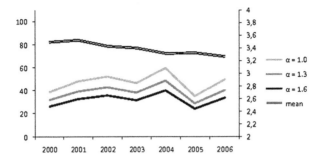

**Fig. 3** ALOS, Hotels and similar establishments: polarization index (left axis) and average value across tourist areas (right axis)

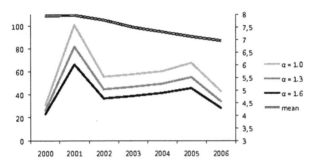

**Fig. 4** ALOS, Other collective accommodations: polarization index (left axis) and average value across tourist areas (right axis)

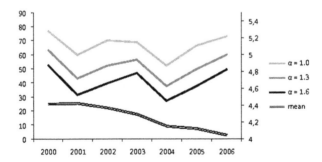

**Fig. 5** ALOS, Italian tourists: polarization index (left axis) and average value across tourist areas (right axis)

(Figs. 3, 4). This denotes that the decrease of the mean ALOS affects all destinations with no relevant average reduction of the gaps.

The slowly decreasing polarization dynamics of the overall set also characterizes the two sets of Italians and foreigners, whose "W" shaped trends (Figs. 5, 6) differ in terms of variability. The wider fluctuation of the Italian tourists trend, and the more stationary time gaps between the clusters of destinations of foreigners, reveal

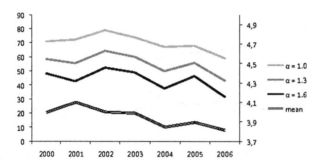

**Fig. 6** ALOS, Foreign tourists: polarization index (left axis) and average value across tourist areas (right axis)

that the latter have been more *constant* in the choice of the length of their vacation across the destinations than the former.

Overall, although a more interesting interpretation of the polarization patterns in time would require a wider dataset, the short term information we provided reveals interesting features concerning the clustering dynamics of tourist destinations. While the interpretation of the index may recall the classical concept of Gini's concentration as also ER state, the axioms of polarization make it different and specifically suited to the measurement of the evolution of the clustering structure. Moreover, our integration of the ER axioms with the hierarchical agglomerative techniques allows to extend the proposed methodology to all the situations when it is required to analyze the dynamics of the grouping deriving from such methods, or simply to compare the *degree of clustering* of two or more sets of data. In addition, the extension of our procedure to inference, which will be the object of future researches, would allow to give our polarization analysis also a predictive value.

**Acknowledgements** This research is part of the PRIN 2007 (Project of Relevant National Interest) *Mobility of regional incoming tourism. Socio-economic aspects of behaviors and motivations*, funded by the Italian Ministry for University and University of Palermo. I thank Prof. Franco Vaccina, the national coordinator of the Project, and Prof. Anna Maria Parroco for their precious suggestions and cooperation. I also thank the anonymous referees for their indications.

# References

Esteban JM, Ray D (1994) On the measurement of polarization. Econometrica 62:819–851
Esteban JM, Gradín C, Ray D (1999) Extensions of a measure of polarization with and application to the income distribution of five OECD countries. In: Luxemb. Income Study Work Pap. Ser 218, Maxwell Sch. Citizsh. Public Aff. Syracuse Univ., Syracuse, New York
EUROSTAT (2010) Occupancy in collective accommodation establishments: domestic and inbound tourism. Reference metadata in euro SDMX metadata structure (ESMS). http://epp.eurostat.ec.europa.eu/portal/page/portal/tourism/data. Cited 3 Aug 2010

Henderson JV (1988) Urban development. Theory, fact and illusion. Oxford University Press, New York

Istat (Italian Statistical Office) (2010) Statistiche per politiche di sviluppo. Risorse turistiche. http://www.istat.it/ambiente/contesto/incipit/turistiche.html. Cited 14 Jun 2010

Scuderi R (2006) Club convergence and cluster persistence: A novel approach to multivariate convergence analysis. In: Proc. XLIII Symp. Ital. Stat. Soc. (SIS), pp 345–348

Ward JH (1963) Hierarchical grouping to optimize an objective function. J Am Stat Assoc 58:236–244

Wolfson MC (1994) When inequality diverge. Am Econ Rev Pap Proc 84:353–358

# Index

W. Gaul et al. (eds.), *Challenges at the Interface of Data Analysis, Computer Science, and Optimization*, Studies in Classification, Data Analysis, and Knowledge Organization, DOI 10.1007/978-3-642-24466-7, © Springer-Verlag Berlin Heidelberg 2012